Recent Advances in Medicinal Plants and Natural Products

Recent Advances in Medicinal Plants and Natural Products

Guest Editors

Ana Paula Coelho Duarte
Ângelo Luís

MDPI
Basel • Beijing • Wuhan • Barcelona • Belgrade • Novi Sad • Cluj • Manchester

Guest Editors

Ana Paula Coelho Duarte
Health Sciences
Research Centre
Universidade da
Beira Interior
Covilha
Portugal

Ângelo Luís
CICS-UBI Health Sciences
Research Centre
University of Beira Interior
Covilhã
Portugal

Editorial Office

MDPI AG
Grosspeteranlage 5
4052 Basel, Switzerland

This is a reprint of the Special Issue, published open access by the journal *International Journal of Molecular Sciences* (ISSN 1422-0067), freely accessible at: https://www.mdpi.com/journal/ijms/special_issues/38U5ZO8P8W.

For citation purposes, cite each article independently as indicated on the article page online and as indicated below:

Lastname, A.A.; Lastname, B.B. Article Title. *Journal Name* **Year**, *Volume Number*, Page Range.

ISBN 978-3-7258-7006-6 (Hbk)
ISBN 978-3-7258-7007-3 (PDF)
https://doi.org/10.3390/books978-3-7258-7007-3

Cover image courtesy of Ângelo Filipe Santos Luís

Contents

Preface

This reprint compiles the scientific contributions originally published within the Special Issue dedicated to advancing molecular knowledge on medicinal plants and natural products. The subject of this work lies at the intersection of natural product chemistry, molecular biology, and therapeutic innovation, reflecting both the complexity and the scientific relevance of bioactive plant-derived compounds. The scope encompasses analytical characterization, mechanistic insights, biological evaluation, and the molecular basis underlying the health-promoting potential of natural substances.

The aim of this reprint is to provide researchers with a consolidated and coherent overview of the diverse studies that shaped the scientific discourse during the active period of the Special Issue. Our motivation in assembling these works is rooted in the rapid expansion of natural product research and the ongoing need for integrated molecular perspectives capable of supporting innovation in drug discovery and functional ingredient development. By gathering these articles into a single reprint, we hope to enhance accessibility to the collective findings and encourage further exploration of the mechanistic foundations that define biological activity.

This work is addressed to scientists, graduate students, and professionals engaged in natural product research, phytochemistry, molecular pharmacology, and related areas. We also extend our appreciation to the authors whose contributions form the core of this reprint, and to the reviewers whose rigorous assessments ensured the high scientific quality of the published studies. Our acknowledgment further includes the Editorial Team, whose assistance supported the successful completion of the Special Issue and the preparation of this reprint.

Ana Paula Coelho Duarte and Ângelo Luís
Guest Editors

International Journal of *Molecular Sciences*

MDPI

Editorial

Special Issue "Recent Advances in Medicinal Plants and Natural Products"

Ângelo Luís * and Ana Paula Duarte *

RISE-Health, Department of Medical Sciences, Faculty of Health Sciences, University of Beira Interior, Av. Infante D. Henrique, 6200-506 Covilhã, Portugal
* Correspondence: afluis27@gmail.com (Â.L.); apcd@ubi.pt (A.P.D.)

Medicinal plants and natural products continue to represent a rich and rapidly evolving field in molecular sciences. Their longstanding importance in traditional medicine, coupled with modern analytical and biotechnological advancements, has accelerated scientific interest in uncovering novel therapeutic agents from natural sources. The Special Issue "Recent Advances in Medicinal Plants and Natural Products," published within the *Bioactives and Nutraceuticals* section of the *International Journal of Molecular Sciences (IJMS)*, was designed to highlight this dynamic research landscape and promote interdisciplinary discussions around natural bioactive compounds. As outlined in the description, this Special Issue showcases studies exploring the therapeutic potential of natural compounds, advances in extraction and chemical characterization techniques, and molecular investigations that clarify mechanisms of action (Figure 1).

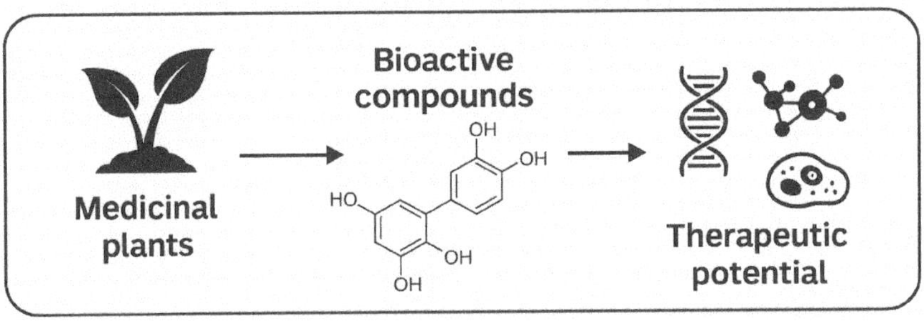

Figure 1. Aims and scope of the Special Issue "Recent Advances in Medicinal Plants and Natural Products."

In recent years, major breakthroughs in analytical chemistry, molecular biology, and biotechnology have transformed how natural compounds are studied. High-resolution chromatographic and spectroscopic techniques have facilitated the discovery of new metabolites, while molecular approaches have illuminated the mechanisms by which these substances exert antioxidant, antimicrobial, and other bioactivities. A distinguishing characteristic of this Special Issue is its strong emphasis on bridging fundamental molecular insights with applied research. The scope explicitly called for contributions that deepen the understanding of molecular mechanisms behind biological activities, allowing researchers to evaluate therapeutic potential through rigorous mechanistic frameworks.

Medicinal plants and natural products remain highly relevant for researchers, clinicians, and policymakers. The contributions collected in this Special Issue demonstrate the broad applicability of natural-compound research across domains ranging from human health and disease management to functional biomaterials and sustainable biotechnologies. In alignment with IJMS editorial values, the Special Issue promotes open-access dissemination, ensuring researchers worldwide can benefit from the findings and engage in global scientific dialog. By offering a curated collection of high-quality, peer-reviewed studies, the Special Issue aims to inform decision-making, stimulate technological development, and reinforce the relevance of natural-products research in modern healthcare and innovation ecosystems.

Natural products continue to occupy a central and enduring role in the search for innovative therapeutic solutions. Their structural diversity, evolutionary optimization, and wide biological activities make them exceptionally rich sources of pharmacologically active molecules. The articles collected in this Special Issue reflect the vibrancy of the field today—combining traditional knowledge, advanced chemical analysis, biological assays, and computational modeling to explore the therapeutic applications and mechanistic complexities of plant-derived compounds.

A recurring theme throughout this Special Issue is the robust characterization and optimization of bioactive compounds. Studies on *Amomyrtus luma* [contribution 1] and *Vaccinium oxycoccos* [contribution 2] provided compelling examples of how phytochemical richness is shaped by environmental factors, ripening stages, and extraction parameters. High phenolic contents, dynamic flavonoid fluctuations, and antioxidant activities reveal how these fruits can be strategically valorized for health-oriented applications. The cinnamaldehyde-rich essential oil of *Cinnamomum zeylanicum* further expands this chemical panorama, displaying strong antioxidant, antimicrobial, immunomodulatory, and antitumor effects [contribution 3]. Meanwhile, the multifaceted chemical composition of *Foeniculum vulgare* extracts—validated through HPLC, GC-MS, and molecular docking approaches—demonstrated their potential as natural antifungal- and antioxidant agents [contribution 4].

Another major axis in this Special Issue is centered on metabolic regulation, inflammation, and tissue homeostasis. Fenugreek extract (Forceterone®) showed strong in vivo and in vitro efficacy in reducing hormone-induced prostate enlargement, modulating inflammatory cytokines, and rebalancing apoptotic pathways [contribution 5]. Research on *Viola yedoensis* extract demonstrated its intriguing potential for skin-barrier repair, upregulating CD44, aquaporin-3, filaggrin, and keratin-10, supported by advanced metabolomic profiling that uncovered numerous previously undescribed compounds [contribution 6]. In a metabolic context, licochalcone D from *Glycyrrhiza uralensis* ameliorated glucose dysregulation and restored mitochondrial function in insulin-resistant hepatocytes, offering promising insights into natural compounds for managing metabolic disorders [contribution 7].

Complementing these biological and analytical studies, structure-based and computational methodologies shed light on molecular mechanisms and multi-target interactions. Phytochemicals from *Detarium senegalense*, such as catechin and epicatechin, exhibited strong binding affinities for PDE5 and other enzymes involved in erectile dysfunction, displaying pharmacokinetic and drug-likeness advantages over synthetic PDE5 inhibitors [contribution 8]. Similarly, T-muurolol showed broad antibacterial potential against *Staphylococcus aureus*, with stable interactions across multiple essential bacterial proteins, coupled with antioxidant, anti-inflammatory, and favorable ADMET properties [contribution 9].

The Special Issue concludes with two comprehensive reviews that expand the scope from specific compounds to broader therapeutic and societal implications. Stilbenes—well-known for their cardioprotective, neuroprotective, anti-inflammatory, and anticancer properties—were discussed in depth, with special attention to bioavailability challenges and emerging formulation strategies [contribution 10]. Finally, *Salvia divinorum* was examined from both a pharmacological and public-health perspective, integrating toxicological evidence, therapeutic potential, patterns of recreational use, and global regulatory complexities [contribution 11].

Together, these contributions illustrate a rich and interconnected landscape. From discovery and characterization to mechanistic elucidation, computational refinement, and translational relevance, they reaffirm natural products as essential pillars in the search for safer, more effective therapeutics and highlight the crucial role of interdisciplinary science in shaping the future of phytochemical research.

The Special Issue "Recent Advances in Medicinal Plants and Natural Products" reflects the richness, diversity, and scientific momentum of this field. By bringing together multidisciplinary insights and highlighting innovative methodologies, the collection offers a valuable resource for advancing natural products research. We hope it will inspire continued innovation, foster new collaborations, and contribute to a deeper understanding of nature's molecular richness as a foundation for future therapeutic development.

Author Contributions: Conceptualization, methodology, writing—original draft preparation, writing—review and editing, Â.L. and A.P.D. All authors have read and agreed to the published version of the manuscript.

Data Availability Statement: No new data were created.

Acknowledgments: We express our deepest appreciation to the authors, reviewers, and collaborators who made this Special Issue possible. Their scientific rigor, innovative methodologies, and commitment to advancing natural-product research provide valuable contributions to the molecular sciences community.

Conflicts of Interest: The authors declare no conflicts of interest.

List of Contributions

1. Gomez-Perez, L.S.; Poblete, J.; García, V.; Vidal, R.L. Cauchao Berry (*Amomyrtus luma*) as a Promising Source of Bioactive Compounds: Optimized Extraction, Phytochemical Characterization, and Assessment of Antioxidant and Antidiabetic Potential. *Int. J. Mol. Sci.* **2025**, *26*, 8391. https://doi.org/10.3390/ijms26178391.
2. Liaudanskas, M.; Šedbarė, R.; Ramanauskas, I.; Janulis, V. Variation in the Chemical Composition of Small Cranberry (*Vaccinium oxycoccos* L.) Fruits Collected from a Bog-Type Habitat in Lithuania. *Int. J. Mol. Sci.* **2025**, *26*, 6956. https://doi.org/10.3390/ijms26146956.
3. Živković, M.; Stanisavljević, I.; Gajović, N.; Pavlović, S.; Simović Marković, B.; Jovanović, I.P.; Cupara, S.; Tadić, V.; Žugić, A.; Milenković, M.T.; et al. Comprehensive Phytochemical Analysis and Evaluation of Antioxidant, Antimicrobial, Cytotoxic, and Immunomodulatory Activities of Commercial Cinnamon Bark Essential Oil (*Cinnamomum zeylanicum* L.). *Int. J. Mol. Sci.* **2025**, *26*, 6482. https://doi.org/10.3390/ijms26136482.
4. Moumen, B.E.; Bouzoubaa, A.; Drioiche, A.; Eddahmouny, M.; Al Kamaly, O.; Shahat, A.A.; Touijer, H.; Hadi, N.; Kharchouf, S.; Cherrat, A.; et al. Unveiling the Chemical Composition, Antioxidant, and Antimicrobial Potentials of *Foeniculum vulgare* Mill: A Combined In Vitro and In Silico Approach. *Int. J. Mol. Sci.* **2025**, *26*, 4499. https://doi.org/10.3390/ijms26104499.
5. Lee, J.Y.; Bang, J.; Kim, J.; Baek, K.-S.; Oh, D.; Lee, Y.-H. Effect of Fenugreek Extract on Testosterone Propionate-Induced Benign Prostatic Hyperplasia. *Int. J. Mol. Sci.* **2025**, *26*, 1261. https://doi.org/10.3390/ijms26031261.
6. Thonthula, S.; Sousa, S.D.; Dubuis, A.; Boudah, S.; Mehta, R.; Singh, A.; Eilstein, J.; Tabet, J.-C.; John, S.; Roy, D.; et al. Improved Skin Barrier Function Along with Hydration Benefits of *Viola yedoensis* Extract, Aesculin, and Schaftoside and LC-HRMS/MS Dereplication of Its Bio-Active Components. *Int. J. Mol. Sci.* **2024**, *25*, 12770. https://doi.org/10.3390/ijms252312770.
7. Lee, Y.G.; Lee, H.M.; Hwang, J.-T.; Choi, H.-K. Licochalcone D from *Glycyrrhiza uralensis* Improves High-Glucose-Induced Insulin Resistance in Hepatocytes. *Int. J. Mol. Sci.* **2024**, *25*, 10066. https://doi.org/10.3390/ijms251810066.

8. Ejeje, J.N.; Agbebi, E.A.; Mathenjwa-Goqo, M.S.; Oje, O.A.; Agboinghale, P.E.; Ebe, I.T.; Obafemi, T.O.; Adewole, E.; Omaka, O.N.; Onikanni, S.A.; et al. Computational Investigation of the Therapeutic Potential of *Detarium senegalense* in the Management of Erectile Dysfunction. *Int. J. Mol. Sci.* **2024**, *25*, 12362. https://doi.org/10.3390/ijms252212362.
9. Bhattacharya, S.; Khanra, P.K.; Dutta, A.; Gupta, N.; Aliakbar Tehrani, Z.; Severová, L.; Šrédl, K.; Dvořák, M.; Fernández-Cusimamani, E. Computational Screening of T-Muurolol for an Alternative Antibacterial Solution against *Staphylococcus aureus* Infections: An In Silico Approach for Phytochemical-Based Drug Discovery. *Int. J. Mol. Sci.* **2024**, *25*, 9650. https://doi.org/10.3390/ijms25179650.
10. Brás, L.P.; Luís, Â.; Chatel, G.; Socorro, S.; Duarte, A.P. Stilbenes from Vine Extracts: Therapeutic Potential and Mechanisms. *Int. J. Mol. Sci.* **2025**, *26*, 8269. https://doi.org/10.3390/ijms26178269.
11. Calado, S.; Pires, B.; Rosendo, L.M.; Rosado, T.; Gallardo, E.; Duarte, A.P. Salvinorin A and *Salvia divinorum*: Toxicology, Pharmacological Profile, and Therapeutic Potential. *Int. J. Mol. Sci.* **2025**, *26*, 5588. https://doi.org/10.3390/ijms26125588.

https://doi.org/10.3390/ijms27031350

International Journal of
Molecular Sciences

MDPI

Article

Cauchao Berry (*Amomyrtus luma*) as a Promising Source of Bioactive Compounds: Optimized Extraction, Phytochemical Characterization, and Assessment of Antioxidant and Antidiabetic Potential

Luis S. Gomez-Perez [1,2,*], Jacqueline Poblete [3], Vivian García [3] and René L. Vidal [1,4,*]

1 Centro de Biología Integrativa, Facultad de Ciencias, Universidad Mayor, Santiago 8380000, Chile
2 Escuela Nutrición y Dietética, Facultad de Medicina y Ciencias de la Salud, Universidad Mayor, Santiago 8580745, Chile
3 Departamento de Ingeniería en Alimentos, Universidad de La Serena, La Serena 1700000, Chile; j.pobletegalleguillos@gmail.com (J.P.); vivian.garcia@userena.cl (V.G.)
4 Centro FONDAP de Gerociencia, Salud Mental y Metabolismo (GERO), Santiago 8380000, Chile
* Correspondence: luis.gomezpe@umayor.cl (L.S.G.-P.); rene.vidal@umayor.cl (R.L.V.)

Abstract

The Cauchao berry (*Amomyrtus luma*), native to southern Chile and Argentina, has been traditionally used in folk medicine, yet scientific evidence supporting its bioactive potential remains limited. This study aimed to optimize the extraction of bioactive compounds and assess their antioxidant and antidiabetic properties. Fresh and freeze-dried samples were compared in terms of proximate composition, dietary fiber, reducing sugars, and fatty acid profiles. Proximate and fiber contents were determined using AOAC methods, while fatty acids were analyzed by gas chromatography, and α-tocopherol levels were measured via HPLC. Extraction optimization was conducted using a Box–Behnken design within a response surface methodology framework, employing freeze-dried samples. Total phenolic (TPC), flavonoid (TFC), and anthocyanin (TAC) contents were quantified spectrophotometrically. Antioxidant potential was assessed by DPPH and ORAC assays, while α-glucosidase inhibition determined antidiabetic activity. Phenolic profiles were characterized by HPLC. Optimal extraction conditions (58% ethanol, 60% ultrasound power, 30 min) enhanced antioxidant response. Results showed high fiber content (~39%), linoleic acid as the predominant fatty acid, and an α-tocopherol concentration of ~95 μg/g. TPC, TFC, and TAC values reached 25.43 ± 0.85, 46.51 ± 1.38, and 5.91 ± 0.40 mg/g d.m., respectively. Antioxidant capacity was 289.54 ± 9.05 μmol TE/g (DPPH) and 451.09 ± 6.04 μmol TE/g (ORAC). The IC_{50} for α-glucosidase inhibition was 0.558 ± 0.015 mg/mL. Phenolic compounds were identified. These findings position the Cauchao berry as a promising source of bioactive compounds with potential health benefits.

Keywords: Cauchao berry; optimized extraction; bio-compounds; antioxidant potential; α-glucosidase inhibition

1. Introduction

Berries are widely acknowledged worldwide as some of the most nutritious foods available, thanks to their rich and diverse content of bioactive compounds. These include phenolic acids, flavonoids, and anthocyanins, which are celebrated for their numerous health-promoting properties [1]. These compounds contribute not only to the vibrant colors of berries but also to the antioxidant, anti-inflammatory, antihypertensive, and antihyperglycemic activities [2].

Chile boasts a rich variety of small fruits, particularly berries, among which the most notable are murta, blueberries, maqui, and calafate, among others. Researchers have increasingly focused on these berries because of their exceptional functional properties, including strong antioxidant activity, anti-inflammatory benefits, and potential positive effects on cardiovascular and metabolic health [1]. Their distinctive composition, abundant in bioactive substances such as polyphenols, vitamins, and dietary fiber, has captured the interest of scientists and healthcare experts globally, making them valuable candidates for use in functional foods and nutraceutical innovations.

However, other wild berries, such as Cauchao berries (*Amomyrtus luma*), remain largely unexplored, presenting a promising opportunity for further scientific research. These fruits could emerge as a new source of bioactive compounds with potential functional properties that may contribute to improved health and wellness. The Cauchao or Chauchao berry is a native and endemic species of southern Chile and Argentina. It thrives in rainforests, particularly near streams and in other high-altitude, humid areas. This globose berry has a pleasant flavor and aroma, transitioning from reddish to blackish as it ripens. Currently, the fruit is primarily consumed fresh or as part of artisanal products, including jams, desserts, juices, pastries, and fermented alcoholic beverages. Historically, it has been used by indigenous peoples for its stimulating and astringent properties [3]. Moreover, studies from ancestral Mapuche medicine report that its leaves are used to lower blood pressure and cholesterol levels and to treat liver diseases [4]. The leaf extract has also been found to contain aromatic compounds such as linalool, 1,8-cineole, and β-caryophyllene, among others [5]. These compounds, widely recognized for their antioxidant, anti-inflammatory, and antimicrobial activities, provide a biochemical rationale for considering *Amomyrtus luma* as a potential source of functional agents. In this context, the fruit, although still underexplored, may also contain bioactive molecules with health-promoting effects, reinforcing the need for scientific investigation into its composition and properties. To date, no peer-reviewed studies in high-impact journals have analyzed the bioactive profile or functional properties of *Amomyrtus luma* fruit.

On the other hand, environmentally friendly extraction methods have been employed to obtain new antioxidant concentrates or natural colorants from berries [6,7]. Technologies such as ultrasound have been explored to reduce extraction time and energy consumption while avoiding the use of high temperatures [8]. Ultrasound-assisted extraction utilizes acoustic cavitation, and when applied at low frequencies (20–100 kHz), it generates large bubbles that implode violently, leading to enhanced cell disruption, improved solvent penetration, and consequently higher extraction yields [9]. This technique is considered an effective alternative for extracting phenolic compounds and anthocyanins from maqui berries [10], as well as from aronia and grapes [11]. Therefore, ultrasound-assisted extraction presents a promising method for obtaining Cauchao berry extracts. Nonetheless, the extraction conditions must be carefully optimized to maximize bio-compound yield.

Berry consumption is currently on the rise, driven by their rich content of bioactive compounds that offer various health benefits for humans. These benefits include antioxidant, anti-inflammatory, and cardioprotective properties, which contribute to reducing the risk of chronic diseases such as cardiovascular conditions, diabetes, and certain types of cancer [12]. However, these properties have been scarcely explored in Cauchao berries. The present study evaluated the antioxidant and antidiabetic potential of extracts rich in bioactive compounds derived from Cauchao berries by investigating these properties. This research aims to provide valuable insights into the health-promoting capabilities of this underexplored fruit.

2. Results

2.1. Proximate Composition and Reducing Sugars

The proximate analysis, shown in Table 1, was conducted on fresh and freeze-dried Cauchao berry samples. The fresh sample exhibited a moisture content comparable to other berries, such as murtilla [13], but higher than that of blueberries and maqui, which have a moisture content of approximately 65 g/100 g on a wet basis [14,15]. The content of ash, crude protein, carbohydrates, crude fiber, and fats was significantly more concentrated in the freeze-dried sample due to the reduced water content in the matrix.

Table 1. Proximate composition and dietary fiber content of Cauchao berries.

Parameters (g/100 g)	Fresh	Freeze-Dried
Moisture	75.31 ^{a} ± 0.41	2.78 ^{b} ± 0.10
Fat	2.15 ^{b} ± 0.15	9.15 ^{a} ± 0.16
Ash	0.57 ^{b} ± 0.03	2.50 ^{a} ± 0.15
Crude protein	1.73 ^{b} ± 0.17	7.15 ^{a} ± 0.23
Crude fiber	3.94 ^{b} ± 0.67	15.25 ^{a} ± 0.50
Carbohydrates	20.24 ^{b} ± 0.32	78.42 ^{a} ± 0.29
Reducing sugars	7.71 ^{b} ± 0.42	39.82 ^{a} ± 1.58
Insoluble dietary fiber *	34.78 ^{a} ± 1.73	34.26 ^{a} ± 2.55
Soluble dietary fiber *	4.60 ^{a} ± 0.50	3.66 ^{a} ± 0.81
Total dietary fiber *	39.37 ^{a} ± 2.23	37.92 ^{a} ± 3.36

Values are expressed as mean ± standard deviation (n = 3). * Dry matter. Different letters in the same row indicate significant differences between treatments ($p < 0.05$), according to the Student's t-test.

As expected, a high concentration of sugars was observed in the freeze-dried Cauchao berries. The fresh value was significantly higher compared to other fruits, such as strawberries, blackberries, raspberries, and cranberries, while being very similar to that of blackcurrants [16].

2.2. Dietary Fiber Content

Dietary fiber refers to the edible components of plants that resist digestion and absorption in the human small intestine. Instead, these fibers remain intact as they pass into the large intestine, where they may undergo complete or partial fermentation [17]. The contents of soluble (SDF), insoluble (IDF), and total (TDF) dietary fiber were determined by the AOAC method. Soluble dietary fiber, which includes components such as non-cellulosic polysaccharides, oligosaccharides, pectins, β-glucans, and gums, is highly valued for its nutritional benefits. IDF includes cellulose, parts of hemicellulose, and lignin [18]. Table 1 shows the SDF, IDF, and TDF for fresh and freeze-dried Cauchao berries. As expected, no differences were observed between the evaluated samples, as both underwent dehydration according to the prescribed methodology. Notably, the TDF value was considerably higher than that reported for comparable fruits, such as murta berries. For instance, murta berry has a TDF value of 17.55 ± 0.37 g/100 g dry matter (d.m.) based on the same determination method [19]. In comparison, the Cauchao berry demonstrated approximately twice the total dietary fiber, highlighting its superior nutritional profile and potential as a rich dietary fiber source.

2.3. Fatty Acids Profile and α-Tocopherol Content

Vegetable fatty acids, which are part of the essential unsaturated fatty acid group, show considerable biological activity. These fatty acids are predominantly found in seeds and oils extracted from fruits such as berries and others [20]. These essential nutrients play a critical role in various bodily functions, including maintaining healthy cell membranes, supporting cardiovascular health, and regulating inflammation [21]. Since the human body cannot synthesize these fatty acids on its own, consuming them through a balanced diet is vital for overall well-being. Table 2 shows the fatty acid profile of the Cauchao berry. Six fatty acids were identified, and no differences were observed between fresh and freeze-dried berries. The profile found is very similar to those reported in murta berries in terms of fatty acids and their quantities [18]. It stands out for its linoleic acid (LA, C18:2n6c) content, which is approximately 80%. LA, an essential fatty acid that must be obtained through the diet [22], is the primary polyunsaturated fatty acid found in Cauchao berries. Oleic acid was found in the second proportion (OA, C18:1n9c), the quantity found was higher than other similar berries such as blueberries, red gooseberries and black currants [23]. The third portion identified was palmitic acid (PA, C16:0), which is one of the most common saturated fatty acids in nature, found in animal and human tissues, as well as in plants, algae, fungi, yeasts, and bacteria [24]. Essential plant-derived fatty acids, including linoleic acid, oleic acid, and palmitic acid, play pivotal roles in human health. Consuming Cauchao berries, which are rich in these fatty acids, may confer substantial health benefits.

Table 2. Fatty acids and α-tocopherol content in Cauchao berries.

Fatty Acids (g/100 g Fatty Acid)	Fresh	Freeze-Dried
C16:0	$6.99^{a} \pm 0.02$	$7.18^{a} \pm 0.13$
C18:0	$2.62^{b} \pm 0.04$	$2.71^{a} \pm 0.02$
C18:1n9c	$8.83^{b} \pm 0.05$	$9.24^{a} \pm 0.17$
C18:2n6c	$80.29^{a} \pm 0.06$	$79.44^{b} \pm 0.03$
C20:0	$0.35^{a} \pm 0.07$	$0.41^{a} \pm 0.01$
C20:1	$0.92^{a} \pm 0.06$	$1.02^{a} \pm 0.04$
α-tocopherol (μg/g)	$95.51^{a} \pm 5.42$	$105.41^{a} \pm 1.39$

Values are expressed as mean ± standard deviation (n = 3). Different letters in the same row indicate significant differences between treatments ($p < 0.05$). Statistical analysis was performed using the Student's t-test for all variables except C20:0, which was analyzed using the Mann–Whitney test due to non-normal distribution.

α-Tocofpherol is one of the primary components of vitamin E, essential for meeting human nutritional requirements, and is primarily obtained through food [25]. Table 2 presents the α-tocopherol content of both fresh and freeze-dried Cauchao berries, with the latter showing a slightly higher concentration, likely due to the concentration effect of solutes caused by the drying process. The α-tocopherol concentration in both forms exceeds that reported in other similar fruits, such as murta berry with values between 1.06 ± 0.02 and 5.80 ± 0.15 mg/100 g [26] and maqui berry with values between 17.1 ± 0.8 μg/g and 31.7 ± 0.5 μg/g [27].

2.4. RSM Method

The response surface modeling method was used to determine the optimal condition of the antioxidant capacity of the Cauchao berry extract. Table 3 presents the factors and levels used in the Box–Behnken model, showing their impact on the antioxidant capacity of Cauchao berry extract, measured by the DPPH method. The results indicated that antioxidant potential varies based on these factors, emphasizing the need to optimize extraction parameters such as time, ultrasonic power, and solvent concentration to maximize antioxidant compound yield and antioxidant capacity.

Figure 1 shows three response surface plots illustrating the impact of the independent variables: solvent (%), power (%), and time (min), on the response variable, DPPH (μmol TE/g). Each plot represents the connection between two independent variables while keeping the third constant, enabling the visualization of individual effects and interactions among the evaluated factors. The color gradient, ranging from blue to brown, indicates response variability, with blue representing lower DPPH values and brown representing higher values.

Table 3. Box–Behnken model, coded factor, and DPPH results.

Factors			Coded Factor			Result
Solvent (%)	Power (%)	Time (min)	X_1	X_2	X_3	DPPH (µmol TE/g d.m)
90	80	25	0	1	1	185.89
50	60	35	1	0	−1	248.70
70	40	35	1	−1	0	246.87
50	40	25	0	−1	−1	254.67
90	60	35	1	0	1	170.16
90	60	15	−1	0	1	169.52
70	60	25	0	0	0	261.32
90	40	25	0	−1	1	191.53
70	60	25	0	0	0	237.68
50	60	15	−1	0	−1	205.62
50	80	25	0	1	−1	277.65
70	60	25	0	0	0	243.76
70	60	25	0	0	0	218.70
70	80	15	−1	1	0	222.84
70	40	15	0	−1	0	254.89
70	80	35	1	1	0	246.09
70	60	25	0	0	0	251.31

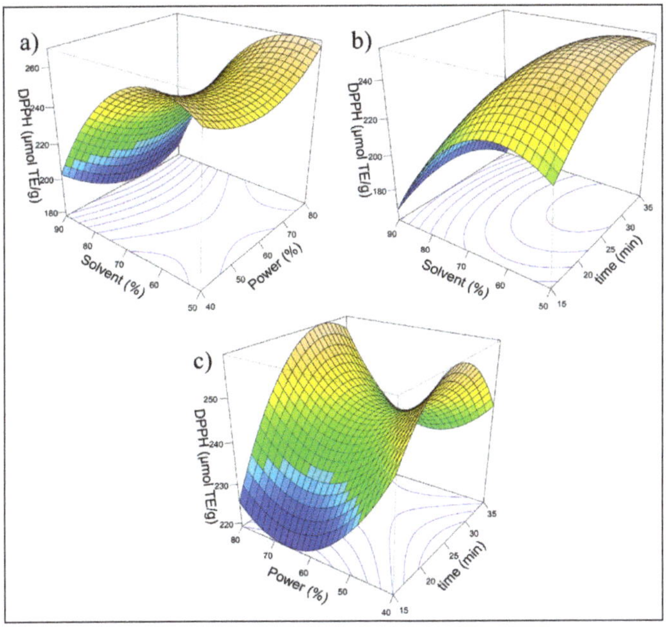

Figure 1. Response surface for antioxidant capacity (DPPH): (**a**) Effect of solvent (%) and power (%) on DPPH; (**b**) effect of solvent (%) and time (min) on DPPH; (**c**) effect of power (%) and time (min) on DPPH. The color gradient, ranging from blue to brown, with blue representing lower DPPH values and brown representing higher values.The response surface model for antioxidant capacity measured by the DPPH assay is described by the following quadratic equation:.

The model fitted using RSM indicates that the linear term of solvent (X_1) was highly significant ($p < 0.001$), demonstrating its strong influence on the response, whereas power

(X_2) and time (X_3) did not show significant effects in the linear model. However, the quadratic terms of X_1, X_2, and X_3 are significant, suggesting a nonlinear relationship with an optimal point. The adjusted R^2 was 0.7855, indicating a good explanatory power, although additional factors may contribute to variability. ANOVA confirms that the first-order effects are significant (p = 0.00264), while interaction effects are not, ruling out synergistic influences. The quadratic effects are significant (p = 0.01292), demonstrating a curvilinear response pattern. The stationary point suggested that maximum antioxidant activity was achieved with 57.58% solvent, 55.96% power, and 29.29 min of extraction, underscoring the importance of these parameters for process optimization. To enhance laboratory reproducibility, the parameters were adjusted to 58% solvent, 60% power, and 30 min of extraction.

$$DPPH = 242.55 - 33.71x_1 - 1.94x_2 + 7.36x_3 - 7.16x_1x_2 - 10.59x_1x_3 + 7.82x_2x_3 - 29.63x_1^2 + 14.51x_2^2 - 14.40x_3^2 \quad (1)$$

This model accounts for linear, quadratic, and interaction effects, providing a mathematical representation of the relationship between the experimental factors and the DPPH response. Figure 2a depicts the model fit by comparing the original experimental DPPH values against the predicted ones (dashed black line), including their range of variability (green shading).

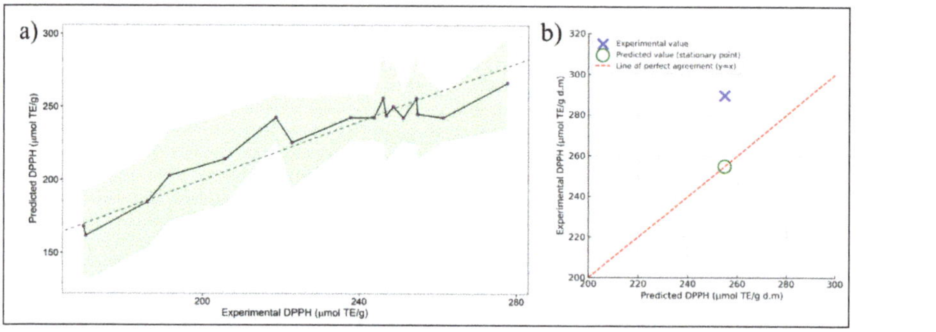

Figure 2. (**a**) Experimental against predicted values of DPPH. The solid black line connects the predicted DPPH values, while the green shaded area represents the confidence interval (uncertainty range) of the prediction. The dashed gray line indicates the line of perfect agreement ($y = x$) for reference. (**b**) Model validation graph comparing predicted and experimental DPPH values at optimal conditions. The large blue cross indicates the experimental value, while the green circle shows the predicted value (stationary point). The red dashed line represents the ideal 1:1 correlation ($y = x$), used to visually assess model accuracy.

2.5. Total Content of Phenolics, Flavonoids, and Anthocyanin

The total bioactive compounds present in the Cauchao berry extract, including total phenolic content, total flavonoid content, and total anthocyanin content, were determined in the extract obtained under optimized conditions according to the RSM results. Table 4 shows the TPC, TFC, and TAC results. Phenolic compounds present in plant-based foods are classified as flavonoids, stilbenes, lignans, phenolic acids, and others, distinguished by their chemical structures [28]. The TPC value of the Cauchao berry extract was 25.43 ± 0.85 mg GAE/g d.m., equivalent to 2543 mg GAE/100 g d.m. This value is considerably higher than those reported for fresh fruits such as strawberry (317–443 mg/100 g), blackberry (411–678 mg/100 g), blueberry (154–585 mg/100 g), blackcurrant (817–1042 mg/100 g) and raspberry (192–529 mg/100 g) [16].

Table 4. Bioactive compound content and antioxidant potential of optimized Cauchao berry extract.

Parameter	Value (Mean ± SD)
TPC (mg GAE/g d.m.)	25.43 ± 0.85
TFC (mg QE/g d.m.)	46.51 ± 1.38
TAC (mg cyanidin-3-glucoside/g d.m.)	5.91 ± 0.40
DPPH (μmol TE/g d.m.)	289.54 ± 9.05
ORAC (μmol TE/g d.m.)	451.09 ± 6.04

Values are expressed as mean ± standard deviation (n = 3).

The TFC of the optimized Cauchao berry extract was 46.51 ± 1.38 mg QE/g d.m., indicating a substantial concentration of these secondary metabolites. When compared to other similar native Chilean berries, such as maqui (*Aristotelia chilensis*) and murta (*Ugni molinae*), which exhibit TFC values ranging from 43.06 ± 0.54 to 55.56 ± 5.60 mg QE/g d.m., respectively [19,29], the Cauchao berry demonstrated comparable levels. The TFC value highlights the potential of Cauchao berries as a competitive source of flavonoids among native berry species.

The TAC value was 5.91 ± 0.40 mg cyanidin-3-glucoside/g d.m. When compared to other berries, such as murta (6.85 mg cya-3-glu eq./g) [1], blackberry (2.10 mg cya-3-glu eq./g), blackcurrant (7.59 mg cya-3-glu eq./g), and different blueberry cultivars (values between 0.03 and 0.66 mg cya-3-glu eq./g) [23]. Cauchao berry presents TAC values higher than most and comparable to those of murta and blackcurrant. This anthocyanin content supports its potential as a promising natural source of compounds beneficial to human health.

2.6. Antioxidant Potential

Table 4 shows the antioxidant potential of optimized Cauchao berry extract evaluated by DPPH and ORAC assays. The antioxidant potential of the Cauchao berry extract was remarkable, with values of 289.54 ± 9.05 μmol TE/g d.m. and 451.09 ± 6.04 μmol TE/g d.m. for DPPH and ORAC, respectively. When compared to another similar berry, such as murta (DPPH: 220 μmol TE/g d.m.; ORAC: 599.5 μmol TE/g d.m.) [29,30], blueberry (DPPH: 125.79 μmol TE/g d.m.; ORAC: 443.01 μmol TE/g d.m.) [31], fresh blackberry (DPPH: 295.8 ± 1.9 μmol TE/g; ORAC: 63.7 ± 0.2 μmol TE/g), and dehydrated blackberry (DPPH: 1203.8 ± 27.9 μmol TE/g d.m.; ORAC: 236.8 ± 1.0 μmol TE/g d.m.) [32], the Cauchao berry presents equivalent values.

2.7. α-Glucosidase Activity

The enzyme α-glucosidase plays a key role in the hydrolysis of complex carbohydrates into simpler, absorbable sugars. Inhibiting its activity is an effective strategy to delay intestinal glucose absorption, thereby reducing postprandial glycemic spikes and potentially slowing the progression of type 2 diabetes mellitus [33]. The Cauchao extract demonstrated a clear, concentration-dependent inhibition of α-glucosidase activity, with a progressive decline in enzymatic function from 67% to nearly 5% at concentrations of 4 mg/mL (Figure 3). The resulting inhibition curve exhibited a decreasing hyperbolic kinetic profile, fitted using a nonlinear exponential model, yielding an IC_{50} of 0.558 ± 0.015 mg/mL, indicative of high inhibitory potency at relatively low concentrations. This value is competitive with other berry extracts reported in the literature, including mulberry fruit (IC_{50} = 0.62 ± 0.06 mg/mL) [34], fermented murta berry juice (IC_{50} = 805 ± 20 μg/mL) [35], and freeze-dried maqui extract (IC_{50} = 0.315 ± 0.046 mg/mL) [27]. Notably, the strong inhibitory effect of the Cauchao extract was observed despite it being a crude, non-fractionated preparation. Although its potency is significantly lower than that of acarbose (IC_{50} = 0.057 ± 0.005 mg/mL), a synthetic reference inhibitor.

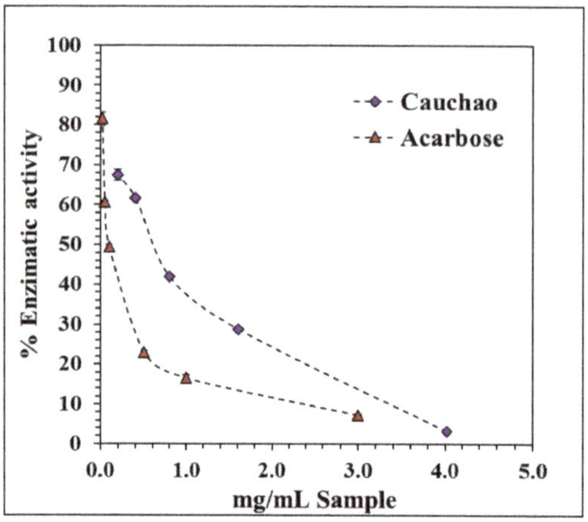

Figure 3. Inhibitory effect of Cauchao berry extract on α-glucosidase activity. Values are averages (n = 3), and error bars are standard deviation.

2.8. Phenolic Compound Profiles

The chromatographic profile of the Cauchao extract at 520 nm (Figure 4) revealed three well-defined peaks at retention times of 12.27, 13.01, and 16.36 min, respectively. The major peak, eluting at 13.01 min and representing over 85% of the total area, was tentatively identified as chlorogenic acid, a hydroxycinnamic acid derivative, with a mean concentration of 3.17 mM. The smaller peaks at 12.27 and 16.36 min were associated with cyanidin, a representative anthocyanin (0.026 mM), and rutin, a flavonol glycoside (0.049 mM), respectively. A fourth compound, sinapic acid (a hydroxycinnamic acid), was also quantified at a relatively high mean concentration (1.52 mM), yet was not clearly visualized in the 520 nm chromatogram. This discrepancy is likely due to its distinct spectral behavior, as hydroxycinnamic acids, such as sinapic acid, exhibit maximal absorbance in the 238–326 nm range [36] and, therefore, are poorly detected at anthocyanin-specific wavelengths. Altogether, these results highlight the co-occurrence of representative phenolic and anthocyanin compounds in the Cauchao extract, supporting its potential as a source of multi-functional bioactive compounds.

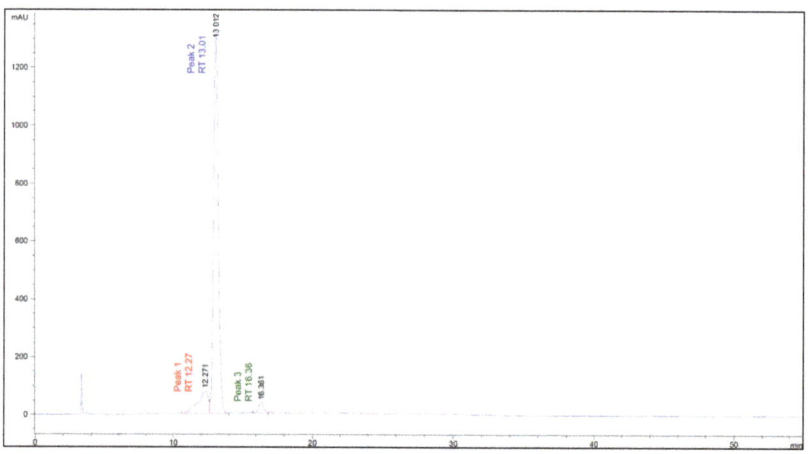

Figure 4. Chromatographic profile of the Cauchao berry extract at 520 nm, where three major peaks were detected at retention times of 12.27, 13.01, and 16.36 min, respectively.

3. Discussion

This discussion interprets the results in light of current scientific evidence, highlighting the physiological, functional, and technological relevance of Cauchao berries. The findings are contextualized through biochemical, nutritional, and health-related perspectives, with comparisons to similar fruit matrices.

3.1. Proximal Composition, Reduces Sugar and Dietary Fiber

The proximal content shows that the freeze-dried Cauchao exhibited a significantly lower moisture content, which is essential for extending the fruit's shelf-life by reducing microbial activity and slowing enzymatic processes. This characteristic highlights freeze-drying as an effective preservation method, particularly for retaining the berries' quality and nutritional value over time. The levels of crude fiber and fats are particularly noteworthy, as they suggest a high dietary fiber content and the presence of relevant fatty acids, respectively. These components are crucial due to their potential health benefits, including promoting digestive health and enhancing the overall nutritional profile. On the other hand, Cauchao berries are a rich source of sugars compared to other fruits with similar characteristics. Sugars are vital to plants, serving both as essential nutrients and as central signaling molecules that regulate gene expression involved in growth, development, metabolism, stress responses, and disease resistance [37].

In terms of dietary fiber, the IDF content was found to be significantly higher than that of SDF. This distinction is particularly noteworthy, as IDF has been widely recognized for its beneficial effects on human health. Studies have linked its consumption to a reduced risk of developing various diseases, including cardiovascular conditions and type 2 diabetes [38]. Additionally, IDF plays a critical role in promoting digestive health by improving bowel regularity, preventing constipation, and improving the production of short-chain fatty acids [39]. On the other hand, soluble dietary fiber is readily available for fermentation by fiber-degrading microorganisms in the intestine. This process results in the production of a variety of beneficial metabolites, such as short-chain fatty acids (SCFAs), which play a crucial role in maintaining gut health and other physiological functions such as increased satiety and reduced energy intake, which leads to control appetite, improve insulin sensitivity, and reduce weight [40]. Also, soluble dietary fibers can bind to metal ions like calcium, magnesium, iron, copper, and zinc, transporting them to the distal colon. As SDFs are degraded by local bacteria, the released ions help inhibit pathogens, enhance gut microbiota diversity, and protect against infections [40]. According to the mentioned fiber properties, Cauchao berries could offer significant benefits that support gut health.

3.2. Fatty Acids Profile and α-Tocopherol Content

The main fatty acids identified in Cauchao berry include linoleic acid, oleic acid, and palmitic acid. These fatty acids are important for human health. Recent studies have suggested that LA may exhibit neuroprotective effects in both in vitro and in vivo models of Parkinson's disease, potentially by modulating lipid droplet dynamics [41]. On the other hand, it has also been shown that the consumption of LA reduces the risk of cardiovascular diseases in healthy individuals [42]. In the brain, OA is a key component of membrane phospholipids and is abundantly present in myelin. A notable reduction in OA levels has been observed in the brains of individuals suffering from major depressive disorders and Alzheimer's disease [43]. This decrease in OA may contribute to the pathophysiology of these conditions, as it plays a crucial role in maintaining neuronal function and integrity. Additionally, it has been suggested that OA may help lower blood pressure by enhancing endothelial function and may also reduce the risk of developing rheumatoid arthritis. Oleic acid is well-known for its health benefits, including its ability to regulate blood glucose levels, lower blood pressure, and provide anti-inflammatory and antimicrobial effects [44]. Palmitic acid is a crucial component of triglycerides in adipose tissue, and its presence in blood carries significant diagnostic implications. High consumption of palmitic acid has been associated with an increased risk of cardiovascular disease (CVD) and elevated levels of LDL and HDL cholesterol. Studies indicate that its intake correlates with a higher risk of heart failure and mortality due to CVD. PA promotes inflammation, contributing to the progression of CVD and cardiac damage [45]. However, recent studies have reported that palmitic acid (PA) has the potential to inhibit the proliferation of cancer cells, including those from prostate, gastric, and breast cancers, suggesting its potential as a basis for developing new therapeutic options for the treatment of various cancer types [46]. Linoleic acid, present in high concentrations in these berries, is not only essential for cardiovascular health but may also exhibit neuroprotective effects. Additionally, oleic acid, known for its antioxidant and anti-inflammatory properties, contributes to the regulation of blood pressure and supports brain health. While excessive consumption of palmitic acid is associated with certain health risks, it is present in relatively low amounts in these berries and has been noted for its potential anti-tumor properties. Including these berries in the diet could represent a beneficial strategy for enhancing overall well-being, mitigating the risk of cardiovascular diseases, and supporting neuronal function.

α-Tocopherol is widely recognized for its antioxidant properties, as it neutralizes reactive oxygen species and prevents lipid peroxidation, thereby protecting cell membranes from oxidative damage [47]. This antioxidant action is essential for maintaining cellular integrity. Moreover, α-tocopherol may help prevent cardiovascular disease by modulating signaling pathways and reducing cell proliferation and oxidative stress. It also inhibits the uptake of oxidized low-density lipoprotein (oxLDL) by immune cells, preventing foam cell formation. Additionally, it decreases inflammation and apoptosis in macrophages. However, some studies suggest that it does not reduce lipid peroxidation in cardiovascular lesions [48]. In terms of its neuroprotective effects, α-tocopherol may contribute to the prevention and management of Alzheimer's disease (AD). Studies have shown that α-tocopherol reduces β-amyloid formation, exhibits antioxidant and anti-inflammatory properties, and supports mitochondrial function, potentially slowing AD progression. Additionally, higher intake has been linked to a reduced risk of dementia and mild cognitive impairment [49]. Given its significant antioxidant, cardiovascular, and neuroprotective benefits, α-tocopherol stands out as a vital bioactive compound with potential therapeutic applications. The high concentration found in Cauchao berries, particularly in their freeze-dried form, underscores their nutritional value as a natural source of vitamin E.

3.3. RSM Method

In the optimization analysis of the extraction process of bio-compounds with RSM; a stationary point was identified at $X_1 = -0.6209$, $X_2 = -0.2019$, and $X_3 = 0.4289$ which, based on eigenanalysis, was characterized as a saddle point. These findings suggest that the response does not exhibit a clear absolute maximum within the range of the factors. However, this stationary point represents the most favorable predicted response within the studied conditions. The experimental value of DPPH at optimal conditions was 289.54 μmol TE/g d.m. (Table 4) and was higher than the value predicted by the model (254.79 μmol TE/g d.m.), as shown in Figure 2b, with an absolute error of 13.64%. This difference can be attributed to natural experimental variability or to the presence of factors not fully accounted for by the quadratic model, such as variations in raw material composition. Nevertheless, the model proved to be a useful predictive tool, pointing toward experimental conditions that led to an even greater antioxidant response. These results confirm the practical applicability of the model for process optimization.

3.4. Total Content of Phenolics, Flavonoids, and Anthocyanin

The high concentration of TPC observed in the Cauchao berry extract could be due to the freeze-drying process, which eliminates water content and concentrates bioactive compounds, as well as to the optimization of extraction conditions, which allowed greater recovery of phenolic compounds. These bioactive compounds are found in various dietary sources, such as tea, fruits, and vegetables, and have been shown to offer potential health benefits. Their antioxidant properties contribute to glycemic control, blood pressure regulation, and improved lipid profiles [50]. Additionally, phenolic compounds, due to their nontoxic nature, play a key role in neuroprotection, acting through antioxidant, anti-inflammatory, and autophagy-modulating mechanisms. In animal models, they have demonstrated the ability to inhibit biomarkers of inflammation, oxidative stress, and neurotoxicity. Furthermore, they regulate various signaling pathways involved in immune and neuronal responses. Their effects include protecting neurons from neurotoxins, reducing neuroinflammation, and improving memory and cognitive function [51]. The TPC value not only positions Cauchao berry as a natural source rich in phenolic compounds but also suggests its possible application in the prevention of diseases associated with oxidative stress and neurodegenerative disorders. Flavonoids constitute a class of polyphenolic compounds widely distributed in the cells and surfaces of various plant tissues. Beyond their role in plant pigmentation, these compounds are distinguished by their importance as bioactive agents involved in multiple physiological functions [52]. Flavonoids exhibit diverse biological activities, including antioxidant, anti-inflammatory, antimicrobial, cardioprotective, and anticancer effects, largely attributed to their hydroxyl groups. Compounds such as quercetin, myricetin, and apigenin contribute to the reduction of oxidative stress, cholesterol, and the risk of cardiovascular and chronic diseases [53]. Some flavonoids are being studied for their neuroprotective properties, associated with their antioxidant and anti-inflammatory properties. These compounds can cross the blood-brain barrier and reach specific brain regions. Their therapeutic potential is highlighted in neurodegenerative and ischemic diseases, such as Alzheimer's, Parkinson's, and stroke [54]. Several natural flavonoids have demonstrated neuroprotective effects by reducing oxidative damage, regulating mitochondrial dysfunction, and modulating cellular pathways, such as Nrf2 (nuclear factor erythroid-related factor 2), ERK (extracellular signal-regulated kinase), and Akt (protein kinase B). Compounds such as catechins, ampelopsin, and pinocembrin stand out for their ability to prevent neuronal death. Therefore, flavonoids are emerging as promising therapeutic agents against neurodegeneration in Parkinson's disease [55]. Anthocyanins are natural pigments that give fruits, leaves, and flowers their characteristic red, blue, and purple colors. In addition, they are notable for their potent antioxidant and anti-inflammatory effects, helping to prevent diseases such as diabetes, obesity, hypertension, and certain types of cancer, especially colon cancer. These benefits are largely due to their ability to block NF-κB (nuclear factor kappa-light-chain-enhancer of activated B cells) mediated inflammatory pathways, thus helping to maintain cardiovascular and metabolic health. However, their effect may vary depending on the specific structure of each compound [56]. Anthocyanins are also notable for their potential to combat neurodegenerative diseases. These pathologies share processes such as brain inflammation, oxidative stress, and neuronal damage due to excessive stimuli, where anthocyanins act by blocking and modulating these pathways. Therefore, they are emerging as promising allies in protecting and improving neurological health [57]. The remarkable anthocyanin content of the Cauchao berry, comparable to that of other recognized fruits, supports its potential as a natural source of bioactive compounds with health benefits. These results not only confirm the bioactive richness of the Cauchao berry, but also open a promising field of research towards its nutraceutical and neuroprotective use in humans.

3.5. Antioxidant Potential

Antioxidant potential comes from key compounds (phenols and flavonoids present in fruits and vegetables) that prevent the oxidation of lipids, proteins, and other biomolecules, protecting both food and the human body from damage caused by reactive oxygen species (ROS). In foods, they extend shelf-life while maintaining flavor, color, and texture. In humans, they reduce the risk of chronic diseases such as cardiovascular disease. Antioxidant efficacy depends on their chemical structure, bioaccessibility, and metabolism [58]. It is important to consider that the evaluation of antioxidant potential using the DPPH and ORAC methods provide complementary information, as both are based on different action mechanisms and, therefore, cannot be compared. The determination of the antioxidant capacity of natural extracts can vary considerably depending on the in vitro method used. This variability is due to the fact that no single assay is capable of considering the complexity of the antioxidant mechanisms involved, the distribution of compounds in heterogeneous matrices, or the influence of other components present in the system. Therefore, the use of more than one complementary analytical method is recommended to obtain a more robust and objective assessment of antioxidant capacity [59].

The DPPH assay measures the capacity of antioxidants to donate electrons or hydrogens to a free radical, reflected by the decrease in absorbance assessed at 515–520 nm. The DPPH radical does not exist in biological systems and can interfere, especially in the presence of certain solvents, which can underestimate antioxidant activity. However, it is simple, inexpensive, rapid, and widely used to compare the antioxidant activity of extracts or pure compounds. On the other hand, the ORAC method evaluates the ability of compounds to neutralize peroxyl radicals generated by AAPH (2,2′-azobis(2-methylpropionamidine) dihydrochloride) through the oxidative degradation of a fluorescent molecule (such as fluorescein). The loss of fluorescence indicates oxidative damage, while the presence of antioxidants reduces this loss [60]. The comparable values obtained in similar fruits in both assays indicate that the Cauchao berry possesses a combination of compounds capable of acting through different mechanisms, which increases its relevance from a nutritional and functional perspective. In this context, the results obtained reinforce the potential application of this berry as a bioactive ingredient in the development of functional foods or nutraceuticals aimed at preventing cellular oxidative damage, with a potentially positive impact on human health.

3.6. α-Glucosidase Activity

A notable inhibitory effect of Cauchao berry extract on the enzyme α-glucosidase was observed. Although the precise mechanism of inhibition is not clarified in this study, this effect may be attributed to synergistic interactions among multiple phenolic constituents in the Cauchao berry matrix, which is consistent with the complex phytochemical profile of native berries. From a functional standpoint, this type of enzyme inhibition is particularly relevant for postprandial glycemic regulation, as early modulation of carbohydrate digestion can help mitigate glucose surges following meals. Accordingly, Cauchao extract emerges as a promising candidate for integration into functional foods or nutraceutical products targeting dietary glycemic control, particularly in individuals at risk of insulin resistance or type 2 diabetes. Although its inhibitory potency is lower than that of acarbose (IC_{50} = 0.057 ± 0.005 mg/mL), a clinically used α-glucosidase inhibitor, the Cauchao extract demonstrated stronger activity than several other natural berry extracts reported in the literature, highlighting its competitive potential despite being a crude, unfractionated preparation.

Furthermore, a strong association has been reported between metabolic regulation and the progression of neurodegenerative diseases, including Parkinson's disease. It has been noted that type 2 diabetes not only increases the risk of developing Parkinson's disease but also accelerates its clinical progression [61]. Given that this condition is amenable to dietary and pharmacological interventions, the inhibitory effect of Cauchao extract on the α-glucosidase enzyme is particularly relevant for future studies exploring potential neuroprotective effects.

3.7. Phenolics Compound Profile

In this study, the phenolic profile of Cauchao was analyzed using high-performance liquid chromatography. The results revealed the presence of several bioactive phenolic compounds, including chlorogenic acid, sinapic acid, cyanidin and its glycosylated derivatives, and rutin. Chlorogenic acid (CGA) was the main component identified in the Cauchao extract. This compound, an ester of caffeic acid and quinic acid, is one of the most abundant phenolic acids in nature and is found in foods such as coffee, eggplant, tomato, and various fruits [62]. CGA has been widely studied for its antioxidant and anti-inflammatory properties. CGA has been shown to modulate the production of inflammatory mediators such as TNF-α, IL-1β, IL-6, IL-8, NO, and PGE2, and regulate key signaling pathways such as NF-κB, MAPK, and Nrf2, providing protective effects against cardiovascular disease and diabetes mellitus [63]. Furthermore, CGA has been reported to exert neuroprotective effects in zebrafish (MPTP) and SH-SY5Y (6-OHDA) cell-based Parkinson's disease models, attenuating the loss of dopaminergic neurons, decreasing apoptosis, and improving locomotion [64]. Likewise, kinetic studies have shown that CGA significantly inhibits β-amyloid (Aβ) aggregation, which is implicated in the pathophysiology of Alzheimer's [65]. This evidence supports the potential of Cauchao extract as a source of bioactive compounds, particularly in the prevention and management of diseases related to oxidative stress and inflammation.

Sinapic acid (SA) was the second major compound identified in the Cauchao extract. This compound is a hydroxycinnamic acid and is present in various plant sources, including fruits such as lemons, oranges, and berries, and is also present in cereals and oilseeds [66]. This compound has been widely studied for its antioxidant, anti-inflammatory, and neuroprotective properties. SA has been reported to block LPS-induced activation of mitogen-activated protein kinases (MAPKs) and protein kinase B (AKT) signaling [67]. Furthermore, it has been shown to exert neuroprotective effects in human neuroblastoma (SHSY5Y), with mitigation of anti-inflammatory markers such as IL-1β and TNF-α and decreased ROS levels [68]. The identification of chlorogenic acid and sinapic acid as the major compounds in the Cauchao extract provides preliminary information on its functional potential. While these findings do not allow direct biological effects to be established, they do suggest that the extract could constitute an interesting natural source of bioactive compounds relevant to health.

Finally, the freeze-dried Cauchao berry extract demonstrates a rich and diverse profile of bioactive compounds, including phenolics, flavonoids, anthocyanins, and essential fatty acids, which collectively contribute to its remarkable antioxidant and metabolic properties. The evidence presented highlights not only the berry's nutritional value but also its promising applicability in the development of functional foods, nutraceuticals, and preventive strategies for oxidative stress-related diseases. Future studies are warranted to further explore the bioavailability, molecular mechanisms, and clinical efficacy of Cauchao-derived compounds, paving the way for their potential integration into therapeutic and dietary interventions aimed at improving human health.

4. Materials and Methods

4.1. Raw Material

The fruits of *Amomyrtus luma* (Cauchao berry), as shown in Figure 5, were obtained from artisanal gatherers located in the rural area of Chiloé Island in southern Chile. These gatherers harvest the fruit directly from wild trees that grow naturally on their ancestral lands. The species is well known and traditionally used by local communities, as documented by Fredes et al. (2024) [69], who describe the cultural significance and ethnobotanical recognition of *A. luma* among women gatherers in the extreme south forests of Chile. The plant was identified based on morphological features and traditional knowledge consistent with the literature. The berries were cleaned and carefully selected to ensure they were free from physical damage. One portion was stored in plastic bags and refrigerated at 4 °C until use, while another portion was freeze-dried and stored in a sealed plastic bag at −18 °C until needed.

Figure 5. *Amomyrtus luma* (Cauchao berries). (**a**) General view of the *Amomyrtus luma* tree bearing ripe fruits, photographed in its natural environment in southern Chile. (**b**) Close-up of ripened Cauchao berries on the plant. Author: Marisol Runiahue. Date: February 2025.

4.2. Proximate Composition and Reducing Sugars

Lipid content was measured through gravimetric analysis after Soxhlet extraction. Crude fiber was determined by performing acid and alkaline hydrolysis on insoluble residues. The crude protein content was assessed using the Kjeldahl method, applying a conversion factor of 6.25. Crude ash content was quantified by incinerating the sample in a muffle furnace at 550 °C. Moisture content was analyzed gravimetrically. All procedures adhered to the guidelines set by the Association of Official Analytical Chemists [70]. Available carbohydrates were calculated by subtracting other components, and all analyses were conducted in triplicate. On the other hand, the reduced sugar content was determined using the Miller method [71], which involves the use of dinitrosalicylic acid (DNS) reagent and spectrophotometric measurement at 540 nm. The analysis was performed in triplicate.

4.3. Dietary Fiber Content

The determination of soluble dietary fiber (SDF) and insoluble dietary fiber (IDF) was conducted using the gravimetric-enzymatic method described in AOAC method No. 991.43. The analysis employed a Total Dietary Fiber Test Kit (TDF100A, Sigma-Aldrich, St. Louis, MO, USA) in conjunction with an Enzymatic Digestion Unit and Filtration System (VELP Scientifica, GDE-CSF6, Usmate, Italy). Total dietary fiber (TDF) was calculated as the sum of SDF and IDF. All analyses were performed in triplicate to ensure accuracy.

4.4. Fatty Acids Profile and α-Tocopherols Content

The oil sample from maqui was extracted following the Bligh and Dyer method (1959) [72]. Fatty acid methyl esters were prepared by saponifying the oil sample with 2 N KOH in methanol, followed by vortex agitation for 1 min. Methylated fatty acids were then extracted using 300 μL of n-hexane with an additional agitation step lasting 3 min. The resulting supernatant was analyzed using a gas chromatograph (Shimatzu, GC-2010 Plus,

Kyoto, Japan) equipped with a flame-ionization detector. The chromatographic conditions for the analysis were as follows: an SP 2560 column was used with a detector temperature set at 260 °C and an injector temperature at 250 °C. The oven temperature program began at 100 °C, held for 5 min, followed by an increase of 4 °C/min until reaching 240 °C, which was maintained for 30 min. Helium was used as the carrier gas with a linear velocity of 20 cm/s. The split ratio was 1:100, and the injection volume was 1 μL.

The α-tocopherol content was analyzed following the method described by Quispe-Fuentes et al. (2020) [27]. The procedure involved weighing 25 mg of extract, which was then mixed with 1 mL of methanol/BHT and stirred for 3 min. The resulting supernatant was filtered using 0.45 μm syringe filters and injected into an HPLC system. Quantification of α-tocopherol was conducted using HPLC with fluorescence detection. The analysis employed a Kromasil 100-5 C18 column (250 × 4.6 mm) with a mobile phase composed of methanol (1:1 *v*/*v*) at a flow rate of 1.2 mL/min. The fluorescence detector was set at excitation and emission wavelengths of 295 nm and 325 nm, respectively. Measurements were performed in triplicate, and the results were expressed as μg of α-tocopherol/g extract.

4.5. Ultrasound/Solvent Extraction

The extraction of bioactive compounds will be conducted using freeze-dried samples and an aqueous methanol solution, following the methodology described by Chandra Singh, Probst, Price, and Kelso (2022) [73], with modifications. The process was carried out using an ultrasonic bath (BIOBASE UC-30si, Jinan, China), operating at a frequency of 40 kHz. For each run, 1.0 g of berry powder was mixed with 12 mL of methanol-water solution at a 1:12 (*w*/*v*) ratio. The extraction was performed in a water bath maintained at 35 °C. Key parameters: extraction time (15, 25, and 35 min), ultrasound power (40, 60, and 80%), and solvent concentration (50, 70, and 90%), were optimized through a response surface methodology, as detailed in Section 4.6. After sonication, the mixture was filtered through Whatman No. 1 paper, and the solid residue was re-extracted under the same conditions. Combined extracts were stored at −80 °C until further analysis.

4.6. Optimization of Extraction Method

The experimental design consists of three factors with three levels each: time extraction (15, 25, and 35 min), ultrasound power (40, 60, and 80%), and solvent concentration (50, 70, and 90%), obtaining 27 combinations, respectively. These combinations were created using Rstudio software (version 2024.04.2) with the AlgDesign library and then optimized using the Box–Behnken model, reducing to 17 representative combinations, as shown in Table 3. The factors were codex by the coded.data function according with coded data = ((factor − center point)/amplitude). The experiments will be carried out in triplicate using freeze-drying Cauchao berry samples. The response surface methodology (RSM), implemented through the RSM library in RStudio, was utilized to determine the optimal combination of factors that maximize the extraction of bioactive compounds, evaluated by their antioxidant capacity using the DPPH assay.

4.7. Total Content of Phenolics, Flavonoids, and Anthocyanins

The total bio-compound content was evaluated in optimized Cauchao berry extract for total phenolic content (TPC), total flavonoid content (TFC), and total anthocyanin content (TAC).

The total phenolic content was determined using the Folin–Ciocalteu assay, following the method described by Rodríguez et al. (2014) [74] with slight modifications. Briefly, 15 μL of the extract was combined with 100 μL of Folin–Ciocalteu reagent (0.2 N) in a microplate. Subsequently, 100 μL of sodium carbonate solution (60 mg/mL) was added, and the mixture was incubated in the dark at room temperature for 90 min. After incubation, the TPC was quantified by measuring the absorbance at 750 nm using a multi-plate reader (Perkin-Elmer, Victor™ X3, Turku, Finland). A calibration curve was constructed using gallic acid as the standard, and TPC results were expressed as milligrams of gallic acid equivalents (GAE) per gram of dry matter (d.m.).

The total flavonoid content was quantified following the methodology outlined by Dini et al. (2010) [75] with minor adjustments. A volume of 500 µL of the optimized pomace extract was mixed with 150 µL of sodium nitrite (5%) and vortexed thoroughly. The mixture was left to rest for 5 min before adding 150 µL of aluminum chloride hexahydrate ($AlCl_3{\cdot}6H_2O$, 10%). It was then incubated at room temperature in the dark for 6 min. To terminate the reaction, 1 mL of sodium hydroxide (NaOH, 1 M) was added, followed by the addition of 1.2 mL of ultrapure water. The absorbance of the final mixture was measured at 415 nm using a spectrophotometer (Thermo Scientific, ORION AQUAMATE 8000, Madison, WI, USA). A calibration curve was constructed using quercetin as the standard, and TFC values were expressed as milligrams of quercetin equivalents (QE) per gram of dry matter (d.m.).

The total anthocyanin content (TAC) was determined using the pH differential method, as described in De Souza et al. (2014) [76]. The red cabbage extract was diluted separately in buffers with pH 1.0 and pH 4.5. Absorbance measurements were taken at 510 nm and 700 nm for both buffer solutions. The difference in absorbance (A) was calculated using Equation (2):

$$A = (A_{510} - A_{700})pH_{1.0} - (A_{510} - A_{700})pH_{4.5} \quad (2)$$

The total anthocyanin content, expressed as cyanidin-3-glucoside equivalents, was computed using the Equation (3):

$$\mathrm{TAC} = (\mathrm{A} \times \mathrm{MW} \times \mathrm{DF} \times \mathrm{Ve} \times 1000)/(\varepsilon \times 1 \times \mathrm{M}) \quad (3)$$

In the Equation (2), MW represents the molecular weight of cyanidin-3-glucoside (449 g/mol), DF denotes the dilution factor, Ve is the volume of the extract, ε corresponds to the molar extinction coefficient of cyanidin-3-glucoside (26,900 $L{\cdot}mol^{-1}{\cdot}cm^{-1}$), and M refers to the mass of the sample used for extraction. This method provides a reliable quantification of anthocyanins by leveraging their pH-dependent color changes, making it a widely used approach for evaluating anthocyanin content in plant-based extracts.

4.8. Antioxidant Potential

The functional properties were determined in optimized Cauchao berry extract. The antioxidant capacity was evaluated using the DPPH assay, following the method described by Grajeda-Iglesias et al. (2016) [77] with minor modifications. For the assay, 20 µL of red cabbage extract was mixed with 180 µL of freshly prepared DPPH solution (120 µM in 80% methanol) in 96-well microplates. The reaction mixture was incubated for 30 min in a multi-plate reader (Perkin-Elmer, Victor™ X3, Hamburg, Germany). Trolox served as the standard for constructing a calibration curve ranging from 1 to 500 µM ($y = -0.001x + 0.5855$; $R^2 = 0.9921$). Absorbance was measured at 510 nm, and the results were expressed as µmol Trolox equivalents (TE) per gram of dry matter (d.m.).

The ORAC (Oxygen Radical Absorbance Capacity) assay was conducted according to the methodology described by Uribe et al. (2014) [78]. In a 96-well microplate reader (Perkin Elmer, Victor™ X3, Hamburg, Germany), 50 µL of red cabbage extract was mixed with 40 µL of phosphate buffer (pH 7.4). Subsequently, 200 µL of fluorescein (100 nM) was added to each well, and the mixture was incubated at 37 °C for 20 min. Following incubation, 35 µL of an AAPH solution (0.36 M) was added to initiate the reaction. Fluorescence readings were performed at excitation (λex) and emission (λem) wavelengths of 485 nm and 535 nm, respectively. The calibration curve for the ORAC assay was generated by plotting Trolox concentrations (5–250 μM) against the area under the fluorescence decay curve, yielding the equation: $y = 0.00002x - 26.664$ ($R^2 = 0.9769$). Results were expressed as Trolox equivalents (μmol TE) per gram of dry matter (d.m.).

4.9. α-Glucosidase Activity

The enzymatic activity of α-glucosidase was assessed following the method described by Quispe-Fuentes et al. (2023) [79], with minor modifications. In a 96-well microplate, 50 μL of optimized Cauchao berry extract at varying concentrations (0.2–2 mg/mL) was mixed with 100 μL of α-glucosidase (0.5 U/mL; Saccharomyces cerevisiae, Sigma G5003, Ronkonkoma, NY, USA) and incubated at 25 °C for 10 min. After the preincubation period, 50 μL of 4-nitrophenyl α-D-glucopyranoside (Sigma N1377, St. Gallen, Switzerland) dissolved in 0.1 M phosphate buffer (pH 6.9) was added. The absorbance was measured at 405 nm using a Victor™ 3 multi-label plate reader, with the readings compared to a control containing 50 μL of buffer solution, representing the maximum enzymatic activity. Measurements were recorded every 30 s for 10 min. Sample and concentration were tested in triplicate, and α-glucosidase activity was calculated as a percentage based on the slope of the resulting exponential curve. Acarbose was used as a reference standard to compare the IC_{50} values of the extract.

4.10. Phenolic Compound Profiles

Three injections (20 μL) of each sample were made in the quantification of phenolic compounds was performed by HPLC-DAD analysis (Shimadzu Corporation, Kyoto, Japan) equipped with a Nucleosil® 300 C18 column (25.0 cm × 4.6 mm; 5 μm particle size) [80]. The mobile phase comprised two solvents: eluent A was a mixture of water and formic acid (99:1, *v*/*v*), and eluent B was acetonitrile. The solvent system started with 8% of B, rising to 15% after 25 min, 22% after 55 min, and 40% after 60 min. The calibration curves were obtained from standard solutions at six different concentrations; the graphs of these showed a good correlation of 0.99 for all standards (cyanidin 3-O-glucoside derivative 1, cyanidin 3-O-glucoside derivative 2, quantified as petunidin 3-O-ruthinoside, and quercetin derivatives were quantified as quercetin 3-O-glucoside).

5. Conclusions

This study identifies Cauchao berry as a novel and promising native source of bioactive compounds with potential antioxidant and antidiabetic properties. The optimized extraction process yielded a phytochemical-rich profile characterized by high levels of total phenolics, flavonoids, anthocyanins, essential fatty acids, and α-tocopherol. These compounds could be responsible for the observed antioxidant activity and α-glucosidase inhibition. Furthermore, the presence of chlorogenic and sinapic acids suggests a potential role in mitigating oxidative stress, metabolic dysfunction, and neurodegenerative conditions, as supported by previous literature. This study provides a pioneering contribution to the scientific characterization of Cauchao berries and highlights their potential as a valuable native resource for the development of functional foods and nutraceuticals derived from underexplored species. Future studies should be directed toward exploring the neuroprotective effects of these compounds, particularly in models of oxidative damage and neurodegeneration.

Author Contributions: Methodology, L.S.G.-P. and R.L.V.; formal analysis, L.S.G.-P., J.P. and V.G.; investigation, L.S.G.-P.; writing—original draft preparation, L.S.G.-P.; writing—review and editing, R.L.V.; project administration, L.S.G.-P.; funding acquisition, R.L.V. and L.S.G.-P. All authors have read and agreed to the published version of the manuscript.

Funding: This work was directly funded by Fondecyt N° 1230980 (R.L.V.). We also thank the support from FONDAP program N° 15150012 and Postdoctoral ANID fellowship N° 3240049 (L.S.G.-P.).

Institutional Review Board Statement: Not applicable.

Informed Consent Statement: Not applicable.

Data Availability Statement: Data is contained within the article.

Acknowledgments: The authors sincerely thank Marisol Runiahue, a traditional gatherer from Chiloé Island, for generously providing access to the fruits of *Amomyrtus luma* (Cauchao) and for sharing her photographic documentation of the species. We deeply value her knowledge, experience, and commitment to preserving the cultural and ecological heritage of native fruits in southern Chile.

Conflicts of Interest: The authors declare no conflicts of interest.

References

1. Vega-Galvez, A.; Rodríguez, A.; Stucken, K. Antioxidant, Functional Properties and Health-promoting Potential of Native South American Berries: A Review. *J. Sci. Food Agric.* **2021**, *101*, 364–378. [CrossRef] [PubMed]
2. Pap, N.; Fidelis, M.; Azevedo, L.; Do Carmo, M.A.V.; Wang, D.; Mocan, A.; Pereira, E.P.R.; Xavier-Santos, D.; Sant'Ana, A.S.; Yang, B.; et al. Berry Polyphenols and Human Health: Evidence of Antioxidant, Anti-Inflammatory, Microbiota Modulation, and Cell-Protecting Effects. *Curr. Opin. Food Sci.* **2021**, *42*, 167–186. [CrossRef]
3. Cordero, S.; Abello, L.; Gálvez, F. *Plantas Silvestres Comestibles y Medicinales de Chile y Otras Partes Del Mundo*; Corporación Chilena de la Madera: Concepción, Chile, 2017.
4. Falkenberg, S.S.; Tarnow, I.; Guzman, A.; Mølgaard, P.; Simonsen, H.T. Mapuche Herbal Medicine Inhibits Blood Platelet Aggregation. *Evid.-Based Complement. Altern. Med.* **2012**, *2012*, 647620. [CrossRef]
5. Weyerstahl, P.; Marschall, H.; Landrum, L.R. Constituents of the Leaf Extract of *Amomyrtus meli* (R. A. Philippi) Legrand et Kausel, *Amomyrtus luma* (Molina) Legrand et Kausel and of *Amomyrtella guili* (Speg.) Kausel. *Flavour. Fragr. J.* **1992**, *7*, 247–251. [CrossRef]
6. Archaina, D.; Leiva, G.; Salvatori, D.; Schebor, C. Physical and Functional Properties of Spray-Dried Powders from Blackcurrant Juice and Extracts Obtained from the Waste of Juice Processing. *Food Sci. Technol. Int.* **2018**, *24*, 78–86. [CrossRef]
7. Gagneten, M.; Corfield, R.; Mattson, M.G.; Sozzi, A.; Leiva, G.; Salvatori, D.; Schebor, C. Spray-Dried Powders from Berries Extracts Obtained upon Several Processing Steps to Improve the Bioactive Components Content. *Powder Technol.* **2019**, *342*, 1008–1015. [CrossRef]
8. Chemat, F.; Zill-e-Huma; Khan, M.K. Applications of Ultrasound in Food Technology: Processing, Preservation and Extraction. *Ultrason. Sonochem.* **2011**, *18*, 813–835. [CrossRef]
9. Dzah, C.S.; Duan, Y.; Zhang, H.; Wen, C.; Zhang, J.; Chen, G.; Ma, H. The Effects of Ultrasound Assisted Extraction on Yield, Antioxidant, Anticancer and Antimicrobial Activity of Polyphenol Extracts: A Review. *Food Biosci.* **2020**, *35*, 100547. [CrossRef]
10. Vázquez-Espinosa, M.; González De Peredo, A.V.; Ferreiro-González, M.; Carrera, C.; Palma, M.; Barbero, G.F.; Espada-Bellido, E. Assessment of Ultrasound Assisted Extraction as an Alternative Method for the Extraction of Anthocyanins and Total Phenolic Compounds from Maqui Berries (*Aristotelia chilensis* (Mol.) Stuntz). *Agronomy* **2019**, *9*, 148. [CrossRef]
11. Watrelot, A.A.; Bouska, L. Optimization of the Ultrasound-Assisted Extraction of Polyphenols from Aronia and Grapes. *Food Chem.* **2022**, *386*, 132703. [CrossRef]
12. López, J.; Vera, C.; Bustos, R.; Florez-Mendez, J. Native Berries of Chile: A Comprehensive Review on Nutritional Aspects, Functional Properties, and Potential Health Benefits. *Food Meas.* **2021**, *15*, 1139–1160. [CrossRef]
13. Lemus-Mondaca, R.; Ah-Hen, K.; Vega-Gálvez, A.; Zura-Bravo, L. Effect of High Hydrostatic Pressure on Rheological and Thermophysical Properties of Murtilla (*Ugni molinae* Turcz) Berries. *J. Food Sci. Technol.* **2016**, *53*, 2725–2732. [CrossRef]
14. Al Hasani, S.; Al-Attabi, Z.; Waly, M.; Al-Habsi, N.; Al-Subhi, L.; Shafiur Rahman, M. Polyphenol and Flavonoid Stability of Wild Blueberry (*Sideroxylon mascatense*) during Air- and Freeze-Drying and Storage Stability as a Function of Temperature. *Foods* **2023**, *12*, 871. [CrossRef]
15. Issis, Q.-F.; Antonio, V.-G.; Elsa, U.; Valeria, V.; Nicole, C.; Jacqueline, P. Vacuum Drying Application to Maqui (*Aristotelia chilensis* [Mol] Stuntz) Berry: Weibull Distribution for Process Modelling and Quality Parameters. *J. Food Sci. Technol.* **2019**, *56*, 1899–1908. [CrossRef]
16. Golovinskaia, O.; Wang, C.-K. Review of Functional and Pharmacological Activities of Berries. *Molecules* **2021**, *26*, 3904. [CrossRef]
17. He, Y.; Wang, B.; Wen, L.; Wang, F.; Yu, H.; Chen, D.; Su, X.; Zhang, C. Effects of Dietary Fiber on Human Health. *Food Sci. Hum. Wellness* **2022**, *11*, 1–10. [CrossRef]

18. Cabrera-Barjas, G.; Quezada, A.; Bernardo, Y.; Moncada, M.; Zúñiga, E.; Wilkens, M.; Giordano, A.; Nesic, A.; Delgado, N. Chemical Composition and Antibacterial Activity of Red Murta (*Ugni molinae* Turcz.) Seeds: An Undervalued Chilean Resource. *Food Meas.* **2020**, *14*, 1810–1821. [CrossRef]
19. Gómez-Pérez, L.S.; Moraga, N.; Ah-Hen, K.S.; Rodríguez, A.; Vega-Gálvez, A. Dietary Fibre in Processed Murta (*Ugni molinae* Turcz) Berries: Bioactive Components and Antioxidant Capacity. *J. Food Sci. Technol.* **2022**, *59*, 3093–3101. [CrossRef]
20. Bederska-Łojewska, D.; Pieszka, M.; Marzec, A.; Rudzińska, M.; Grygier, A.; Siger, A.; Cieślik-Boczula, K.; Orczewska-Dudek, S.; Migdał, W. Physicochemical Properties, Fatty Acid Composition, Volatile Compounds of Blueberries, Cranberries, Raspberries, and Cuckooflower Seeds Obtained Using Sonication Method. *Molecules* **2021**, *26*, 7446. [CrossRef]
21. Waehler, R. Fatty Acids: Facts vs. Fiction. *Int. J. Vitam. Nutr. Res.* **2023**, *93*, 268–288. [CrossRef]
22. Marangoni, F.; Agostoni, C.; Borghi, C.; Catapano, A.L.; Cena, H.; Ghiselli, A.; La Vecchia, C.; Lercker, G.; Manzato, E.; Pirillo, A.; et al. Dietary Linoleic Acid and Human Health: Focus on Cardiovascular and Cardiometabolic Effects. *Atherosclerosis* **2020**, *292*, 90–98. [CrossRef] [PubMed]
23. Zorzi, M.; Gai, F.; Medana, C.; Aigotti, R.; Morello, S.; Peiretti, P.G. Bioactive Compounds and Antioxidant Capacity of Small Berries. *Foods* **2020**, *9*, 623. [CrossRef] [PubMed]
24. Murru, E.; Manca, C.; Carta, G.; Banni, S. Impact of Dietary Palmitic Acid on Lipid Metabolism. *Front. Nutr.* **2022**, *9*, 861664. [CrossRef]
25. Lee, G.Y.; Han, S.N. The Role of Vitamin E in Immunity. *Nutrients* **2018**, *10*, 1614. [CrossRef] [PubMed]
26. López, J.; Shun Ah-Hen, K.; Vega-Gálvez, A.; Morales, A.; García-Segovia, P.; Uribe, E. Effects of Drying Methods on Quality Attributes of Murta (*Ugni molinae* Turcz) Berries: Bioactivity, Nutritional Aspects, Texture Profile, Microstructure and Functional Properties. *J. Food Process Eng.* **2017**, *40*, e12511. [CrossRef]
27. Quispe-Fuentes, I.; Vega-Gálvez, A.; Aranda, M.; Poblete, J.; Pasten, A.; Bilbao-Sainz, C.; Wood, D.; McHugh, T.; Delporte, C. Effects of Drying Processes on Composition, Microstructure and Health Aspects from Maqui Berries. *J. Food Sci. Technol.* **2020**, *57*, 2241–2250. [CrossRef]
28. Belščak-Cvitanović, A.; Durgo, K.; Huđek, A.; Bačun-Družina, V.; Komes, D. Overview of Polyphenols and Their Properties. In *Polyphenols: Properties, Recovery, and Applications*; Elsevier: Amsterdam, The Netherlands, 2018; pp. 3–44, ISBN 978-0-12-813572-3.
29. Rodríguez, K.; Ah-Hen, K.S.; Vega-Gálvez, A.; Vásquez, V.; Quispe-Fuentes, I.; Rojas, P.; Lemus-Mondaca, R. Changes in Bioactive Components and Antioxidant Capacity of Maqui, *Aristotelia chilensis* [Mol] Stuntz, Berries during Drying. *LWT* **2016**, *65*, 537–542. [CrossRef]
30. López, J.; Vega-Gálvez, A.; Ah-Hen, K.S.; Rodríguez, A.; Quispe-Fuentes, I.; Delporte, C.; Valenzuela-Barra, G.; Arancibia, Y.; Zambrano, A. Evaluation of the Antioxidant, Anti-Inflammatory, and Anti-Tumoral Properties of Bioactive Compounds Extracted from Murta Berries (*Ugni molinae* T.) Dried by Different Methods. *Front. Plant Sci.* **2023**, *14*, 1095179. [CrossRef]
31. Uribe, E.; Vega-Galvez, A.; Pasten, A.; Ah-Hen, K.S.; Mejias, N.; Sepúlveda, L.; Poblete, J.; Gomez-Perez, L.S. Drying: A Practical Technology for Blueberries (*Vaccinium corymbosum* L.)—Processes and Their Effects on Selected Health-Promoting Properties. *Antioxidants* **2024**, *13*, 1554. [CrossRef]
32. Lutz, M.; Hernández, J.; Henríquez, C. Phenolic Content and Antioxidant Capacity in Fresh and Dry Fruits and Vegetables Grown in Chile. *CyTA J. Food* **2015**, *13*, 541–547. [CrossRef]
33. Mechchate, H.; Es-safi, I.; Louba, A.; Alqahtani, A.S.; Nasr, F.A.; Noman, O.M.; Farooq, M.; Alharbi, M.S.; Alqahtani, A.; Bari, A.; et al. In Vitro Alpha-Amylase and Alpha-Glucosidase Inhibitory Activity and In Vivo Antidiabetic Activity of *Withania frutescens* L. Foliar Extract. *Molecules* **2021**, *26*, 293. [CrossRef] [PubMed]
34. Wattanathorn, J.; Kawvised, S.; Thukham-mee, W. Encapsulated Mulberry Fruit Extract Alleviates Changes in an Animal Model of Menopause with Metabolic Syndrome. *Oxid. Med. Cell. Longev.* **2019**, *2019*, 5360560. [CrossRef]
35. Escobar-Beiza, N.; Pérez-Correa, J.R.; Franco, W. Fermentation of Murta (*Ugni molinae*) Juice: Effect on Antioxidant Activity and Control of Enzymes Associated with Glucose Assimilation. *Int. J. Mol. Sci.* **2023**, *24*, 15197. [CrossRef]
36. Robbins, R.J. Phenolic Acids in Foods: An Overview of Analytical Methodology. *J. Agric. Food Chem.* **2003**, *51*, 2866–2887. [CrossRef]
37. Khatri, D.; Chhetri, S.B.B. Reducing Sugar, Total Phenolic Content, and Antioxidant Potential of Nepalese Plants. *BioMed Res. Int.* **2020**, *2020*, 7296859. [CrossRef] [PubMed]
38. Zhang, G.; Wang, D.; Ding, Y.; Zhang, J.; Ding, Y.; Lyu, F. Effect and Mechanism of Insoluble Dietary Fiber on Postprandial Blood Sugar Regulation. *Trends Food Sci. Technol.* **2024**, *146*, 104354. [CrossRef]
39. Bai, X.; He, Y.; Quan, B.; Xia, T.; Zhang, X.; Wang, Y.; Zheng, Y.; Wang, M. Physicochemical Properties, Structure, and Ameliorative Effects of Insoluble Dietary Fiber from Tea on Slow Transit Constipation. *Food Chem. X* **2022**, *14*, 100340. [CrossRef] [PubMed]
40. Guan, Z.-W.; Yu, E.-Z.; Feng, Q. Soluble Dietary Fiber, One of the Most Important Nutrients for the Gut Microbiota. *Molecules* **2021**, *26*, 6802. [CrossRef]
41. Alarcon-Gil, J.; Sierra-Magro, A.; Morales-Garcia, J.A.; Sanz-SanCristobal, M.; Alonso-Gil, S.; Cortes-Canteli, M.; Niso-Santano, M.; Martínez-Chacón, G.; Fuentes, J.M.; Santos, A.; et al. Neuroprotective and Anti-Inflammatory Effects of Linoleic Acid in Models of Parkinson's Disease: The Implication of Lipid Droplets and Lipophagy. *Cells* **2022**, *11*, 2297. [CrossRef]

42. Froyen, E.; Burns-Whitmore, B. The Effects of Linoleic Acid Consumption on Lipid Risk Markers for Cardiovascular Disease in Healthy Individuals: A Review of Human Intervention Trials. *Nutrients* **2020**, *12*, 2329. [CrossRef]
43. Santa-María, C.; López-Enríquez, S.; Montserrat-de La Paz, S.; Geniz, I.; Reyes-Quiroz, M.E.; Moreno, M.; Palomares, F.; Sobrino, F.; Alba, G. Update on Anti-Inflammatory Molecular Mechanisms Induced by Oleic Acid. *Nutrients* **2023**, *15*, 224. [CrossRef] [PubMed]
44. Bhattacharjee, B.; Pal, P.K.; Chattopadhyay, A.; Bandyopadhyay, D. Oleic Acid Protects against Cadmium Induced Cardiac and Hepatic Tissue Injury in Male Wistar Rats: A Mechanistic Study. *Life Sci.* **2020**, *244*, 117324. [CrossRef] [PubMed]
45. Shramko, V.S.; Polonskaya, Y.V.; Kashtanova, E.V.; Stakhneva, E.M.; Ragino, Y.I. The Short Overview on the Relevance of Fatty Acids for Human Cardiovascular Disorders. *Biomolecules* **2020**, *10*, 1127. [CrossRef] [PubMed]
46. Wang, X.; Zhang, C.; Bao, N. Molecular Mechanism of Palmitic Acid and Its Derivatives in Tumor Progression. *Front. Oncol.* **2023**, *13*, 1224125. [CrossRef]
47. Engwa, G.A.; Nweke, F.N.; Nkeh-Chungag, B.N. Free Radicals, Oxidative Stress-Related Diseases and Antioxidant Supplementation. *Altern. Ther. Health Med.* **2022**, *28*, 114–128.
48. Sozen, E.; Demirel, T.; Ozer, N.K. Vitamin E: Regulatory Role in the Cardiovascular System. *IUBMB Life* **2019**, *71*, 507–515. [CrossRef]
49. Pelczarski, M.; Wolaniuk, S.; Zaborska, M.; Sadowski, J.; Sztangreciak-Lehun, A.; Bułdak, R.J. The Role of α-Tocopherol in the Prevention and Treatment of Alzheimer's Disease. *Mol. Cell Biochem.* **2025**, *480*, 3385–3398. [CrossRef]
50. Rana, A.; Samtiya, M.; Dhewa, T.; Mishra, V.; Aluko, R.E. Health Benefits of Polyphenols: A Concise Review. *J. Food Biochem.* **2022**, *46*, e14264. [CrossRef]
51. Tavan, M.; Hanachi, P.; De La Luz Cádiz-Gurrea, M.; Segura Carretero, A.; Mirjalili, M.H. Natural Phenolic Compounds with Neuroprotective Effects. *Neurochem. Res.* **2024**, *49*, 306–326. [CrossRef]
52. Mutha, R.E.; Tatiya, A.U.; Surana, S.J. Flavonoids as Natural Phenolic Compounds and Their Role in Therapeutics: An Overview. *Futur. J. Pharm. Sci.* **2021**, *7*, 25. [CrossRef]
53. Ekalu, A.; Habila, J.D. Flavonoids: Isolation, Characterization, and Health Benefits. *Beni-Suef Univ. J. Basic Appl. Sci.* **2020**, *9*, 45. [CrossRef]
54. Hasan, S.; Khatri, N.; Rahman, Z.N.; Menezes, A.A.; Martini, J.; Shehjar, F.; Mujeeb, N.; Shah, Z.A. Neuroprotective Potential of Flavonoids in Brain Disorders. *Brain Sci.* **2023**, *13*, 1258. [CrossRef] [PubMed]
55. Kim, T.Y.; Leem, E.; Lee, J.M.; Kim, S.R. Control of Reactive Oxygen Species for the Prevention of Parkinson's Disease: The Possible Application of Flavonoids. *Antioxidants* **2020**, *9*, 583. [CrossRef]
56. Alam, M.A.; Islam, P.; Subhan, N.; Rahman, M.M.; Khan, F.; Burrows, G.E.; Nahar, L.; Sarker, S.D. Potential Health Benefits of Anthocyanins in Oxidative Stress Related Disorders. *Phytochem. Rev.* **2021**, *20*, 705–749. [CrossRef]
57. Zaa, C.A.; Marcelo, Á.J.; An, Z.; Medina-Franco, J.L.; Velasco-Velázquez, M.A. Anthocyanins: Molecular Aspects on Their Neuroprotective Activity. *Biomolecules* **2023**, *13*, 1598. [CrossRef]
58. Gulcin, İ. Antioxidants and Antioxidant Methods: An Updated Overview. *Arch. Toxicol.* **2020**, *94*, 651–715. [CrossRef]
59. Lu, W.; Shi, Y.; Wang, R.; Su, D.; Tang, M.; Liu, Y.; Li, Z. Antioxidant Activity and Healthy Benefits of Natural Pigments in Fruits: A Review. *Int. J. Mol. Sci.* **2021**, *22*, 4945. [CrossRef] [PubMed]
60. Munteanu, I.G.; Apetrei, C. Analytical Methods Used in Determining Antioxidant Activity: A Review. *Int. J. Mol. Sci.* **2021**, *22*, 3380. [CrossRef]
61. Athauda, D.; Evans, J.; Wernick, A.; Virdi, G.; Choi, M.L.; Lawton, M.; Vijiaratnam, N.; Girges, C.; Ben-Shlomo, Y.; Ismail, K.; et al. The Impact of Type 2 Diabetes in Parkinson's Disease. *Mov. Disord.* **2022**, *37*, 1612–1623. [CrossRef]
62. Wang, L.; Pan, X.; Jiang, L.; Chu, Y.; Gao, S.; Jiang, X.; Zhang, Y.; Chen, Y.; Luo, S.; Peng, C. The Biological Activity Mechanism of Chlorogenic Acid and Its Applications in Food Industry: A Review. *Front. Nutr.* **2022**, *9*, 943911. [CrossRef]
63. Huang, J.; Xie, M.; He, L.; Song, X.; Cao, T. Chlorogenic Acid: A Review on Its Mechanisms of Anti-Inflammation, Disease Treatment, and Related Delivery Systems. *Front. Pharmacol.* **2023**, *14*, 1218015. [CrossRef] [PubMed]
64. Gao, X.; Zhang, B.; Zheng, Y.; Liu, X.; Rostyslav, P.; Finiuk, N.; Sik, A.; Stoika, R.; Liu, K.; Jin, M. Neuroprotective Effect of Chlorogenic Acid on Parkinson's Disease like Symptoms through Boosting the Autophagy in Zebrafish. *Eur. J. Pharmacol.* **2023**, *956*, 175950. [CrossRef] [PubMed]
65. Majid, A.; Garg, S. Modeling Inhibitory Effects of Chlorogenic Acid on Amyloid Beta Aggregation. *Ind. Eng. Chem. Res.* **2024**, *63*, 7636–7645. [CrossRef]
66. Pandi, A.; Kalappan, V.M. Pharmacological and Therapeutic Applications of Sinapic Acid—An Updated Review. *Mol. Biol. Rep.* **2021**, *48*, 3733–3745. [CrossRef]
67. Yasser, M.B.; Hagag, R.S.; El-Sayed, N.M.; Hazem, R.M. Sinapic Acid: A Brief Review of Its Therapeutic Potential and Molecular Targets in Parkinson's Disease. *Rec. Pharm. Biomed. Sci.* **2025**, *9*, 1–7. [CrossRef]
68. Prabhakar, P.; Ahmed, B.A.; Chidambaram, S.B.; Kumar, A.; Pandian, A. In Vitro Ameliorative Effects of Sinapic Acid on Parkinson Related Neurotoxicity in SHSY5Y Cell Lines. *Int. J. Nutr. Pharmacol. Neurol. Dis.* **2023**, *13*, 16–24. [CrossRef]

69. Fredes, C.; Parada, A.; Aguirre, C.; Rojas, L.; Robert, P.; Bernales, M. We Only Collect What We Need: Women's Experiences on Collecting Cauchao (Amomyrtus Luma Molina) in Food Systems of Extreme South Forests in Chile. *Sustainability* **2024**, *16*, 9460. [CrossRef]
70. AOAC. *Official Method of Analysis*, 15th ed.; Association of Official Analytical Chemists: Arlington, VA, USA, 1990.
71. Miller, G.L. Use of Dinitrosalicylic Acid Reagent for Determination of Reducing Sugar. *Anal. Chem.* **1959**, *31*, 426–428. [CrossRef]
72. Bligh, E.G.; Dyer, W.J. A rapid method of total lipid extraction and purification. *Can. J. Biochem. Physiol.* **1959**, *37*, 911–917. [CrossRef]
73. Chandra, M.; Probst, Y.; Price, W.; Kelso, C. Relative comparisons of extraction methods and solvent composition for Australian blueberry anthocyanins. *J. Food Compos. Anal.* **2022**, *105*, 104232. [CrossRef]
74. Rodríguez, K.; Ah-Hen, K.; Vega-Gálvez, A.; López, J.; Quispe-Fuentes, I.; Lemus-Mondaca, R.; Gálvez-Ranilla, L. Changes in Bioactive Compounds and Antioxidant Activity during Convective Drying of Murta (*Ugni molinae* T.) Berries. *Int. J. Food Sci. Technol.* **2014**, *49*, 990–1000. [CrossRef]
75. Dini, I.; Tenore, G.C.; Dini, A. Antioxidant Compound Contents and Antioxidant Activity before and after Cooking in Sweet and Bitter Chenopodium Quinoa Seeds. *LWT-Food Sci. Technol.* **2010**, *43*, 447–451. [CrossRef]
76. Souza, V.B.D.; Fujita, A.; Thomazini, M.; Da Silva, E.R.; Lucon, J.F.; Genovese, M.I.; Favaro-Trindade, C.S. Functional Properties and Stability of Spray-Dried Pigments from Bordo Grape (*Vitis labrusca*) Winemaking Pomace. *Food Chem.* **2014**, *164*, 380–386. [CrossRef] [PubMed]
77. Grajeda-Iglesias, C.; Salas, E.; Barouh, N.; Baréa, B.; Panya, A.; Figueroa-Espinoza, M.C. Antioxidant Activity of Protocatechuates Evaluated by DPPH, ORAC, and CAT Methods. *Food Chem.* **2016**, *194*, 749–757. [CrossRef]
78. Uribe, E.; Lemus-Mondaca, R.; Vega-Gálvez, A.; Zamorano, M.; Quispe-Fuentes, I.; Pasten, A.; Di Scala, K. Influence of Process Temperature on Drying Kinetics, Physicochemical Properties and Antioxidant Capacity of the Olive-Waste Cake. *Food Chem.* **2014**, *147*, 170–176. [CrossRef] [PubMed]
79. Quispe-Fuentes, I.; Uribe, U.E.; Vega-Gálvez, A.; Poblete, G.J.; Olmos, C.A.; Pasten, C.A. Solar Drying of Flame Seedless (*Vitis vinifera* L.) Grape after Different Pretreatments: Characterization of Raisin's Physicochemical and Functional Properties. *Food Meas.* **2023**, *17*, 2755–2766. [CrossRef]
80. Martins, M.S.; Azevedo, R.; Alves, G.; Almeida, A.; De Pinho, P.G.; Garcia-Viguera, C.; Moreno, D.A.; Silva, L.R.; Gonçalves, A.C. Assessment of Chemical Composition of Blackberries and Mulberries from Covilhã Region, Portugal. *J. Food Compos. Anal.* **2025**, *137*, 106832. [CrossRef]

International Journal of
Molecular Sciences

MDPI

Article

Variation in the Chemical Composition of Small Cranberry (*Vaccinium oxycoccos* L.) Fruits Collected from a Bog-Type Habitat in Lithuania

Mindaugas Liaudanskas [1,2,*], Rima Šedbarė [1], Irmantas Ramanauskas [3] and Valdimaras Janulis [1]

[1] Department of Pharmacognosy, Faculty of Pharmacy, Lithuanian University of Health Sciences, LT-50162 Kaunas, Lithuania; rima.sedbare@lsmu.lt (R.Š.); valdimaras.janulis@lsmu.lt (V.J.)
[2] Institute of Pharmaceutical Technologies, Faculty of Pharmacy, Lithuanian University of Health Sciences, LT-50162 Kaunas, Lithuania
[3] Department of Languages and Education, Faculty of Medicine, Lithuanian University of Health Sciences, LT-50275 Kaunas, Lithuania; irmantas.ramanauskas2@lsmu.lt
* Correspondence: mindaugas.liaudanskas@lsmu.lt

Abstract

This study revealed variations in the composition and in vitro antioxidant activity of proanthocyanidins, hydroxycinnamic acid derivatives, flavonols, anthocyanins, and triterpene compounds in small cranberry fruit samples collected from a bog-type natural habitat in Lithuania during intensive ripening of the fruit. The highest total amounts of identified flavonols were determined at the beginning of fruit ripening on September 10 (1232.84 ± 31.73 µg/g). The highest total amounts of proanthocyanidins (1.85 ± 0.02 mg EE/g, $p < 0.05$), anthocyanins (4096.79 ± 5.93 µg/g, $p < 0.05$), and triterpene compounds (8248.46 ± 125.60 µg/g, $p < 0.05$) were detected in small cranberry fruit samples collected in the middle of the ripening period (September 16–18). The most potent in vitro antiradical and reducing activity was found in extracts of small cranberry fruit collected on September 10 (95.25 ± 1.15 µmol TE/g and 159.26 ± 0.77 µmol/g, respectively). The strongest correlation was found between the total content of hydroxycinnamic acid derivatives in the small cranberry fruit samples and the in vitro reducing activity of their extracts (0.858, $p < 0.01$). Among the individual compounds, the strongest correlation was observed between the amounts of isoquercitrin and guaijaverin in *V. oxycoccos* fruit samples and the in vitro reducing activity as assessed by the CUPRAC method (0.844, $p < 0.01$ and 0.769, $p < 0.05$, respectively).

Keywords: anthocyanins; antioxidant activity in vitro; flavonols; hydroxycinnamic acid derivatives; proanthocyanidins; ripening period; triterpene compounds; *Vaccinium oxycoccos* L.

1. Introduction

The use of herbal medicines and dietary supplements has increased significantly over the last decades [1]. This trend has led to the growing popularity of medicinal plant materials, which accumulate biologically active compounds with a wide range of pharmacological effects. Such compounds protect structural molecules in the human body from damage caused by various exogenous and endogenous agents. The growing interest in plant materials encourages a wider application of modern methods for studying their chemical composition and biological effects. It also promotes the development of research-based recommendations for the use of botanical preparations. Plant materials and products derived from plants of the family *Ericaceae* Juss. have strong antioxidant properties [2], making them important for practical medicine, healthy eating, cosmetics, and the food industry.

Small cranberry (*Vaccinium oxycoccos* L.), a member of the *Ericaceae* family and the genus *Vaccinium* L., is widespread in the cold and temperate zones of Eurasia and North America [3]. In Lithuania, the largest cranberry habitats are located in areas of raised bogs and intermediate wetlands. In Lithuania, land reclamation in the second half of the 20th century drained a large number of raised bogs and intermediate wetlands, which reduced the areas of the natural habitats of small cranberries [4]. Conservation of wetlands is relevant for the preservation and development of small cranberry habitats. In Lithuania, wetlands are distributed throughout the country; most of them are small and cover areas of up to 50 ha [5]. Small cranberry fruits have been used for food and folk medicine in the Baltic Sea region since ancient times and have attracted the attention of scientists as well as health and nutrition experts. In food processing, fruits are used to produce jams, juices, sauces, jellies, and soft drinks [6]. In folk medicine, the medicinal plant material of small cranberries is used to treat diseases of the digestive and urinary systems, while the juice is used to treat rheumatism, throat diseases, general body weakness, and avitaminosis [7]. Small cranberry fruits have been found to contain flavonols (quercetin and myricetin glycosides), proanthocyanidins, hydroxycinnamic acid derivatives, anthocyanins, triterpene compounds, organic acids, and vitamins [4,8]. Triterpene compounds of small cranberry fruits exhibit pronounced anti-inflammatory activity [9]. Flavonols, proanthocyanidins, phenolic acids, and anthocyanins have been found to have antioxidant effects [10]. Plant materials and preserves rich in phenolic compounds have been shown to be of value in the treatment of oncological, cardiovascular, and neurodegenerative diseases [11].

Thus, it is relevant to investigate the qualitative and quantitative variation in the composition of biologically active compounds in plant materials by applying modern techniques used for the analysis of plant secondary metabolites. Changes in the qualitative and quantitative composition of secondary metabolites during the ripening of small cranberry fruit influence their biological effects, determining the qualitative composition and the organoleptic properties of the botanical material [12]. It is also important to determine and evaluate the chemical composition of small cranberry fruits growing under Lithuanian climatic conditions in order to identify (i) the compounds that determine their antioxidant effect and (ii) the patterns of variation in the in vitro antiradical and reducing activity of the bioactive compounds in small cranberry fruit extracts. Detailed studies on the phytochemical composition of small cranberry fruit will ensure the preparation of high-quality small cranberry plant material, will allow for the rational use of plant resources by determining the optimum timing of small cranberry fruit picking and will provide research-based recommendations for the rational harvesting of small cranberry fruit.

The aim of the study was to determine the variation in the phytochemical composition and in vitro antioxidant activity of small cranberry fruit samples collected in a bog-type natural habitat in Lithuania during intensive ripening of fruit.

2. Results and Discussion

2.1. Determination of the Composition of Total Proanthocyanidins, Hydroxycinnamic Acid Derivatives, Individual Flavonols, and Chlorogenic Acid

Changes in the qualitative and quantitative composition of secondary metabolites in small cranberry fruits during their development and ripening determine their quality and may influence their effects on biological systems in the human body. It is thus relevant to investigate the variation in the qualitative and quantitative composition of phenolic compounds in small cranberry fruits growing in natural habitats in the territory of the Republic of Lithuania during their development and ripening.

Anthocyanins, flavonols, proanthocyanidins, and phenolic acids are the main groups of biologically active compounds in small cranberry plant materials [10]. Kylli et al. [13], using ultra-high-performance liquid chromatography coupled with photodiode array and fluorescence detectors (UHPLC-PDA-FL), found that proanthocyanidins accounted for 63%, anthocyanins for 16%, flavonols for 14%, hydroxycinnamic acids for 7%, and hydroxybenzoic acids for 0.05% of the total amount of phenolic compounds in *V. oxycoccos* fruit samples.

The proanthocyanidins found in small cranberry fruit are important for the prevention and complementary treatment of urinary tract infections [3-Jurikova]. A complex of type A trimeric proanthocyanidins inhibits the adhesion of a uropathogenic urinary-tract-infection-causing *Escherichia coli* strain to urinary epithelial cell receptors [14]. Herbal preparations of some small cranberry species have traditionally been used to alleviate symptoms of mild recurrent lower urinary tract infections [7]. Proanthocyanidins are found in plants of the genus *Vaccinium*, but some proanthocyanidins are analytical markers of the chemical composition of small cranberry fruit raw material [15].

In our study, we used spectrophotometric analysis to evaluate the proanthocyanidin content of small cranberry fruit samples collected during their development and ripening. The total proanthocyanidin content of the samples was expressed as the (−)-epicatechin equivalent (EE) and ranged from 1.33 mg EE/g to 1.85 mg EE/g. The mean total proanthocyanidin content of the samples analyzed was calculated to be 1.55 ± 0.17 mg EE/g. The highest total proanthocyanidin content (1.85 ± 0.02 mg EE/g, $p < 0.05$) was found in the small cranberry fruit samples collected on September 18. The conditions during the fruit ripening period that influenced the changes in proanthocyanidin content were primarily related to the intensive fruit ripening stage [16,17], which is characterized by significant physiological and biochemical changes in the fruit. During the intensive fruit ripening stage, the metabolic activity of the plant increases, leading to enhanced synthesis and accumulation of secondary metabolites, including proanthocyanidins. Environmental factors such as temperature [18], light intensity [19], and water availability [20] during ripening may also contribute to the regulation of proanthocyanidin biosynthesis by affecting enzymatic activities and gene expression related to flavonoid pathways. Therefore, the observed increase in proanthocyanidin content reflects the combined effect of the fruit's developmental stage and the prevailing environmental conditions. The lowest total proanthocyanidin content (1.33 ± 0.04 mg EE/g) was found in the *V. oxycoccos* fruit samples collected on October 3, but it was not statistically significantly different from the total proanthocyanidin content of small cranberry fruit samples collected on September 11 or September 16.

The data presented by Jungfer et al. [21] indicate that proanthocyanidins accumulate in larger amounts in large cranberry fruit samples than in small cranberry fruit samples. Šedbarė et al. [22] reported that total amounts of proanthocyanidins found in small cranberry fruit samples collected in protected areas of Lithuania ranged from 0.92 mg EE/g to 3.04 mg EE/g. The highest total proanthocyanidin content (3.04 ± 0.14 mg EE/g) was found in fruit samples collected from mesotrophic-type wetland in the southern part of Lithuania (Žuvintas Biosphere Reserve) [22]. The levels found were 1.6–2.3 times higher than the total amount of proanthocyanidins found in our samples collected during the period of intense fruit ripening. The trends of the changes in the quantitative composition of proanthocyanidins found in the studies by Vvedenskaya et al. [16] and Wang et al. [23] differ from our results. These researchers indicate that the concentration of proanthocyanidins decreases most intensively at the beginning of small cranberry fruit ripening and increases later on in the ripening period [16,23]. During our study, the highest total amount of proanthocyanidins was determined in the middle of the intensive ripening period of the small cranberries. In order to assess the range of variation in the total proanthocyanidin content, a coefficient of variation (10.91%) was calculated, which shows a slight variation in the total proanthocyanidin content during the ripening period of the *V. oxycoccos* fruit. The variation in the total proanthocyanidin content in the small cranberry fruit samples during the ripening period is presented in Figure 1.

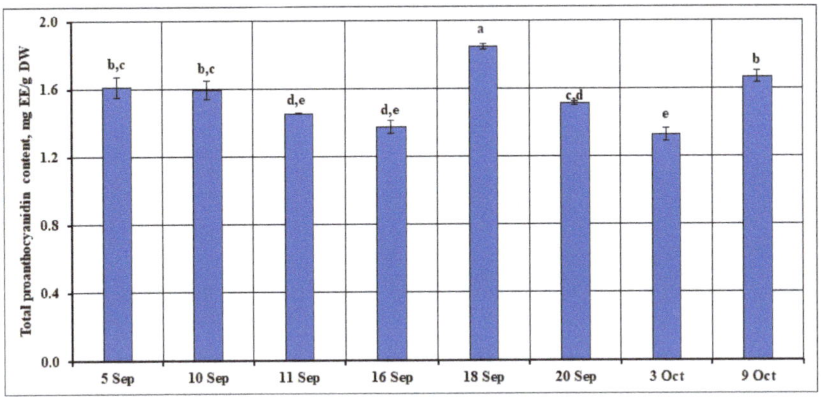

Figure 1. Variation in the total amount of proanthocyanidins in small cranberry fruit samples during the ripening period. Different letters indicate statistically significant differences in the total amount of proanthocyanidins between the *V. oxycoccos* fruit samples tested ($p < 0.05$).

Hydroxycinnamic acids identified in medicinal plant materials have a wide range of biological effects: antioxidant, antimicrobial, anticancer, cardioprotective, hepatoprotective, anti-inflammatory, neuroprotective, antidiabetic, anti-aging, and other biological effects [24].

The analysis using the spectrophotometric method showed that the total content of hydroxycinnamic acid derivatives in small cranberry fruit samples ranged from 11.85 mg ChAE (chlorogenic acid equivalent)/g to 19.27 mg ChAE/g. The calculated mean total amount of hydroxycinnamic acid derivatives in small cranberry fruit samples was 15.04 ± 2.43 mg ChAE/g. The highest total amounts of hydroxycinnamic acid derivatives (19.27 ± 2.01 mg ChAE/g) were found in samples of *V. oxycoccos* fruit collected at the beginning of the ripening period on September 10. At the beginning of fruit ripening, the biosynthetic pathways responsible for the production of hydroxycinnamic acid derivatives are highly active, as these compounds play a role in plant defense mechanisms and in regulating oxidative processes during early fruit development [25]. As ripening progresses, the concentrations of phenolic acids mostly decrease [26], and the plant's metabolism shifts toward the synthesis of other secondary metabolites, such as anthocyanins and proanthocyanidins, which contribute to coloration and other ripening-related traits [27]. Therefore, the higher levels of hydroxycinnamic acid derivatives observed at the onset of ripening reflect their functional role in the early stages of fruit maturation and their subsequent decline as other compounds accumulate. As the fruit ripened, the total amount of hydroxycinnamic acid derivatives decreased. The lowest total amount of hydroxycinnamic acid derivatives (11.85 ± 0.44 mg ChAE/g) was detected in samples of small cranberry fruit collected at the end of the ripening period on October 3. The data from studies conducted by other researchers are consistent with our findings, which show a decrease in the amount of phenolic acids in small cranberry fruit samples during the ripening period. Oszmiański et al. [28] report that the content of phenolic acids decreased remarkably (by 12.3–36.5%) in fruit samples of different *V. macrocarpon* cultivars. The highest amounts of hydroxycinnamic acid derivatives at the beginning of the ripening period were also found in fruit samples of *Vaccinium corymbosum* L. [29]. The calculated coefficient of variation (16.12%) showed a slight variation in the total amount of hydroxycinnamic acid derivatives during the ripening of small cranberry fruits. The variation in the total amount of hydroxycinnamic acid derivatives during the ripening of *V. oxycoccos* fruits is presented in Figure 2.

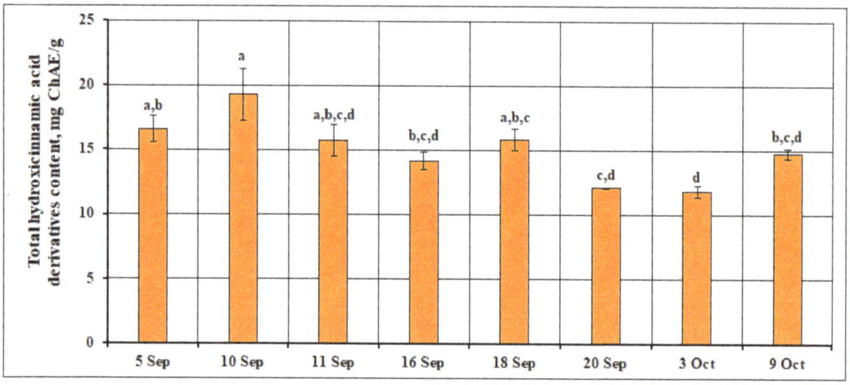

Figure 2. Variation in the total amount of hydroxycinnamic acid derivatives in small cranberry fruit samples during the ripening period. Different letters indicate statistically significant differences in the total amount of hydroxycinnamic acid derivatives between the *V. oxycoccos* fruit samples tested ($p < 0.05$).

In the fruit samples of *V. oxycoccos*, we identified and quantified chlorogenic acid, which belongs to the group of phenolic acids, as well as other compounds of the flavonol group: myricetin and its glycoside myricetin-3-galactoside, as well as quercetin and its glycosides hyperoside (quercetin-3-galactoside), isoquercitrin (quercetin-3-glucoside), guaijaverin (quercetin-3-arabinopyranoside), avicularin (quercetin-3-arabinofuranoside), and quercitrin (quercitrin-3-rhamnoside).

The highest total amount of identified flavonols was determined in the samples of *V. oxycoccos* fruits collected on September 10 and September 18 (1232.84 ± 31.73 µg/g and 1159.85 ± 13.77 µg/g, respectively). The percentage of myricetin-3-galactoside content among the total amount of the identified flavonol compounds ranged from 25.66% to 39.41%, and the percentage of hyperoside content ranged from 31.76% to 37.27%. The percentage of avicularin in small cranberry fruit samples ranged from 10.42% to 14.64%. The highest content of this compound (180.51 ± 4.64 µg/g, $p < 0.05$) was found in fruit samples of *V. oxycoccos* collected on September 10, while the lowest content (96.89 ± 1.48 µg/g, $p < 0.05$) was found in fruit samples of *V. oxycoccos* collected on September 16. The percentage of hyperoside in small cranberry fruit samples collected in northern Lithuania (Kamanos State Strict Nature Reserve) and southern Lithuania (Žuvintas Biosphere Reserve) was found to be slightly higher (38.33%) than in the samples we analyzed. Meanwhile, the percentage of myricetin-3-galactoside (31.38%) and avicularin (12.37%) corresponded to the percentage of this compound in the fruit samples of *V. oxycoccos* collected in our study [22]. The percentage of avicularin (12.37%) in *V. oxycoccos* fruit samples collected in northern Lithuania (Kamanos State Strict Nature Reserve) and southern Lithuania (Žuvintas Biosphere Reserve) is in line with the results of our study (12.37%). Wang et al. [23] reported that the hyperoside content in *V. macrocarpon* fruit samples accounted for 31% to 46%, and myricetin-3-galactoside content accounted for 19% to 32% of the total quantified flavonol content. The highest amounts of myricetin-3-galactoside were detected in *V. oxy-*

coccos fruit samples collected on September 5 and September 18 (442.80 ± 15.61 μg/g and 454.67 ± 5.38 μg/g, respectively). The highest hyperoside content (459.49 ± 11.82 μg/g, $p < 0.05$) was found in small cranberry fruit samples collected on September 10. The lowest levels of myricetin-3-galactoside and hyperoside were found in *V. oxycoccos* fruit samples collected on September 16 (252.93 ± 3.87 μg/g and 247.70 ± 3.78 μg/g, respectively). The coefficient of variation for myricetin-3-O-galactoside was 20.37%, and for hyperoside, it was 17.97%. Oszmiański et al. [28] reported higher hyperoside (276.2–443.3 mg/100 g) and myricetin-3-O-galactoside (143.1–392.0 mg/100 g) contents in samples of large cranberry fruit. The flavonoid biosynthesis process can be influenced by temperature [30], soil water content [31], and other factors, which were not evaluated in this study. Urbstaite et al. [32] found higher levels of avicularin (293.79–613.80 μg/g) in *V. macrocarpon* fruit samples than those detected in small cranberry samples analyzed in this study. The percentage of hyperoside (34.9%), myricetin-3-O-galactoside (30.8%), and avicularin (12.3%) detected in small cranberry fruit samples collected in southern Lithuania (Čepkeliai State Strict Nature Reserve) were within the range of the results of our study [33]. The variation in the amount of myricetin-3-galactoside, hyperoside, and avicularin in small cranberry fruit samples during the ripening period is presented in Figure 3.

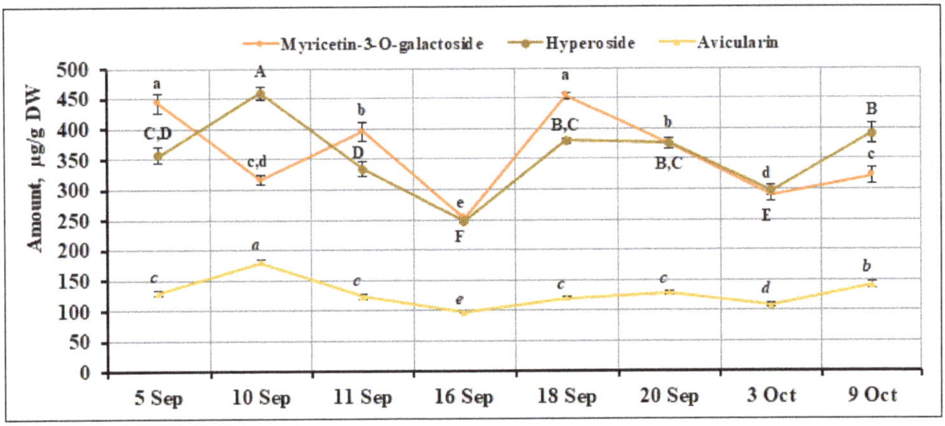

Figure 3. Variation in the quantitative composition of myricetin-3-O-galactoside, hyperoside, and avicularin in small cranberry fruit samples during the ripening period. Different letters (a–e for myricetin-3-O-galactoside, A–F for hyperoside, *a–e* for avicularin) indicate statistically significant differences in the content of these compounds between the *V. oxycoccos* fruit samples tested ($p < 0.05$).

We analyzed the quantitative composition of other flavonol compounds in small cranberry fruit samples. The percentage of isoquercitrin content varied from 5.26% to 6.91% of the total amount of the identified and quantified flavonol compounds. The highest content (81.44 ± 2.10 μg/g, $p < 0.05$) of isoquercitrin was found in fruit samples of *V. oxycoccos* collected on September 10, while the lowest content was found in fruit samples collected on September 16 and October 3 (53.07 ± 0.83 μg/g and 48.04 ± 1.57 μg/g, respectively, Figure 4). The percentage of quercitrin among the flavonol group compounds ranged from 1.39% to 4.57%. Its highest amount (56.32 ± 1.45 μg/g, $p < 0.05$) was detected in *V. oxycoccos* fruit samples collected on September 10, while the lowest amount (14.64 ± 0.38 μg/g, $p < 0.05$) was detected in fruit samples collected on September 20 (Figure 4). The coefficient of variation for quercitrin was calculated to be 37.82%, demonstrating a significant variation in the quantitative composition of this compound during the ripening period. The highest level of myricetin (27.34 ± 0.34 μg/g, $p < 0.05$) was detected on September 18, while the lowest levels were detected in fruit samples collected on September 5 and

October 3 (18.59 ± 0.65 µg/g and 19.12 ± 0.63 µg/g, respectively, Figure 4). The percentage of myricetin content ranged from 1.65% to 2.95% among all the flavonol compounds detected. The coefficient of variation was calculated and showed a marginal variation of 12.63% for myricetin. The percentage of quercitrin (4.87%) in *V. oxycoccos* fruit samples collected in northern Lithuania (Kamanos State Strict Nature Reserve) and southern Lithuania (Žuvintas Biosphere Reserve) was slightly higher, and the percentage of myricetin (1.82%) was in line with the results of our study. In the fruit samples of small cranberries harvested in southern Lithuania (Čepkeliai State Strict Nature Reserve), the percentage of guaijaverin was slightly lower (4.7%), and the percentage of quercitrin and myricetin was higher (4.6% and 3.1%, respectively) than the percentage of these compounds in the fruit samples analyzed in our study [33]. Wang et al. [23] reported that in samples of large cranberries of different cultivars, the percentage of hyperoside was 31–46%, that of myricetin-3-galactoside was 19–32%, that of avicularin was 7–17%, and that of quercitrin was 7–14% of the total amount of flavonols found. Urbstaite et al. [32] reported higher levels of quercitrin (227.49–413.68 µg/g) in large cranberry fruit samples than those detected in *V. oxycoccos* fruit samples analyzed in this study. The variation in the amount of isoquercitrin, quercitrin, and myricetin in small cranberry fruit samples during the ripening period is presented in Figure 4.

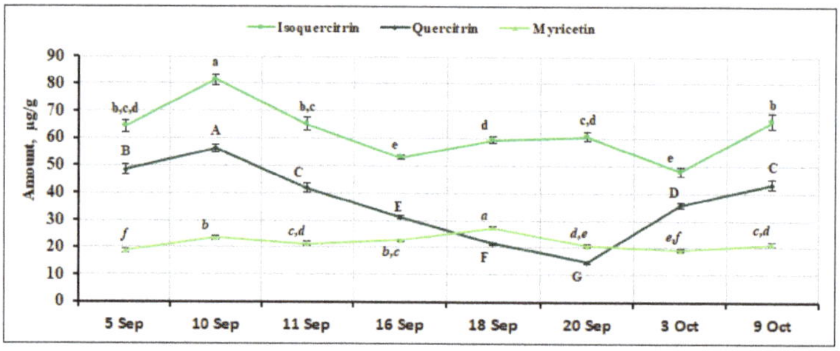

Figure 4. Variation in the amount of isoquercitrin, quercitrin, and myricetin in small cranberry fruit samples during the ripening period. Different letters (a–e for isoquercitrin, A–G for quercitrin, *a–f* for myricetin) indicate statistically significant differences in the content of these compounds between the *V. oxycoccos* fruit samples tested ($p < 0.05$).

The percentage of the amount of guaijaverin among the flavonols detected ranged from 4.86% to 7.13%. The highest content (87.91 ± 2.26 µg/g, $p < 0.05$) was found in small cranberry fruit samples collected on September 10. The quantitative composition of guaijaverin varied quite significantly during the ripening of small cranberry fruit (Figure 5), with the calculated coefficient of variation being 25.84%. The highest levels of quercetin were found in *V. oxycoccos* fruit samples collected on September 10 and October 9 (27.47 ± 0.72 µg/g and 25.95 ± 1.10 µg/g, respectively). The lowest amounts of quercetin were found in small cranberry fruit samples collected on September 5 and September 20 (15.73 ± 0.55 µg/g and 17.02 ± 0.43 µg/g, respectively). The percentage of quercetin content found in the samples ranged from 1.40% to 2.64%. The amount of quercetin in the studied small cranberry fruit samples varied significantly, with the calculated coefficient of variation being 19.38%. Our results confirm those of Šedbarė et al. [22]. Survey data on small cranberries collected in northern Lithuania (Kamanos State Strict Nature Reserve) and southern Lithuania (Žuvintas Biosphere Reserve) show that quercetin accounted for 2.02% [22]. Šedbarė et al. [22] reported that the percentage of guaijaverin (4.57%) was slightly lower than that in the fruit samples analyzed in our study. In small cranberry fruit samples collected in southern Lithuania (Čepkeliai State Strict Nature Reserve), the percentage of quercetin was higher

(4.4%) and the percentage of guaijaverin was slightly lower (4.7%) than those found in *V. oxycoccos* fruit samples collected during the ripening period that were analyzed in our study [33]. The variation in the quantitative composition of guaijaverin and quercetin is shown in Figure 5. The variation in the percentage composition of flavonol group compounds identified in all *V. oxycoccos* fruit samples is presented in the Supplementary Materials in Table S1.

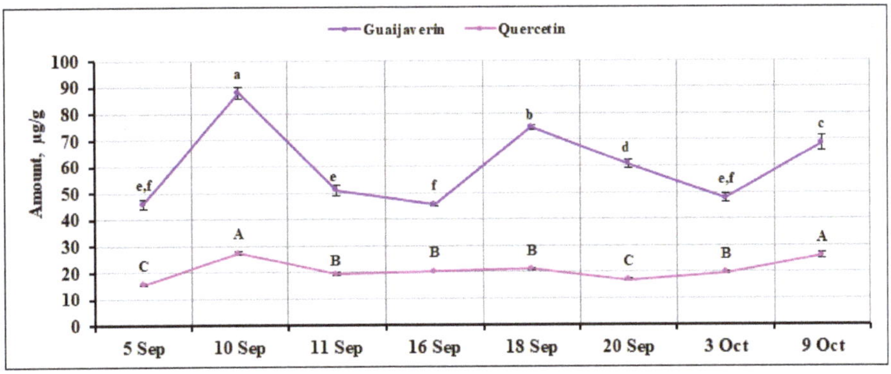

Figure 5. Variation in the quantitative composition of guaijaverin and quercetin in small cranberry fruit samples during the ripening period. Different letters (a–f for guaijaverin, A–C for quercetin) indicate statistically significant differences in the content of these compounds between the *V. oxycoccos* fruit samples tested ($p < 0.05$).

Chlorogenic acid, a member of the group of phenolic acids, was also detected in small cranberry fruit samples. The highest amounts of this acid were found in small cranberry fruit samples collected on September 16 and October 9 (124.66 ± 1.90 µg/g and 126.55 ± 5.34 µg/g, respectively), while the lowest (47.04 ± 1.22 µg/g, $p < 0.05$) amounts were found in samples collected on September 10. The percentage of chlorogenic acid ranged from 3.68% to 13.93% of the total phenolic compounds identified and quantified. The coefficient of variation in the quantitative composition of chlorogenic acid in the samples collected during the ripening period of small cranberries showed a significant variation (coefficient of variation, CV = 34.80%) for this compound. Šedbarė et al. [22] reported that the chlorogenic acid content of *V. oxycoccos* fruit samples collected in northern Lithuania (Kamanos State Strict Nature Reserve) and southern Lithuania (Žuvintas Biosphere Reserve) ranged from 17 µg/g to 1224 µg/g. The highest amount of chlorogenic acid (1224 ± 41 µg/g) was found in fruit samples of small cranberries collected in the mesotrophic type wetland of northern Lithuania (Kamanos State Strict Nature Reserve) [22], which was 9.7 to 26 times higher than the chlorogenic acid content of the ripening fruit samples of *V. oxycoccos* analyzed in our study. Stobnicka and Gniewosz [34] tested fruit samples of *V. oxycoccos* extracted with different extractants and reported a higher content of chlorogenic acid (610–963 µg/g) than that found in the samples we tested. Higher amounts of chlorogenic acid (720.0–1296.2 µg/g) were also reported by Oszmiański et al. [17] in fruit samples of large cranberries. These results could be due to interspecies and intraspecies differences in the phytochemical composition of the plant organs, climatic conditions, soil geochemical composition, genetic [4,8,35,36], and other factors, which were not assessed in this study. The detailed data on the variation in the quantitative composition of chlorogenic acid in *V. oxycoccos* fruit samples during the ripening period are presented in Figure 6.

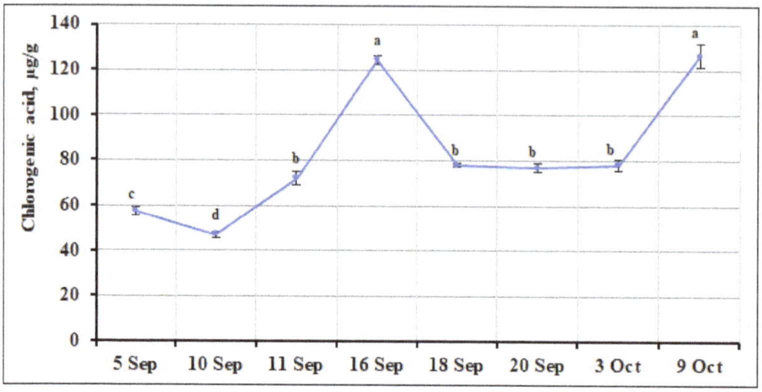

Figure 6. Variation in the quantitative composition of chlorogenic acid in small cranberry fruit samples during the ripening period. Different letters indicate statistically significant differences in the content of this compound between the *V. oxycoccos* fruit samples tested ($p < 0.05$).

Studies on the phytochemical composition of small cranberry fruit in selected habitats to determine the qualitative and quantitative composition of flavonols, proanthocyanidins, proanthocyanidins, and hydroxycinnamic acid derivatives, as well as studies on the trends in the content of these compounds during the ripening period, are important for the preparation of high-quality small cranberry fruit raw material with a known chemical composition of bioactive compounds.

2.2. Determination of the Quantitative Composition of Anthocyanins

Anthocyanins found in berries, fruits, and vegetables are an important group of secondary metabolites found in plants. They protect plant organs from the harmful effects of UV radiation and adverse environmental conditions (frost, drought, oxidative processes, etc.) and give plant organs different colors [37]. Studies of the qualitative and quantitative composition of anthocyanins are important in order to assess the variation in anthocyanin composition of cranberry fruits during their development and ripening. Anthocyanin group compounds have hypoglycemic, antibacterial, anti-inflammatory, anticancer, memory-enhancing, antioxidant, and other biological properties [38].

UHPLC analysis of anthocyanins allowed the identification and quantification of 12 compounds in this group: 9 glycosides (delphinidin-3-O-galactoside, cyanidin-3-O-galactoside, cyanidin-3-O-glucoside, cyanidin-3-O-arabinoside, peonidin-3-O-galactoside, peonidin-3-O-glucoside, peonidin-3-O-arabinoside, malvidin-3-O-galactoside, and malvidin-3-O-glucoside) and 3 aglycones (anthocyanidins)—cyanidin, peonidin, and malvidin. The collection of small cranberry fruit samples during the ripening period was performed by using UHPLC according to the methodology described by Urbstaite et al. [32]. The highest total amount of identified anthocyanins (4096.79 ± 5.93 µg/g, $p < 0.05$) was determined in the samples of *V. oxycoccos* fruits collected on September 18.

Three peonidin glycosides were identified and quantified in the small cranberry fruit samples tested: peonidin-3-O-galactoside, peonidin-3-O-glucoside, and peonidin-3-O-arabinoside. Peonidin-3-O-galactoside was the dominant compound among all the anthocyanin group compounds quantified. In 2011, Česonienė et al. published similar research data [39]. The calculated percentage of peonidin-3-O-galactoside ranged from 22.36% to 37.70%. The highest amount of glycoside (1486.04 ± 18.11 µg/g, $p < 0.05$) was detected in *V. oxycoccos* fruit samples collected on September 18, while the lowest amount was detected in fruit samples collected on September 5 and 11 (638.48 ± 15.73 µg/g and 680.76 ± 30.33 µg/g, respectively). The coefficient of variation (CV = 29.95%) showed a significant variation in peonidin-3-O-galactoside content. The percentage of peonidin-3-O-galactoside (26.4–30.4%) in small cranberry fruit samples collected in southern Lithuania (Čepkeliai State Strict Nature Reserve) reported by Šedbarė et al. [33] was in agreement with our results. Vorsa and Polashock [40] reported that the percentage of peonidin-3-O-galactoside in *V. oxycoccos* fruits (24.5%) was similar to that found in our samples. The percentage of peonidin-3-O-galactoside (29.15%) in cultured *V. oxycoccos* fruit samples found in a study by Česonienė et al. [39] is in line with the results obtained in our study. The evaluation of the quantitative composition of other peonidin glycosides in small cranberry fruit samples showed that other peonidin glycosides accounted for a lower percentage. The percentage of peonidin-3-O-arabinoside ranged from 13.94% to 20.38%. Its highest levels in *V. oxycoccos* fruit samples were found at the end of the ripening period, on October 9 (674.12 ± 28.42 μg/g, $p < 0.05$), and the lowest levels were found at the beginning of the ripening period, on September 5 and September 12 (402.22 ± 14.40 and 366.47 ± 14.47 μg/g, respectively). The peonidin-3-O-arabinoside content showed a moderate variation during the ripening of small cranberries with an estimated coefficient of variation of 21.97%. The percentage of peonidin-3-arabinoside (17.4–18.4%) in *V. oxycoccos* fruit samples collected in southern Lithuania (Čepkeliai State Strict Nature Reserve) was consistent with the results of our study [33]. The percentage of peonidin-3-O-arabinoside in small cranberry fruit reported by US scientists was 17.3% and is consistent with the percentage found in our samples [40]. The percentage of peonidin-3-arabinoside (19.64%) content in cultured *V. oxycoccos* fruit samples found by Česonienė et al. [39] is in agreement with the results of our study. The percentage of peonidin-3-O-glucoside content in small cranberry fruit samples collected during the ripening period ranged from 1.93% to 23.45%. The highest levels were found in small cranberry fruit samples collected on September 5 and September 11 (605.95 ± 21.82 μg/g and 616.42 ± 24.38 μg/g, respectively). The lowest levels of peonidin-3-O-glucoside were found in fruit samples of *V. oxycoccos* collected on September 10, September 18, and September 20 (63.00 ± 2.01 μg/g, 91.61 ± 1.10 μg/g, and 66.49 ± 1.65 μg/g, respectively). The calculated coefficient of variation (98.26%) demonstrates a very significant variation in the amount of peonidin-3-O-glucoside during the ripening of small cranberry fruits. The percentage of peonidin-3-glucoside (14.4%) in fruit samples of *V. oxycoccos* collected in southern Lithuania (Čepkeliai State Strict Nature Reserve) was within the range of our results [33]. The percentage of peonidin-3-O-glucoside in *V. oxycoccos* fruit reported by US scientists (4.7%) is in line with that found in our study [40]. Česonienė et al. [39] reported that the average percentage of peonidin-3-O-glucoside found in fruit samples of cultivated small cranberries of different cultivars was 6.21%. The variation in the peonidin glycoside content in small cranberry fruit samples during the ripening period is presented in Figure 7.

We qualitatively and quantitatively evaluated the cyanidin glycosides cyanidin-3-O-galactoside, cyanidin-3-O-arabinoside, and cyanidin-3-O-glucoside in *V. oxycoccos* fruit samples. Cyanidin-3-O-galactoside is the predominant anthocyanin among the quantitatively identified compounds, which is in line with the results obtained by other researchers [40,41]. Its percentage ranged from 13.84% to 22.26%. The highest content of cyanidin-3-O-galactoside (911.88 ± 10.71 μg/g, $p < 0.05$) was found in fruit samples collected on September 18, while the lowest content (318.59 ± 4.69 μg/g, $p < 0.05$) was found in fruit samples collected on September 16. The coefficient of variation showed a significant (33.86%) variation in cyanidin-3-O-galactoside content. The percentage of cyanidin-3-O-galactoside (16.2–20.1%) found in fruit samples of *V. oxycoccos* collected in southern Lithuania (Čepkeliai State Strict Nature Reserve) was consistent with the results of our study [33]. The percentage of cyanidin-3-O-galactoside in *V. oxycoccos* fruit samples reported by Vorsa and Polashock [40] (26.7%) was higher than that found in the small cranberry samples we examined. Česonienė et al. [39] reported that the percentage of cyanidin-3-O-galactoside in small cranberry samples was 20.44%, which is similar to that found in our samples. Other cyanidin glycosides were detected in lower amounts. The percentage of cyanidin-3-O-arabinoside ranged from 13.77% to 18.29%. The stages of fruit ripening cause variations in the chemical composition of small cranberries due to changes in metabolic activity during development. Early ripening stages are often characterized by the accumulation of certain phenolic compounds, such as phenolic acids, which serve protective and antioxidant roles [42]. As ripening progresses, the synthesis shifts toward other compounds like anthocyanins and proanthocyanidins, which contribute to fruit coloration, flavor, and defense [43]. Environmental factors and enzymatic activities also interact with these developmental processes, leading to dynamic changes in the chemical profile throughout ripening. Therefore, the chemical composition reflects the fruit's physiological and biochemical status at each ripening stage [42]. Our study was limited solely to investigating the variation of biologically active compounds in cranberry fruits during the period of intensive fruit ripening; the assessment of the influence of enzymatic activity and environmental conditions was not the aim of this research. Its highest amount of cyanidin-3-O-arabinoside (726.36 ± 9.03 μg/g, $p < 0.05$) was detected in small cranberry fruit samples collected on September 18, while the lowest amounts were found in fruit samples collected on September 11 and 16 (362.04 ± 16.11 μg/g and 347.85 ± 5.17 μg/g, respectively). In our study, cyanidin-3-O-arabinoside showed a moderate variation of CV = 25.55% during the ripening of small cranberry fruit. The percentage of cyanidin-3-O-arabinoside (17.0–19.0%) found in fruit samples of *V. oxycoccos* collected in southern Lithuania (Čepkeliai State Strict Nature Reserve) was within the range of our results [33]. In the USA, researchers reported that the percentage of cyanidin-3-O-arabinoside in small cranberry fruit samples was 13.1% [40]. Česonienė et al. [39], who studied fruit samples of *V. oxycoccos* in Lithuania, reported a percentage of cyanidin-3-O-arabinoside (21.33%) that was slightly higher than that found in our samples. The percentage of cyanidin-3-O-glucoside was the lowest of all the cyanidin glycosides detected, ranging from 0.49% to 6.02%. The highest level of cyanidin-3-O-glucoside was detected in fruit samples collected on September 5 (171.93 ± 4.68 μg/g, $p < 0.05$), while the lowest levels of cyanidin-3-O-glucoside were detected in fruit samples collected on September 10 and 20 (15.60 ± 0.45 μg/g and 16.91 ± 0.45 μg/g, respectively). The quantitative composition of cyanidin-3-O-glucoside varied over a wide range, with an estimated CV = 102.80%. The percentage of cyanidin-3-O-glucoside (3.9%) found in fruit samples of *V. oxycoccos* collected in southern Lithuania (Čepkeliai State Strict Nature Reserve) was in line with the results of our study [33]. The percentage of cyanidin-3-O-glucoside content (3.23%) reported by Česonienė et al. [39] is consistent with our results. The variation in the quantitative composition of cyanidin glycosides in small cranberry fruit samples during the ripening period is shown in Figure 8.

Two malvidin glycosides, malvidin-3-O-arabinoside and malvidin-3-O-galactoside, were identified and quantified in small cranberry fruit samples. The percentage of malvidin-3-O-arabinoside in *V. oxycoccos* fruit samples during the ripening period ranged from 0.70% to 2.07%. The highest amounts of this compound were found in fruit samples collected on September 20 and October 3 (64.32 ± 1.60 μg/g and 63.35 ± 2.06 μg/g, respectively), while the lowest amount (19.97 ± 0.70 μg/g, $p < 0.05$) was found in samples collected on September 5. The quantitative composition of malvidin-3-O-arabinoside varied significantly during the ripening of small cranberry fruit (CV = 39.64%). The percentage of malvidin-3-galactoside ranged from 0.40% to 1.21% during the ripening period. The highest amount of this compound (41.60 ± 0.93 μg/g, $p < 0.05$) was detected in the fruit samples collected on September 20, while the lowest amounts were found in the fruit samples collected on September 10, September 16, and October 9 (13.62 ± 0.42 μg/g, 13.30 ± 0.25

μg/g, and 13.36 ± 0.57 μg/g, respectively). The quantitative composition of malvidin-3-galactoside varied widely during the ripening of the fruit, with an estimated CV = 51.65%. Delphinidin-3-O-galactoside was also identified and quantified in small cranberry samples. In our study, the percentage of this compound ranged from 0.50% to 0.78%. The highest levels of delphinidin-3-O-galactoside were found in fruit samples collected on September 18 and 20 (25.32 ± 0.30 μg/g and 26.76 ± 0.82 μg/g, respectively), while the lowest level (14.40 ± 0.47 μg/g) was found in fruit samples collected on September 5. The coefficient of variation was calculated and showed a mean variation (CV = 23.01%) for delphinidin-3-O-galactoside. Šedbarė et al. [44], who studied the fruit of large cranberries, reported a percentage of malvidin-3-O-galactoside (0.45%) that was similar to that found in our samples, with the percentage of malvidin-3-O-arabinoside (0.61%) being slightly lower and the percentage of delphinidin-3-O-galactoside (2.04%) being higher than that found in the samples we studied. The variation in the quantitative composition of malvidin glycosides and delphinidin-3-O-galactoside in small cranberry fruit samples during the ripening period is shown in Figure 9.

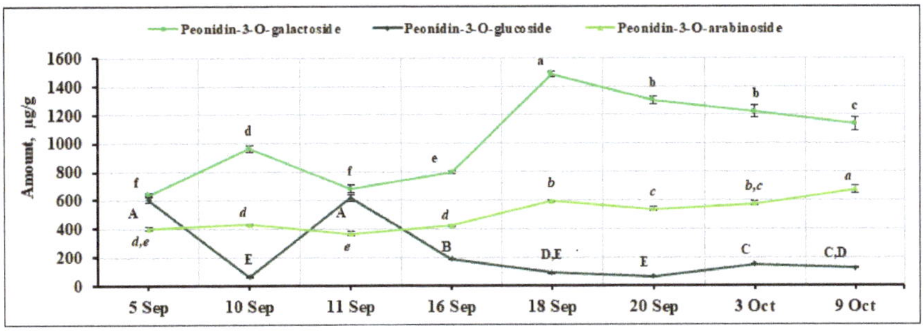

Figure 7. Variation in the quantitative composition of peonidin glycosides in small cranberry fruit samples during the ripening period. Different letters (a–f for peonidin-3-O-galactoside, A–E for peonidin-3-O-glucoside, *a–e* for peonidin-3-arabinoside) indicate statistically significant differences in the amount of anthocyanin compounds detected between the *V. oxycoccos* fruit samples tested ($p < 0.05$).

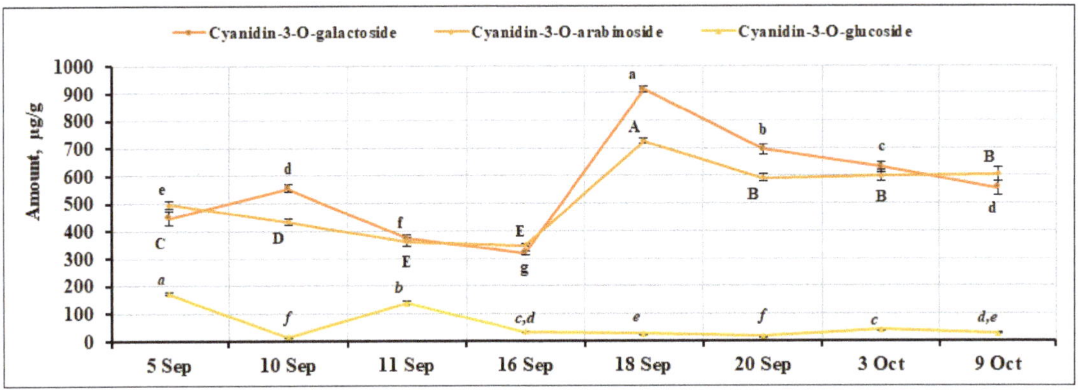

Figure 8. Variation in the quantitative composition of cyanidin glycosides in small cranberry fruit samples during the ripening period. Different letters (a–g for cyanidin-3-O-galactoside, A–E for cyanidin-3-O-arabinoside, *a–f* for cyanidin-3-O-glucoside) indicate statistically significant differences in the amount of anthocyanin compounds detected between the *V. oxycoccos* fruit samples tested ($p < 0.05$).

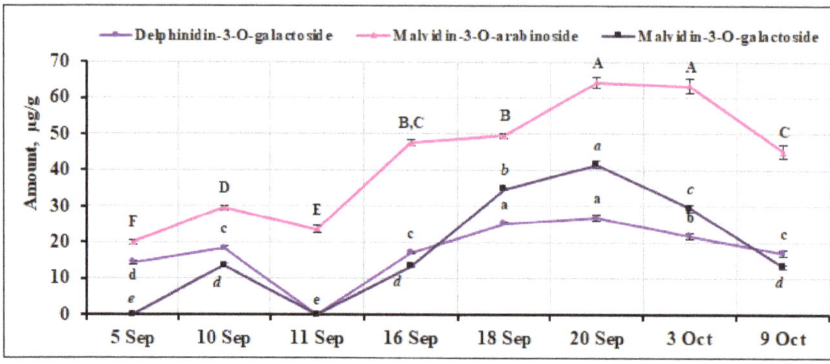

Figure 9. Variation in the quantitative composition of malvidin glycosides and delphinidin-3-O-galactoside in small cranberry fruit samples during the ripening period. Different letters (a–e for delphinidin-3-O-galactoside, A–F for malvidin-3-O-arabinoside, *a–e* for malvidin-3-O-galactoside) indicate statistically significant differences in the amount of anthocyanin compounds detected between the *V. oxycoccos* fruit samples tested ($p < 0.05$).

Three anthocyanin aglycones—anthocyanidins peonidin, malvidin, and cyanidin—were identified and quantified in small cranberry fruit samples. The percentage of peonidin in *V. oxycoccos* fruit samples during the ripening period ranged from 0.95% to 2.00%. The highest amount of this aglycone (70.64 ± 0.84 μg/g, $p < 0.05$) was detected in fruit samples collected on September 18, while the lowest amount (27.01 ± 0.95 μg/g, $p < 0.05$) was detected in fruit samples collected on September 5. Peonidin content varied significantly during the ripening period, with an estimated CV = 31.35%. The percentage of malvidin in *V. oxycoccos* fruit samples ranged from 0.53% to 2.12% during the ripening period. The highest amounts of this compound were found in fruit samples collected on September 16 and October 3 (48.85 ± 0.75 μg/g and 48.30 ± 1.92 μg/g, respectively), while the lowest amount was found in samples collected on September 5 (15.14 ± 0.53 μg/g, $p < 0.05$). The quantitative composition of malvidin in small cranberry fruit samples during the ripening period varied quite widely, with an estimated CV = 38.66%. The percentage of cyanidin in *V. oxycoccos* fruit samples during the ripening period ranged from 0.55% to 0.86%. The highest amounts of this aglycone (34.72 ± 0.85 μg/g, $p < 0.05$) were found in fruit samples collected on September 18, while the lowest amounts were found in fruit samples collected on September 5 and September 11 (15.60 ± 0.58 μg/g and 15.51 ± 0.96 μg/g, respectively. The variation in the quantitative composition of anthocyanidins identified and quantified in small cranberry fruit samples during the ripening period is shown in Figure 10. The variation in the percentage composition of all anthocyanins identified in all *V. oxycoccos* fruit samples is presented in the Supplementary Materials in Table S2.

We chose a site where the samples were collected and where it was appropriate to determine the qualitative and quantitative composition of bioactive compounds. The results of the study showed an increasing trend in the content of most of the identified compounds of the anthocyanin group during the ripening of small cranberry fruit, which provides insight into the quantitative changes of individual compounds and allows for the timing of small cranberry fruit harvesting when the concentration of anthocyanins is highest. Chemical composition studies on small cranberry fruit samples allow for the qualitative and quantitative assessment of the composition of individual anthocyanins and for the targeted preparation of small cranberry raw material with the highest content of anthocyanins, which are natural pigments with strong antioxidant activity.

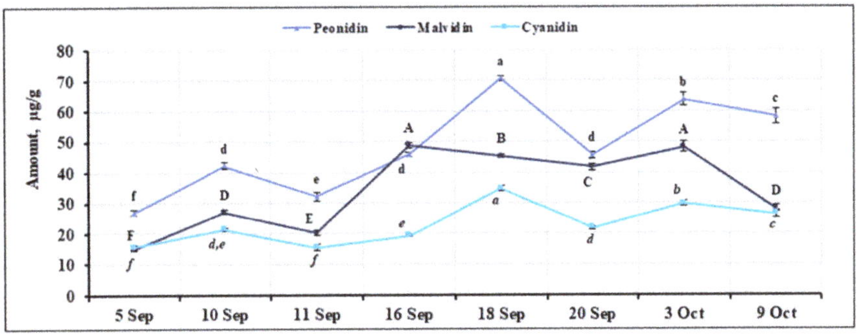

Figure 10. Variation in the quantitative composition of anthocyanidins in small cranberry fruit samples during the ripening period. Different letters (a–f for peonidin, A–F for malvidin, *a–f* for cyanidin) indicate statistically significant differences in the content of anthocyanidin group compounds between the *V. oxycoccos* fruit samples tested ($p < 0.05$).

2.3. Determination of the Quantitative Composition of Triterpene Compounds

The outer surface of the fruit peel is covered with a waxy layer that protects the fruit from water evaporation, temperature changes, UV radiation, and microorganisms [45]. The waxy layer of cranberry berries has been found to contain triterpene compounds [46], which have a biological effect. Triterpene compounds have anti-inflammatory, antibacterial, antiviral, hepatoprotective, gastroprotective, cardioprotective, hypolipidemic, immunoregulatory, antiatherosclerotic, anticancer, and antioxidant pharmacological activities [47].

Using UHPLC, nine compounds of the triterpene group were identified and quantified in small cranberry fruit samples: maslinic acid, corosolic acid, oleanolic acid, ursolic acid, campesterol, α-amyrin, β-sitosterol, and squalene. The highest total amount of identified triterpene compounds (8248.46 ± 125.60 µg/g, $p < 0.05$) was determined in the samples of *V. oxycoccos* fruits collected on September 16.

Four triterpenic acids were detected in small cranberry fruit samples: maslinic acid, corosolic acid, oleanolic acid, and ursolic acid. Ursolic acid was the predominant acid in the fruit samples of *V. oxycoccos* collected during the ripening period. Data published by other researchers confirm our results, showing that ursolic acid was the most abundant triterpene compound in small cranberry fruit samples [45]. The percentage of ursolic acid ranged from 63.83% to 67.49% of the total amount of triterpenic compounds detected in the samples. Its highest amount (5495.43 ± 83.68 µg/g, $p < 0.05$) was detected in fruit samples of *V. oxycoccos* collected on September 16. The coefficient of variation (6.61%) showed a slight variation in ursolic acid in the fruit samples tested. The variation in triterpene compounds content during cranberry fruit ripening is influenced by multiple factors, including the developmental stage of the fruit [17], which affects biosynthetic activity; the enzymatic processes involved in triterpenoid synthesis that fluctuate throughout ripening; genetic differences among cultivars; and various environmental conditions such as climate and other factors [48]. Additionally, biotic and abiotic stresses can modulate triterpenoid production as part of the plant's defense mechanisms [49]. Together, these factors contribute to the dynamic changes observed in triterpene compounds levels during the ripening process. The percentage of ursolic acid was higher in fruit samples of *V. oxycoccos* collected from northern Lithuania (Kamanos State Strict Nature Reserve) and southern Lithuania (Žuvintas Biosphere Reserve), accounting for 76.24% of the total amount of triterpene compounds quantified [22]. A higher percentage of ursolic acid (76.8%) than in our study was also found in small cranberry fruit samples collected in southern Lithuania (Čepkeliai State Strict Nature Reserve) [33]. Zhang et al. [50] analyzed cranberry fruit samples and found 65.9 mg/100 g of ursolic acid. The variation in the amount of ursolic acid in small cranberry fruit samples collected during the ripening period is presented in Figure 11.

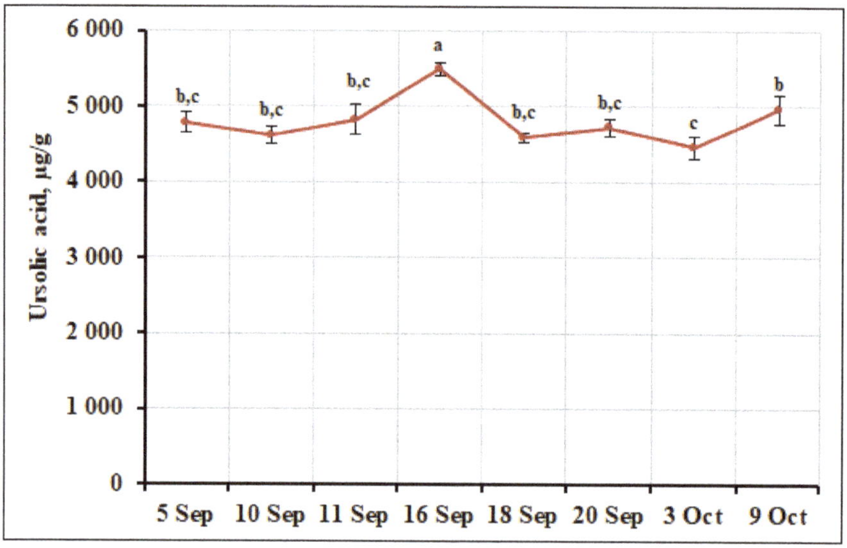

Figure 11. Variation in the amount of ursolic acid in small cranberry fruit samples during the ripening period. Different letters indicate statistically significant differences in ursolic acid content between the *V. oxycoccos* fruit samples tested ($p < 0.05$).

The percentage of oleanolic acid in our fruit samples ranged from 14.68% to 18.16%. The highest amounts were found in fruit samples collected on September 10 and 16 (1314.54 ± 32.95 µg/g and 1381.18 ± 21.03 µg/g, respectively). The lowest amounts were found in fruit samples collected on September 18 and October 3 (997.99 ± 11.81 µg/g and 1093.18 ± 36.29 µg/g, respectively). The calculated coefficient of variation (10.62%) reflects a slight variation in the oleanolic acid content of the fruit samples. The percentage of oleanolic acid (17.44%) in small cranberry fruits collected in northern Lithuania (Kamanos State Strict Nature Reserve) and southern Lithuania (Žuvintas Biosphere Reserve) reported by Šedbarė et al. [22] was in line with the results of our study. The percentage of oleanolic acid (17.12%) found in samples of small cranberries collected in southern Lithuania (Čepkeliai State Strict Nature Reserve) was within the range of our results [33]. The percentage of β-sitosterol in the total amount of triterpene compounds detected in small cranberry fruits ranged from 11.97% to 14.11%. The highest amount of β-sitosterol (1034.65 ± 15.75 µg/g) was detected in small cranberry fruit samples collected on September 16, but it was not statistically significantly different from the amount of this compound in the fruit samples collected on September 5, September 10, September 11, or September 20 (1002.39 ± 30.20 µg/g, 1020.98 ± 25.60 µg/g, 1020.23 ± 42.85 µg/g, and 990.48± 23.74 µg/g, respectively). The lowest amount of β-sitosterol (890.34 ± 33.83 µg/g) was detected in small cranberry fruit samples collected on October 9, but it was not statistically significantly different from the amount of β-sitosterol detected in the fruit samples collected on September 18 or October 3 (943.54 ± 11.17 µg/g and 922.52 ± 30.62 µg/g, respectively). The quantitative composition of β-sitosterol in small cranberry fruit samples during the ripening period showed the least variation of all the phenolic compounds identified and quantified, with an estimated coefficient of variation of 5.40%. Wu et al. [51] indicate that β-sitosterol is the major compound in the phytosterol group of cranberry fruits, which is consistent with our results. Sedbarė et al. [52] reported slightly higher β-sitosterol content (1068.3 ± 16.02

μg/g) in small cranberry fruit samples than that found in our samples. The variation in the quantitative composition of β-sitosterol and oleanolic acid in samples of small cranberry fruit collected during the ripening period is shown in Figure 12.

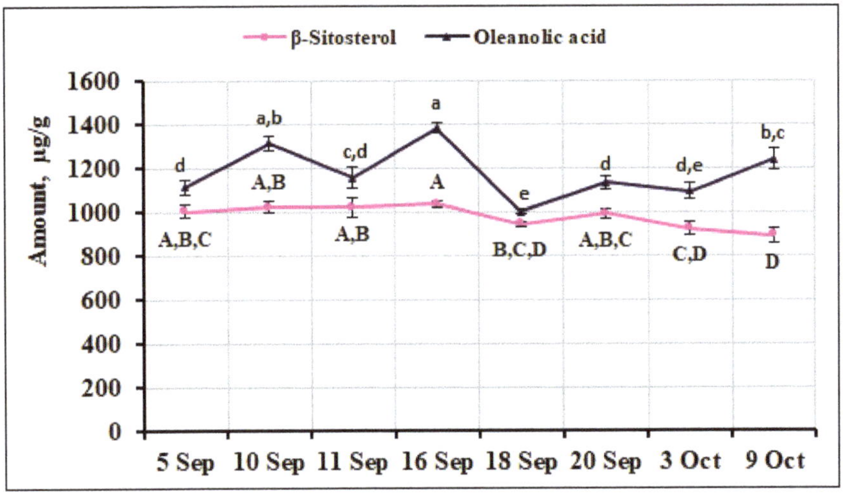

Figure 12. Variation in the quantitative composition of β-sitosterol and oleanolic acid in small cranberry fruit samples during the ripening period. Different letters (a–e for oleanolic acid, A–D for β-sitosterol) indicate statistically significant differences in the content of these compounds between the *V. oxycoccos* fruit samples tested ($p < 0.05$).

The percentage of corosolic acid in the small cranberry fruit samples ranged from 1.37% to 1.87%. Its highest amount (138.56 ± 2.11 μg/g) was detected in *V. oxycoccos* fruit samples collected on September 16, but it was not statistically significantly different from the levels of corosolic acid detected in fruit samples collected on September 5, September 10, or September 11 (134.52 ± 4.07 μg/g, 130.61 ± 3.29 μg/g, and 137.07 ± 5.76 μg/g, respectively). The lowest amounts of corosolic acid were found in small cranberry fruit samples collected on September 18 and September 20 (96.42 ± 1.14 μg/g and 97.62 ± 2.34 μg/g, respectively). The calculated coefficient of variation (14.56%) reflects a slight variation in the quantitative composition of this acid in the small cranberry samples analyzed. The percentage of α-amyrin among the total amount of triterpene compounds detected ranged from 1.11% to 2.26%. The highest amount of this compound (153.24 ± 5.09 μg/g, $p < 0.05$) was detected in the fruit samples collected on October 3, while the lowest amount (80.57 ± 2.02 μg/g, $p < 0.05$) was detected in the fruit samples collected on September 10. The quantitative composition of α-amyrin in the small cranberry fruit samples showed a slight variation during the ripening period, with an estimated coefficient of variation of 17.20%. β-Amyrin was only quantified in the small cranberry fruit samples collected on October 9, and its amount was found to be 24.76 ± 0.94 μg/g. The percentage corosolic acid (7%) in the fruit samples of *V. oxycoccos* collected in northern Lithuania (Kamanos State Strict Nature Reserve) was significantly higher than that found in our samples [22]. The percentage of corosolic acid (3.2%) found in samples of small cranberries collected in southern Lithuania (Čepkeliai State Strict Nature Reserve) was higher than that found in our study [33]. The percentage of α-amyrin content (1.98%) in *V. oxycoccos* fruit samples collected in northern Lithuania (Kamanos State Strict Nature Reserve) and southern Lithuania (Žuvintas Biosphere Reserve) was within the range of our results [22]. The percentage of α-amyrin (1.65%) found in fruit samples of small cranberries collected in southern Lithuania (Čepkeliai State Strict Nature Reserve) was in line with the results of our study [33]. Sedbarė et al. [52] found a lower amount of α-amyrin (57.1 ± 0.86 μg/g) in *V. oxycoccos* fruit samples. The variation in α-amyrin and corosolic acid contents of small cranberry fruit samples collected during the ripening period is presented in Figure 13.

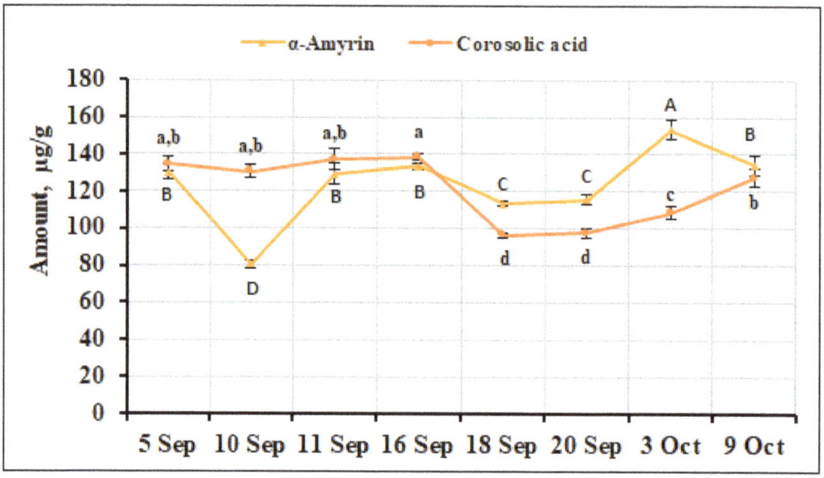

Figure 13. Variation in the quantitative composition of α-amyrin and corosolic acid in small cranberry fruit samples during the ripening period. Different letters (a–d for corosolic acid, A–D for α-amyrin) indicate statistically significant differences in the content of these compounds between the *V. oxycoccos* fruit samples tested ($p < 0.05$).

The percentage of maslinic acid of the total amount of triterpene compounds detected in *V. oxycoccos* fruit samples during the ripening period ranged from 0.31% to 0.51%. The highest amount of maslinic acid (37.22 ± 1.61 μg/g) was found in fruit samples collected on September 11, but it was not statistically significantly different from that found in samples collected on September 5 or September 16 (34.49 ± 1.13 μg/g and 37.04 ± 0.71 μg/g, respectively). The lowest amount of maslinic acid was found in *V. oxycoccos* fruit samples collected on September 18 and 20 (22.09 ± 0.34 μg/g and 22.05 ± 0.53 μg/g, respectively). The quantitative composition of maslinic acid in small cranberry fruit samples collected during the ripening period showed the greatest variation of all the triterpenic acids detected, with a coefficient of variation calculated at 20.69%. The percentage of maslinic acid (2%) in the fruit samples of *V. oxycoccos* collected in northern Lithuania (Kamanos State Strict Nature Reserve) was significantly higher than that found in our samples [22]. The percentage of maslinic acid (0.83%) found in samples of small cranberries collected in southern Lithuania (Čepkeliai State Strict Nature Reserve) was higher than that found in our study [33]. The amount of campesterol detected in the small cranberry samples tested was significantly lower, with percentages ranging from 0.16% to 0.43% of the total amount of triterpene compounds quantified. The highest amount of campesterol (30.92 ± 0.78 μg/g, $p < 0.05$) was found in fruit samples collected on September 10, while the lowest amount was found in fruit samples collected on October 3 and 9 (11.28 ± 0.38 μg/g and 12.24 ± 0.46 μg/g, respectively). Small cranberry fruit samples collected during the ripening period showed the presence of a non-cyclic triterpene compound, squalene. Its percentage varied from 0.06% to 0.20% among all the identified and quantified triterpene compounds. The highest amount of squalene (13.29 ± 0.44 μg/g, $p < 0.05$) was detected in fruit samples collected on October 3, while the lowest amount was detected in fruit samples collected on September 11 and September 20 (4.32 ± 0.19 μg/g and 4.73 ± 0.12 μg/g, respectively). The quantitative composition of squalene in small cranberry fruit samples collected during the ripening period showed the greatest variation of all the triterpene compounds detected, with a calculated coefficient of variation of 40.93%, reflecting a significant variation in the content of this compound among the samples analyzed. Sedbare et al. [52] reported higher squalene content (48.6 ± 0.73 μg/g) in small cranberry fruit samples than that found in our samples. The variation in the amount of maslinic acid, campesterol, and squalene in small cranberry fruit samples collected during the ripening period is shown in Figure 14. The variation in the percentage composition of all triterpene compounds identified in all *V. oxycoccos* fruit samples is presented in the Supplementary Materials in Table S3.

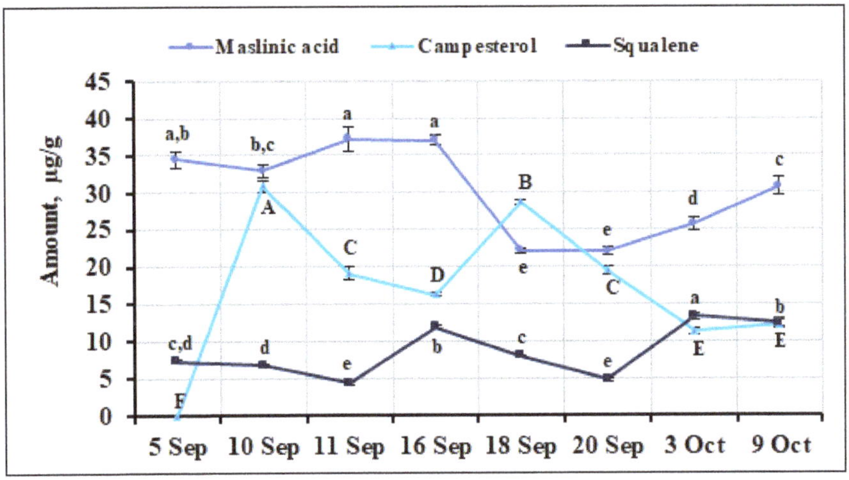

Figure 14. Variation in the quantitative composition of maslinic acid, campesterol, and squalene in small cranberry fruit samples during the ripening period. Different letters (a–e for maslinic acid, A–F for campesterol, *a–e* for squalene) indicate statistically significant differences in the content of these compounds between the *V. oxycoccos* fruit samples tested ($p < 0.05$).

Small cranberry fruit plant material accumulates triterpene compounds with a wide range of biologically active properties, and therefore, studies on the qualitative and quantitative composition of these compounds are relevant. The triterpene compounds detected in small cranberry fruit can be used as markers for the detection of triterpenes in *V. oxycoccos* fruit samples during the ripening period, as well as in the preparation of small cranberry fruit. Studies on the chemical composition of the fruit allow for the provision of high-quality plant material with a known qualitative and quantitative composition of triterpene compounds.

2.4. Determination of Antioxidant Activity of V. oxycoccos Fruit Extracts

Recently, studies have been carried out to identify compounds in plant matrices with strong antioxidant effects. Antioxidants inhibit inflammatory processes, slow down aging, and protect against oxidative stress [53]. For these reasons, it is worthwhile to investigate the variation in the quantitative composition of phenolic and triterpene compounds during the ripening of small cranberries and to assess the dynamics of the variation of antiradical and reductive activities during ripening in order to determine the fruit harvesting time at which the selected small cranberry fruit extracts exhibit the strongest antioxidant activity in vitro. It is thus of interest to assess the correlation between the amounts of bioactive compounds detected and the in vitro antioxidant activity.

The in vitro antiradical activity of *V. oxycoccos* fruit sample extracts ranged from 75.16 μmol TE (Trolox equivalents)/g to 95.25 μmol TE/g. The mean in vitro antiradical activity of all the extracts tested was found to be 83.71 ± 7.15 μmol TE/g. The strongest in vitro antiradical activity was observed for the extracts of small cranberry fruit collected on September 10 and September 18 (95.25 ± 1.15 μmol TE/g and 92.16 ± 0.20 μmol TE/g, respectively). The weakest in vitro antiradical activity (75.16 ± 2.42 μmol TE/g) was found in extracts of small cranberry fruit collected on September 16, but it was not statistically significantly different from that found in extracts of small cranberry fruit collected on September 11, September 20, or October 3 (79.98 ± 0.77 μmol TE/g, 76.31 ± 3.51 μmol TE/g, and 81.34 ± 2.34 μmol TE/g, respectively). The coefficient of variation, which reflects the range of variation in the in vitro antiradical activity of the tested extracts, was found to be insignificant (CV = 8.54%). The in vitro antiradical activity of the extracts of small cranberry fruit collected from different locations of Lithuania, as assessed by the ABTS (2,2′-azino-bis(3-ethylbenzothiazoline-6-sulfonic acid) assay, varied over a wider range (CV = 18.86%), from 37.38 μmol TE/g to 141.28 μmol TE/g [54].

The in vitro reducing activity of the extracts of small cranberry samples collected during the ripening period was found to range from 75.33 μmol TE/g to 159.26 μmol TE/g. The mean calculated reducing activity of all *V. oxycoccos* fruit sample extracts tested was 117.45 ± 24.82 μmol TE/g. The strongest in vitro reducing activity (159.26 ± 0.77 μmol TE/g, $p < 0.05$) was observed in extracts of small cranberry fruit collected on September 10, and the weakest (75.33 ± 0.78 μmol TE/g, $p < 0.05$) was found in extracts of fruit collected on October 3. The calculated coefficient of variation showed a moderate variation (CV = 21.13%) in the in vitro reducing activity of the extracts of *V. oxycoccos* fruit samples collected during the ripening period. The in vitro reducing activity of fruit extracts of large cranberries grown in Lithuania, as assessed by the CUPRAC (CUPric Reducing Antioxidant Capacity) method, was stronger (215.15–493.87 μmol TE/g) than that observed in the present study. Namiesnik et al. [55], who investigated the in vitro reducing activity of large cranberry fruit extracts using the CUPRAC methodology, reported a weaker activity (49.38 ± 4.4 μM TE/g) than what we observed in our tested small cranberry fruit extracts. The variation in the in vitro antiradical and reducing activities of the extracts of small cranberry fruit collected in natural habitats during the ripening period is presented in Figure 15.

A statistical correlation analysis was performed to assess the relationship between the quantitative composition of anthocyanins, proanthocyanidins, hydroxycinnamic acid derivatives, flavonols, chlorogenic acid, and triterpene compounds collected from *V. oxycoccos* fruit samples during the ripening period and the in vitro antiradical and reducing activity of *V. oxycoccos* fruit extracts.

Statistical analysis showed a strong correlation between the total amount of hydroxycinnamic acid derivatives in small cranberry fruit samples and the in vitro reducing and antiradical activity of their extracts (r = 0.858, $p < 0.01$ and r = 0.771, $p < 0.05$, respectively). There was a strong correlation (r = 0.711, $p < 0.01$) between the total amount of proanthocyanidins in *V. oxycoccos* fruit samples and the in vitro antiradical activity of their extracts. A moderate correlation (r = 0.609, $p < 0.05$) was found between the total amount of proanthocyanidins in the tested small cranberry fruit samples and the in vitro reducing activity of their extracts.

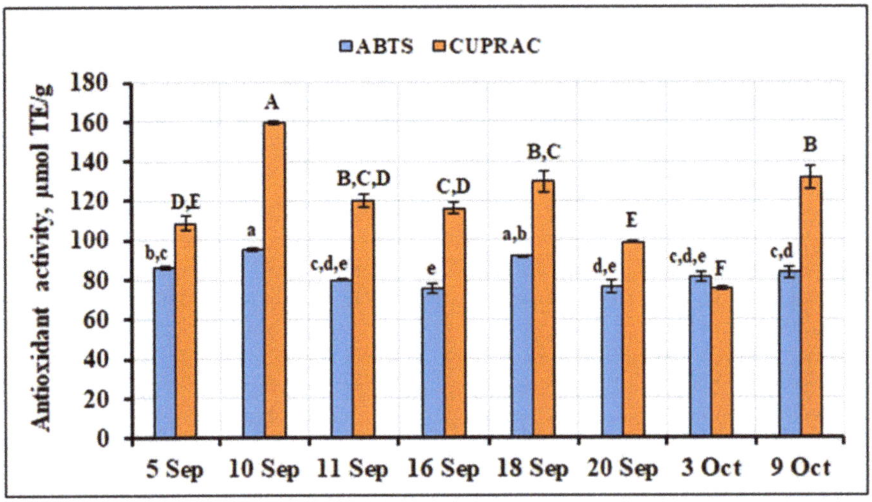

Figure 15. Variation in the in vitro antioxidant activity of the extracts of *V. oxycoccos* fruit samples collected during the ripening period. Different letters (a–e for ABTS, A–F for CUPRAC) indicate statistically significant differences between the antioxidant activity estimates of the tested small cranberry fruit sample extracts ($p < 0.05$).

In the group of the identified flavonol compounds, the strongest positive correlation between individual compounds was found between the amounts of isoquercitrin (r = 0.844, $p < 0.01$) and guaijaverin (r = 0.769, $p < 0.05$) and the in vitro reducing activity assessed by the CUPRAC method. The levels of avicularin and quercitrin detected in small cranberry fruit samples correlated strongly with the in vitro reducing activity of the tested sample extracts (r = 0.721, $p < 0.05$ and r = 0.731, $p < 0.05$, respectively). The amount of guaijaverin and hyperoside in small cranberry fruit samples correlated strongly with the in vitro antiradical activity assessed by the ABTS assay (r = 0.754, $p < 0.05$ and r = 0.747, $p < 0.05$, respectively). All the results of the Pearson's correlation analysis we performed are presented in the Supplementary Material (Table S4). In our previous study, we found a moderately strong correlation between the ABTS-assessed in vitro antiradical activity of *V. oxycoccos* fruit sample extracts and the amounts of myricetin-3-O-galactoside and hyperoside (r = 0.660, $p < 0.01$ and r = 0.606, $p < 0.01$, respectively) [54].

Oszmiański et al. [17] reported a positive moderate correlation between the amount of flavonols and the total content of all the identified phenolic compounds in large cranberry fruit extracts and the in vitro antiradical activity of these extracts assessed by the ABTS assay (r = 0.646 and r = 0.508, accordingly), a strong positive correlation (r = 0.728) between the flavonol content and the in vitro reducing activity assessed by the FRAP method, and a moderate positive correlation (r = 0.591) between the total content of phenolic compounds and the in vitro reducing activity. These authors reported a strong positive correlation between the in vitro antiradical and reducing activity of large cranberry fruit extracts and the total amount of triterpene compounds (r = 0.852 and r = 0.736, respectively), as well as a moderate positive correlation between the amount of anthocyanins and the in vitro antiradical and reducing activity of these extracts (r = 0.675 and r = 0.614, respectively) [17].

Small cranberry fruits are a valuable plant material accumulating flavonols, phenolic acids, proanthocyanidins, anthocyanins, and triterpene compounds. In plant matrices, it is often not possible to attribute biological effects to a specific group of compounds or to a specific compound, as the complex of compounds they accumulate act synergistically [56]. As the quantitative composition of the biologically active compounds of *V. oxycoccos* fruits changes during the ripening period, so does the antioxidant activity of small cranberry fruit extracts, and its evaluation is important to ensure the quality of the *V. oxycoccos* fruits and the biological effects of the preparations made from them.

3. Materials and Methods

3.1. Plant Material

The samples of small cranberry fruits (200 g each) growing under Lithuanian climatic conditions used in this study were hand-picked in the village of Janoniai, Molėtai district, from a wetland near Lake Pagulbis (55°22′59.0″ N 25°13′58.4″ E), during September–October 2021. During the collection of *V. oxycoccos* fruit samples in the Molėtai district, where the Pagulbis Lake bog is located, the average air temperature was 10.6 °C in September and 7.6 °C in October. The average precipitation in Molėtai was 37.0 mm in September and 31.9 mm in October. *V. oxycoccos* fruit samples were frozen at −20 °C temperature and stored until lyophilization. Small cranberry fruits were lyophilized using the technique described by Šedbarė et al. [22]. All the presented data were recalculated for dry weight.

3.2. Extraction Procedure

The extraction of biologically active compounds from *V. oxycoccos* fruit samples was performed by applying a technique described by Liaudanskas et al. [57], using a 2% (*v*/*v*) solution of hydrogen chloride in 70% (*v*/*v*) ethanol as an extractant.

3.3. Spectrophotometric Analysis

All the spectrophotometric measurements were carried out with an i3 UV-VIS spectrophotometer (Hanon instrument Co., Ltd., Jinan, China).

3.3.1. Determination of the Total Amount of Proanthocyanidins and Hydroxycinnamic Acid Derivatives

In the first stage of our study, we applied spectrophotometry to assess the variation in the quantitative composition of the different groups of phenolic compounds—proanthocyanidins and hydroxycinnamic acid derivatives—in small cranberry fruit samples. The total amount of proanthocyanidins in the extracts of *V. oxycoccos* fruit samples was evaluated by the DMCA (4-(dimethylamino)cinnamaldehyde) assay according to the methodology described by Šedbarė et al. [22] with minor modifications. A 50 µL measure of small cranberry fruit extract was used for the analysis. The total amount of proanthocyanidins was expressed as (–)-epicatechin equivalents (EEs) in mg/g of the small cranberry fruit samples. Meanwhile, the total amount of hydroxycinnamic acid derivatives in the extracts of *V. oxycoccos* fruit samples was determined according to the technique described by Zymonė et al. [58] and expressed as chlorogenic acid equivalents (ChAEs) in mg/g of the *V. oxycoccos* fruit samples.

3.3.2. Determination of Antioxidant Activity In Vitro

The ABTS and CUPRAC assays were performed using the techniques described by Balciunaitiene et al. [59]. The in vitro antioxidant activity of the tested extracts was expressed as Trolox equivalents (TE) in µmol/g of the *V. oxycoccos* fruit samples.

3.4. Chromatographic Methods

The evaluation of anthocyanin content in the extracts of *V. oxycoccos* fruit samples was performed according to the technique described by Vilkickyte et al. [60]. Qualitative and quantitative analysis of flavonol compounds and chlorogenic acid in small cranberry fruit samples collected during the ripening period was performed by using UHPLC according to the methodology described by Urbstaite et al. [32]. The analysis of triterpenic compounds was performed according to the technique described by Sedbare et al. [52].

3.5. Data Analysis

Statistical analyses were performed using SPSS Statistics 21 (IBM, Chicago, IL, USA) and Microsoft Excel 2016 (Microsoft, Redmond, WA, USA) software. The sample analyses were performed in triplicate. All results are expressed as mean ± standard deviation. A single-factor analysis of variance (ANOVA) along with a post hoc Tukey test was employed for statistical analysis. Differences at $p < 0.05$ were considered to be significant. The coefficient of variation was calculated to determine the variation in the quantitative composition of the studied bioactive compounds between samples of *V. oxycoccos* fruit collected during their ripening period. The correlation was determined by Pearson's statistical analysis. Correlation coefficients were as follows: 0.00–0.10 was regarded as a negligible correlation, 0.10–0.39 was regarded as a weak correlation, 0.40–0.69 was regarded as a moderate correlation, 0.70–0.89 was regarded as a strong correlation, and 0.90–1.00 was regarded as a very strong correlation [61].

4. Conclusions

The study provides new information on the qualitative and quantitative composition and accumulation patterns of biologically active compounds in *V. oxycoccos* fruits growing in natural wetland-type habitats in Lithuanian climate conditions.

The highest total amounts of identified flavonols were determined at the beginning of fruit ripening on September 10. Among individual flavonol compounds, myricetin-3-galactoside and hyperoside predominated. The highest amounts of myricetin-3-galactoside were detected in *V. oxycoccos* fruit samples collected on September 5 and September 18. The highest hyperoside content was found in small cranberry fruit samples collected on September 10.

Meanwhile, the highest total amounts of proanthocyanidins, anthocyanins, and triterpenes were detected in fruit samples collected in the middle of the ripening period (September 16–18). Among anthocyanin group compounds, peonidin-3-O-galactoside and cyanidin-3-O-galactoside predominated. The highest amounts of these compounds were determined in *V. oxycoccos* fruit samples collected on September 18. Among triterpene group compounds, ursolic acid clearly predominated. The highest amount of this acid was determined in small cranberry fruit samples collected on September 16. The in vitro antioxidant activity was found to be strongest in extracts of cranberry fruit collected on September 10.

Studies on the qualitative and quantitative composition of biologically active compounds in cranberry fruit are important for the chemotyping of cranberries growing in natural cenopopulations in Lithuania. Detailed studies on the phytochemical composition of cranberry fruits ensure the preparation of high-quality cranberry raw material, protect cranberry plantations, and allow for a rational use of plant resources. Phytochemical studies on cranberry fruit are important in order to determine the optimum timing for harvesting cranberries and to provide research-based recommendations for the rational harvesting of cranberry fruit in natural habitats.

Supplementary Materials: The following supporting information can be downloaded at https://www.mdpi.com/article/10.3390/ijms26146956/s1.

Author Contributions: Conceptualization, V.J.; methodology, M.L. and R.Š.; software, R.Š.; formal analysis, R.Š.; investigation, R.Š.; resources, R.Š.; data curation, V.J.; writing—original draft preparation, M.L.; writing—review and editing, M.L., I.R. and V.J.; visualization, M.L.; supervision,

V.J.; project administration, V.J.; funding acquisition, V.J. All authors have read and agreed to the published version of the manuscript.

Funding: This research received no external funding.

Institutional Review Board Statement: Not applicable.

Informed Consent Statement: Not applicable.

Data Availability Statement: All data generated during this study are included in this article.

Conflicts of Interest: The authors declare no conflicts of interest.

References

1. Martins, E. The growing use of herbal medicines: Issues relating to adverse reactions and challenges in monitoring safety. *Front. Pharmacol.* **2014**, *4*, 177. [CrossRef] [PubMed]
2. Martău, G.A.; Bernadette-Emőke, T.; Odocheanu, R.; Soporan, D.A.; Bochiș, M.; Simon, E.; Vodnar, D.C. *Vaccinium* species (*Ericaceae*): Phytochemistry and biological properties of medicinal plants. *Molecules* **2023**, *28*, 1533. [CrossRef] [PubMed]
3. Jurikova, T.; Skrovankova, S.; Mlcek, J.; Balla, S.; Snopek, L. Bioactive compounds, antioxidant activity, and biological effects of European cranberry (*Vaccinium oxycoccos*). *Molecules* **2018**, *24*, 24. [CrossRef] [PubMed]
4. Viskelis, P.; Rubinskienė, M.; Jasutienė, I.; Šarkinas, A.; Daubaras, R.; Česonienė, L. Anthocyanins, Antioxidative, and Antimicrobial Properties of American Cranberry (*Vaccinium macrocarpon* Ait.) and Their Press Cakes. *J. Food Sci.* **2009**, *74*, C157–C161. [CrossRef] [PubMed]
5. Baškytė, R.; Raščius, G.; Kavaliauskas, P.; Tukačiauskas, T. *Protected Areas in Lithuania*; Lututė: Vilnius, Lithuania, 2019; pp. 13–37.
6. Adamczak, A.; Gąbka, M.; Buchwald, W. Fruit yield of European cranberry (*Oxycoccus palustris* Pers.) in different plant communities of peatlands (Northern Wielkopolska, Poland). *Acta Agrobot.* **2012**, *62*, 97–105. [CrossRef]
7. Raal, A.; Kõiva, M.; Kuperjanov, A.; Vilbaste, K.; Vlasova, I.; Koshovyi, O. Multi-use of cranberries (*Vaccinium* spp.): Heritage and pharmaceutical results. *Folklore* **2023**, *89*, 107–142. [CrossRef]
8. Česonienė, L.; Daubaras, R.; Jasutiene, I.; Miliauskiene, I.; Zych, M. Investigations of anthocyanins, organic acids, and sugars show great variability in nutritional and medicinal value of European cranberry (*Vaccinium oxycoccos*) fruit. *J. Appl. Bot. Food Qual.* **2015**, *88*, 295–299.
9. Xue, L.; Liu, C.; Ma, H.; Seeram, N.P.; Neto, C.C. Anti-Inflammatory activities of cranberry fruit extracts in human THP-1 monocytes are influenced by their phytochemical composition. *ACS Food Sci. Technol.* **2022**, *2*, 75–83. [CrossRef]
10. Nemzer, B.V.; Al-Taher, F.; Yashin, A.; Revelsky, I.; Yashin, Y. Cranberry: Chemical composition, antioxidant activity and impact on human health: Overview. *Molecules* **2022**, *27*, 1503. [CrossRef] [PubMed]
11. Pandey, K.B.; Rizvi, S.I. Plant polyphenols as dietary antioxidants in human health and disease. *Oxid. Med. Cell. Longev.* **2009**, *2*, 270–278. [CrossRef] [PubMed]
12. Pott, D.M.; Vallarino, J.G.; Osorio, S. Metabolite changes during postharvest storage: Effects on fruit quality traits. *Metabolites* **2020**, *10*, 187. [CrossRef] [PubMed]
13. Kylli, P.; Nohynek, L.; Puupponen-Pimiä, R.; Westerlund-Wikstrom, B.; Leppanen, T.; Welling, J.; Moilanen, E.; Heinonen, M. Lingonberry (*Vaccinium vitis-idaea*) and European cranberry (*Vaccinium microcarpon*) proanthocyanidins: Isolation, identification, and bioactivities. *J. Agric. Food Chem.* **2011**, *59*, 3373–3384. [CrossRef] [PubMed]
14. Salerno-Paestum, I. *WHO Monographs on Selected Medicinal Plants*; WHO: Geneva, Switzerland, 2006; Volume 4, pp. 149–166.
15. Neto, C.C.; Krueger, C.G.; Lamoureaux, T.L.; Kondo, M.; Vaisberg, A.J.; Hurta, R.A.; Curtis, S.; Matchett, M.D.; Yeung, H.; Sweeney, M.I. MALDI-TOF MS characterization of proanthocyanidins from cranberry fruit (*Vaccinium macrocarpon*) that inhibit tumor cell growth and matrix metalloproteinase expression in vitro. *J. Sci. Food Agric.* **2006**, *86*, 18–25. [CrossRef]
16. Vvedenskaya, I.O.; Vorsa, N. Flavonoid composition over fruit development and maturation in American cranberry, *Vaccinium macrocarpon* Ait. *Plant Sci.* **2004**, *167*, 1043–1054. [CrossRef]
17. Oszmiański, J.; Kolniak-Ostek, J.; Lachowicz, S.; Gorzelany, J.; Matlok, N. Phytochemical compounds and antioxidant activity in different cultivars of cranberry (*Vaccinium macrocarpon* L.). *J. Food Sci.* **2017**, *82*, 2569–2575. [CrossRef] [PubMed]
18. Poudel, P.R.; Koyama, K.; Goto-Yamamoto, N. Evaluating the Influence of Temperature on Proanthocyanidin Biosynthesis in Developing Grape Berries (*Vitis vinifera* L.). *Mol. Biol. Rep.* **2020**, *47*, 3501–3510. [CrossRef] [PubMed]
19. Guo, X.L.; Hu, J.B.; Wang, D.L. Effect of Light Intensity on Blueberry Fruit Coloration, Anthocyanin Synthesis Pathway Enzyme Activity, and Gene Expression. *Russ. J. Plant Physiol.* **2023**, *70*, 136. [CrossRef]
20. Cáceres-Mella, A.; Talaverano, M.I.; Villalobos-González, L.; Ribalta-Pizarro, C.; Pastenes, C. Controlled Water Deficit during Ripening Affects Proanthocyanidin Synthesis, Concentration and Composition in Cabernet Sauvignon Grape Skins. *Plant Physiol. Biochem.* **2017**, *117*, 34–41. [CrossRef] [PubMed]

21. Jungfer, E.; Zimmermann, B.F.; Ruttkat, A.; Galensa, R. Comparing procyanidins in selected *Vaccinium species* by UHPLC-MS^2 with regard to authenticity and health effects. *J. Agric. Food Chem.* **2012**, *60*, 9688–9696. [CrossRef] [PubMed]
22. Šedbarė, R.; Sprainaitytė, S.; Baublys, G.; Viskelis, J.; Janulis, V. Phytochemical composition of cranberry (*Vaccinium oxycoccos* L.) fruits growing in protected areas of Lithuania. *Plants* **2023**, *12*, 1974. [CrossRef] [PubMed]
23. Wang, Y.; Johnson-Cicalese, J.; Singh, A.P.; Vorsa, N. Characterization and quantification of flavonoids and organic acids over fruit development in American cranberry (*Vaccinium macrocarpon*) cultivars using HPLC and APCI-MS/MS. *Plant Sci.* **2017**, *262*, 91–102. [CrossRef] [PubMed]
24. Sova, M. Antioxidant and antimicrobial activities of cinnamic acid derivatives. *Mini Rev. Med. Chem.* **2012**, *12*, 749–767. [CrossRef] [PubMed]
25. Singh, R.; Rastogi, S.; Dwivedi, U.N. Phenylpropanoid metabolism in ripening fruits. *Compr. Rev. Food Sci. Food Saf.* **2010**, *9*, 398–416. [CrossRef] [PubMed]
26. Gruz, J.; Ayaz, F.A.; Torun, H.; Strnad, M. Phenolic acid content and radical scavenging activity of extracts from medlar (*Mespilus germanica* L.) fruit at different stages of ripening. *Food Chem.* **2011**, *124*, 271–277. [CrossRef]
27. Kapoor, L.; Simkin, A.J.; George Priya Doss, C.; Siva, R. Fruit ripening: Dynamics and integrated analysis of carotenoids and anthocyanins. *BMC Plant Biol.* **2022**, *22*, 27. [CrossRef] [PubMed]
28. Oszmiański, J.; Lachowicz, S.; Gorzelany, J.; Matłok, N. The Effect of different maturity stages on phytochemical composition and antioxidant capacity of cranberry cultivars. *Eur. Food Res. Technol.* **2018**, *244*, 705–719. [CrossRef]
29. Castrejón, A.D.R.; Eichholz, I.; Rohn, S.; Kroh, L.W.; Huyskens-Keil, S. Phenolic profile and antioxidant activity of highbush blueberry (*Vaccinium corymbosum* L.) during fruit maturation and ripening. *Food Chem.* **2008**, *109*, 564–572. [CrossRef]
30. Wang, G.; Cao, F.; Chang, L.; Guo, X.; Wang, J. Temperature has more effects than soil moisture on biosynthesis of flavonoids in ginkgo (*Ginkgo biloba* L.) leaves. *New For.* **2014**, *45*, 797–812. [CrossRef]
31. Hodaei, M.; Rahimmalek, M.; Arzani, A.; Talebi, M. The Effect of water stress on phytochemical accumulation, bioactive compounds and expression of key genes involved in flavonoid biosynthesis in *Chrysanthemum morifolium* L. *Ind. Crops Prod.* **2018**, *120*, 295–304. [CrossRef]
32. Urbstaite, R.; Raudone, L.; Liaudanskas, M.; Janulis, V. Development, validation, and application of the UPLC-DAD methodology for the evaluation of the qualitative and quantitative composition of phenolic compounds in the fruit of American cranberry (*Vaccinium macrocarpon* Aiton). *Molecules* **2022**, *27*, 467. [CrossRef] [PubMed]
33. Šedbarė, R.; Grigaitė, O.; Janulis, V. Peculiarities of the variation of biologically active compounds in fruit of *Vaccinium oxycoccos* L. growing in the Čepkeliai state strict nature reserve. *Molecules* **2023**, *28*, 5888. [CrossRef] [PubMed]
34. Stobnicka, A.; Gniewosz, M. Antimicrobial protection of minced pork meat with the use of swamp cranberry (*Vaccinium oxycoccos* L.) fruit and pomace extracts. *J. Food Sci. Technol.* **2018**, *55*, 62–71. [CrossRef] [PubMed]
35. Harnly, J.; Geng, P.; Polashock, J.; Chen, P.; Johnson, J.; Vorsa, N. Impact of genetics and environment on cranberry fruit metabolites. *J. AOAC Int.* **2025**, qsaf056. [CrossRef] [PubMed]
36. Zhou, Y.; Liu, Y.; Zhang, X.; Gao, X.; Shao, T.; Long, X.; Rengel, Z. Effects of soil properties and microbiome on highbush blueberry (*Vaccinium corymbosum*) growth. *Agronomy* **2022**, *12*, 1263. [CrossRef]
37. Landi, M.; Tattini, M.; Gould, K.S. Multiple functional roles of anthocyanins in plant-environment interactions. *Environ. Exp. Bot.* **2015**, *119*, 4–17. [CrossRef]
38. Li, F.; Sun, Q.; Chen, L.; Zhang, R.; Zhang, Z. Unlocking the health potential of anthocyanins: A structural insight into their varied biological effects. *Crit. Rev. Food Sci. Nutr.* **2025**, *65*, 2134–2154. [CrossRef] [PubMed]
39. Česonienė, L.; Daubaras, R.; Jasutienė, I.; Venclovienė, J.; Miliauskienė, I. Evaluation of the biochemical components and chromatic properties of the juice of *Vaccinium macrocarpon* Aiton and *Vaccinium oxycoccos* L. *Plant Foods Hum. Nutr.* **2011**, *66*, 238–244. [CrossRef] [PubMed]
40. Vorsa, N.; Polashock, J.J. Alteration of anthocyanin glycosylation in cranberry through interspecific hybridization. *J. Am. Soc. Hortic. Sci.* **2005**, *130*, 711–715. [CrossRef]
41. Česonienė, L.; Jasutienė, I.; Šarkinas, A. Phenolics and anthocyanins in berries of European cranberry and their antimicrobial activity. *Medicina* **2009**, *45*, 992–999. [CrossRef] [PubMed]
42. Fawole, O.A.; Opara, U.L. Changes in physical properties, chemical and elemental composition and antioxidant capacity of pomegranate (cv. Ruby) fruit at five maturity stages. *Sci. Hortic.* **2013**, *150*, 37–46. [CrossRef]
43. de Pascual-Teresa, S.; Sanchez-Ballesta, M.T. Anthocyanins: From plant to health. *Phytochem. Rev.* **2008**, *7*, 281–299. [CrossRef]
44. Šedbarė, R.; Pašakinskienė, I.; Janulis, V. Changes in the composition of biologically active compounds during the ripening period in fruit of different large cranberry (*Vaccinium macrocarpon* Aiton) cultivars grown in the Lithuanian collection. *Plants* **2023**, *12*, 202. [CrossRef] [PubMed]
45. Klavins, L.; Klavins, M. Cuticular wax composition of wild and cultivated northern berries. *Foods* **2020**, *9*, 587. [CrossRef] [PubMed]

46. Klavins, L.; Viksna, A.; Kviesis, J.; Klavins, M. Lipids of cultivated and wild *Vaccinium* Spp. Berries from Latvia. In Proceedings of the 13th Baltic Conference on Food Science and Technology "Food. Nutrition. Well-Being." Foodbalt 2019 Conference Proceeding, Jelgava, Latvia, 2–3 May 2019; pp. 198–203.
47. Rascón-Valenzuela, L.A.; Torres-Moreno, H.; Velázquez-Contreras, C.; Garibay-Escobar, A.; Robles-Zepeda, R.E. Triterpenoids: Synthesis, uses in cancer treatment and other biological activities. In *Advances in Medicine and Biology*; Berhardt, L.V., Ed.; Nova Science Publishers: New York, NY, USA, 2017; Volume 106, pp. 139–182.
48. Xue, L.; Otieno, M.; Colson, K.; Neto, C. Influence of the growing Region on the phytochemical composition and antioxidant properties of North American cranberry fruit (*Vaccinium macrocarpon* Aiton). *Plants* **2023**, *12*, 3595. [CrossRef] [PubMed]
49. Jan, R.; Asaf, S.; Numan, M.; Lubna; Kim, K.M. Plant secondary metabolite biosynthesis and transcriptional regulation in response to biotic and abiotic stress conditions. *Agronomy* **2021**, *11*, 968. [CrossRef]
50. Zhang, F.; Daimaru, E.; Ohnishi, M.; Kinoshita, M.; Tokuji, Y. Oleanolic acid and ursolic acid in commercial dried fruits. *Food Sci. Technol. Res.* **2013**, *19*, 113–116. [CrossRef]
51. Wu, X.; Xue, L.; Tata, A.; Song, M.; Neto, C.; Xiao, H. Bioactive components of polyphenol-rich and non-polyphenol-rich cranberry fruit extracts and their chemopreventive effects on colitis-associated colon cancer. *J. Agric. Food Chem.* **2020**, *68*, 6845–6853. [CrossRef] [PubMed]
52. Sedbare, R.; Raudone, L.; Zvikas, V.; Viskelis, J.; Liaudanskas, M.; Janulis, V. Development and validation of the UPLC-DAD methodology for the detection of triterpenoids and phytosterols in fruit samples of *Vaccinium macrocarpon* Aiton and *Vaccinium oxycoccos* L. *Molecules* **2022**, *27*, 4403. [CrossRef] [PubMed]
53. Gulcin, İ. Antioxidants and Antioxidant Methods: An Updated Overview. *Arch. Toxicol.* **2020**, *94*, 651–715. [CrossRef] [PubMed]
54. Liaudanskas, M.; Šedbarė, R.; Janulis, V. Determination of biologically active compounds and antioxidant capacity in vitro in fruit of small cranberries (*Vaccinium oxycoccos* L.) growing in natural habitats in Lithuania. *Antioxidants* **2024**, *13*, 1045. [CrossRef] [PubMed]
55. Namiesnik, J.; Vearasilp, K.; Nemirovski, A.; Leontowicz, H.; Leontowicz, M.; Pasko, P.; Martinez-Ayala, A.L.; Gonzalez-Aguilar, G.A.; Suhaj, M.; Gorinstein, S. In vitro studies on the relationship between the antioxidant activities of some berry extracts and their binding properties to serum albumin. *Appl. Biochem. Biotechnol.* **2014**, *172*, 2849–2865. [CrossRef] [PubMed]
56. de Las Heras, N.; Galiana, A.; Ballesteros, S.; Quintela, J.C.; Bonilauri, I.; Lahera, V.; Martín-Fernández, B. Polyphenols and triterpenes combination in an in vitro model of cardiac damage: Protective effects. *Int. J. Mol. Sci.* **2023**, *24*, 7977. [CrossRef] [PubMed]
57. Liaudanskas, M.; Viškelis, P.; Jakštas, V.; Raudonis, R.; Kviklys, D.; Milašius, A.; Janulis, V. Application of an optimized HPLC method for the detection of various phenolic compounds in apples from Lithuanian cultivars. *J. Chem.* **2014**, *2014*, 542121. [CrossRef]
58. Zymonė, K.; Liaudanskas, M.; Lanauskas, J.; Nagelytė, M.; Janulis, V. Variability in the Qualitative and Quantitative Composition of Phenolic Compounds and the In Vitro Antioxidant Activity of Sour Cherry (*Prunus cerasus* L.) Leaves. *Antioxidants* **2024**, *13*, 553. [CrossRef] [PubMed]
59. Balciunaitiene, A.; Zehra, S.H.; Liaudanskas, M.; Zvikas, V.; Viskelis, J.; Nuapia, Y.B.; Siukscius, A.; Singh, P.K.; Janulis, V.; Viskelis, P. Biosynthesis of silver nanoparticles via *Medusomyces gisevii* fermentation with *Origanum vulgare* L. extract: Antimicrobial properties, antioxidant properties, and phytochemical analysis. *Molecules* **2025**, *30*, 1706. [CrossRef] [PubMed]
60. Vilkickyte, G.; Motiekaityte, V.; Vainoriene, R.; Liaudanskas, M.; Raudone, L. Development, validation, and application of UPLC-PDA method for anthocyanins profiling in *Vaccinium* L. berries. *J. Berry Res.* **2021**, *11*, 583–599. [CrossRef]
61. Schober, P.; Boer, C.; Schwarte, L.A. Correlation coefficients: Appropriate use and interpretation. *Anesth. Analg.* **2018**, *126*, 1763–1768. [CrossRef] [PubMed]

International Journal of
Molecular Sciences

MDPI

Article

Comprehensive Phytochemical Analysis and Evaluation of Antioxidant, Antimicrobial, Cytotoxic, and Immunomodulatory Activities of Commercial Cinnamon Bark Essential Oil (*Cinnamomum zeylanicum* L.)

Milja Živković [1], Isidora Stanisavljević [2], Nevena Gajović [2], Slađana Pavlović [2], Bojana Simović Marković [2], Ivan P. Jovanović [2], Snežana Cupara [1], Vanja Tadić [3], Ana Žugić [3], Marina T. Milenković [4] and Ana Barjaktarević [1,*]

[1] Department of Pharmacy, Faculty of Medical Sciences, University of Kragujevac, Svetozara Markovica 69, 34000 Kragujevac, Serbia; zivkovicmilja@gmail.com (M.Ž.); snezanacupara@yahoo.com (S.C.)

[2] Center for Molecular Medicine and Stem Cell Research, Faculty of Medical Sciences, University of Kragujevac, Svetozara Markovića 69, 34000 Kragujevac, Serbia; isidorastanisavljevic97@gmail.com (I.S.); gajovicnevena@yahoo.com (N.G.); sladjadile@gmail.com (S.P.); bojana.simovic@gmail.com (B.S.M.); ivanjovanovic77@gmail.com (I.P.J.)

[3] Department for Pharmaceutical Research and Development, Institute for Medicinal Plant Research "Dr. Josif Pančić", Tadeuša Koscuška 1, 11000 Belgrade, Serbia; vtadic@mocbilja.rs (V.T.); azugic@mocbilja.rs (A.Ž.)

[4] Department of Microbiology and Immunology, Faculty of Pharmacy, University of Belgrade, Vojvode Stepe 450, 11221 Belgrade, Serbia; marina.milenkovic@pharmacy.bg.ac.rs

* Correspondence: ana.radovanovic@medf.kg.ac.rs

Abstract

The essential oil derived from the bark of *Cinnamomum zeylanicum* L., *Lauraceae*, has gained significant attention because of its numerous biological benefits. This study aimed to perform a phytochemical analysis of commercially available *Cinnamomum zeylanicum* bark essential oil and to evaluate its antioxidant, antimicrobial, immunomodulatory, and antitumor properties. GC–MS analysis was employed to determine the phytochemical composition. The major component of the total essential oil composition was (*E*)-cinnamaldehyde, constituting 77.93%, followed by eugenol (4.34%), *E*-caryophyllene (3.68%), and linalool (2.79%). The antioxidant activity was confirmed by DPPH, ABTS, CUPRAC, and TAC assays. In the broth microdilution assay, cinnamon essential oil demonstrated strong antimicrobial activity, with MIC values ranging from 7.37 to 29.50 μg/mL. Furthermore, cinnamon essential oil demonstrated selective antitumor activity by inducing apoptosis and cell-cycle arrest in human colorectal cancer cells (HCT116) while sparing non-cancerous cells (MRC-5). In HCT116 cells, cinnamon essential oil induced apoptosis, downregulated Cyclin D and p-AKT, and caused G1-phase arrest. Additionally, cinnamon essential oil modulated immune responses by reducing pro-inflammatory cytokine production in activated splenocytes and enhancing pro-inflammatory activity in naïve cells. These findings highlight the great potential of the cinnamon bark essential oil in the development of new therapeutic agents.

Keywords: *Cinnamomum zeylanicum*; bark essential oil; antioxidant; antimicrobial; antitumor; immunomodulatory

1. Introduction

Cinnamon (*Cinnamomum verum* J. Presl. or *Cinnamomum zeylanicum* L., *Lauraceae*), commonly known as 'true cinnamon,' is an aromatic spice that has been traditionally used for its medicinal and culinary properties [1]. Over the past twenty years, there has been a growing tendency to develop new herbal medicines from natural sources [2]. Recently, the essential oil derived from cinnamon bark has gained significant attention for its diverse biological activities, including antioxidant, antimicrobial, immunomodulatory, and anticancer effects [3]. These properties are largely attributed to the unique chemical composition of the essential oil, particularly the presence of bioactive compounds such as cinnamaldehyde, eugenol, cinnamyl alcohol, linalool, and various terpenoids, which play crucial roles in inhibiting harmful microorganisms, neutralising oxidative stress, modulating the immune response, and suppressing tumor development [1,3]. Cinnamaldehyde was found to be a major bioactive compound, making up 60–80% of the cinnamon bark essential oil (CEO) [4,5]. The specific composition of the oil can vary depending on factors such as the plant's geographical origin and the method of extraction. Particularly, cinnamaldehyde, as a major component, is responsible for the antioxidant and antimicrobial properties of essential oil, among other effects [4,6]. The mechanism behind the CEO's antioxidant activity is mainly described as its ability to neutralise free radicals and reduce oxidative stress [7]. Previous studies have shown that cinnamaldehyde and eugenol play crucial roles in inhibition of lipid peroxidation and protection against oxidative damage [3,7]. A study conducted by Denkova-Kostova R et al. [8] showed the efficacy of bark CEO as a natural antioxidant and antimicrobial agent. The oil demonstrated high antimicrobial activity against spore-forming bacteria, saprophytic microorganisms, fungi, and yeast. The essential oil was more effective against Gram-positive bacteria, compared to Gram-negative bacteria, because of the evident difference in the structure of the bacterial cell wall [8]. Additionally, CEO has demonstrated efficacy against drug-resistant strains of *Escherichia coli*, *Staphylococcus aureus*, and *Candida albicans* [9,10]. Along with its antioxidant and antimicrobial properties, CEO has been found to possess immunomodulatory and antitumor activity. Recent research indicates that compounds like cinnamaldehyde can modulate the immune response by regulating the production of pro-inflammatory cytokines and reducing inflammation. This makes cinnamon oil a promising agent for managing immune-related conditions, such as autoimmune diseases and chronic inflammatory disorders [11]. Previously published studies have reported the cytotoxic activity of CEO on several human and animal cancer cell lines, including lung cancer, breast adenocarcinoma, chronic myelogenous erythroleukemia, mouse breast cancer, and mouse lymphocytic leukemia cell lines [12,13]. It has been found that cinnamaldehyde induces apoptosis and inhibits the proliferation of cancer cells in various cancer models, including breast, lung, and colorectal cancers [14].

The present study aimed to perform a phytochemical analysis of commercially available *C. zeylanicum* bark essential oil and to expand the existing knowledge of its antioxidant, antimicrobial, immunomodulatory, and antitumor properties.

2. Results

2.1. GC–MS Analysis of the CEO

The gas chromatography–mass spectrometry (GC–MS) analysis of the bark CEO identified 27 constituents, which together represented 99.66% of the essential oil composition (Table 1). All identified components have been categorized into five categories: monoterpene hydrocarbons, oxygenated monoterpenes, sesquiterpenes, aromatic compounds, and cinnamic derivatives. A quantity of nine components within the monoterpene hydrocarbons category severally constituted 0.05–1.45%. Linalool was the most abundant component among the oxygenated monoterpenes (2.79%); *E*-caryophyllene (3.68%) among the sesquiterpene hydrocarbons, and eugenol (4.34%) among the aromatic compounds. However, the dominant component of the total essential oil composition was (*E*)-cinnamaldehyde, within the cinnamic derivatives category, with a share of 77.93%.

Table 1. Chemical composition of *C. zeylanicum* bark essential oil.

N°	Class/Compound	Kovats Index	Percentage (%)
	Monoterpene Hydrocarbons		
1	α-thujene	924	0.06
2	α-pinene	932	0.65
3	Camphene	946	0.05
4	β-pinene	974	0.21
5	α-phellandrene	1002	0.58
6	α-terpinene	1014	0.10
7	*p*-cymene	1020	1.38
8	Limonene + β-phellandrene	1025 + 1025	1.45
9	Terpinolene	1086	0.06
	Oxygenated Monoterpenes		
10	1,8-cineol	1026	0.83
11	Linalool	1095	2.79
12	Terpinen-4-ol	1174	0.30
13	α-terpineol	1186	0.58
	Aromatic Compounds		
14	Benzaldehyde	952	0.09
15	Phenyl ethyl alcohol	1106	0.10
16	Eugenol	1356	4.34
17	Eugenol acetate	1521	0.11
18	Benzyl benzoate	1759	0.75
	Sesquiterpenes		
19	α-copaene	1374	0.14
20	(*E*)-caryophyllene	1417	3.68
21	α-humulene	1452	0.24
22	*Cis*-muurola-4(14),5-diene	1465	0.09
23	Caryophyllene oxide	1582	0.16
	Cinnamic Derivatives		
24	(*Z*)-cinnamaldehyde	1217	0.05
25	(*E*)-cinnamaldehyde	1267	77.93
26	(*E*)-cinnamylacetate	1443	2.74
27	(*E*)-allyl cinnamate	1548	0.20
Total			99.66

2.2. Antioxidant Activity

In this study, the antioxidant activity of *C. zeylanicum* bark essential oil was assessed using four different in vitro assays: the DPPH Radical Scavenging Activity Assay (DPPH), ABTS Radical Scavenging Activity Assay (ABTS), Cupric Ion Reducing Antioxidant Capacity Assay (CUPRAC), and Total Antioxidant Capacity Assay (TAC) assays. The results of antioxidant activity are presented in Table 2. According to IC_{50} values determined using the DPPH, ABTS, and CUPRAC assays (IC_{50} = 9.53 ± 0.2 mg/mL; 0.35 ± 0.01 mg/mL; and 0.14 ± 0.002 mg/mL, respectively), CEO has shown a weak antioxidant activity compared to BHT (IC_{50} = 0.011 mg/mL; 0.006 mg/mL; and 0.02 mg/mL, respectively). The results of the TAC assay showed a value of 0.01 mg AAE/g for CEO, compared to 9.80 mg AAE/g for BHT (Butylated hydroxytoluene). This indicates that the CEO has a low capacity with respect to the reduction of molybdenum ions (Mo^{+6} into Mo^{+5}).

Table 2. Antioxidant activity of C. *zeylanicum* bark essential oil and a standard antioxidant, based on the DPPH, ABTS, CUPRAC, and TAC assays.

	IC_{50} DPPH (mg/mL)	IC_{50} ABTS (mg/mL)	IC_{50} CUPRAC (mg/mL)	TAC (mg AAE/g)
CEO	9.53 ± 0.20	0.35 ± 0.01	0.14 ± 0.02	0.01 ± 0.00
BHT	0.011 ± 0.001	0.006 ± 0.001	0.02 ± 0.00	9.80 ± 0.15

The results are expressed as means (n = 3) ± SD. CEO—Cinnamon essential oil; BHT—Butylated hydroxytoluene.

2.3. Antimicrobial Activity

The in vitro antimicrobial activity of the cinnamon bark essential oil was evaluated using the broth microdilution method, and according to the CLSI recommendations [15].

The obtained MIC values for CEO were in the range of 7.37–29.50 μg/mL (Table 3). The most pronounced antimicrobial activity was observed against *S. aureus*, *S. epidermidis*, and *C. albicans*, with identical MIC values of 7.37 μg/mL. The CEO was less active against *B. subtilis*, *E. coli*, *K. pneumoniae*, and *S. enterica* serovar Abony, with a MIC value of 14.75 μg/mL. However, *P. aeruginosa* was the most resistant to CEO (MIC = 29.50 μg/mL). According to the results presented in Table 3, Gram-positive bacteria, except *B. subtilis*, were more susceptible to CEO, compared to the tested Gram-negative bacteria. Generally, all MIC values for tested microbial strains are very low (significantly lower than 100 μg/mL), and that fact can classify CEO in the category of essential oils with strong antimicrobial activity. However, the observed MIC values of CEO were notably higher than those for the standard antibiotics and the antifungal agent, indicating lower antibacterial potency compared to synthetic agents.

Table 3. Antimicrobial activity of CEO and standard antibiotics, expressed as MIC (μg/mL).

Bacterial Strain	ATCC	CEO MIC (μg/mL)	MIN MIC μg/mL	AMK MIC μg/mL	NYS MIC μg/mL
Staphylococcus aureus	6538	7.37	0.40	1.00	N.t.
Staphylococcus epidermidis	1228	7.37	0.25	N.t.	N.t.
Bacillus subtilis	6633	14.75	0.05	3.25	N.t.
Escherichia coli	8739	14.75	N.t.	12.00	N.t.
Klebsiella pneumoniae	13,883	14.75	N.t.	12.50	N.t.
Pseudomonas aeruginosa	9027	29.50	N.t.	15.00	N.t.
Salmonella enterica subsp. *enterica* serovar Abony	6017	14.75	N.t.	12.50	N.t.
Yeast strain					
Candida albicans	10,231	7.37	N.t.	N.t.	0.85

N.t.—Not tested. MIN—Minocycline; AMK—Amikacin; NYS—Nystatin.

2.4. Evaluation of In Vitro Cytotoxicity of CEO

The cytotoxic effects of the CEO were assessed using the MTT assay. This analysis was performed on several cell lines, including mouse (4T1) and human (MDA-MB-468) breast cancer cells and mouse (CT26) and human (HCT116) colon cancer cells, as well as the non-cancerous human fibroblast cell line (MRC-5). The cell lines were exposed to varying concentrations of the CEO (ranging from 100 to 0.78 μg) for 48 h. The findings indicate that the CEO significantly affected the viability of the 4T1, MDA-MB-468, HCT116, and CT26 cancer cell lines (Figure 1), while having minimal impact on the non-cancerous MRC-5 cells (Figure 1). The CEO was shown to reduce MRC-5 cell viability in a dose-dependent manner (Figure 1). However, at concentrations ranging from 0.78 to 100 μg, CEO exhibited a lower cytotoxic effect on MRC-5 cells, compared to its effects on cancerous cell lines (4T1, MDA-MB-468, HCT116, and CT26). Furthermore, cisplatin (*cis*-diamminedichloroplatinum(II),

CDDP) exhibited stronger cytotoxic effects on mouse 4T1 and CT26 cell lines, as well as human MDA-MB 468 and HCT116 cell lines, in comparison to CEO, under the same experimental conditions. It also reduced the viability of non-cancerous MRC-5 cells more than did CEO (Figure 1).

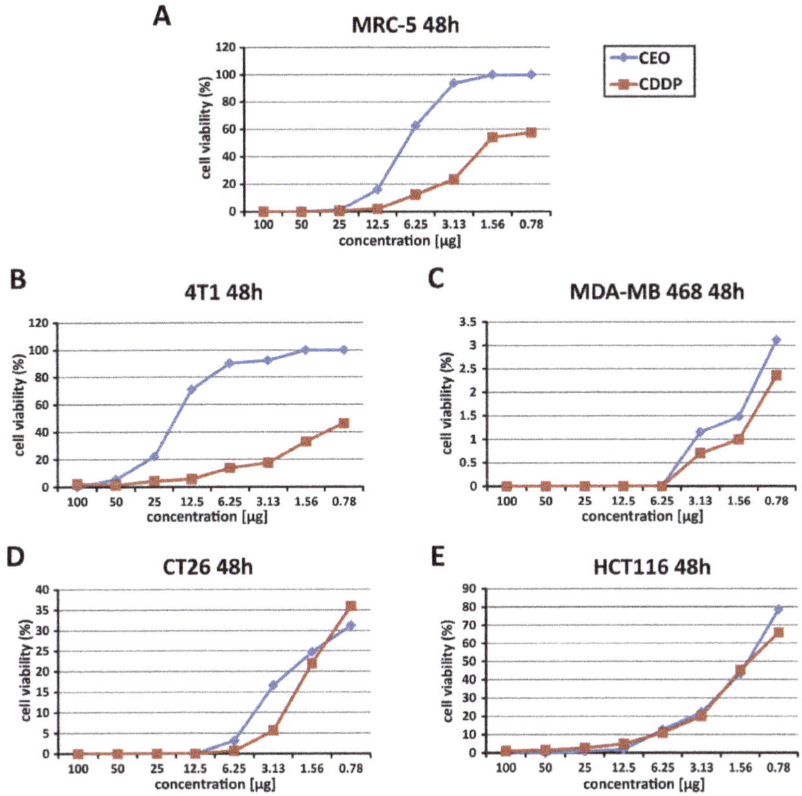

Figure 1. Dose-dependent cytotoxic effects of CEO. This figure illustrates the dose-dependent cytotoxicity of the CEO on various cell lines, as assessed by the MTT assay, following 48 h of exposure. The graphs show the survival rates of human fibroblasts (MRC-5) (**A**), murine and human breast carcinoma cells (4T1 and MDA-MB 468) (**B**,**C**), and murine and human colon cancer cells (CT26 and HCT116) (**D**,**E**), treated with the CEO and cisplatin (CDDP). The data presented represent mean values from three independent experiments, each conducted in triplicate.

In the context of the study of the cytotoxic effectiveness of CEO, a detailed assessment of the half-maximal inhibitory concentration (IC_{50}) values was conducted. This analysis focused on breast cancer cell lines (mouse 4T1 and human MDA-MB-468) as well as colon cancer cell lines (mouse CT26 and human HCT116). The IC_{50} values, representing the concentration needed to inhibit 50% of cell viability, are listed in full in Table 4. Upon examining these values, it was found that the CEO demonstrated more potent cytotoxic effects against the human HCT116 cells, compared to other cell lines. Also, CDDP is one of the most commonly used chemotherapeutic agents and is known for its cytotoxic effects on both cancerous and non-cancerous cells. However, CEO exhibited stronger cytotoxicity toward HCT116 cells, compared to CDDP, while simultaneously showing lower toxicity towards non-cancerous MRC-5 cells, as shown in Table 4.

The therapeutic selectivity of the cinnamon essential oil was further assessed by calculating the selectivity index, which is presented in Table 5. One of the key findings of this study was the notably high selectivity index for the CEO, with the HCT116 cell line exhibiting the highest selectivity index among all tested cell lines. This suggests that CEO is significantly more cytotoxic to the HCT116 cancer cells than to other cancerous and non-cancerous cells. Moreover, although CDDP exhibited greater cytotoxicity, it displayed lower selectivity towards non-cancerous MRC-5 cells, confirming its stronger but less selective effect compared to CEO (Table 5). Based on these findings, further research was conducted using the HCT116 cell line.

Table 4. IC_{50} Values for CEO and cisplatin (CDDP) on the MRC-5, 4T1, MDA-MB 468, CT26, and HCT116 cell lines, as determined by the MTT assay.

Cell Lines	IC_{50} (μg) 48 h	
	CEO	CDDP
MRC-5	7.21 ± 0.25	4.08 ± 2.11
4T1	12.43 ± 4.55	1.33 ± 0.74
MDA-MB 468	3.78 ± 0.93	3.44 ± 1.34
CT26	<0.78	<0.78
HCT116	1.18 ± 0.93	5.16 ± 1.82

Table 5. Selectivity index (SI) for CEO and cisplatin (CDDP) on the MRC-5, 4T1, MDA-MB 468, CT26, and HCT116 cell lines.

Cell Lines	Selectivity Index (IC_{50} MRC-5/IC_{50})	
	CEO	CDDP
4T1	0.6	3.1
MDA-MB 468	1.9	1.2
CT26	N/A	N/A
HCT116	6.1	0.8

2.5. Assessment of Apoptotic Cell Death Induced by CEO

The evaluation of apoptosis in HCT116 cells treated with CEO was performed using the Annexin V and propidium iodide double-staining assay. In this analysis, cells were treated with the CEO and then stained with Annexin V-FITC and PI. The results, shown in Figure 2A, demonstrated that a significant portion of HCT116 cells underwent both early and late apoptosis after 24 h of treatment with the CEO. Notably, necrosis does not appear to be the mechanism of cell death through which the CEO exerts its effects, as HCT116 cells did not undergo necrosis (Figure 2A). These findings highlight the dual mechanism of the CEO, which promotes cell death via apoptotic pathways, by suggesting its potential as a powerful anti-carcinoma agent. The results of this study indicate that treatment with IC_{50} concentrations of the CEO significantly increased the percentage of Bax+ HCT116 cells, while simultaneously reducing the percentage of Bcl-2+ HCT116 cells (Figure 2B,C). Furthermore, the CEO was shown to enhance the percentage of caspase-3+ HCT116 cells (Figure 2D). The observed increase in caspase-3 expression, coupled with a decrease in Bcl-2 expression, points to potential mechanisms of action for the CEO, particularly in inducing apoptosis. Additionally, the marked increase in Bax, a pro-apoptotic protein, following treatment with the CEO, underscores the role of CEO in promoting apoptosis in HCT116 cells.

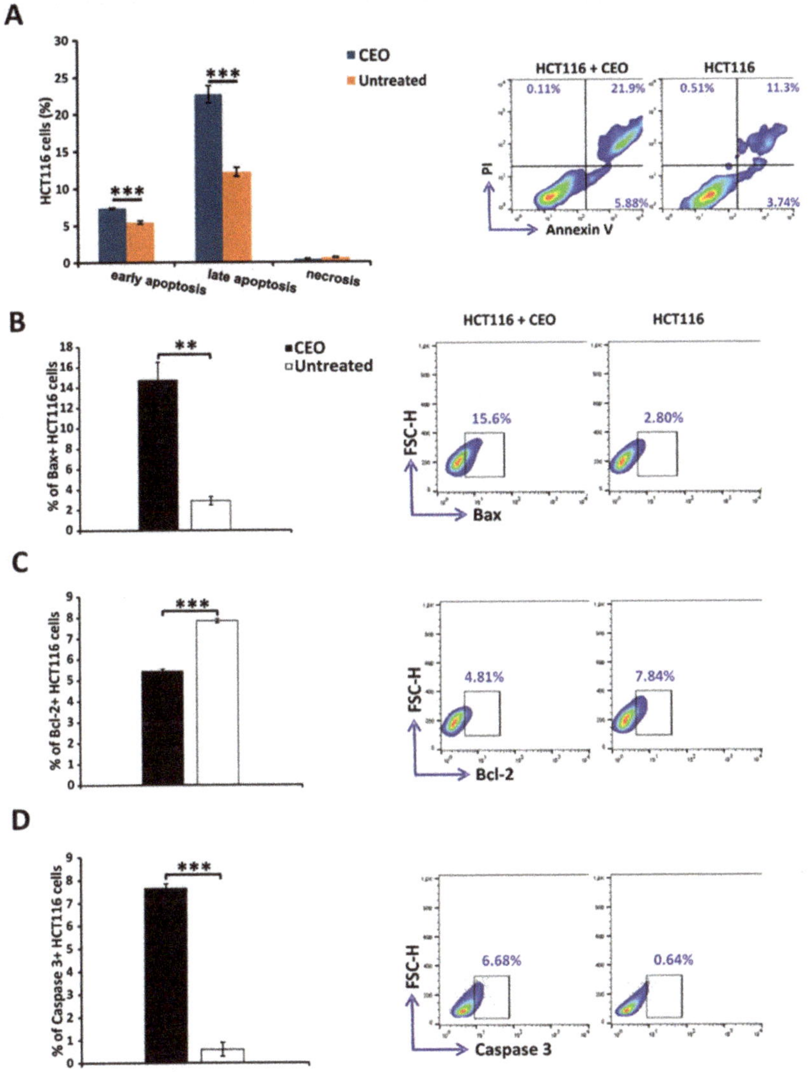

Figure 2. Induction of apoptosis and modulation of apoptosis-related proteins in HCT116 colorectal cancer cells by the CEO. Apoptosis was assessed using flow cytometry with Annexin V (FITC) and propidium iodide (PI) double staining to differentiate between viable (AnnV−PI−), early apoptotic (AnnV+PI−), late apoptotic (AnnV+PI+), and necrotic (AnnV−PI+) populations (**A**). Additionally, the expression levels of key apoptosis-related proteins, including Bax (**B**), Bcl-2 (**C**), and caspase-3 (**D**), were evaluated in both untreated and CEO-treated HCT116 cells after 24 h. Statistical analysis was performed using the non-parametric Mann–Whitney test. Statistical significance is indicated by ** $p < 0.01$, and *** $p < 0.001$, comparing CEO-treated cells to the untreated control group.

2.6. Effects of CEO on Cell Cycle and Proliferation in HCT116 Cancer Cells

In the investigation of CEO's antiproliferative effects, Ki67 expression levels were analyzed in HCT116 cells. The results demonstrated a significant reduction in Ki67 expression in CEO-treated HCT116 cells compared to untreated control cells (Figure 3A), indicating a notable decrease in cell proliferation. To assess the effect of the CEO on Cyclin D, HCT116 cells were treated with an IC_{50} concentration of the essential oil for 24 h. The results, shown in Figure 3B, reveal that the CEO significantly decreased the percentage of Cyclin D$^+$ cells, suggesting a reduction in the number of cells progressing from the G1 phase to the S phase. This selective downregulation of Cyclin D indicates that the CEO specifically interferes with cell-cycle progression at the G1 phase, potentially inducing cell-cycle arrest and preventing further proliferation of HCT116 cells.

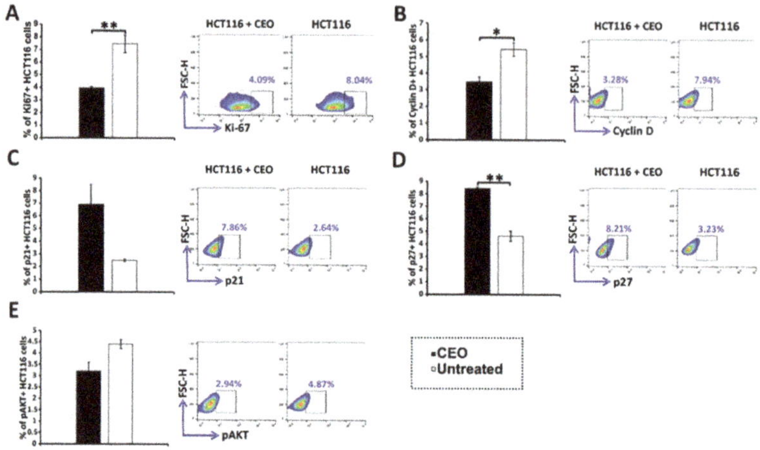

Figure 3. Expression of cell-cycle and signaling proteins in HCT116 cells treated with CEO. The expression of Ki67 (**A**), Cyclin D (**B**), p21 (**C**), p27 (**D**), and p-AKT (**E**) in HCT116 cells treated with the CEO is displayed. Representative FACS plots show the expression of key proteins following 24 h of exposure to the CEO. The data, obtained from flow cytometry, are presented as the mean ± SEM from three independent experiments. Statistical analysis was performed using the non-parametric Mann–Whitney test. Statistical significance is indicated by * $p < 0.05$ and ** $p < 0.01$, highlighting significant differences in protein expression between CEO-treated cells and the untreated control group.

Treatment with CEO induced significant alterations in the expression of key cell-cycle regulators in HCT116 cells. Specifically, there was an increase in the proportion of cells expressing the tumor suppressors p21 and p27 (Figure 3C,D). This upregulation is likely to contribute to CDK suppression, leading to cell-cycle arrest. Consistent with these findings, the increased expression of p21 and p27 was accompanied by a reduction in cyclin D levels, further reinforcing the disruption of cell-cycle progression. Additionally, CEO treatment led to a significant decrease in the levels of phosphorylated AKT (p-AKT) (Figure 3E). The reduction in p-AKT, a critical component of oncogenic signaling pathways that support cell survival and proliferation, suggests a potential antitumor effect of the CEO. These changes in p21, p27, and p-AKT expression highlight the CEO's ability to disrupt key processes involved in cell-cycle regulation and survival signaling, underscoring its potential as an effective anticancer agent.

2.7. *In Vitro, CEO Reduces the Concentration of Pro-Inflammatory Cytokines*

In further research, the immunomodulatory effects of CEO were examined. Splenocytes freshly isolated from healthy mice were cultured with medium, ConA, CEO, or both CEO and ConA simultaneously. Treatment with ConA only, a poly-clonal T cell activator (21), notably elevated the levels of IL-1β, TNF-α, IFN-γ, IL-17, and IL-10 in the cell supernatant after 24 h, compared to untreated cells (Figure 4A–E). CEO by itself also significantly raised the concentrations of IL-1β, TNF-α, IFN-γ, IL-17, and IL-10, compared to splenocytes incubated in medium (Figure 4A–E). However, the combined treatment of ConA and CEO significantly lowered the levels of IL-1β, TNF-α, IFN-γ, IL-17, and IL-10, compared to ConA stimulation (Figure 4A–E). Furthermore, the CEO appears to promote a predominance of pro-inflammatory cytokines over IL-10, which is shown in Figure 5A–D. This effect is more pronounced in non-activated cells, and likely demonstrates components of the innate immune response. Interestingly, while CEO exerts an immunosuppressive effect on activated cells, it simultaneously activates naïve cells, thereby exhibiting a dual effect on immune-cell functionality.

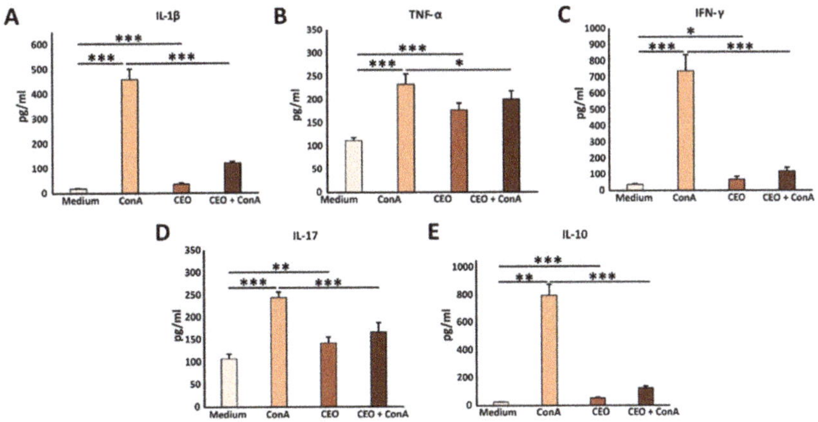

Figure 4. CEO decreases cytokine production in vitro. The graphs display the concentrations of IL-1β (**A**), TNF-α (**B**), IFN-γ (**C**), IL-17 (**D**), and IL-10 (**E**) in the supernatants of splenocytes from healthy mice after 24 h of incubation with medium only, ConA, CEO, or a combination of CEO and ConA. The data are presented as the mean ± SEM. Statistical analysis was performed using the non-parametric Kruskal–Wallis test. Statistical significance is indicated by * $p < 0.05$, ** $p < 0.01$, and *** $p < 0.001$, highlighting significant differences in cytokine concentration between groups.

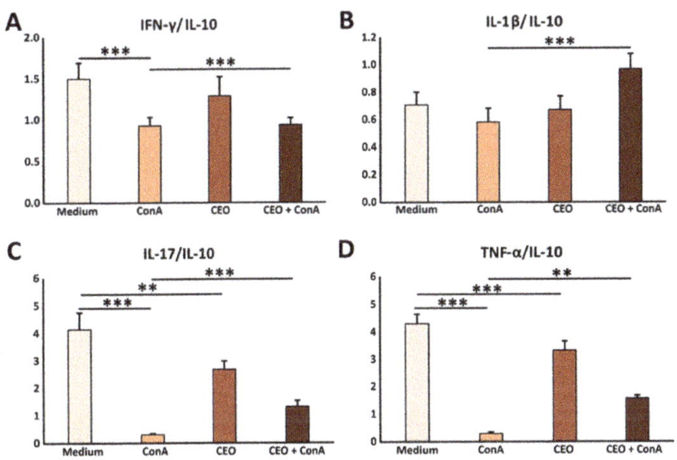

Figure 5. CEO facilitates the predominance of the pro-inflammatory cytokine milieu in vitro. The graphs display the ratios of IFN-γ/IL-10 (**A**), IL-1β/IL-10 (**B**), IL-17/IL-10 (**C**), and TNF-α/IL-10 (**D**) measured in the supernatants of splenocytes from healthy mice after 24 h of incubation with medium only, ConA, CEO, or a combination of CEO and ConA. The data are presented as mean ± SEM. Statistical analysis was performed using the non-parametric Kruskal–Wallis test. Statistical significance is indicated by ** $p < 0.01$, and *** $p < 0.001$, highlighting significant differences in cytokine concentration between groups.

3. Discussion

Plants provide an abundant source of potent medicinal compounds called phytochemicals, many of which play multiple roles in folk medicine worldwide. The application of different essential oils is associated with several biological properties, including antioxidant, antimicrobial, antiviral, anticancer, anti-inflammatory, immunomodulatory and antiprotozoal activities [16].

3.1. GC–MS Analysis

Generally, the data obtained by GC–MS analysis in this study were consistent with reports in the literature on the chemical composition of bark CEO. (*E*)-cinnamaldehyde is a dominant component of the bark CEO, present at a high concentration, according to several authors [8,10]. The concentrations of (*E*)-cinnamaldehyde in the bark CEO were in the range of 46–91.82%, depending on the plant's habitat [17–19]. When comparing the results of this study with findings in the literature, we observed that our sample of CEO showed a higher concentration of (*E*)-cinnamaldehyde (77.93%) than the Iranian sample (71.5%), and the concentration reported by Elgendy et al. (64.84%) [18]. On the other hand, the concentration of cinnamaldehyde in the bark CEO obtained in this study was lower than in the Indian sample (91.82%) [19]. The other quantitatively most representative components in our CEO sample were as follows: eugenol (4.34%), (*E*)-caryophyllene (3.68), and linalool (2.79), respectively. Additionally, differences were evident in the levels of linalool and sesquiterpenes in the study conducted by Behbahani et al., while our study demonstrated greater concentrations of benzyl benzoate and other cinnamic derivatives [17]. (*E*)-cinnamaldehyde, as the major bioactive compound, is considered to be responsible for the biological activities of bark CEO, including the antimicrobial, antioxidant, and anti-inflammatory properties of the oil [1]. Its dominance suggests a central role in defining the overall activity of the CEO. The high percentage of (*E*)-cinnamaldehyde, combined with the contributions of oxygenated monoterpenes and eugenol, suggested significant potential applications in antimicrobial and antioxidant contexts.

Minor components, including monoterpene hydrocarbons (α-pinene, p-cymene, and limonene + β-phellandrene), and oxygenated monoterpenes (linalool and 1,8-cineole) contribute to the CEO's bioactivity, particularly its antimicrobial and antioxidant effects [2,3]. Sesquiterpene hydrocarbons, namely, (*E*)-caryophyllene and α-humulene, add anti-inflammatory and analgesic properties [5].

3.2. Antioxidant Activity

It is well established that antioxidant activity cannot be accurately assessed using a single test; therefore, four commonly used assays, which differ in their underlying principles, have been employed [20]. The antioxidant activity of essential oils is likely due to the synergistic interactions among their constituents, with the major components playing a key role in this beneficial biological effect [17]. The antioxidant activity of (*E*)-cinnamaldehyde, α-pinene, eugenol, and β-caryophyllene has been reported in the literature [20–22]. However, based on the results of the DPPH, ABTS, and CUPRAC assays, the antioxidant activity of the CEO was weak, with an IC_{50} value higher than 100 μg/mL. The CEO sample was the most effective in reducing cupric ions, which was confirmed in the CUPRAC assay ($IC_{50} = 0.14 \pm 0.002$ mg/mL). Our results were consistent with previous studies (Gheorghe-Irimia et al. and Elgamma et al.), which also found IC_{50} values above 0.1 mg/mL in the DPPH assay [9,23]. However, most studies reported the antioxidant activity of CEO as DPPH radical scavenging activity, expressed as the percent of inhibition, which prevents direct comparison with our results. These results ranged from 21.3% to 71.12% in DPPH inhibition [8,9,16,23,24]. Factors such as the plant's geographical origin, plant parts used, and extraction methods may have influenced variations in chemical profile and, more directly, the antioxidant properties of the essential oil. The qualitative profiles of the CEO samples from different sources were similar, but quantitative differences were observed in the key bioactive compounds. Kallel et al. (2019) reported weak antioxidant activity (21.3% DPPH inhibition) for CEO, despite a cinnamaldehyde content determination similar to our sample (77.34% vs. 77.93%). In contrast to this study, our research confirmed the presence of compounds such as linalool and (*E*)-caryophyllene [25]. The classification of samples as moderate or strong antioxidants based on high percentages of DPPH radical inhibition was related to their higher contents of linalool (7.00%), β-caryophyllene (6.40%), and 1,8-cineole (5.40%), which enhance antioxidant activity via hydrogen donation [17]. Moreover, it was observed that CEO, with a high percentage of DPPH radical scavenging, which corresponds to high antioxidant activity, contains $83.71 \pm 0.69\%$ of *cis*-cinnamaldehyde, as its major compound. In our sample, the dominant compound was (*E*)-cinnamaldehyde (77.93%, Table 1), which is the *trans* stereoisomer of cinnamaldehyde. Differences between *cis* and *trans* isomers can influence the biological activity of essential oil, including its radical-scavenging ability. Additionally, the presence of terpinolene and D-limonene may have contributed to the higher antioxidant activity compared to our sample [8]. There is a lack of specific, comparable data on the antioxidant potential of CEO, as evaluated by the CUPRAC and ABTS assays, in the literature. The result obtained for CEO total antioxidant capacity, as determined by the TAC assay, was found to be 0.01 mg AAE/g (equivalent to 10 μg AAE/g). However, reports in the literature have determined higher values of CEO total antioxidant capacity (91.57 μg AAE/mL; 0.27 mg AAE/mg) [23,25]. These differing results highlight the importance of using multiple assays to obtain a comprehensive understanding of a sample's antioxidant capacity.

In addition, the differences in the chemical composition and antioxidant activity of *C. zeylanicum* Blume could be attributed to the two different chemotypes identified by Farias et al. [26,27]. Chemotype differences significantly influence the functional properties of essential oils, and their potential applications in the pharmaceutical and food industries.

3.3. *Antimicrobial Activity*

The available in vitro and in vivo studies suggest that CEO exhibits broad-spectrum antimicrobial and antifungal activity. The suggested antimicrobial mechanism involves the ability of hydrophobic essential oils to disrupt the bacterial cell membrane and its structures [28]. Studies in the literature have proposed a classification of the plant materials, including the essential oils, which is based on MIC values. There are slight differences within the available literature data. Generally, essential oils with MIC values lower than 100 μg/mL are considered to be strong antimicrobials; those between 100 and 500 μg/mL are considered to be moderate inhibitors; and those between 500 and 1000 μg/mL are considered to be weak antimicrobial agents [28–30]. According to this classification, our sample of bark CEO expressed extremely strong antimicrobial activity against all tested microbial strains, with MIC values much lower than 100 μg/mL (7.37–29.50 μg/mL). Gram-negative bacteria, including *E. coli*, *K. pneumoniae*, *P. aeruginosa*, and *S. enterica* serovar Abony, have shown higher MIC values, indicating greater resistance. This can be associated with the additional layer of lipopolysaccharides (LPS) in their outer membrane, which serves as a barrier to the entry of many antimicrobial agents. Our finding of greater resistance of Gram-negative bacteria to essential oil was consistent with data in the literature [31–33]. The antimicrobial activity of CEO is well-documented in the literature. The study conducted by Elgammal et al. [23] confirmed the antimicrobial activity of bark CEO, using the agar well diffusion method, in line with our findings. They observed that bark CEO inhibits the growth of both Gram-positive and Gram-negative bacteria, with greater susceptibility observed in Gram-positive bacteria and fungi [23]. In tension with the results of Unlu et al. [34], our sample demonstrated greater antimicrobial activity, with significantly lower MIC values against all mutual microbial strains, although their study included a broader microbial spectrum [34]. Furthermore, our findings were supported by the results of Behbahani et al. [17], who reported the valuable antimicrobial activity of bark CEO, especially against Gram-positive bacteria. The comprehensive study of Saki et al. revealed the very potent antimicrobial effects of bark CEO against 150 extensively drug-resistant isolates, (60 Gram-positive and 90 Gram-negative bacteria strains) [35]. The low effective concentrations of CEO indicated the need for further investigations and highlight the great potential of CEO in the synthesis of new antimicrobial agents, particularly against multidrug-resistant bacteria [35]. However, the variations across studies may stem from differences in testing methods, extraction techniques, essential oil composition, and microbial strains.

According to the results of this study, the CEO exhibited lower antimicrobial activity compared to the standard antibiotics minocycline and amikacin and the antifungal agent nystatin. However, the great antimicrobial potential of CEO as a plant-derived product could be useful in adjuvating an antibiotic effect. There is evidence that a combination of CEO with colistine and ciprofloxacin demonstrated synergistic activity against multidrug resistant *P. aeruginosa* [36,37].

The antimicrobial activity of bark CEO is largely attributed to its chemical composition, particularly the high content of (*E*)-cinnamaldehyde (77.93%). Minor constituents, such as eugenol (4.34%) and *E*-caryophyllene (3.68%), further enhance this effect by disrupting microbial cell membranes and inhibiting vital enzymes.

3.4. In Vitro Evaluation of the Cytotoxic Effects of CEO

Nowadays, there is an increasing interest in exploring the antitumor and immunomodulatory activities of essential oils, which are primarily influenced by the major phytochemicals but also by the total composition of the oil, due to potential synergistic, additive, or antagonistic effects [38]. The findings of this study indicated that the CEO significantly affected the viability of the 4T1, MDA-MB-468, HCT116, and CT26 cancer cell lines, while having minimal impact on the non-cancerous MRC-5 cells. The IC_{50} values obtained for all evaluated cancer cell lines indicate that our CEO sample could be considered as a potential candidate in the development of anticancer drugs. This assessment is based on the criteria established by the U.S. National Cancer Institute, which recognizes plant-derived compounds with IC_{50} values below 30 μg/mL as potential anticancer agents. [12]. The cytotoxic effect of CEO against the human HCT116 cells was higher, when compared to cisplatin (CDDP), a widely used chemotherapeutic agent [39]. The antitumor effects of CEO were validated in numerous in vitro and in vivo studies [12,39,40]. In a study conducted by Nile et al. [41], an IC_{50} of 13.5 μg for cinnamon bark extract against human HCT116 cells was reported, while our study showed a significantly greater potency associated with CEO (IC_{50} = 1.18 μg). These differences might be attributed to the different cinnamon bark-derived products (extract vs. essential oil) and the varying concentrations of cinnamaldehyde [39]. Moreover, CEO has shown anticancer potential against several human cancer cell lines, including breast, lung, and prostate carcinomas [12,39].

Additionally, Kallel et al. [25] reported the pronounced cytotoxic effects of CEO against cervical cancer (HeLa) and Burkitt's lymphoma (Raji) cell lines, with IC_{50} values below 1 μg/mL, further supporting its potential as an effective anticancer agent.

When developing low-toxicity drugs, it is essential to assess their anti-proliferative effects on normal, non-cancerous cells, such as the MRC-5 cell line. Our results showed a dose-dependent reduction of MRC-5 cell viability, and a lower cytotoxic effect of CEO on

MRC-5 cells compared to its effects on cancerous cell lines (4T1, MDA-MB-468, HCT116, and CT26). This selective reduction in cytotoxicity toward MRC-5, combined with its significant impact on tumor cell viability, suggested that the CEO may offer selective cytotoxicity in targeting cancer cells. Our findings were consistent with previous studies demonstrating the anti-proliferative effects of cinnamon bark on normal cells, including CCD-112CoN normal colon cells and primary mouse lymphocytes [42]. One of the key findings of this study was the notably high selectivity index for the CEO, with the HCT116 cell line exhibiting the highest selectivity index among all tested cell lines. This suggested that the CEO exhibited pronounced cytotoxicity toward HCT116 cancer cells, along with strong selectivity over other cancerous and healthy cells, highlighting its potential for targeted cancer therapy. These findings prompted further research on the HCT116 cell line.

These findings expand the existing knowledge of CEO's antitumor activity and support further investigations into its potential as an anticancer drug.

3.5. Assessment of Apoptotic Cell Death Induced by CEO

Anexin V FITC staining, which occurs before membrane integrity is lost, indicates early apoptotic cells as Annexin V FITC positive and PI negative; viable cells as both Annexin V FITC negative and PI negative; and late apoptotic or dead cells as both Annexin V FITC positive and PI positive [43]. The results of this study demonstrated that a significant portion of HCT116 cells underwent both early and late apoptosis after 24 h of treatment with the CEO. Notably, necrosis does not appear to be the mechanism of cell death through which the CEO exerts its effects, as HCT116 cells did not undergo necrosis. These findings highlight the dual mechanism of the CEO, which promotes cell death via apoptotic pathways, suggesting its potential as a powerful anti-carcinoma agent. Kubatka et al. [13] reported similar findings in human adenocarcinoma cell lines (MDA-MB-231 and MCF-7) treated with CEO for 24 to 72 h. Significant increases in the proportions of MDA-MB-231 cells in early and late apoptotic stages were observed, whereas MCF-7 cells showed minimal presence in the late apoptotic or necrotic phase.

Apoptosis is mainly initiated through two pathways: the extrinsic pathway, which involves death receptors activating caspase-3, and the intrinsic or mitochondrial pathway [40]. The intrinsic apoptotic pathway is initiated by an imbalance between pro-apoptotic and anti-apoptotic proteins (e.g., Bax and Bcl-2), leading to mitochondrial membrane permeabilization, cytochrome c release, and subsequent activation of caspase-3, culminating in apoptosis [42]. Bax (Bcl-2-associated X protein), a pro-apoptotic member of the Bcl-2 protein family, induces cell death by enhancing the release of cytochrome c from the mitochondria [44]. Alternatively, Bcl-2 (B-cell lymphoma 2) acts as an anti-apoptotic protein that prevents apoptosis by inhibiting the activity of pro-apoptotic proteins [45]. Caspase-3, a key executioner caspase in the apoptosis process, is activated in apoptotic cells through both the extrinsic (death ligand) and intrinsic (mitochondrial) pathways. It plays a central role in cleaving several critical cellular proteins, driving the morphological and biochemical alterations that define apoptosis [46]. The analysis of apoptotic markers in this study indicated that treatment with IC_{50} concentrations of the CEO significantly increased the percentage of Bax+ HCT116 cells, while simultaneously reducing the percentage of Bcl-2+ HCT116 cells. Moreover, the treatment with CEO resulted in a higher percentage of caspase-3+ HCT116 cells. The observed increase in caspase-3 expression, coupled with a decrease in Bcl-2 expression, points to potential mechanisms of action for the CEO, particularly in inducing apoptosis. Additionally, the marked increase in Bax, a pro-apoptotic protein, following treatment with the CEO underscored its role in promoting apoptosis in HCT116 cells. According to Nile et al. [41], cinnamon extract induced apoptosis in HCT116 cells through activation of the caspase cascade. Moreover, there was growing evidence supporting the regulatory role of phytochemicals in modulating the Bax/Bcl-2/caspase-3 signaling pathway in breast cancer. A significant increase in the Bax/Bcl-2 ratio and caspase-3 expression observed in mammary tumors of rats treated with CEO, as reported by Kubatka et al. [13], aligned with our findings in HCT116 cells.

3.6. *Effects of CEO on Cell Cycle and Proliferation in HCT116 Cancer Cells*

Ki67 is a well-established proliferation marker, expressed in actively dividing cells (G1, S, G2, and M phases), but absent during the resting phase (G0) cell cycle. Its strong correlation with cell division makes it valuable for assessing tumor growth and evaluating anticancer therapy efficacy in both clinical and research settings [47]. The results of this study revealed a significant reduction in Ki67 expression in CEO-treated HCT116 cells, compared to untreated control cells, indicating a significant decrease in cell proliferation. Studies have indicated that CEO effectively inhibited tumor cell proliferation by suppressing the expression of the nuclear proliferation marker Ki67, which was in line with our results. Additionally, a significant reduction in Ki67 expression following CEO treatment, compared to control groups, was demonstrated in in vivo models of rat mammary carcinoma and Ehrlich ascites carcinoma [13,48].

Cyclins are essential regulators of the cell cycle, a family of proteins that manage the cell's progression by activating cyclin-dependent kinases (CDKs). The levels of cyclins fluctuate throughout the various stages of the cell cycle, and their precise expression is crucial for the accurate timing and smooth transition between phases [49,50]. Cyclin D forms complexes with CDK4 and CDK6, which are key in facilitating the shift from the G1 phase to the S phase by preparing the cell for DNA replication. Cyclin D is often considered a responder to external mitogenic signals and plays a significant role in determining whether a cell should enter the division cycle [51]. The regulated expression of Cyclin D is crucial for proper cell-cycle control, and disruptions in its expression are frequently observed in various cancers [52]. The results showed that the CEO significantly decreased the percentage of Cyclin D^+ cells, indicating a reduction in the number of cells progressing from the G1 phase to the S phase. This selective downregulation of Cyclin D pointed out that the CEO specifically interferes with cell-cycle progression at the G1 phase, potentially inducing cell-cycle arrest and inhibiting further proliferation of HCT116 cells. Nagle et al. [53] demonstrated that cinnamaldehyde, the primary active component of cinnamon essential oil, induced cell-cycle arrest at the G2/M phase in HCT116 colon cancer cells. Similarly, a study on cinnamon extract, also rich in cinnamaldehyde, reported a dose-dependent decrease in the G1 phase, accompanied by an increase in the sub-G1 population or G2/M phase arrest in HCT116 cells [41]. Furthermore, CEO treatment led to a delayed accumulation of MDA-MB-231 cells in the G0/G1 phase, along with an increase in the S phase of the cell cycle [13].

The p21, p27, and phospho-AKT (p-AKT) play crucial roles in regulating the cell cycle. The p21 and p27, cyclin-dependent kinase (CDK) inhibitors, regulate cell-cycle control by blocking the activity of CDK–Cyclin complexes, including those involving Cyclin D. These proteins can be activated in response to various stimuli, such as DNA damage or other stressors, serving as regulatory factors that arrest cell-cycle progression, particularly at the G1 checkpoint, by inhibiting the Cyclin D/CDK4/6 complex [54,55]. This inhibition is critical for the cell's response to damage, preventing the replication of damaged DNA [55]. In contrast, p-AKT, the phosphorylated form of protein kinase B (AKT), is a key oncogenic factor involved in regulating cell proliferation and survival. The phosphorylation of AKT activates downstream pathways that promote cell growth and survival, making it a crucial target for cancer therapy. Dysregulation of AKT signaling is commonly observed in various cancers; it contributes to tumorigenesis and resistance to treatment [56]. Treatment with CEO induced significant alterations in the expression of key cell-cycle regulators in HCT116 cells. Specifically, there was an increase in the proportion of cells expressing the tumor suppressors p21 and p27. This upregulation likely contributed to CDK suppression, leading to cell-cycle arrest. Consistent with these findings, the increased expression of p21 and p27 was associated with a reduction in cyclin D levels, further supporting the disruption of cell-cycle progression. Moreover, CEO treatment significantly decreased the levels of phosphorylated AKT (p-AKT). The reduction in p-AKT, a crucial component of the oncogenic signaling pathways that support cell survival and proliferation, might be connected with a potential antitumor effect of the CEO. These changes in p21, p27 and p-AKT expression emphasized the CEO's ability to disrupt key processes involved in cell-cycle regulation and survival signaling, highlighting its potential as an effective anticancer agent. Li et al. [57] proposed that the mechanism of action of cinnamaldehyde involved the modulation of genes associated with apoptosis, invasion, and adhesion, primarily through the inhibition of the PI3K/Akt signaling pathway.

3.7. *In Vitro, CEO Reduces the Concentration of Pro-Inflammatory Cytokines*

Generally, essential oils have been found to suppress inflammation and reduce cytokine production by disrupting key components of inflammatory pathways [38]. Furthermore, the literature indicates the effects of essential oils on the production of interleukins, tumor necrosis factor, thromboxane, and leukotrienes [58]. The evaluation of the immunomodulatory effects of CEO in this research showed that treatment with ConA only, a poly-clonal T cell activator, notably elevated the levels of IL-1β, TNF-α, IFN-γ, IL-17, and IL-10 in the cell supernatant after 24 h, compared to untreated cells [59]. Also, CEO by itself significantly raised the concentrations of IL-1β, TNF-α, IFN-γ, IL-17, and IL-10, compared to splenocytes incubated in medium. However, the combination of the ConA and CEO treatments significantly reduced the levels of IL-1β, TNF-α, IFN-γ, IL-17, and IL-10, compared to ConA stimulation alone. Furthermore, the CEO appeared to promote a predominance of pro-inflammatory cytokines over IL-10. This effect was more pronounced in non-ConA primed cells, which represented likely components of the innate immune response. Interestingly, while CEO exerted an immunosuppressive effect on activated cells, it simultaneously activated naïve cells, demonstrating a dual impact on immune-cell function. To the best of our knowledge, only a limited number of studies have investigated the effects of CEO on the production of inflammatory biomarkers. Cinnamon essential oil has been shown to significantly reduce the expression of several pro-inflammatory mediators, including vascular cell adhesion molecule-1 (VCAM-1), intercellular adhesion molecule-1 (ICAM-1), monocyte chemoattractant protein-1 (MCP-1), interferon gamma-induced protein 10 (IP-10), interferon-inducible T-cell alpha chemoattractant (I-TAC), and monokine induced by gamma interferon (MIG) in a human skin disease model [60]. Additionally, CEO suppressed IL-2 secretion in the Caco-2 human intestinal epithelial cell model and reduced TNF-α levels in lipopolysaccharide (LPS)-induced mice [42,44].

4. Materials and Methods

4.1. *Essential Oil Material*

CEO was purchased from Fares, Timisoara, Romania. According to the manufacturer, the essential oil was obtained by steam distillation from the bark of the species *Cinnamomum zeylanicum* Blume. The sample was stored in the original package, a dark glass bottle, at room temperature and away from sunlight before the analysis.

4.2. GC and GC–MS Analysis of the CEO

The chemical composition of the cinnamon bark *C. zeylanicum* essential oil was determined by gas chromatography–mass spectrometry (GC–MS) using a Shimadzu GCM-SQP2010 ultra mass spectrometer (Shimadzu Corporation, Kyoto, Japan) fitted with a flame ionic detector coupled with a GC2010 gas chromatograph (Shimadzu Corporation, Kyoto, Japan). The InertCap5 capillary column (60.0 m × 0.25 mm × 0.25 μm) was employed for separation. Helium (He) was used as a carrier gas at a split ratio 1:5 and a linear velocity of 35.2 cm/s. The ion source temperature was 200 °C, the injector temperature was 250 °C, and the detector temperature 300 °C, while the column temperature was linearly programmed from 40 to 260 °C, at the rate of 4 °C/min; then from 260 to 310, at a rate 10 °C/min; and after reaching 310 °C, kept isothermally for 10 min. TPEO was dissolved in EtOH (*v*/*v* 96%) and consecutively injected in the amount of 1 μL. The contents of different compounds were determined based on the normalized area of chromatograms and defined as each content according to the corresponding peak area (the mean of three determinations). The identification of the constituents was performed by comparing their mass spectra and retention indices (RIs) with those obtained from authentic samples and/or listed in the NIST/Wiley mass-spectra libraries, using different types of search (PBM/NIST/AMDIS) and available data in the literature [61].

4.3. Antioxidant Activity

4.3.1. DPPH Radical Scavenging Activity Assay

The antioxidant activity of CEO was evaluated by a DPPH radical (2,2-diphenyl-1-picrylhydrazyl) scavenging activity assay, which was performed according to the method described by Blois, with some modifications [62]. The reaction mixture consisted of 20 μL of an ethanolic solution of essential oil in increasing concentrations and 280 mL of 0.1 mM ethanolic solution of DPPH (0.04 mg/mL). After the 30 min' incubation at room temperature and in a dark place, the absorbance values of the reaction mixtures were determined spectrophotometrically at 517 nm, using the BioTek Epoch 2 Microplate Spectrophotometer (Agilent Technologies, Santa Clara, CA, USA). Butylated hydroxytoluene (BHT) was used as a positive control, while pure ethanol served as a negative control. The antioxidant activity, expressed as the percentage of inhibition, was determined according to the following equation:

$$\mathrm{DPPH}(\%) = \left[\left(\mathrm{A}_{\mathrm{control}} - \mathrm{A}_{\mathrm{sample}}\right)/\mathrm{A}_{\mathrm{control}}\right] \times 100$$

where $A_{control}$ is the absorbance of the negative control and A_{sample} is the absorbance of the sample.

The results were expressed as an IC_{50} value (the sample concentration able to neutralise 50% of the DPPH free radicals), using a linear regression method.

All the measurements were performed in triplicate, and the results are expressed as the mean values for the concentration expressed as mg/mL ± standard deviation.

4.3.2. Cupric Ion Reducing Antioxidant Capacity Assay (CUPRAC Assay)

The CUPRAC assay was conducted using the procedure described by Milošević et al. [63]. A quantity of 60 μL of CEO in increasing concentrations diluted in ethanol was added to a microplate in a series of wells. This was followed by the addition to each well of 240 μL of a mixture consisting of 75 μL of aqueous solution of $CuCl_2 \cdot 2H_2O$ (1.7 μg/mL), 75 μL of methanolic solution of neocuproine (7.5 mM), and 90 μL of ammonium acetate buffer (pH ≈ 7). The reaction mixture was incubated for 30 min at an ambient temperature. The absorbance was recorded at 450 nm using a BioTek Epoch 2 Microplate Spectrophotometer (Agilent

Technologies, Santa Clara, CA, USA), with measurements conducted in triplicate. The same procedure was conducted with BHT as a positive control. The reducing activity was calculated using the following equation:

$$(\%)\text{ reducing power} = \left[\left(A_{sample} - A_{control}\right)/A_{sample}\right] \times 100$$

where A_{sample} is the absorbance of the sample, and $A_{control}$ is the absorbance of the negative control (20 μL of pure ethanol). The results are expressed as IC_{50} values, using a linear regression method. The IC_{50} values were calculated as the concentration (mg/mL) which provides 50% of the reducing power. All the measurements were performed in triplicate, and the results are expressed as the mean value ± standard deviation.

4.3.3. ABTS Radical Scavenging Activity Assay

A 2,2′-azino-bis-(3-ethylbenzothiazoline-6-sulfonic acid) (ABTS) radical cation decolorization assay was applied to determine the antioxidant activity of the CEO. The assay was performed according to the method previously described by Čutović et al., with some modifications [64]. The ABTS solution was prepared by dissolving 20 mg $ABTS^{\bullet+}$ radical cation (7.8 mmol/L) in 5 mL deionized water; this was followed by the addition of 88 μL of aqueous solution of potassium persulfate. The incubation of the mixture lasted 16–20 h at 4 °C in the dark. The control solution was the mixture of 280 μL $ABTS^{\bullet+}$ radical cation solution and 20 μL of ethanol. The reaction mixture consisted of 20 μL of the sample, and 280 μL of the ABTS solution was added into microplates and incubated for 6 min at 25 ± 5 °C in the dark. The absorbance was recorded at 734 nm using a BioTek Epoch 2 Microplate Spectrophotometer (Agilent Technologies, Santa Clara, CA, USA), with measurements conducted in triplicate. The percentage of inhibition was calculated using the following equation:

$$\text{Inhibition}(\%) = \left[\left(A_{control} - A_{sample}\right)/A_{control}\right] \times 100$$

where $A_{control}$ is the absorbance of the negative control (pure ethanol) and A_{sample} is the absorbance of the sample. The obtained results were transformed using a linear regression method. The IC_{50} values were calculated as the concentration (mg/mL) that provides 50% of the inhibition.

4.3.4. Total Antioxidant Capacity Assay (TAC)

The total antioxidant capacity assay was carried out according to a modified method, based on the reduction of Mo(VI) to Mo(V) previously described by Milošević et al. [63]. An aliquot of 20 μL of essential oil dissolved in ethanol was combined with 200 μL of reagent solution (0.6 M sulfuric acid, 28 mM sodium phosphate, and 4 mM ammonium molybdate). The tubes with reaction mixture were incubated for 90 min at 60 °C. For the blank, 100 μL of deionized water was used in place of the sample. After the samples were cooled at room temperature, the absorbance of each sample was measured at 695 nm in the spectrophotometer. The calibration curve was made with L-ascorbic acid solution as a positive control (0.3125–10 mmol/L). The results are expressed as L-Ascorbic Acid Equivalents (mg AAE/g) of three replicates.

4.4. Antimicrobial Activity

4.4.1. Microbial Strains

Standard strains of microorganisms sourced from the American Type Culture Collection (ATCC, Microbiologics, St. Cloud, MN, USA) were utilized in this research. The antimicrobial activity of the essential oil was tested on seven bacterial strains and one yeast strain. The Gram-positive bacteria tested were *S. aureus* (ATCC 6538), *S. epidermidis* (ATCC 1228), and *B. subtilis* (ATCC 6633). The Gram-negative bacteria included *E. coli* (ATCC 8739), *P. aeruginosa* (ATCC 9027), *K. pneumoniae* (ATCC 13883), and *S. enterica* serovar Abony (ATCC 6017). Moreover, the antimicrobial effect was evaluated on *C. albicans* (ATCC 10231) as a yeast strain.

4.4.2. Determination of Minimum Inhibitory Concentration (MIC)

The minimum inhibitory concentration (MIC) of the essential oil was determined using the broth microdilution method in standard sterile 96-well flat-bottomed microtiter plates according to the Clinical and Laboratory Standards Institute guidelines [15]. Fresh microbial cultures were established by incubation on Müller–Hinton agar (TSA; Torlak, Belgrade, Serbia) for 24 h at 37 °C. Individual colonies were then suspended in saline solution to a density of approximately 1.5×10^8 CFU/mL and measured using a DEN-1 McFarland densitometer (Biosan, Riga, Latvia). Bacterial suspensions and *C. albicans* yeast were diluted with Müller–Hinton broth to final concentrations of 1.5×10^6 CFU/mL for bacteria and 2×10^6 CFU/mL for *C. albicans*, and 100 µL of the suspension was added per well. Dilutions of CEO were prepared by dissolving it in dimethyl sulfoxide (DMSO); this was followed by serial dilutions in broth. These dilutions were added in equal volumes to each well, yielding final concentrations between 7.37 and 118 µg/mL. The microtiter plates were incubated at 37 °C for 24 h. Each concentration was tested in duplicate and the broth microdilution test was repeated three times. The standard antibiotics minocycline and amikacin, as broad-spectrum antibiotics commonly used in clinical practice for the treatment of various types of Gram-positive and Gram-negative bacterial infections, along with the antifungal agent nystatin, were used as positive controls. The results are expressed as minimum inhibitory concentrations (MIC)—the lowest concentrations of the tested essential oil at which no microbial growth is observed. MIC values indicate bacteriostatic and fungistatic activities.

4.5. Cytotoxic Activity

4.5.1. Cell Culture

In this study, a variety of cell lines were carefully cultured, including mouse (4T1) and human (MDA-MB-468) breast carcinoma cells, mouse (CT26) and human (HCT116) colon carcinoma cells, and a human fibroblast cell line (MRC-5). These cell lines were obtained from the American Type Culture Collection (ATCC), Microbiologics, St. Cloud, MN, USA. All cell lines were cultured in Dulbecco's modified Eagle's medium (DMEM) supplemented with 10% fetal bovine serum (FBS) from Sigma Aldrich, St. Louis, MO, USA. The cultures were maintained in a controlled environment with 5% CO_2 under standard laboratory conditions. For the experiments, only cell suspensions with over 95% viability were used. Cell viability and counts were carefully determined using trypan blue staining, ensuring the accuracy and reliability of the experimental outcomes.

4.5.2. In Vitro Evaluation of the Cytotoxic Effects of CEO

To assess the cytotoxic activity of CEO, an MTT (3-(4,5-dimethylthiazol-2-yl)-2,5-diphenyltetrazolium bromide) assay was performed, in line with previously established protocols [65,66]. The MRC5, 4T1, MDA-MB 468, CT26, and HCT116 cell lines, all in the exponential growth phase, were harvested from culture flasks, counted, and seeded into 96-well plates at a density of 5×10^3 cells per well. After allowing 24 h for cell adhesion, 100 µL of serially diluted CEO and cisplatin (ranging from 0.78 to 100 µg) were added to each well. The cells were incubated for 48 h at 37 °C in a 5% CO_2 incubator. Following incubation, the plates were centrifuged, the supernatant was removed, and 100 µL of MTT solution (5 mg/mL in PBS) was added to each well for a 4 h incubation. The plates were then centrifuged again, the cell-free supernatants were discarded, and 150 μL of DMSO, along with 20 μL glycine buffer, was added to dissolve the formazan crystals. After shaking the plates for 10 min, the optical density of each well was measured at 595 nm using a Zenyth 3100 microplate reader (Anthos Labtec Instruments GmbH, Salzburg, Austria). The percentage of cytotoxicity was calculated in accordance with the procedures described in a previous study [43].

4.6. Evaluation of Apoptosis

To evaluate apoptosis in HCT116 cells treated with CEO, the Annexin V and propidium iodide double staining assay was used, following protocols from previous studies [39,64]. This method enabled the quantification of cells in early and late apoptosis, as well as those undergoing necrosis. In additional experiments, cells were fixed and permeabilized with a permeabilization buffer from BD Bioscience, Heidelberg, Germany. They were then incubated with antibodies specific to Bcl-2, Bax, and caspase-3, following previously established procedures [43]; these antibodies were provided by Thermo Fisher Scientific Inc., Waltham, MA, USA. The treated cells were analyzed using a FACS Calibur flow cytometer (BD Biosciences, San Jose, CA, USA), and the data were processed with FlowJo software (v10.8.2). This method allowed for a detailed and comprehensive evaluation of the apoptotic responses in cells exposed to the various treatments, providing deeper insight into the apoptotic mechanisms activated under the experimental conditions.

4.7. The Impact of CEO on Cell-Cycle Dynamics

The potential effects of the CEO on the cell-cycle dynamics of HCT116 tumor cells were further investigated. Specifically, the Ki67 expression in tumor cells treated with the CEO was evaluated and compared to untreated cells. The experimental setup included a treatment group exposed to IC_{50} concentrations of the CEO, and a control group of untreated cells; all were cultured in 25 mL flasks. After a 24 h incubation post-treatment, the cells were trypsinized, washed three times with PBS, and counted. Each sample was then stained with Ki67-specific antibodies (eBioscience, San Diego, CA, USA) and 1 μL of Vybrant® DyeCycleTM Ruby stain (Thermo Fisher Scientific, Inc.). The samples were analyzed using a FACS Calibur flow cytometer (BD Biosciences, San Jose, CA, USA), with at least 15,000 events per sample being analyzed. Data analysis was performed using FlowJo vX.0.7 software, providing a detailed evaluation of the cell-cycle phases.

4.8. Analysis of the Effects of CEO on Cell-Cycle Regulators

HCT116 tumor cells were treated with the IC_{50} concentration of the CEO or maintained in culture medium alone as a control for 24 h, after which they underwent a series of cellular analyses. The cells were first fixed and permeabilized using a permeabilization buffer from BD Bioscience, Heidelberg, Germany. They were then incubated with specific antibodies targeting key cell-cycle regulators, including Cyclin D, p21, p27, and phospho-AKT (p-AKT), all supplied by Thermo Fisher Scientific, followed by secondary FITC-conjugated Dnk anti-rabbit IgG mAb (Abcam Limited, Cambridge, UK). The cells were analyzed using a FACS Calibur flow cytometer (BD Biosciences, San Jose, CA, USA), and the data were processed and interpreted using FlowJo software (v10.8.2).

4.9. In Vitro Assessment of the Immunomodulatory Impacts of CEO

The impact of CEO on cytokine production by splenocytes was assessed. Splenocytes from healthy BALB/C mice were isolated as per a previously established method [67]. Cell viability was evaluated using a trypan blue exclusion assay, and only those with over 95% viability were used for subsequent experiments. The splenocytes were placed in 96-well plates (1×10^5 cells per well) and divided into four groups: a control group incubated with complete medium, cells stimulated with concanavalin A (ConA) (0.5 μg/mL), cells treated with CEO (IC_{50}), and cells treated with both CEO and ConA. The cells were incubated for 24 h at 37 °C in a humidified environment with 5% CO_2. After incubation, cell viability was confirmed using the trypan blue assay, showing that viability was above 90% in all groups, with no significant differences. The cells were then centrifuged, and the supernatants were stored at −80 °C for later analysis. Cytokine levels of IL-1β, TNF-α, IFN-γ, IL-17, and IL-10 were quantified using high-sensitivity ELISA kits from R&D Systems (Minneapolis, MN, USA) as specified by the manufacturer [68].

4.10. Statistical Analysis

All measurements of antioxidant activity evaluation were taken in triplicate, and the results are expressed as mean values with standard deviation (SD). Calculation of IC_{50} values was performed using Microsoft Excel 365. The antimicrobial assay involved the testing of each concentration in duplicate across three independent experiments, and the results are expressed as MIC values. The data were analyzed using SPSS software, version 26.0. Statistical comparisons were performed using the nonparametric Mann–Whitney U test and Kruskal–Wallis test. The normality of data distribution was assessed using the Kolmogorov–Smirnov test. A p-value of <0.05 was considered statistically significant, p-values of <0.01 were regarded as highly significant, and p-values of <0.001 were considered very highly significant.

5. Conclusions

The present study revealed 99.66% of the total chemical composition of commercial *C. zeylanicum* bark essential oil, with (*E*)-cinnamaldehyde being a major component (77.93%), followed by eugenol (4.34%), *E*-caryophyllene (3.68%), and linalool (2.79%). The essential oil has shown considerable antioxidant activity and strong antimicrobial activity, with a more pronounced effect on Gram-positive bacteria and *C. albicans* than on Gram-negative bacteria. In addition, the essential oil exhibited pronounced antitumor effects, particularly against HCT116 colorectal cancer cells, in which it induced apoptosis through modulation of Bax, Bcl-2, and caspase-3 expression, and triggered G1-phase cell-cycle arrest via downregulation of Cyclin D and p-AKT. The oil also significantly reduced the proliferation marker Ki67. Importantly, CEO demonstrated selectivity by sparing non-cancerous MRC-5 fibroblasts. Furthermore, the essential oil exerted immunomodulatory effects by reducing the production of pro-inflammatory cytokines (IL-1β, TNF-α, IFN-γ, and IL-17) in activated splenocytes, while enhancing pro-inflammatory responses in non-activated, naïve immune cells.

These findings highlighted the great potential of cinnamon bark essential oil in the development of new therapeutic agents.

Author Contributions: Conceptualization, A.B., I.P.J., B.S.M., S.C., M.T.M. and V.T.; methodology, V.T., M.T.M., I.P.J. and B.S.M.; formal analysis and investigation, M.Ž., B.S.M., I.S., N.G., S.P., M.T.M. and A.Ž.; data curation, A.B. and I.P.J.; writing—original draft preparation, M.Ž., A.B. and I.S.; writing—review and editing, I.P.J., B.S.M., V.T. and M.T.M.; project administration, A.B. and M.Ž. All authors have read and agreed to the published version of the manuscript.

Funding: This research was supported by the Ministry of Science, Technological Development and Innovation of the Republic of Serbia (contract No. 451-03-137/2025-03/200111) and Junior Projects of Faculty of Medical Sciences, University of Kragujevac, Serbia (JP 09/25).

Institutional Review Board Statement: The study was conducted in accordance with the Declaration of Helsinki, and the protocol was approved by the Ethics Committee of Faculty of Medical Sciences, University of Kragujevac, Serbia (protocol code No. 01-8353/2 from 12 September 2024).

Informed Consent Statement: Not applicable.

Data Availability Statement: All data generated for this study are included in the article.

Conflicts of Interest: The authors declare no conflicts of interest.

Abbreviations

The following abbreviations are used in this manuscript:

CEO	Cinnamon essential oil
GC-MS	Gas chromatography–mass spectrometry
DPPH	2,2-diphenyl-1-picrylhydrazyl
ABTS	2,2′-azino-bis-(3-ethylbenzothiazoline-6-sulfonic acid)
CUPRAC	Cupric ion reducing antioxidant capacity
TAC	Total antioxidant capacity
BHTMIC	Butylated hydroxytolueneMinimum inhibitory concentration
MIN	Minocycline
AMK	Amikacin
NYS	Nystatine
IC50	Concentration that provides 50% of inhibition
MTT	(3-(4,5-dimethylthiazol-2-yl)-2,5-diphenyltetrazolium bromide)
4T1	Mouse breast cancer cells
MDA-MB-468	Human breast cancer cell line
CT26	Mouse colon cancer cell line
HCT116	Human colon cancer cell line
MRC-5	Human fibroblast cell line
CDDP	cis-diamminedichloroplatinum(II) or cisplatin
Ki67	Marker of proliferation
p21	Cyclin-dependent kinase inhibitor 1A
p27	Cyclin-dependent kinase inhibitor 1B
p-AKT	Phosphorylated AKT
ConA	Concanavalin A
IL-1β	Interleukin 1 beta
TNF-α	Tumor necrosis factor alpha
IFN-γ	Interferon gamma
IL-17	Interleukin 17
IL-10	Interleukin 10

References

1. Liñán-Atero, R.; Aghababaei, F.; García, S.R.; Hasiri, Z.; Ziogkas, D.; Moreno, A.; Hadidi, M. Clove Essential Oil: Chemical Profile, Biological Activities, Encapsulation Strategies, and Food Applications. *Antioxidants* **2024**, *13*, 488. [CrossRef] [PubMed]
2. Raut, J.S.; Karuppayil, S.M. A Status Review on the Medicinal Properties of Essential Oils. *Ind. Crops Prod.* **2014**, *62*, 250–264. [CrossRef]
3. Bassolé, I.H.N.; Juliani, H.R. Essential Oils in Combination and Their Antimicrobial Properties. *Molecules* **2012**, *17*, 3989–4006. [CrossRef] [PubMed]
4. Pramod, K.; Ansari, S.H.; Ali, J. Eugenol: A Natural Compound with Versatile Pharmacological Actions. *Nat. Prod. Commun.* **2010**, *5*, 1999–2006. [CrossRef]
5. Scandiffio, R.; Geddo, F.; Cottone, E.; Querio, G.; Antoniotti, S.; Gallo, M.P.; Maffei, M.E.; Bovolin, P. Protective Effects of (E)-β-Caryophyllene (BCP) in Chronic Inflammation. *Nutrients* **2020**, *12*, 3273. [CrossRef]
6. Friedman, M. Chemistry, Antimicrobial Mechanisms, and Antibiotic Activities of Cinnamaldehyde against Pathogenic Bacteria in Animal Feeds and Human Foods. *J. Agric. Food Chem.* **2017**, *65*, 10406–10423. [CrossRef]
7. Mutlu, M.; Bingol, Z.; Uc, E.M.; Köksal, E.; Goren, A.C.; Alwasel, S.H.; Gulcin, İ. Comprehensive Metabolite Profiling of Cinnamon (*Cinnamomum zeylanicum*) Leaf Oil Using LC-HR/MS, GC/MS, and GC-FID: Determination of Antiglaucoma, Antioxidant, Anticholinergic, and Antidiabetic Profiles. *Life* **2023**, *13*, 136. [CrossRef]

8. Denkova-Kostova, R.; Teneva, D.; Tomova, T.; Goranov, B.; Denkova, Z.; Shopska, V.; Slavchev, A.; Hristova-Ivanova, Y. Chemical Composition, Antioxidant and Antimicrobial Activity of Essential Oils from Tangerine (*Citrus reticulata* L.), Grapefruit (*Citrus paradisi* L.), Lemon (*Citrus lemon* L.) and Cinnamon (*Cinnamomum zeylanicum* Blume). *Z. Naturforsch C J. Biosci.* **2021**, *76*, 175–185. [CrossRef]
9. Gheorghe-Irimia, R.-A.; Tăpăloagă, D.; Tăpăloagă, P.-R.; Ghimpețeanu, O.-M.; Tudor, L.; Militaru, M. Spicing Up Meat Preservation: *Cinnamomum zeylanicum* Essential Oil in Meat-Based Functional Foods—A Five-Year Review. *Foods* **2024**, *13*, 2479. [CrossRef]
10. Didehdar, M.; Chegini, Z.; Tabaeian, S.P.; Razavi, S.; Shariati, A. Cinnamomum: The New Therapeutic Agents for Inhibition of Bacterial and Fungal Biofilm-Associated Infection. *Front. Cell Infect. Microbiol.* **2022**, *12*, 930624. [CrossRef]
11. Hua, K.-F.; Lin, Y.-B.; Chiu, H.-W.; Wong, W.-T.; Ka, S.-M.; Wu, C.-H.; Lin, W.-Y.; Wang, C.-C.; Hsu, C.-H.; Hsu, H.-T.; et al. Cinnamaldehyde Inhibits the NLRP3 Inflammasome by Preserving Mitochondrial Integrity and Augmenting Autophagy in Shigella Sonnei-Infected Macrophages. *J. Inflamm.* **2024**, *21*, 18. [CrossRef] [PubMed]
12. Najar, B.; Shortrede, J.E.; Pistelli, L.; Buhagiar, J. Chemical Composition and in Vitro Cytotoxic Screening of Sixteen Commercial Essential Oils on Five Cancer Cell Lines. *Chem. Biodivers.* **2020**, *17*, e1900478. [CrossRef] [PubMed]
13. Kubatka, P.; Kello, M.; Kajo, K.; Samec, M.; Jasek, K.; Vybohova, D.; Uramova, S.; Liskova, A.; Sadlonova, V.; Koklesova, L.; et al. Chemopreventive and Therapeutic Efficacy of *Cinnamomum zeylanicum* L. Bark in Experimental Breast Carcinoma: Mechanistic In Vivo and In Vitro Analyses. *Molecules* **2020**, *25*, 1399. [CrossRef]
14. Banerjee, S.; Banerjee, S. Anticancer Potential and Molecular Mechanisms of Cinnamaldehyde and Its Congeners Present in the Cinnamon Plant. *Physiologia* **2023**, *3*, 173–207. [CrossRef]
15. Clinical and Laboratory Standards Institute (CLSI). *Methods for Dilution Antimicrobial Susceptibility Tests for Bacteria That Grow Aerobically: Approved Standard*, 10th ed.; CLSI Document M07-A10; CLSI: Wayne, PA, USA, 2015.
16. Bakkali, F.; Averbeck, S.; Averbeck, D.; Idaomar, M. Biological Effects of Essential Oils—A Review. *Food Chem. Toxicol.* **2008**, *46*, 446–475. [CrossRef]
17. Alizadeh Behbahani, B.; Falah, F.; Lavi Arab, F.; Vasiee, M.; Tabatabaee Yazdi, F. Chemical Composition and Antioxidant, Antimicrobial, and Antiproliferative Activities of *Cinnamomum zeylanicum* Bark Essential Oil. *Evid. Based Complement. Altern. Med.* **2020**, *2020*, 5190603. [CrossRef]
18. Elgendy, E.M.; Ibrahim, H.S.; Elmeherry, H.F.; Sedki, A.G.; Mekhemer, F.U. Chemical and Biological Comparative in Vitro Studies of Cinnamon Bark and Lemon Peel Essential Oils. *Food Nutr. Sci.* **2016**, *8*, 110–125. [CrossRef]
19. Amran, D.A.A.; Al-Derawi, K.H.; Al-Saadi, S. Phytochemical Constituents of Bark Essential Oils of *Cinnamomum zeylanicum* Blume and Effects on Liver Tissue of Rats. *Univ. Thi-Qar J. Sci.* **2023**, *10*, 165–170. [CrossRef]
20. Zugic, A.; Jeremic, I.; Isakovic, A.; Arsic, I.; Savic, S.; Tadic, V. Evaluation of Anticancer and Antioxidant Activity of a Commercially Available CO_2 Supercritical Extract of Old Man's Beard (*Usnea barbata*). *PLoS ONE* **2016**, *11*, e0146342. [CrossRef]
21. Dahham, S.S.; Tabana, Y.M.; Iqbal, M.A.; Ahamed, M.B.K.; Ezzat, M.O.; Majid, A.S.A.; Majid, A.M.S.A. The Anticancer, Antioxidant and Antimicrobial Properties of the Sesquiterpene β-Caryophyllene from the Essential Oil of *Aquilaria crassna*. *Molecules* **2015**, *20*, 11808–11829. [CrossRef]
22. Sihoglu Tepe, A.; Ozaslan, M. Anti-Alzheimer, Anti-Diabetic, Skin-Whitening, and Antioxidant Activities of the Essential Oil of *Cinnamomum zeylanicum*. *Ind. Crops Prod.* **2020**, *145*, 112069. [CrossRef]
23. Elgammal, E.; El Gendy, A.E.N.; Elgamal, A.E.-B. Mechanism of Action and Bioactivities of *Cinnamomum zeylanicum* Essential Oil against Some Pathogenic Microbes. *Egypt. Pharm. J.* **2020**, *19*, 162. [CrossRef]
24. Gülçin, İ. Antioxidant Activity of Eugenol: A Structure–Activity Relationship Study. *J. Med. Food* **2011**, *14*, 975–985. [CrossRef]
25. Kallel, I.; Hadrich, B.; Gargouri, B.; Chaabane, A.; Lassoued, S.; Gdoura, R.; Bayoudh, A.; Ben Messaoud, E. Optimization of Cinnamon (*Cinnamomum zeylanicum* Blume) Essential Oil Extraction: Evaluation of Antioxidant and Antiproliferative Effects. *Evid.-Based Complement. Altern. Med.* **2019**, *2019*, 6498347. [CrossRef]
26. Azad, R.; Jayaprada, N.V.T.; Ranaweera, S.A.; Ranawaka, R.A.A.K.; Jayasekara, L.; Senanayake, G.; Hirotoshi, T.; Geekiyanage, S. Diversity and Morpho-Chemo Correlations of Cinnamon (*Cinnamomum verum* J. Presl) Bark Oil from Sri Lanka. *J. Agric. Food Res.* **2024**, *18*, 101500. [CrossRef]
27. Farias, A.P.P.; Monteiro, O.d.S.; da Silva, J.K.R.; Figueiredo, P.L.B.; Rodrigues, A.A.C.; Monteiro, I.N.; Maia, J.G.S. Chemical Composition and Biological Activities of Two Chemotype-Oils from *Cinnamomum verum* J. Presl Growing in North Brazil. *J. Food Sci. Technol.* **2020**, *57*, 3176–3183. [CrossRef]
28. Ranasinghe, P.; Pigera, S.; Premakumara, G.S.; Galappaththy, P.; Constantine, G.R.; Katulanda, P. Medicinal Properties of 'True' Cinnamon (*Cinnamomum zeylanicum*): A Systematic Review. *BMC Complement. Altern. Med.* **2013**, *13*, 275. [CrossRef]
29. Aligiannis, N.; Kalpoutzakis, E.; Mitaku, S.; Chinou, I.B. Composition and Antimicrobial Activity of the Essential Oils of Two Origanum Species. *J. Agric. Food Chem.* **2001**, *49*, 4168–4170. [CrossRef]

30. Crevelin, E.J.; Caixeta, S.C.; Dias, H.J.; Groppo, M.; Cunha, W.R.; Martins, C.H.G.; Crotti, A.E.M. Antimicrobial Activity of the Essential Oil of Plectranthus Neochilus against Cariogenic Bacteria. *Evid. Based Complement. Altern. Med.* **2015**, *2015*, 102317. [CrossRef]
31. Raeisi, M.; Tajik, H.; Yarahmadi, A.; Sanginabadi, S. Antimicrobial Effect of Cinnamon Essential Oil Against *Escherichia coli* and *Staphylococcus aureus*. *Health Scope* **2015**, *4*, e21808. [CrossRef]
32. Marongiu, B.; Piras, A.; Porcedda, S.; Tuveri, E.; Sanjust, E.; Meli, M.; Sollai, F.; Zucca, P.; Rescigno, A. Supercritical CO_2 Extract of *Cinnamomum zeylanicum*: Chemical Characterization and Antityrosinase Activity. *J. Agric. Food Chem.* **2007**, *55*, 10022–10027. [CrossRef] [PubMed]
33. Mirković, S.; Tadić, V.; Milenković, M.T.; Ušjak, D.; Racić, G.; Bojović, D.; Žugić, A. Antimicrobial Activities of Essential Oils of Different *Pinus* Species from Bosnia and Herzegovina. *Pharmaceutics* **2024**, *16*, 1331. [CrossRef] [PubMed]
34. Unlu, M.; Ergene, E.; Unlu, G.V.; Zeytinoglu, H.S.; Vural, N. Composition, Antimicrobial Activity and in Vitro Cytotoxicity of Essential Oil from *Cinnamomum zeylanicum* Blume (Lauraceae). *Food Chem. Toxicol.* **2010**, *48*, 3274–3280. [CrossRef] [PubMed]
35. Saki, M.; Seyed-Mohammadi, S.; Montazeri, E.A.; Siahpoosh, A.; Moosavian, M.; Latifi, S.M. In Vitro Antibacterial Properties of *Cinnamomum zeylanicum* Essential Oil against Clinical Extensively Drug-Resistant Bacteria. *Eur. J. Integr. Med.* **2020**, *37*, 101146. [CrossRef]
36. Utchariyakiat, I.; Surassmo, S.; Jaturanpinyo, M.; Khuntayaporn, P.; Chomnawang, M.T. Efficacy of Cinnamon Bark Oil and Cinnamaldehyde on Anti-Multidrug Resistant *Pseudomonas aeruginosa* and the Synergistic Effects in Combination with Other Antimicrobial Agents. *BMC Complement. Altern. Med.* **2016**, *16*, 158. [CrossRef]
37. Abdelatti, M.A.I.; Abd El-Aziz, N.K.; El-Naenaeey, E.Y.M.; Ammar, A.M.; Alharbi, N.K.; Alharthi, A.; Zakai, S.A.; Abdelkhalek, A. Antibacterial and Anti-Efflux Activities of Cinnamon Essential Oil against Pan and Extensive Drug-Resistant *Pseudomonas aeruginosa* Isolated from Human and Animal Sources. *Antibiotics* **2023**, *12*, 1514. [CrossRef]
38. Sandner, G.; Heckmann, M.; Weghuber, J. Immunomodulatory Activities of Selected Essential Oils. *Biomolecules* **2020**, *10*, 1139. [CrossRef]
39. Ćorović, K.; Stojković, D.L.; Petrović, Đ.S.; Milić, S.S.; Đukić, M.B.; Radojević, I.D.; Raković, I.; Jurišević, M.; Gajović, N.; Jovanović, M.; et al. Newly Synthesized Palladium(Ii) Complexes with Dialkyl Esters of (S, S)-Propylenediamine-N, N′-Di-(2,2′-Di-(4-Hydroxy-Benzil))Acetic Acid: In Vitro Investigation of Biological Activities and HSA/DNA Binding. *Dalton Trans.* **2024**, *53*, 7922–7938. [CrossRef]
40. Jan, R.; Chaudhry, G.-S. Understanding Apoptosis and Apoptotic Pathways Targeted Cancer Therapeutics. *Adv. Pharm. Bull.* **2019**, *9*, 205–218. [CrossRef]
41. Nile, A.; Shin, J.; Shin, J.; Park, G.S.; Lee, S.; Lee, J.-H.; Lee, K.-W.; Kim, B.G.; Han, S.G.; Saini, R.K.; et al. Cinnamaldehyde-Rich Cinnamon Extract Induces Cell Death in Colon Cancer Cell Lines HCT 116 and HT-29. *Int. J. Mol. Sci.* **2023**, *24*, 8191. [CrossRef]
42. Zhou, L.; Wang, S.; Cao, L.; Ren, X.; Li, Y.; Shao, J.; Xu, L. Lead Acetate Induces Apoptosis in Leydig Cells by Activating PPARγ/Caspase-3/PARP Pathway. *Int. J. Environ. Health Res.* **2021**, *31*, 34–44. [CrossRef] [PubMed]
43. Arsenijevic, M.; Milovanovic, M.; Jovanovic, S.; Arsenijevic, N.; Markovic, B.S.; Gazdic, M.; Volarevic, V. In Vitro and in Vivo Anti-Tumor Effects of Selected Platinum(IV) and Dinuclear Platinum(II) Complexes against Lung Cancer Cells. *J. Biol. Inorg. Chem.* **2017**, *22*, 807–817. [CrossRef]
44. Kale, J.; Osterlund, E.J.; Andrews, D.W. BCL-2 Family Proteins: Changing Partners in the Dance towards Death. *Cell Death Differ.* **2018**, *25*, 65–80. [CrossRef]
45. He, Y.; Liu, D.; Zhu, C.; Chen, X.; Ye, H.; Ye, M. Prevention of the *Lachnum* Polysaccharide and Its Selenium Derivatives on Cisplatin-Induced Acute Kidney Injury in Mice. *Bioorganic Med. Chem. Lett.* **2024**, *97*, 129192. [CrossRef]
46. McIlwain, D.R.; Berger, T.; Mak, T.W. Caspase Functions in Cell Death and Disease. *Cold Spring Harb. Perspect. Biol.* **2013**, *5*, a008656. [CrossRef]
47. Deutsch, T.M.; Fischer, C.; Riedel, F.; Haßdenteufel, K.; Michel, L.L.; Sütterlin, M.; Riethdorf, S.; Pantel, K.; Wallwiener, M.; Schneeweiss, A.; et al. Relationship of Ki-67 Index in Biopsies of Metastatic Breast Cancer Tissue and Circulating Tumor Cells (CTCs) at the Time of Biopsy Collection. *Arch. Gynecol. Obstet.* **2024**, *309*, 235–248. [CrossRef]
48. Larasati, Y.A.; Meiyanto, E. Revealing the Potency of Cinnamon as an Anti-Cancer and Chemopreventive Agent. *Indones. J. Cancer Chemoprev.* **2018**, *9*, 47–62. [CrossRef]
49. Ahmad, A.; Tiwari, R.K.; Saeed, M.; Ahmad, I.; Ansari, I.A. Glycyrrhizin Mediates Downregulation of Notch Pathway Resulting in Initiation of Apoptosis and Disruption in the Cell Cycle Progression in Cervical Cancer Cells. *Nutr. Cancer* **2022**, *74*, 622–639. [CrossRef]
50. Bertoli, C.; Skotheim, J.M.; de Bruin, R.A.M. Control of Cell Cycle Transcription during G1 and S Phases. *Nat. Rev. Mol. Cell Biol.* **2013**, *14*, 518–528. [CrossRef]
51. Topacio, B.R.; Zatulovskiy, E.; Cristea, S.; Xie, S.; Tambo, C.S.; Rubin, S.M.; Sage, J.; Kõivomägi, M.; Skotheim, J.M. Cyclin D-Cdk4,6 Drives Cell-Cycle Progression via the Retinoblastoma Protein's C-Terminal Helix. *Mol. Cell* **2019**, *74*, 758–770.e4. [CrossRef]

52. Wang, Y.; Li, W.; Wang, Z.; Ren, H.; Li, Y.; Zhang, Y.; Yang, P.; Pan, S. Genistein Upregulates Cyclin D1 and CDK4 Expression and Promotes the Proliferation of Ovarian Cancer OVCAR-5 Cells. *Clin. Chim. Acta* **2021**, *512*, 100–105. [CrossRef] [PubMed]
53. Nagle, A.A.; Gan, F.-F.; Jones, G.; So, C.-L.; Wells, G.; Chew, E.-H. Induction of Tumor Cell Death through Targeting Tubulin and Evoking Dysregulation of Cell Cycle Regulatory Proteins by Multifunctional Cinnamaldehydes. *PLoS ONE* **2012**, *7*, e50125. [CrossRef] [PubMed]
54. Maddika, S.; Ande, S.R.; Panigrahi, S.; Paranjothy, T.; Weglarczyk, K.; Zuse, A.; Eshraghi, M.; Manda, K.D.; Wiechec, E.; Los, M. Cell Survival, Cell Death and Cell Cycle Pathways Are Interconnected: Implications for Cancer Therapy. *Drug Resist. Updat.* **2007**, *10*, 13–29. [CrossRef] [PubMed]
55. Al Bitar, S.; Gali-Muhtasib, H. The Role of the Cyclin Dependent Kinase Inhibitor P21cip1/Waf1 in Targeting Cancer: Molecular Mechanisms and Novel Therapeutics. *Cancers* **2019**, *11*, 1475. [CrossRef]
56. He, K.; Xu, T.; Xu, Y.; Ring, A.; Kahn, M.; Goldkorn, A. Cancer Cells Acquire a Drug Resistant, Highly Tumorigenic, Cancer Stem-like Phenotype through Modulation of the PI3K/Akt/β-Catenin/CBP Pathway. *Int. J. Cancer* **2014**, *134*, 43–54. [CrossRef]
57. Li, J.; Teng, Y.; Liu, S.; Wang, Z.; Chen, Y.; Zhang, Y.; Xi, S.; Xu, S.; Wang, R.; Zou, X. Cinnamaldehyde Affects the Biological Behavior of Human Colorectal Cancer Cells and Induces Apoptosis via Inhibition of the PI3K/Akt Signaling Pathway. *Oncol. Rep.* **2016**, *35*, 1501–1510. [CrossRef]
58. Valdivieso-Ugarte, M.; Plaza-Diaz, J.; Gomez-Llorente, C.; Lucas Gómez, E.; Sabés-Alsina, M.; Gil, Á. *In Vitro* Examination of Antibacterial and Immunomodulatory Activities of Cinnamon, White Thyme, and Clove Essential Oils. *J. Funct. Foods* **2021**, *81*, 104436. [CrossRef]
59. Fan, X.; Men, R.; Huang, C.; Shen, M.; Wang, T.; Ghnewa, Y.; Ma, Y.; Ye, T.; Yang, L. Critical Roles of Conventional Dendritic Cells in Autoimmune Hepatitis via Autophagy Regulation. *Cell Death Dis.* **2020**, *11*, 23. [CrossRef]
60. Han, X.; Parker, T.L. Antiinflammatory Activity of Cinnamon (*Cinnamomum zeylanicum*) Bark Essential Oil in a Human Skin Disease Model. *Phytother. Res.* **2017**, *31*, 1034–1038. [CrossRef]
61. Adams, D.R.P. *Identification of Essential Oil Components by Gas Chromatography/Mass Spectrometry*, 4th ed.; Allured Publ. Crop: Carol Steam, IL, USA, 2017.
62. Blois, M.S. Antioxidant Determinations by the Use of a Stable Free Radical. *Nature* **1958**, *181*, 1199–1200. [CrossRef]
63. Milošević, M.D.; Marinković, A.D.; Petrović, P.; Klaus, A.; Nikolić, M.G.; Prlainović, N.Ž.; Cvijetić, I.N. Synthesis, Characterization and SAR Studies of Bis(Imino)Pyridines as Antioxidants, Acetylcholinesterase Inhibitors and Antimicrobial Agents. *Bioorganic Chem.* **2020**, *102*, 104073. [CrossRef] [PubMed]
64. Čutović, N.; Batinić, P.; Marković, T.; Radanović, D.; Marinković, A.; Bugarski, B.; Jovanović, A.A. Physicochemical Properties of Winter Savory Extracts Prepared Using Ultrasound-Assisted Extraction. *Lek. Sirovine* **2022**, *42*, 34–42. [CrossRef]
65. Pavlović, S.; Petrović, B.; Ćoćić, D.; Schreurer, A.; Sretenović, S.; Nešić, M.D.; Nišavić, M.; Maric, Z.; Stanisavljević, I.; Ćorović, I.; et al. New Pd(II)-Pincer Type Complexes as Potential Antitumor Drugs: Synthesis, Nucleophilic Substitution Reactions, DNA/HSA Interaction, Molecular Docking Study and Cytotoxic Activity. *Dalton Trans.* **2024**, *53*, 18560–18574. [CrossRef] [PubMed]
66. Vučelj, S.; Hasić, R.; Ašanin, D.; Šmit, B.; Caković, A.; Bogojeski, J.; Serafinović, M.Ć.; Marković, B.S.; Stojanović, B.; Pavlović, S.; et al. Modes of Interactions with DNA/HSA Biomolecules and Comparative Cytotoxic Studies of Newly Synthesized Mononuclear Zinc(II) and Heteronuclear Platinum(II)/Zinc(II) Complexes toward Colorectal Cancer Cells. *Int. J. Mol. Sci.* **2024**, *25*, 3027. [CrossRef] [PubMed]
67. Stanisavljević, I.; Živković, M.; Rajković, S.; Obradović, M.; Jurišević, M.; Pavlović, S.; Simović-Marković, B.; Gajović, N.; Ćorović, I.; Jocić, M.; et al. Immunomodulatory Effects of Mononuclear 5,6-Epoxy-5,6-Dihydro-1,10-Phenanthroline Platinum(II) Complex. *Kragujev. J. Sci.* **2024**, *46*, 73–84. [CrossRef]
68. Jovanovic, M.; Simovic Markovic, B.; Gajovic, N.; Jurisevic, M.; Djukic, A.; Jovanovic, I.; Arsenijevic, N.; Lukic, A.; Zdravkovic, N. Metabolic Syndrome Attenuates Ulcerative Colitis: Correlation with Interleukin-10 and Galectin-3 Expression. *World J. Gastroenterol.* **2019**, *25*, 6465–6482. [CrossRef]

International Journal of *Molecular Sciences*

MDPI

Article

Unveiling the Chemical Composition, Antioxidant, and Antimicrobial Potentials of *Foeniculum vulgare* Mill: A Combined In Vitro and In Silico Approach

Bouchra El Moumen [1], Amal Bouzoubaa [1], Aziz Drioiche [1,2,*], Mohamed Eddahmouny [1], Omkulthom Al Kamaly [3], Abdelaaty Abdelaziz Shahat [4], Hanane Touijer [1], Nadia Hadi [1], Samira Kharchouf [1], Ali Cherrat [1], Kamal Fadili [1], Hajar El Ouadni [5], Amina Bari [6] and Touriya Zair [1,*]

1 Research Team of Chemistry of Bioactive Molecules and the Environment, Laboratory of Innovative Materials and Biotechnology of Natural Resources, Faculty of Sciences, Moulay Ismaïl University, B.P. 11201 Zitoune, Meknes 50070, Morocco; elmoumen.bouchra4@gmail.com (B.E.M.); bouzoubaa.amal@yahoo.fr (A.B.); mo.eddahmouny@edu.umi.ac.ma (M.E.); hananetou@gmail.com (H.T.); nadia.hadi1@gmail.com (N.H.); samirakharchouf@yahoo.fr (S.K.); alicherrat2016@gmail.com (A.C.); k.fadili@edu.umi.ac.ma (K.F.)

2 Higher Institute of Nursing Professions and Health Techniques of Fez, Regional Health Directorate Fez-Meknes, EL Ghassani Hospital, Fez 30000, Morocco

3 Department of Pharmaceutical Sciences, College of Pharmacy, Princess Nourah bint Abdulrahman University, P.O. Box 84428, Riyadh 11671, Saudi Arabia; omalkmali@pnu.edu.sa

4 Pharmacognosy Department, College of Pharmacy, King Saud University, P.O. Box 2457, Riyadh 11451, Saudi Arabia; ashahat@ksu.edu.sa

5 Pharmacodynamics Research Team ERP, Laboratory of Pharmacology and Toxicology, Faculty of Medicine and Pharmacy, University Mohammed V in Rabat, Rabat B.P. 6203, Morocco; elouadni.hajar@gmail.com

6 Laboratory of Biotechnology, Environment Agrifood and Health, Faculty of Sciences Dhar El Mahraz, Sidi Mohamed Ben Abdellah University, BP 1796 Atlas, Fez 30000, Morocco; aminabari3@gmail.com

* Correspondence: drioichelabo@gmail.com (A.D.); t.zair@umi.ac.ma (T.Z.); Tel.: +212-600394841 (A.D.)

Abstract: This study on *Foeniculum vulgare* Mill., derived from seeds collected in Meknes (Morocco), evaluates in vitro and in silico the therapeutic potential of its extracts and essential oil through a comprehensive analysis of its phytochemical composition, as well as its antioxidant and antimicrobial activities. Aqueous extracts (E0), hydroethanolic extract (E1) obtained via Soxhlet, decoction (E2), and essential oil (EO) obtained through hydrodistillation were analyzed using HPLC/UV-ESI-MS and GC-MS, revealing a richness in phenolic and terpenic compounds. The quantification of total polyphenols, flavonoids, and tannins in aqueous and organic extracts was performed using spectrophotometric methods. Antioxidant activity was assessed through three methods: DPPH, FRAP, and Total Antioxidant Capacity (TAC). The antimicrobial activity of the essential oil and decoction was evaluated by microdilution in microplate assays. The aqueux extract was dominated by butyl ferulate (14.33%), while hydroethanolic extract contained chlorogenic acid (14.79%) and quercetin-3-glucuronide (13%). The extract (E_2) was characterized by dihydrocaffeic acid (11.25%) and 3-O-caffeoylshikimic acid (11.08%), whereas the EO was primarily composed of fenchone (24.72%), trans-anethole (22.22%), and limonene (20.48%). Antioxidant assays (DPPH/FRAP/TAC) demonstrated decreasing efficacy as follows: EO exhibited the highest efficiency (IC_{50} = 51.45 μg/mL), followed by E_1 (93.71 μg/mL), E_0 (212.86 μg/mL), and E_2 (397.41 μg/mL), confirming a correlation between phenolic composition and antioxidant activity. Furthermore, antimicrobial tests highlighted a pronounced fungicidal effect against *Candida albicans* (MIC = 3.13 mg/mL) and *Aspergillus niger* (6.25 mg/mL), contrasting with a more moderate inhibition of *Escherichia coli* and *Staphylococcus aureus*. Molecular docking simulations identified stable interactions between chlorogenic acid, quercetin-3-glucuronide, and microbial proteases, suggesting a synergistic inhibitory mechanism. This research validates the potential of *F. vulgare* as a source of bioactive molecules with promising applications in phytotherapy for managing oxidative stress and fungal infections, while emphasizing the need for clinical studies to confirm these effects in vivo.

Keywords: *Foeniculum vulgare* Mill; fenchone; anethole; butyl ferulate; chlorogenic acid; quercetin-3-glucuronide; antioxidant activity; antimicrobial activity

1. Introduction

Foeniculum vulgare Mill., commonly known as fennel, is a herbaceous plant belonging to the Apiaceae family, widely recognized for its numerous medicinal, culinary, and industrial applications. The genera within this family are characterized by strong flavors and aromas due to the presence of schizogenous ducts containing oil, mucilage, and resins, which can be found in both the aerial parts (leaves, stems, and fruits) and the roots [1]. This distinctive feature endows the plants of the Apiaceae family, including *Foeniculum vulgare*, with a wealth of secondary metabolites, such as coumarins, flavonoids, saponins, and terpenoids, making them highly suitable for various sectors as follows: food (nutrition, beverages, and spices), pharmaceutical, and cosmetic industries [2]. Furthermore, many species within this family are used in traditional medicine for the treatment of gastrointestinal, reproductive, and respiratory disorders [3,4].

Foeniculum vulgare is a perennial plant with delicate, feathery, and almost filamentous foliage, capable of reaching up to 2 m in height. It bears a morphological resemblance to dill. Its leaves, striated and three to four times pinnate, feature filiform segments measuring up to 4 cm in length. The small yellow flowers are grouped into broad, flat umbels, with the flowering period extending from July to October. The fruits, oblong to ovoid in shape, measure 3 to 5 mm in length, and the seeds mature between September and October [1,5]. These dry fruits are widely used in culinary applications [6]. This species grows naturally in coastal regions of the Mediterranean but has become extensively naturalized in various parts of the world, thanks to its adaptability to diverse climates and its resilience to dry, sunlit soils [7]. Its status as a cosmopolitan plant is also attributed to its historical use in traditional medicine and gastronomy, notably as a seasoning.

The use of *F. vulgare* dates back to antiquity, where it was cultivated by Egyptians, Chinese, Indians, and Romans, particularly for its aromatic seeds and edible shoots [8]. The Romans utilized fennel, not only for its seeds but also for its succulent stems, which remain a commonly consumed vegetable in southern Italy [9]. All parts of the plant, including its roots, stems, leaves, and fruits, are aromatic and can be employed in various ways. Fennel fruits, as well as its essential oils, serve as flavoring agents in food products, such as liqueurs, bread, pickles, pastries, and cheeses [10–12], while also being used in the formulation of cosmetic and pharmaceutical products [13,14].

Studies have demonstrated that *F. vulgare* is effective in addressing a wide range of infectious disorders of bacterial, fungal [15,16], viral, mycobacterial, and protozoal origins [17]. Fennel essential oils have been reported to possess anti-inflammatory, antispasmodic, antiseptic, antithrombotic, antitumor, chemopreventive, cytoprotective, hepatoprotective, hypoglycemic, estrogenic, carminative, diuretic, and analgesic properties [18–20], along with beneficial effects in the treatment of gastrointestinal [21] and neurological disorders [22]. The essential oil of *F. vulgare* exhibits significant antimicrobial activity against both Gram-positive and Gram-negative bacteria [23], as well as against yeasts, although its efficacy varies depending on the phase (liquid or vapor) and the concentration used [24]. Moreover, this essential oil possesses strong antioxidant properties, including the ability to scavenge free radicals and enhance the activity of antioxidant enzymes, such as superoxide dismutase, catalase, and glutathione peroxidase [25,26]. Additionally, the essential oil of *F. vulgare* has shown hypoglycemic effects in diabetic rats, significantly reducing blood glucose levels and improving the activity of antioxidant enzymes [27]. Furthermore, certain publications have highlighted that *F. vulgare* possesses a particular effect on memory enhancement and can reduce stress [28].

Although the phytochemistry and antimicrobial properties of fennel, particularly its essential oils, have been studied in a fragmented manner, the available data remain insufficient to fully exploit the potential of this plant. In this context, our study stands out for its in-depth and innovative approach, aiming to uncover the richness of Moroccan *F. vulgare*. By combining a detailed analysis of its phytochemical compounds with a rigorous evaluation of its antioxidant and antimicrobial capacities, both for its essential oils and three distinct types of extracts, we make a significant contribution to the understanding and valorization of this species in key fields, such as health, cosmetics, and the food industry.

2. Results

2.1. Quality Control of Plant Material

Table 1 summarizes the results of the plant material quality control. *F. vulgare* seeds have a moisture content of around 25.12%, an acid pH of 5.5, and contain 6.4% mineral matter and 93.6% organic matter. The dosage of metallic trace elements showed fairly low contents of arsenic (0.0058 mg/g), chromium (0.0008 mg/g), antimony (0.0023 mg/g), copper (0.003 mg/g) while lead, cadmium and titanium are undetectable against a slightly high iron content (0.271 mg/g) (Table 2). It should be noted that these results are below the limit values for each metallic trace element.

Table 1. Quality control of plant matter, including moisture content (MC), pH, acidity, organic material (OM), and mineral matter (MM).

Species	MC (%)	pH	Acidity	MM (%)	MO (%)
F. vulgare	25.12 ± 0.001	5.5 ± 0.00	0.11 ± 0.00	6.4 ± 0.067	93.6

Table 2. Concentration of heavy metals (mg/L) (ICP) and FAO/WHO Maximum Limit (2009).

Species	Arsenic (As)	Chrome (Cr)	Antimoine (Sb)	Plomb (Pb)	Cadmium (Cd)	Iron (Fe)	Copper (Cu)	Titanium (Ti)
F. vulgare	0.0058	0.0008	0.0023	Undetectable	Undetectable	0.271	0.003	Undetectable
Maximum Limit (mg/L)	0.05	0.05	0.005	0.05	0.005	20	1	-

2.2. Phytochemical Screening of Plant Material

The phytochemical screening highlighted the richness of this plant in vitamins, minerals, primary and secondary metabolites. Table 3 displays the composition of primary and secondary metabolites in *F. vulgare*. It was observed that *F. vulgare* is abundant in lipids (sterols and triterpenes), proteins, reducing sugars, and carbohydrates (oses and holosides). Concerning secondary metabolites, the species is rich in polyphenols, such as flavonoids (leucoanthocyanins, flavones) and tannins alkaloids, mucilages, as well as saponosides. These results are in line with those reported by Kooti et al. (2015) [29] and demonstrate the plant's richness in components known for their medicinal, culinary, and dietary benefits.

Table 3. Chemical families present in *F. vulgare*.

Chemical Group			*F. vulgare*
Secondary Metabolites	Lipids (Lieberman–Burchard reaction)		++
	Protein	Biuret reaction	+
		Xanthoprotein reaction	++
	Reducing sugar		+
	Polysaccharide		+
Secondary Metabolites	Tannins		+++
	Catechic tannins		+++
	Gallic tannins		+
	Flavonoids		++
	Cyanidin reaction		Flavones
	Leucoanthocyanins		++
	Saponosides		+
	Alkaloids		+
	Reducing compounds		++
	Monosaccharides and holosides		++
	Mucilages		++
	Sterols and triterpenes		++

+: weak positive test; ++: positive test; +++: strongly positive test.

2.3. Contents of Polyphenols, Flavonoids and Condensed Tannins

The yields of the various fennel seed extracts using the three methods E (0), E (1) and E (2), are shown in Figure 1a. It can be seen that the extracts obtained by the methods used gave fairly high and almost similar yields, of the order of 26.45%, 25.78%, and 24.26%, respectively. Our results concerning the yield are much higher than those obtained by Anwar et al. (2009) [30], which brought in maximum yields of 15.63% and 6.21%, respectively, from fennel seeds using the ethanol and methanol. Figure 1b–d also present the composition of *F. vulgare* in total polyphenols, flavonoids and condensed tannins. Figure 1b illustrates the total polyphenol content, which are 36.91, 27.09 and 25.38 mg EQ AG/g ES, respectively, for E (2), E (1) and E (0). While for flavonoids, the highest contents were 19.71; 14.78 mg EQ AG/g ES and the lowest were of the order of 5.08 mg EQ AG/g ES recorded, respectively, by E (2), E (0) and E (1) (Figure 1c). Whereas for condensed tannins, the contents are much lower, and were around 0.142; 0.135; 0.131 mg EQC/g ES, respectively, in E (0), E (2) and E (1) (Figure 1d). These differences in the quantity of total polyphenols, flavonoids and catechin tannins, may be due to the varying effectiveness of solvent extraction [31].

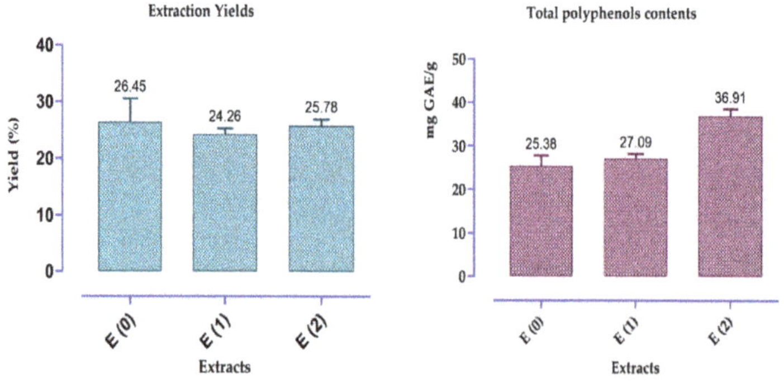

(a) (b)

Figure 1. *Cont.*

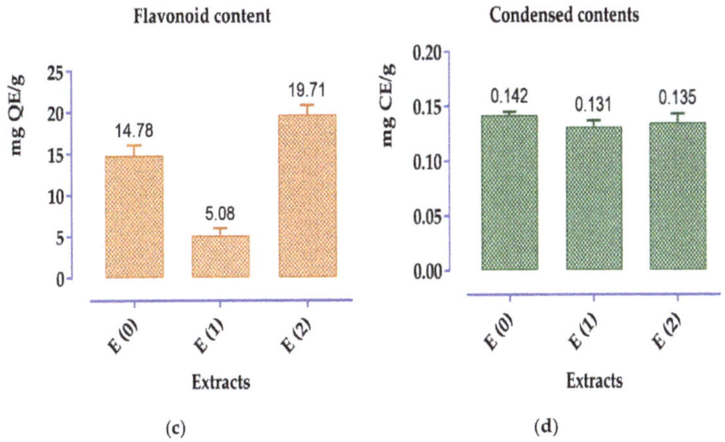

Figure 1. Extraction yield (**a**) and contents of polyphenols (**b**), flavonoids (**c**) and catechic tannins (**d**) in *F. vulgare* extracts. Values denote the mean ± standard deviation based on three separate experiments.

2.4. Identification of the Chemical Composition of Phenolic Compounds from the Extracts of F. vulgare by LC/UV

The chemical analysis of *F. vulgare* extracts was carried out using the HPLC/UV-ESI-MS technique, enabling a detailed characterization of the compounds present in the chromatograms shown in Figure 2. In-depth interpretation of the mass spectra in negative ionization mode, coupled with examination of the chromatograms, led to the identification of 65 molecules, listed in Supplementary Table S1. The three extraction methods used—aqueous Soxhlet extraction (E (0)), hydroethanolic Soxhlet extraction (E (1)), and decoction (E (2))—revealed distinct phytochemical profiles. The aqueous extract E (0) showed the presence of 28 bioactive compounds, dominated by phenolic acids (Table 4), which represent 49.96% of the total content. The major compound identified in E (0) is butyl ferulate, which accounts for 14.33% of the extract. The hydroethanolic extract E (1), obtained with an ethanol/water mixture (70:30), was distinguished by its greater molecular diversity, with 34 compounds identified. Phenolic acids also dominate this extract, representing 33.39% of the total content, while flavonoids make up a notable proportion of 42.91% (Table 4). Among the major compounds, chlorogenic acid (14.79%) and quercetin-3-glucuronide (13%) are particularly abundant. Finally, the decocted extract E (2) allowed the identification of 37 bioactive compounds, with a predominance of phenolic acids (37.3%). However, this extract stands out due to its richness in lignans (7.82%), phenolic diterpenes (8.08%), and dipeptides (5.42%), suggesting that boiling promotes the extraction of thermostable or weakly polar compounds. The major compounds in this extract include dihydrocaffeic acid (11.25%) and 3-O-caffeoylshikimic acid (11.08%). These results highlight the influence of extraction conditions on the phytochemical profile of the extracts. They particularly emphasize the effectiveness of the hydroalcoholic solvent with intermediate polarity in extracting a diverse range of secondary metabolites, while also illustrating the complementarity of the different approaches.

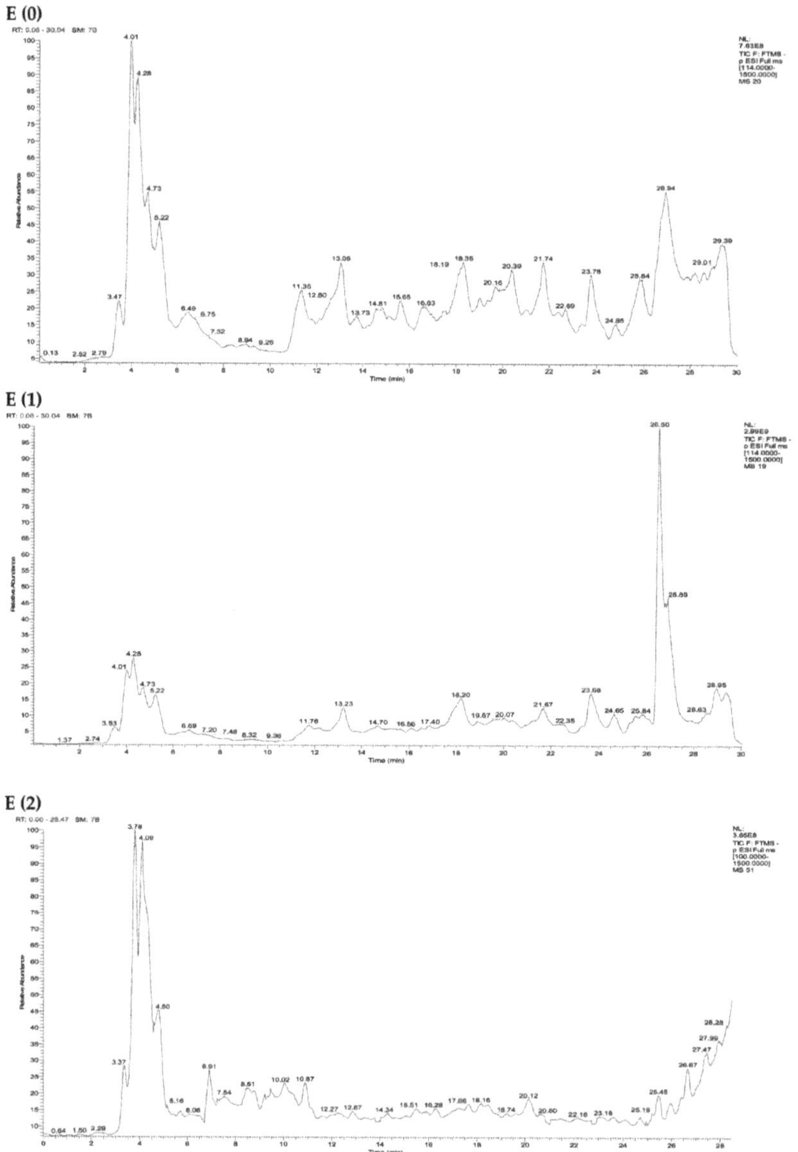

Figure 2. HPLC/UV-ESI-MS chromatogram of *F. vulgare* extracts.

The combined HPLC/UV-ESI-MS analysis of *F. vulgare* extracts allowed for the fine and structured characterization of the fragmentation profiles of bioactive compounds, providing a deep understanding of their molecular architecture. The observed fragmentation mechanisms are specific to each chemical class and reveal essential diagnostic signatures for structural validation. Lignans, exemplified by medioresinol ($[M-H]^-$ = 387), fragment into ions at m/z 207 and 179, indicating a classical breakage of ether bonds between phenylpropanoid units. Phenolic acids, such as embelic acid, dihydrocaffeic acid ($[M-H]^-$ = 181), and 3-O-caffeoylshikimic acid ($[M-H]^-$ = 335), show typical losses of CO_2 (m/z 137) or caffeic residues (m/z 291, 197), characteristic of carboxylic chain or ester bond cleavages. Chlorogenic acid ($[M-H]^-$ = 353) and iso-chlorogenic acid A ($[M-H]^-$ = 515) generate fragments at m/z 191 and 179 (quinic acid and caffeic acid, respectively), confirming their nature as acylquinates. Phenolic diterpenes like rosmanol ($[M-H]^- = 345$) produce an ion at m/z 283, suggesting isoprene loss or dehydration, while sesquiterpenoids like emmotine A ($[M-H]^- = 277$) reveal fragments at m/z 243 due to terpene skeleton degradation. Glycosylated polyphenols, such as resveratrol-3-glucoside ($[M-H]^- = 389$) and quercetin-3-glucuronide ($[M-H]^- = 477$), undergo losses of 162 u (glucose) and 176 u (glucuronide), revealing fragment ions at m/z 227, 301, and 151, typical of flavonoid fragmentation. Aglycone flavonoids, such as apigenin ($[M-H]^- = 269$) and 7-methoxy-2-methylisoflavone ($[M-H]^- = 265$), show breakage of the C-ring and decarboxylations, leading to ions at m/z 227, 237, 179, and 159. Esters like butyl ferulate ($[M-H]^- = 249$) lose a butyl group (m/z 193), while more complex polyphenols like salvianolic acid K ($[M-H]^- = 555$) reveal multi-stage fragmentation at m/z 493, 313, and 179, indicating a structure rich in caffeic units. These fragmentation patterns, specific and reproducible, serve as powerful tools for structural identification, chemical classification, and confirmation of annotations in complex plant extracts. They also highlight the metabolic richness of *F. vulgare* and reinforce the pharmacognostic interest of this medicinal species. Figure 3 illustrates the structures of the major identified compounds, providing essential information for a thorough structural analysis.

Table 4. Percentages of the compounds classes identified in the *F. vulgare*.

Classes	R.A (%)		
	E (0)	E (1)	E (2)
Carboxylic ester	1.04	0	0.66
Dipeptide	0	0	5.42
Ester	0	0.55	0
Lignan	0	1.21	7.82
Phenolic compound	7.12	3.54	2.36
Fatty acid	0	0	1.67
Phenolic acid	49.96	33.39	37.3
Flavonoid	24.4	42.91	19.72
Phenolic diterpene	0	1.29	8.08
Polyphenol	11.2	9.94	7.84
Quinone	0	0	0.49
Terpenoid	4.42	6.58	6.46
Vitamin	1.82	0.56	1.84

F. vulgare extracts are distinguished by their richness in bioactive compounds, particularly polyphenols and phenolic acids, which provide potent antioxidant and antimicrobial properties. These findings support the work of [32], as well as [33,34], which highlighted the key role of these compounds in the biological activities of this plant. The ethanolic extract, known for its effectiveness in extracting flavonoids and polyphenols, stands out for its ability to concentrate bioactive compounds, such as quercetin-3-glucuronide, quercetin-3-D-xyloside, apigenin, and rutin. These molecules, widely documented for their protective properties at the cellular level, have been highlighted in various studies [35]. Additionally, the notable presence of chlorogenic acid and iso-chlorogenic acid A provides promising potential for antimicrobial applications, as confirmed by the studies of Khlood et al. (2023) [36] and Mingsan et al. (2020) [37]. Although the decoction extract is less concentrated in flavonoids, it remains effective for extracting certain polyphenols and phenolic acids while offering better bioavailability for certain molecules. Thus, these results emphasize the relevance of *F. vulgare* extracts for various applications in the pharmaceutical and food industries, due to their remarkable functional properties.

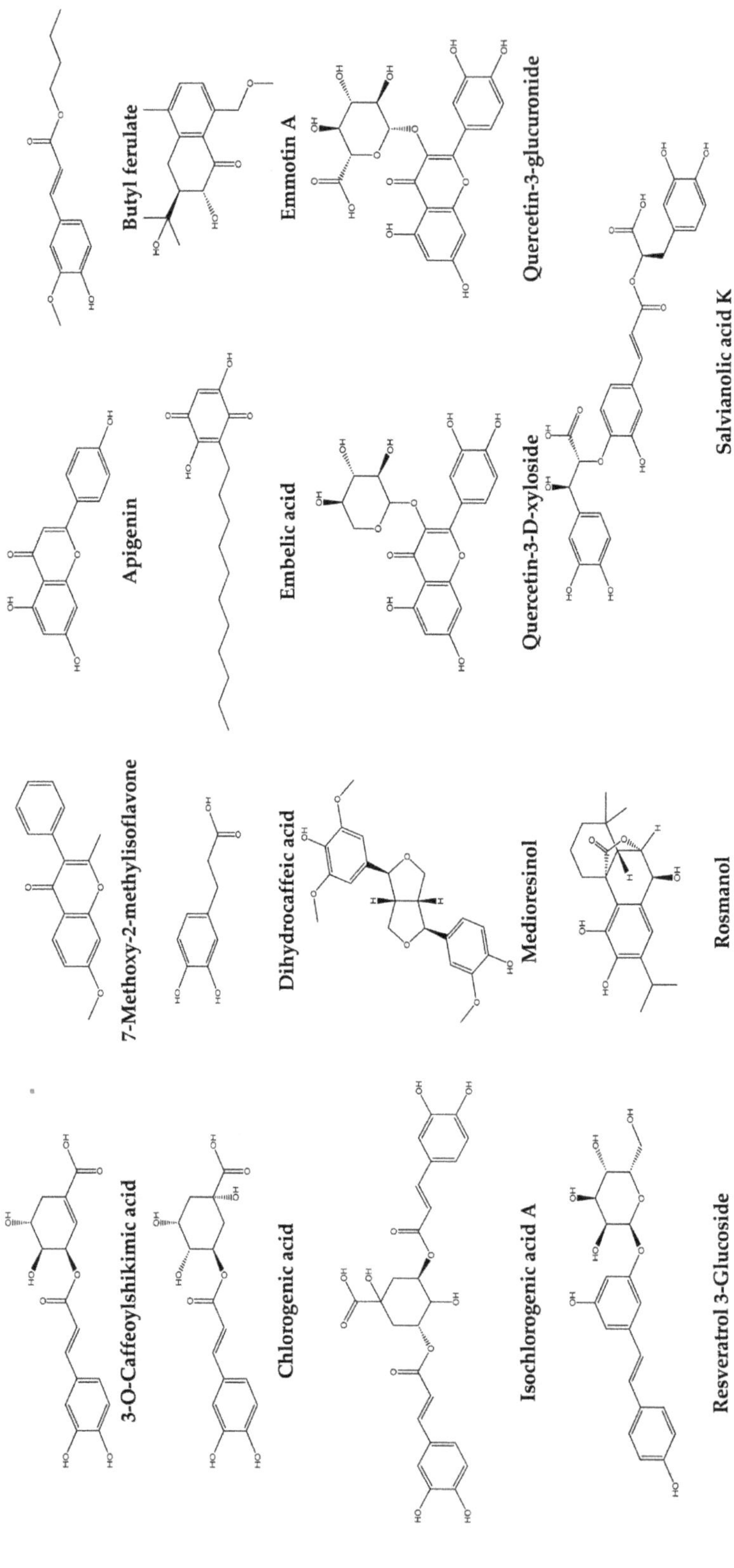

Figure 3. Structures of the compounds identified in the extracts of *F. vulgare*.

2.5. Yields and Quality Control of EO

The yield of fennel seed EO is shown in Table 5. The yield of hydro-distilled fennel seed essential oil is 2.5%. The results obtained in the present study are in line with those reported by other researchers, who found yields between 2.82 and 3.38% using hydro-distillation [31,38]. This yield is higher than that obtained by Diao et al., 2014 [23], which recorded a yield of 1.74%, while the yield of EO and ethanol extract of fennel seeds from Portugal were 0.1% and 6.9%, respectively [30].

Table 5. Yield and density of *F. vulgare* oils.

Species	Yield (%)	Density (g/mL)
***F. vulgare* Mill**	2.500 ± 0.067	0.964 ± 0.002

2.6. Chemical Composition of *F. vulgare* Seed EO

The results of the chemical composition of fennel essential oil are shown in Table 6. A total of 25 chemical compounds representing approximately 99.99% of the total EO composition of *F. vulgare* leaves were identified using gas chromatography-mass spectrometry (GC-MS) (Figure 4). The main constituents of the EO tested were fenchone (24.72%), followed by trans-anethole (22.22%), limonene (20.48%), cis-anethole (19.18%) and Methylchavicol (estragole) (8.79%) (Figure 5, Table 6). The illustrated results also showed the presence of two chemical families at high levels, specifically oxygenated monoterpenes (77.02%) and monoterpenes (22.90%). In addition, the fennel EO tested also contained considerable quantities of various minor constituents whose contribution was less than 1%, such as sesquiterpenes (0.07%). Regarding the chemical constituent groups represented, cis/trans-anethole, fenchone, and estragole were the major oxygenated monoterpenes, while limonene was the major monoterpene. In this regard, a comparative profile has been reported by several researchers who revealed the presence of trans-anethole, fenchone, estragole and limonene as the main components of the EOs of *F. vulgare* fennel seeds with percentages depending on the region (Podgorica, central Montenegro, Turkey) [15,39,40]. The variation in the chemical composition of EO from one country to another could be due to agro-climatic conditions (climate, season, geographic) among many more factors that influence the adaptive metabolism of plants [41].

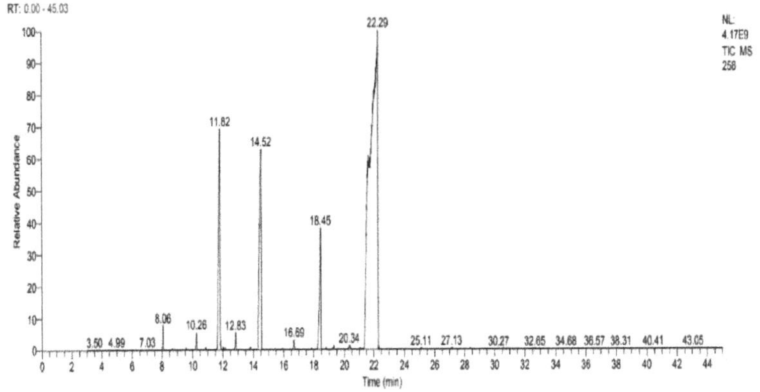

Figure 4. Chromatogram (GC-MS) of the EO of *F. vulgare* seeds.

Table 6. Relative percentage composition of *F. vulgare* EO by GC-MS analysis.

Compound	RA %	KI
α–Pinene	1.1	939
Camphene	0.08	954
Sabinene	0.1	975
β–Pinene	0.03	979
Myrcene	0.82	990
α–Phellandrene	0.13	1002
β–Phellandrene	0.05	1029
Limonene	20.48	1029
1,8-Cineole	0.11	1031
β–cis-Ocimene	0.11	1037
γ–Terpinene	0.85	1059
Fenchone	24.72	1086
Terpinolene	0.13	1088
Cis-Thujone	0.03	1102
Trans-Pinene hydrate	0.05	1122
Camphor	0.52	1146
Terpinen-4-ol	0.06	1177
Methyl chavicol (estragole)	8.79	1196
Fenchyl acetate <endo->	0.07	1220
Fenchyl acetate <exo->	0.18	1232
Cis-anethole	19.18	1252
Trans-anethole	22.22	1284
Anisyl methyl ketone	0.06	1382
Germacrene D	0.07	1481
Trans-Methyl isoeugenol	0.05	1492
Identified compounds (%)	99.99	
Monoterpenes (%)	22.90	
Oxygenated monoterpenes (%)	77.02	
Sesquiterpenes (%)	0.07	
Oxygenated sesquiterpenes (%)	0.0	

RA: Relative abundance (%); KI: Kovats Index.

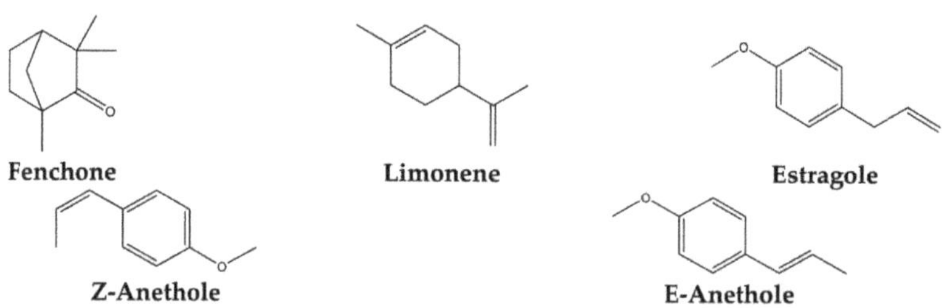

Figure 5. Structures of the major compounds identified in the EO of *F. vulgare*.

Due to its hypolipidemic and antiatherogenic activities, this plant could be used to control cardiovascular disorders as stated by Garga et al. (2009) [42]. Fennel's high polyphenol content means that it may play an essential role in cancer chemoprevention. The anethole contained in its seeds has an inhibitory effect on the activation of TNF-α by the transcription factor NF-KB. Studies have shown that anethole also inhibits cellular responses induced by such cytokines, which could explain its role in cancer protection [31]. As with green anise, Chinese star anise and tarragon, the main active ingredient in both fennel varieties is anethole, which makes up around 80% of the EO [43]. Bitter fennel is also rich in fenchone, while sweet fennel contains more estragole [41]. Laboratory experiments have demonstrated anethole's antibacterial and antimycotic properties (against microscopic fungi) [44,45].

Fenchone is a monoterpene occurring in the EOs of various plants, including *F vulgare* [46,47]. The studies conducted by Pessoa et al. (2020), Araruna et al. (2025), and Aćimović et al. (2025) have substantiated the anti-inflammatory, antioxidant, wound-healing, antidiarrheal, antifungal, antinociceptive, and bronchodilatory properties of fenchone [48–50]. It is thought to possess antispasmodic properties, helping to relieve stomach aches and painful menstruation [51]. Additionally, the anti-tumorigenic effects of D-limonene have been widely studied across various cancer types [52].

A comprehensive analysis of the chemical composition of the essential oil (EO) revealed that *F. vulgare* EO is predominantly composed of oxidized ethers (41.58%), followed by ketones (24.81%) and hydrocarbons (23.93%), with alcohols present in lower amounts (9.67%). Notably, other chemical groups, such as aldehydes and epoxides, are entirely absent from the EO of this plant (Figure 6).

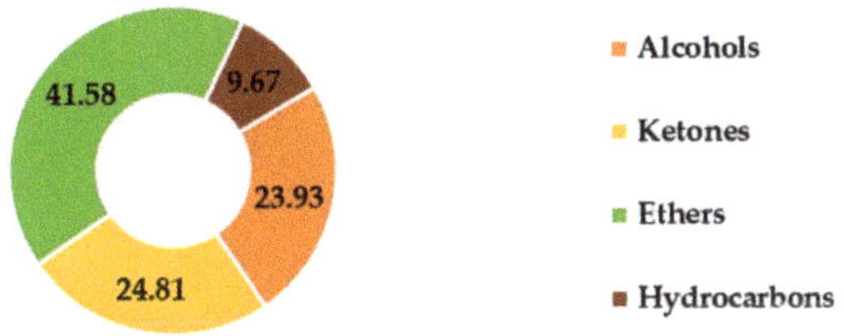

Figure 6. Distribution of families of chemical compounds of *F. vulgare* EO.

2.7. Antioxidant Activities

We examined the free radical neutralizing activity and lipid oxidation inhibition of fennel seed essential oil and the three extracts. Free radicals, which are involved in the lipid peroxidation process, are considered to play a major role in many chronic pathologies, such as cancer and cardiovascular disease among others, hence the need to neutralize them. The free radical neutralizing activities of EO (Figure 7a) and *F. vulgare* extracts (Figure 7b) were measured by the DPPH assay. *F. vulgare* seed extracts showed excellent radical-neutralizing activity, with IC_{50} values (the extract concentration providing 50% of inhibition) of 212.86; 93.71; 397.41 μg/mL for E (0), E (1), E (2) extracts, respectively (Figure 7). Still according to the IC_{50} values grouped together in Figure 7a, the antioxidant power of *F. vulgare* EO is 51.45 μg/mL, which is higher than that of the standard antioxidant ascorbic acid. In comparison with the antioxidant activity of the essential oil and extracts, the antioxidant power of the E (2) extract is lower than that of the other extracts and the EO. Moreover, the E (1) extract has a higher antioxidant activity than the E (0) and E (2) extracts. However, Anwar et al. (2009) [30] and Goswami et al., 2014 [53], reported IC_{50} values between 23.61 and 83 μg/mL, using methanol and ethanol extracts, which are considerably lower than the results achieved with the extracts used. The free radical scavenging activity of fennel extracts could be due to their higher content of phenolic components. These hydroxyl phenolic compounds can donate hydrogen atoms to DPPH and trap it [54].

The reduction activity of ferric iron to ferrous iron was carried out on the E (0), E (1) and E (2) extracts of *F. vulgare*, as well as on ascorbic acid, the standard used as a reference. The results show that the reducing power of iron is proportional to the increase in concentration of the three extracts studied. The concentrations that provide 50% inhibition (IC_{50}) were calculated from the curve in Figure 7c. The results showed that the IC_{50} value of ascorbic acid has a higher antioxidant power (5.32 μg/mL). Then the IC_{50} values of the extracts: E (1) (153.6 μg/mL), E (2) (170 μg/mL) and E (0) (254.12 μg/mL), revealed a less significant antioxidant power (Figure 7c). This antioxidant activity of the extracts may have a link with the chemical constituents of these extracts. Indeed, butyl ferulate (14.33%), chlorogenic acid (14.79%) and Dihydrocaffeic acid (11.25%), which are the major compounds in the extracts of *F. vulgare*, revealed a strong correlation with the antioxidant activity.

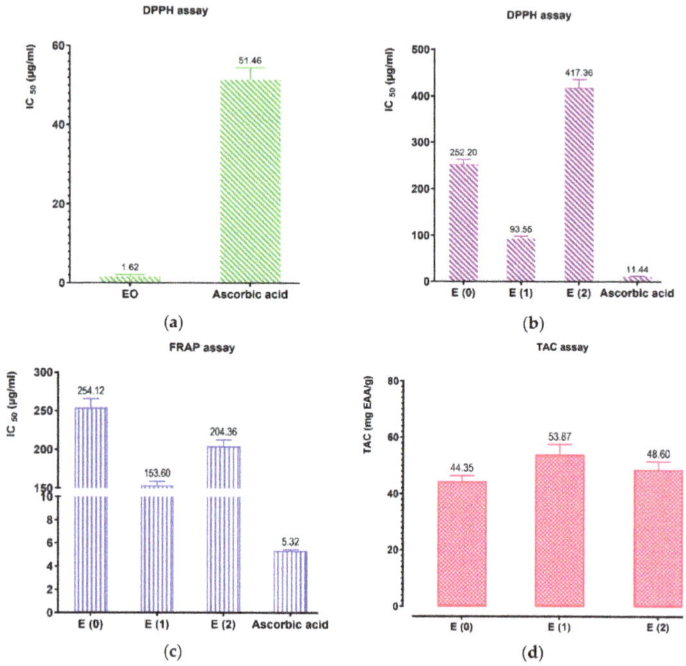

Figure 7. Antioxidant activity of the EO and extracts by DPPH assay (**a**) and (**b**), respectively. Antioxidant activity of the extracts by FRAP assay (**c**) and TAC assay (**d**). Mean values ± standard deviations of determinations performed in triplicate are reported. Means are significantly different ($p < 0.001$).

The ferric-to-ferrous iron reduction assay was performed on three extracts of *F. vulgare*: the aqueous extract obtained via Soxhlet (E_0), the hydroethanolic extract obtained via Soxhlet (E_1), and the decoction extract (E_2), alongside ascorbic acid, which served as the reference standard. The results indicate that the iron-reducing capacity of the extracts increases proportionally with their concentration. The half-maximal inhibitory concentration (IC_{50}) values were determined from the dose–response curve shown in Figure 7c. Ascorbic acid demonstrated the strongest antioxidant activity with an IC_{50} value of 5.32 µg/mL. In comparison, the IC_{50} values of the extracts: E (1) (153.6 µg/mL), E (2) (170 µg/mL), and E (0) (254.12 µg/mL), demonstrate a relatively modest antioxidant capacity. The observed antioxidant activity of the extracts is likely attributable to their chemical composition. Notably, key compounds, such as butyl ferulate (14.33%), chlorogenic acid (14.79%), and dihydrocaffeic acid (11.25%) in *F. vulgare* extracts, have been shown to exhibit a strong correlation with antioxidant activity [55].

The results of the antioxidant activity, obtained through the total antioxidant capacity test described in Figure 7d, indicate that the hydroethanolic extract obtained by Soxhlet presents the highest antioxidant activity (53.87 mg EAA/g), followed by the decoction extract (48.60 mg EAA/g), and finally, the aqueous extract obtained by Soxhlet, which shows the lowest activity (44.35 mg EAA/g). Therefore, the hydroethanolic extract (E (1)) stands out as the most effective in terms of antioxidant activity, likely due to specific compounds or conditions associated with this extract.

2.8. Antimicrobial Activity of Essential Oils and Extracts of *F. vulgare*

For several years, many studies have focused on EOs and medicinal plant extracts to inhibit the growth of microbes [31]. In this study, the antimicrobial activity of fennel (*Foeniculum vulgare*) essential oil and extracts was assessed against a panel of ten clinically relevant pathogenic strains of public health importance, namely the following: *Enterobacter cloacae, Klebsiella pneumoniae, Escherichia coli, Staphylococcus aureus* (BLACT), *Staphylococcus epidermidis, Candida albicans, Candida dubliniensis, Candida tropicalis, Candida parapsilosis,* and *Aspergillus niger*. The results of the minimum inhibitory concentrations (MICs) for fennel essential oil ranged from 3.13 to 50 mg/mL across the tested strains, underscoring its antimicrobial potential (Table 7). Notably, the lowest MICs were observed against *Candida albicans* (3.13 mg/mL) and *Aspergillus niger* (6.25 mg/mL), whereas higher concentrations, such as 50 mg/mL, were required to inhibit *Staphylococcus aureus* (BLACT) and *Staphylococcus epidermidis*. These findings suggest that antimicrobial efficacy depends on both the targeted microorganism and on the complex chemical composition of the EO. These results are consistent with those previously reported by Roby et al., 2013 [31], who observed comparable antimicrobial effects against *Escherichia coli, Candida albicans,* and *Staphylococcus aureus*. Furthermore, Anwar et al., 2009 [30] also highlighted the antifungal activity of fennel essential oil against *Aspergillus* species, thereby supporting the findings presented herein. Regarding the extracts, the hydroethanolic extract obtained by Soxhlet extraction (E1) exhibited broad-spectrum activity against all tested bacterial and fungal strains. The lowest MICs were recorded against *Candida parapsilosis* (0.78 mg/mL) and *Enterobacter cloacae* (12.5 mg/mL), followed by moderate activity against *Aspergillus niger* and *Klebsiella pneumoniae* (25 mg/mL). The E0 extract also demonstrated antimicrobial activity, with MICs of 50 mg/mL for most strains, except for *Klebsiella pneumoniae* and *Escherichia coli*, which were found to be resistant. In contrast, extract E2 showed limited activity, being effective only against *Enterobacter cloacae, Staphylococcus aureus* (BLACT), and *Candida albicans*, with an MIC of 50 mg/mL (Table 7).

Table 7. Antimicrobial activity of *F. vulgare* EO and extracts.

Strains	EO (mg/mL)		Extracts (mg/mL)						Gentamicin	Terbinafin
			E (0)		E (1)		E (2)			
	MIC	MBC/MFC	MIC	MBC/MFC	MIC	MBC/MFC	MIC	MBC/MFC	MIC (µg/mL)	MIC (µg/mL)
Enterobacter cloacae	25	25	50	50	12.5	25	50	50	>4	-
Klebsiella pneumoniae	25	50	>50	>50	25	25	>50	>50	<=1	-
Escherichia coli sauvage	25	25	>50	>50	50	50	>50	>50	2	-
Staphylococcus aureus BLACT	50	50	50	50	50	50	50	50	<0.5	-
Staphylococcus epidermidis	50	100	50	50	50	50	>50	>50	2	-
Candida albicans	3.13	6.25	50	50	50	50	50	50	-	12.500
Candida dubliniensis	25	25	50	50	50	50	>50	>50	-	3.125
Candida tropicalis	12.5	25	50	50	12.5	12.5	>50	>50	-	12.500
Candida parapsilosis	25	50	50	50	0.78	0.78	>50	>50	-	6.250
Aspergillus niger	6.25	6.25	50	50	25	25	>50	>50	-	3.125

MIC: minimum inhibitory concentration; MBC: minimum bactericidal concentration; MFC: minimum fungicidal concentration.

Food spoilage and foodborne pathogens can compromise the nutritional quality of foods, causing biochemical changes, weight loss, as well as toxicity that harm human health [56]. EOs, which are volatile, odoriferous compounds from plant metabolism, find several applications in food flavoring and preservation [57]. Recent research has emphasized the antimicrobial potential of monoterpene and sesquiterpene hydrocarbons, along with their oxygenated derivatives, which are the primary constituents of EOs [58–60]. Such findings support our study, as fennel seed EO, which contains these components, demonstrates capacity as a natural antimicrobial agent. Fennel is traditionally well known to treat various infectious diseases caused by bacteria, fungi, viruses, and mycobacteria [61,62]. Several studies have already supported the antimicrobial effects of EOs [63]. The data here showed that the extracts and fennel EO were effective against the tested microorganisms, confirming fennel's antimicrobial and antifungal properties. This consequently suggests its extract could be used to combat multiple antibiotic-resistant bacteria. Such findings are in line with previous studies on fennel EOs results [64]. Several researchers have linked these activities to the presence of terpenes, phenols, aldehydes and ketones as the main components of EOs and, based on the GC-MS results. Our results are confirmatory and attribute the antimicrobial activity obtained in this study to the presence of such compounds [65]. In addition to antimicrobial activities, EO and various fennel extracts have shown good antioxidant activities [65]. Fennel is known as an excellent source of natural antioxidants, owing to its high polyphenol and flavonoid content. Phenolic compounds, such as caffeoylquinic acid and rosmarinic acid, have been proven to have very good antioxidant potential [66,67].

2.9. Molecular Docking

This computational study thoroughly examined the antimicrobial and antioxidant properties of the primary chemical compounds found in fennel. The results, summarized in Table 8, highlighted significant differences between phenolic extracts and essential oil components. The antimicrobial potential of fennel compounds was tested against the following ten critical microbial proteins: 7TI1, 3RAE, 4DUH, 2W9S, 1JIJ, 3KP5, 7RJB, 5V5Z, 4YBF, and 4ZA5. Among the phenolic extracts, isochlorogenic acid A demonstrated outstanding affinity, especially with 3KP5 (−10.3 kcal/mol), followed by salvianolic acid K and quercetin-3-D-xyloside, which also exhibited strong interactions with 3KP5 (−9.5 kcal/mol) and 3RAE (−9.9 kcal/mol), respectively. These findings underscore the remarkable antimicrobial potential of phenolic compounds, particularly against targets like 3KP5 and 3RAE. In contrast, monoterpenes from the essential oil, such as trans-anethole, limonene, and fenchone, demonstrated weaker affinities, with binding scores ranging between −4.2 and −6.9 kcal/mol. Chlorogenic acid, while showing moderate interactions with 3RAE (−9 kcal/mol) and 3KP5 (−8.5 kcal/mol), was less effective compared to other phenolic compounds. Rosmanol displayed strong selectivity, with a high affinity for 3RAE (−9.9 kcal/mol) but a weaker interaction with 3KP5 (−6.8 kcal/mol). Lastly, quercetin-3-glucuronide exhibited similar activity to quercetin-3-D-xyloside, though slightly less effective, particularly against 3RAE and 7RJB. These results further highlight the prominent role of flavonoids in the antimicrobial activity of fennel.

The antioxidant activities were evaluated against the following five oxidative targets: 5QJ2, 3NRZ, 1OG5, 1N8Q, and 2CDU. Salvianolic acid K showed a particularly strong affinity for 3NRZ ($\Delta G = -10.9$ kcal/mol) and 1OG5 ($\Delta G = -10.0$ kcal/mol), suggesting an exceptional ability to interact with proteins involved in oxidative stress pathways. Apigenin and rosmanol also demonstrated remarkable performance, with scores below −9 kcal/mol for several targets, thereby confirming their crucial role in antioxidant activity. Additionally, mediorsinol showed a marked selectivity for 5QJ2 ($\Delta G = -9.0$ kcal/mol), suggesting a specific mechanism of action. In contrast, the terpenic compounds present in the essential oil, such as fenchone, limonene, and trans-anethole, exhibited significantly lower affinities ($\Delta G \geq -6.5$ kcal/mol), indicating that they contribute secondarily to the overall antioxidant activity of fennel. These results highlight that phenolic extracts, particularly salvianolic acid K and apigenin, play a key role in neutralizing free radicals and modulating redox pathways, while the terpenes in the essential oil, though effective, exert a more moderate antioxidant effect.

Table 8. Docking scores of EO and extract compounds from *F. vulgare* against various target proteins (binding energies in kcal/mol).

	Molecules\Proteins	Antimicrobial Activities											Antioxidant Activities			
		7TI1	**3RAE**	**4DUH**	**2W9S**	**1JIJ**	**3KP5**	**7RJB**	**5V5Z**	**4YBF**	**4ZA5**	**5qj2**	**3nrz**	**1og5**	**1n8q**	**2cdu**
Extracts	3-O-Caffeoylshikimic acid	−7.4	−8.4	−6.4	−7.9	−7.4	−8.5	−8.5	−7.6	−7.1	−7.9	−8.4	−8.6	−8.7	−8.8	−8.3
	7-Methoxy-2-methylisoflavone	−6.9	−7.9	−6	−7	−6.9	−7.7	−9.5	−7.3	−6.9	−8.2	−7.9	−8.1	−8.5	−8.6	−8.2
	Apigenin	−7	−8.4	−6.6	−8.1	−7	−8.2	−8.6	−7.2	−7	−8.4	−8.2	−9.1	−8.8	−9.1	−7.9
	Butyl ferulate	−6.2	−6.4	−4.8	−6.1	−6.2	−6.4	−7.1	−6.1	−5.9	−6.8	−6.1	−7.2	−6.8	−7.3	−7.5
	Chlorogenic acid	−7.2	−9	−6.5	−7.9	−7.2	−8.5	−8.5	−7.6	−7.2	−8.4	−8	−8.9	−8.8	−8.7	−8.2
	Dihydrocaffeic acid	−5.9	−6.2	−5.1	−6.6	−5.9	−6.3	−6.3	−5.8	−5.7	−6.3	−6	−7.2	−6.3	−6.6	−6.4
	Embelic acid	−5.5	−6.4	−4.6	−6.3	−5.5	−7	−6.9	−6.3	−5.4	−6.7	−6.3	−7.3	−6.9	−7.6	−7.3
	Isochlorogenic acid A	−8.1	−9.3	−8.2	−8.6	−8.1	−10.3	−9.7	−8.5	−8.6	−9.2	−8.7	−9.9	−9.1	−9	−8.3
	Medioresinol	−7.2	−8.6	−6.5	−7.9	−7.2	−8.9	−8.3	−7.6	−7.2	−7.9	−9	−7.8	−7.8	−8.4	−8.7
	Quercetin-3-D-xyloside	−7.6	−9.9	−7.1	−8	−7.6	−8.5	−9.1	−7.6	−7.7	−8.5	−9.2	−8.1	−8.8	−9.8	−8.1
	Quercetin-3-glucuronide	−7.6	−9.3	−7.1	−8	−7.6	−8.5	−9.1	−7.6	−7.7	−8.6	−9.1	−8.1	−8.8	−9.1	−8.1
	Rosmanol	−7.8	−9.9	−7.2	−7.5	−7.8	−6.8	−8.3	−7.7	−7.8	−8.2	−9.1	−7.9	−9.4	−8.4	−8.6
	Salvianolic acid K	−8	−9.4	−7.3	−9	−8	−9.5	−8.7	−8.6	−8	−9.5	−9	−10.9	−10	−9.6	−7.9
EO	Fenchone	−5.6	−5.1	−4.8	−5.5	−5.6	−6.5	−5.6	−5.7	−5.4	−6	−5.4	−5.7	−5.9	−6.1	−6.3
	Trans-anethole	−5.7	−5.1	−4.3	−5.2	−5.7	−5.7	−6.8	−5.3	−5	−5.6	−5	−6.5	−6	−6.1	−6.4
	Limonene	−5.8	−5	−4.3	−5	−5.8	−6.1	−6.9	−5.5	−5.1	−5.8	−5.4	−6.5	−5.8	−6	−6.5
	Cis-anethole	−5.7	−5	−4.2	−5.2	−5.7	−6.1	−6.6	−5.7	−4.9	−5.5	−5	−6.3	−5.8	−5.9	−6.5
	Estragole	−5.8	−5.2	−4.2	−5.2	−5.8	−5.9	−6.6	−5.4	−4.8	−5.6	−5.2	−6.2	−5.7	−6	−6.1

2.9.1. Interaction with Antibacterial Proteins

This research identifies specific molecular interactions between several phenolic compounds and two main bacterial targets as follows: DNA topoisomerase 4 (3RAE) and the transcriptional regulator TcaR (3KP5), as shown in Table 9. Chlorogenic acid and iso-chlorogenic acid A form hydrogen bonds with topoisomerase residues DG H:1 and DT F:7, while Quercetin-3-D-xyloside creates two hydrogen bonds (ARG A:28, ASP C:510) along with a hydrophobic interaction (HIS A:76). Quercetin-3-glucuronide interacts with DT F:7 and SER A:80, and both Rosmanol and Salvianolic acid K bind through hydrophobic interactions (ALA A:115/PRO A:113 and ALA A:29, respectively), impairing bacterial replication. Additionally, these compounds influence TcaR: chlorogenic and iso-chlorogenic acids via hydrophobic interactions (ALA B:24/ALA B:38), quercetins through hydrogen bonds (ASN B:20), and Rosmanol/Salvianolic acid K through hydrophobic interactions (ALA B:38 and ILE A:9/LEU A:12/VAL B:19/MET B:114/ALA B:118), which may alter bacterial gene expression. These findings highlight a dual action mechanism: enzymatic inhibition and transcriptional modulation, demonstrating the potential of these compounds as broad-spectrum antimicrobial agents.

2.9.2. Interaction with Antioxidant Proteins

The study reveals specific molecular interactions between phenolic compounds and two key oxidative stress enzymes, as described in Table 10. Regarding xanthine oxidase (3NRZ), apigenin forms hydrogen bonds with ARG L:606 and ARG J:32, as well as π-Alkyl interactions with LEU J:41 and PRO L:675, while chlorogenic acid and iso-chlorogenic acid A establish hydrophobic interactions with VAL C:591/VAL L:591 and ARG J:37, respectively. Rosmanol combines a hydrogen bond (THR A:24) and a π-Anion interaction (ASP A:21), whereas salvianolic acid K interacts via VAL L:591 and PRO L:597. For cytochrome P450 2C9 (1OG5), apigenin shows hydrogen bonds (SER A:343, LYS B:421), a π-Anion interaction (ASP A:349), and a π-Alkyl bond (LYS B:420). Chlorogenic and iso-chlorogenic acids present Alkyl interactions with LYS B:423 and ALA A:439, rosmanol displays π-T-shaped contacts (PHE A:419) and π-Alkyl interactions (LYS A:421), and salvianolic acid K forms a π-Alkyl bond with LYS B:420. These findings demonstrate a dual mechanism of xanthine oxidase inhibition and cytochrome P450 2C9 modulation, suggesting that these compounds could act as natural antioxidants by reducing free radical production and neutralizing reactive oxygen species, thus offering promising prospects for the prevention of oxidative stress-related diseases.

The antioxidant mechanism of *F. vulgare* extracts and essential oil primarily relies on their capacity to neutralize free radicals and to reduce transition metals, thereby preventing oxidative chain reactions. The predominant phenolic compounds found in the extracts, such as chlorogenic acid, quercetin-3-glucuronide, and dihydrocaffeic acid, have been identified as key contributors to this mechanism. These molecules act as electron or hydrogen atom donors, enabling them to stabilize free radicals, such as DPPH$^{\cdot}$ and to convert ferric ions (Fe^{3+}) into ferrous ions (Fe^{2+}). Moreover, the significant correlation observed between the concentration of polyphenols and the outcomes of the DPPH, FRAP, and TAC assays underscores the pivotal role of these compounds in the antioxidant potential. This observation is consistent with the findings of Panda et al. (2019) [68], who demonstrated that the chemical modification of biomolecules like chitin by phenolic acids enhances both their water solubility and antioxidant activity. Additionally, oxygenated monoterpenes, such as fenchone and trans-anethole, although contributing to a lesser extent, further augment the overall effect by enhancing the scavenging of reactive oxygen species (ROS). These combined mechanisms highlight the potential of *F. vulgare* as a natural source of antioxidants capable of counteracting oxidative stress.

Table 9. Two-dimensional and three-dimensional interactions of *F. vulgare* compounds with target proteins associated with antimicrobial activities.

Molecules\Proteins	3RAE		3KP5	
	2D	3D	2D	3D
Chlorogenic acid		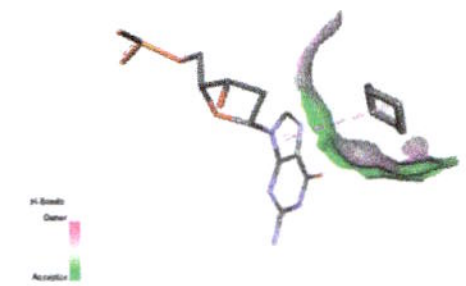	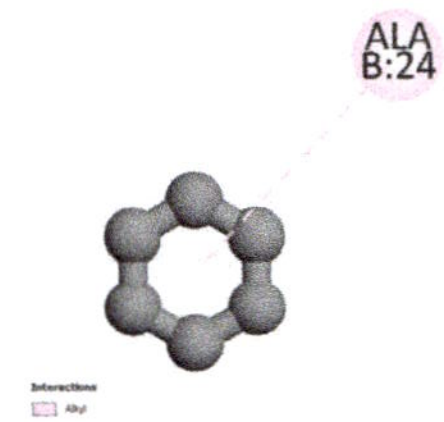	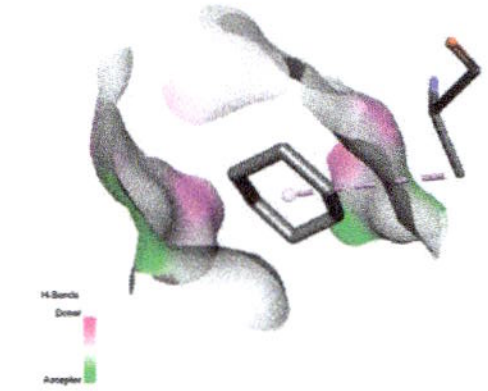
Isochlorogenic acid A	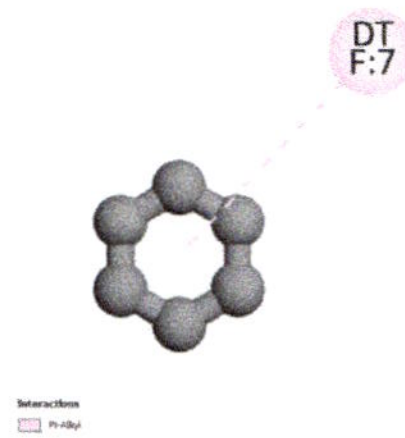	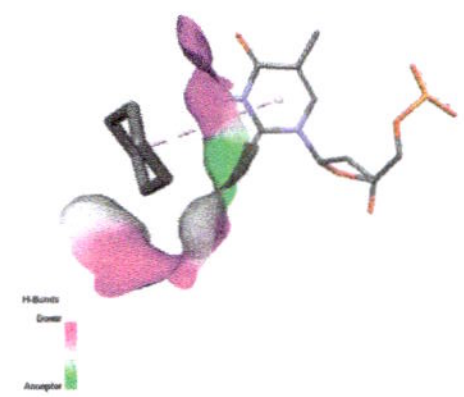	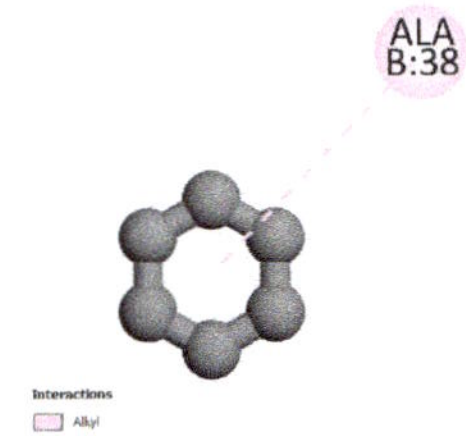	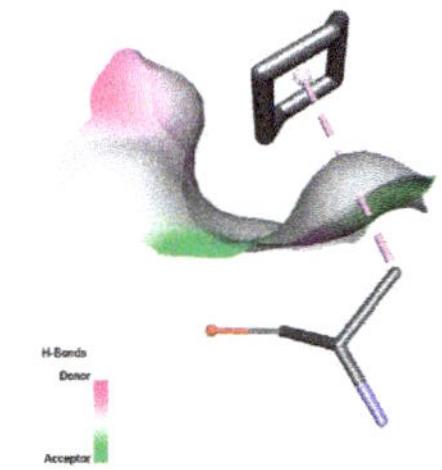
Quercetin-3-D-xyloside	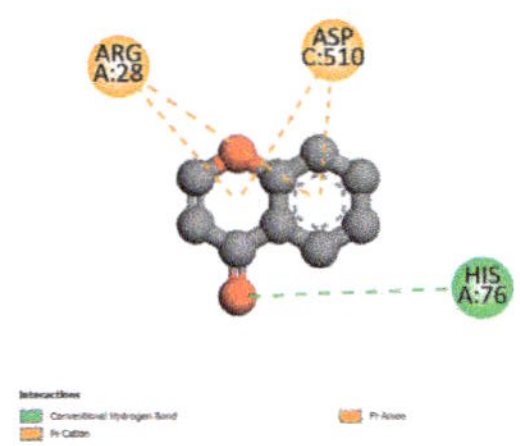	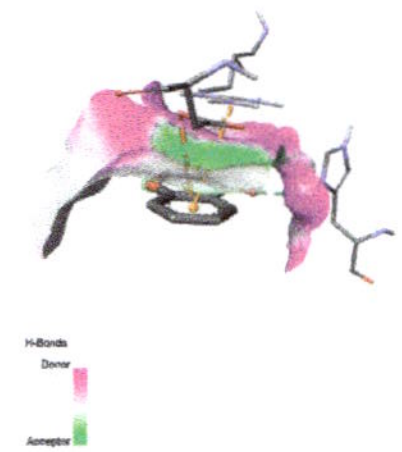	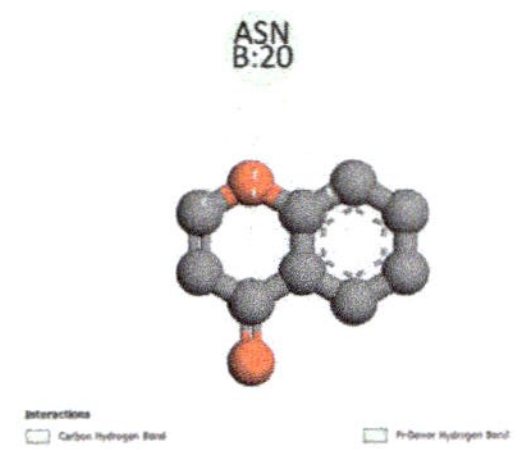	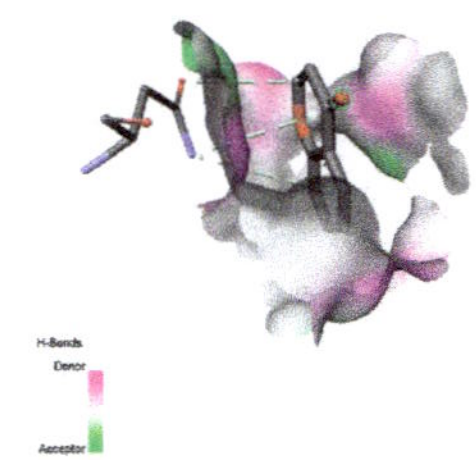

Table 9. *Cont.*

Molecules\Proteins	3RAE		3KP5	
	2D	3D	2D	3D
Quercetin-3-glucuronide	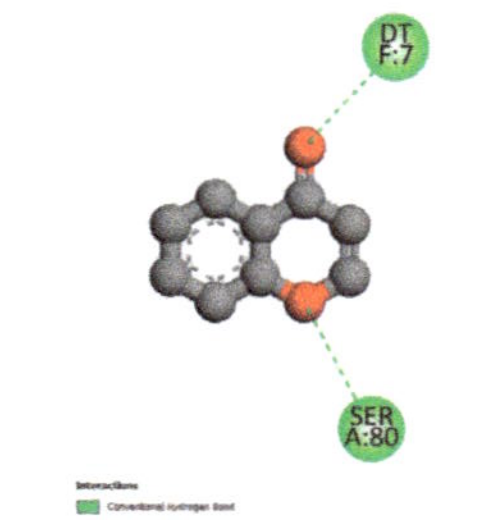	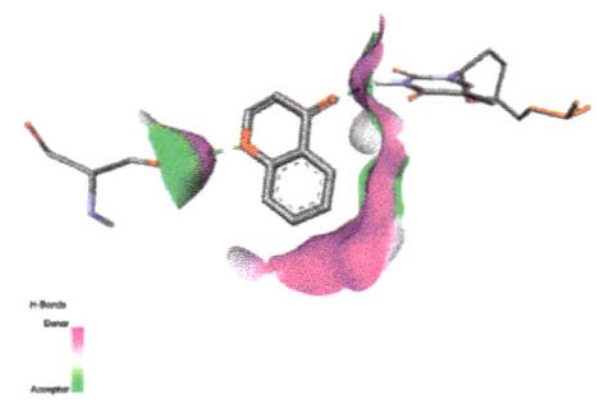		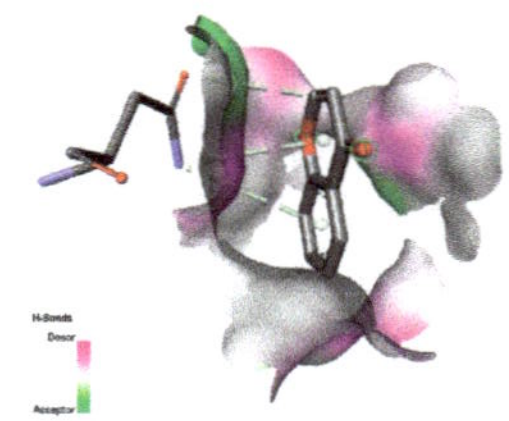
Rosmanol	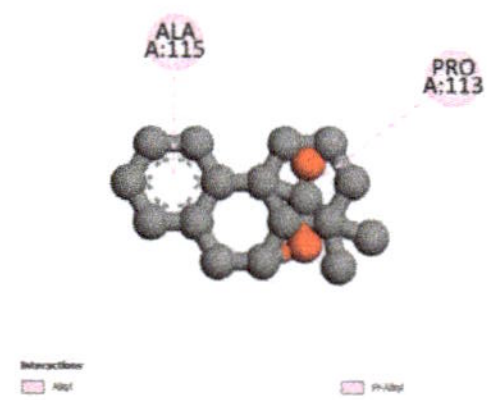	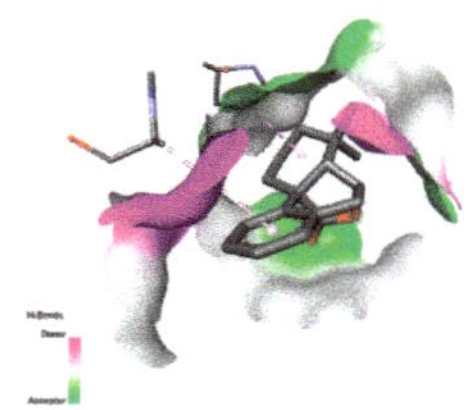	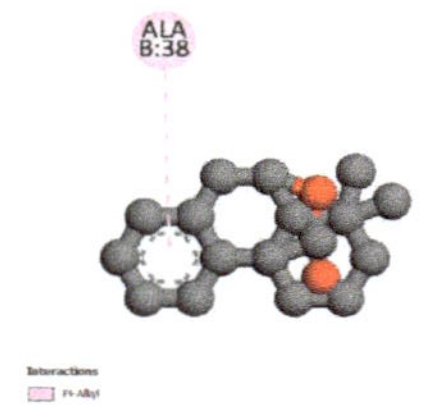	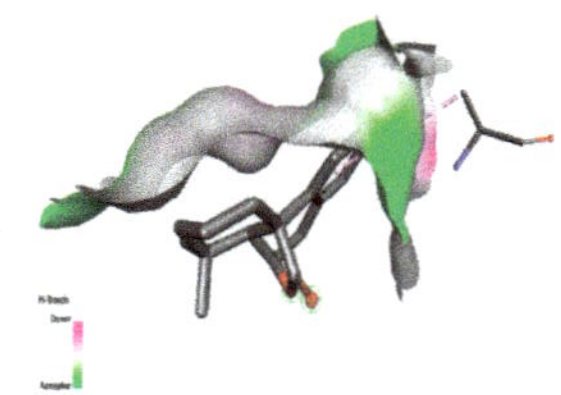
Salvianolic acid K	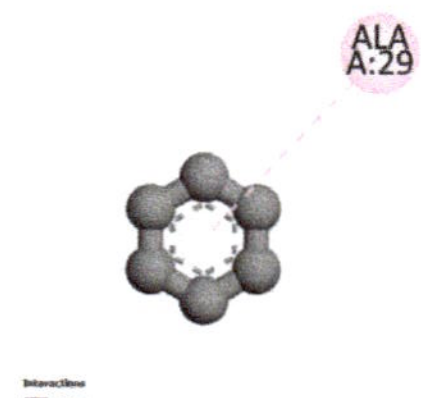	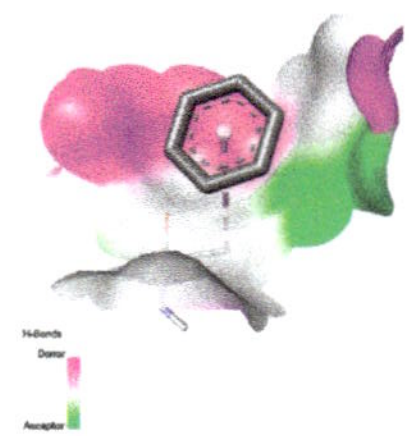	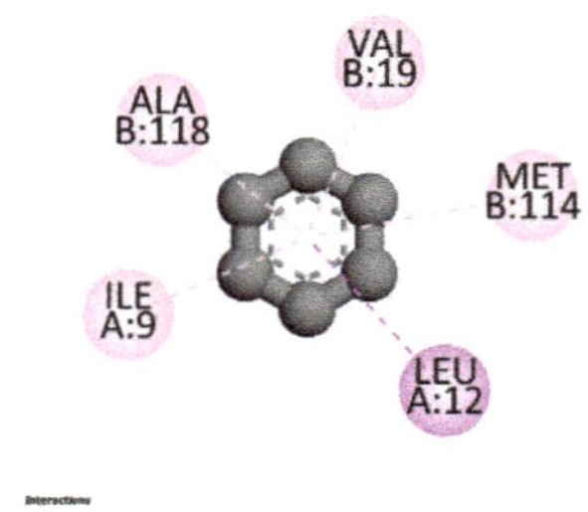	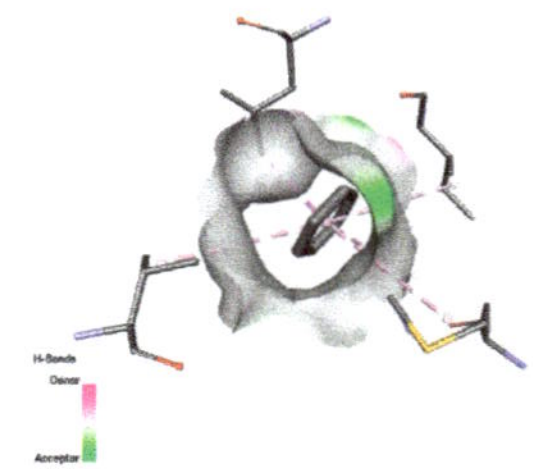

Table 10. Two-dimensional and three-dimensional interactions of *F. vulgare* compounds with target proteins involved in antioxidant activities.

Molecules\Proteins	3NRZ		1OG5	
	2D	3D	2D	3D
Apigenin	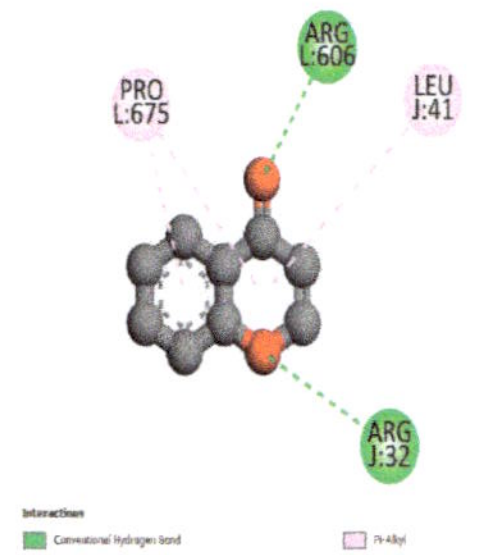	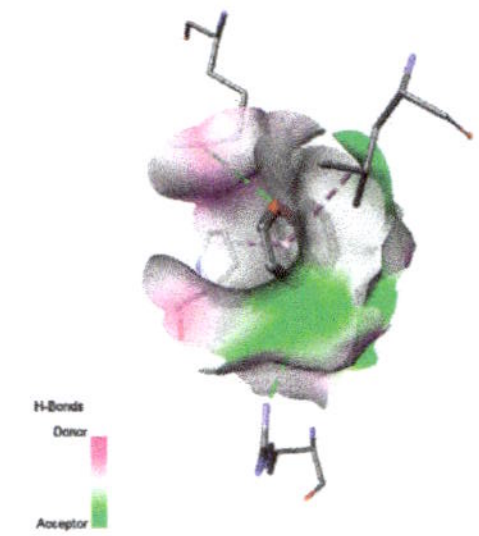	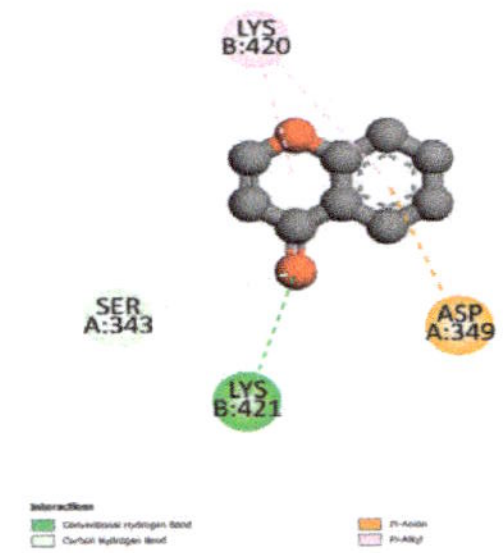	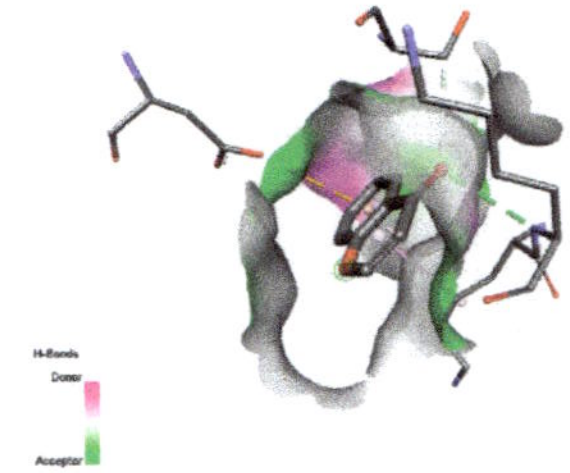
Chlorogenic acid		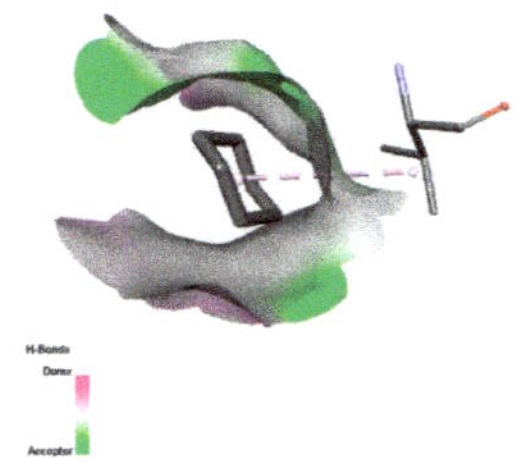		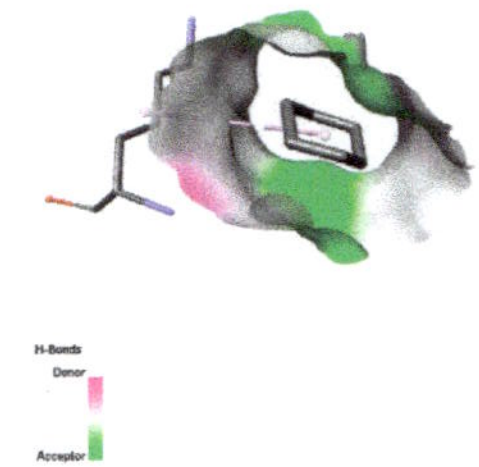
Isochlorogenic acid A	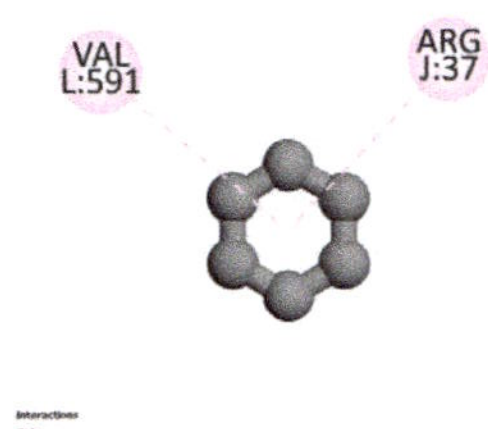	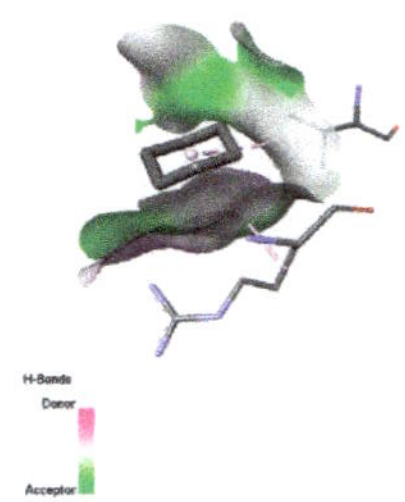	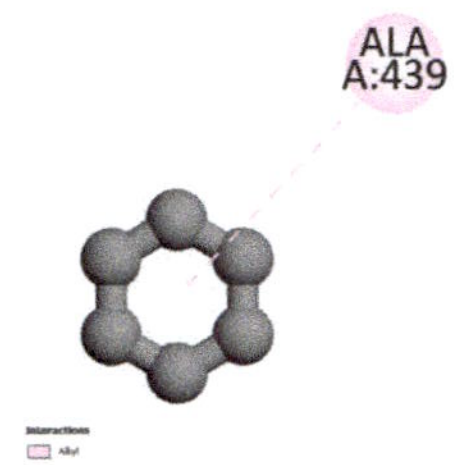	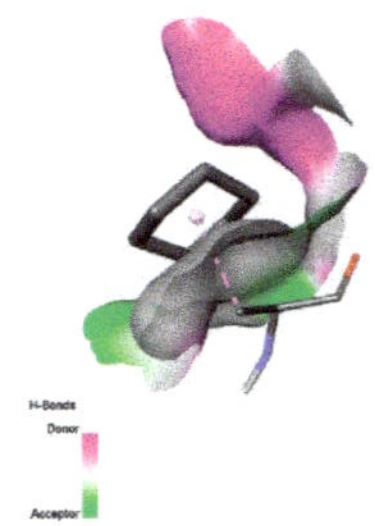

Table 10. *Cont.*

Molecules\Proteins	3NRZ		1OG5	
	2D	3D	2D	3D
Rosmanol	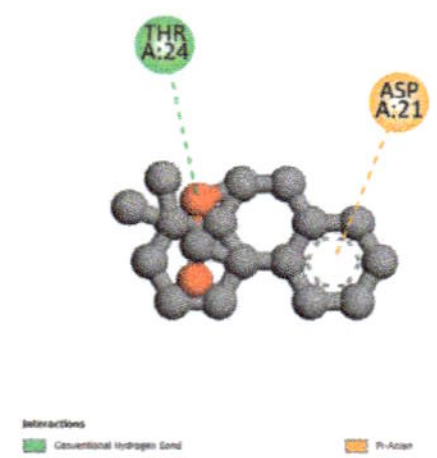	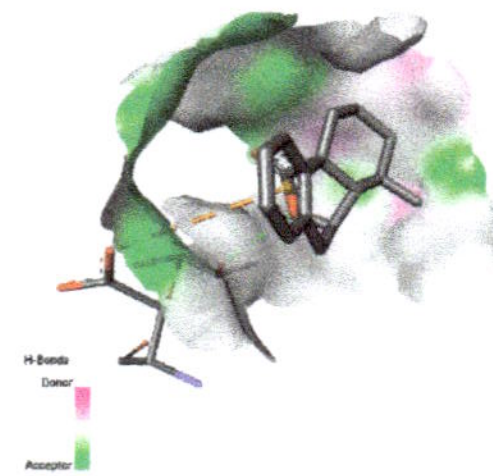	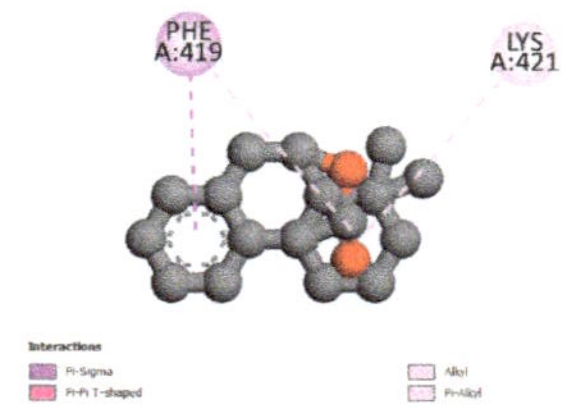	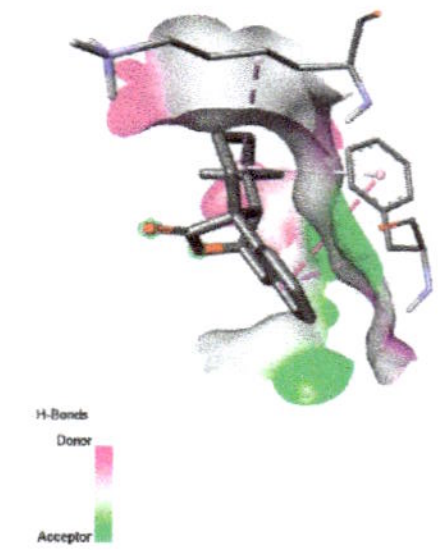
Salvianolic acid K	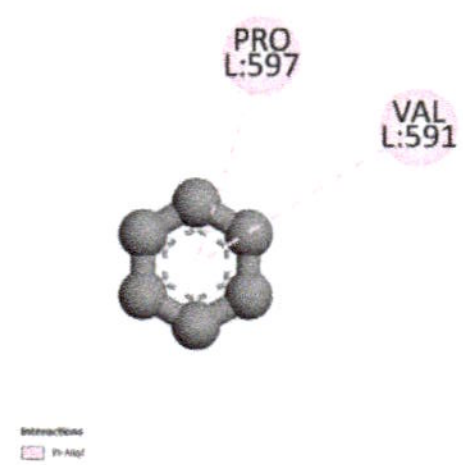	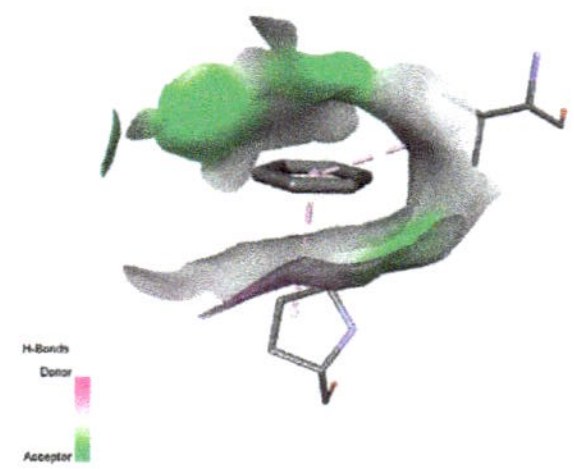		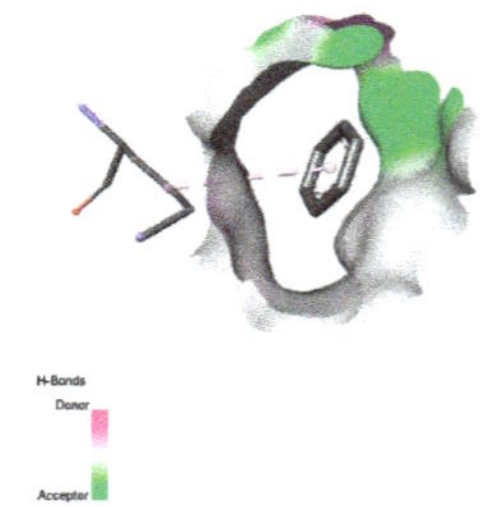

The comparative analysis reveals that phenolic extracts exhibit significantly higher antimicrobial and antioxidant activity than essential oils. This difference can be explained by the presence of several hydroxyl groups in polyphenols, which facilitate hydrogen and π-π interactions with the active sites of proteins, thereby enhancing their biological activity, as demonstrated by the research of Zhou et al. (2020) [69] and Chen et al. (2024) [70]. More specifically, the remarkable results observed for iso-chlorogenic acid A and salvianolic acid K, both derivatives of caffeic acid, highlight the importance of catechol motifs in modulating these biological activities, which has been confirmed by the work of Razaviamri et al. (2021) [71] and Dangles et al. (2012) [72]. In comparison, phenolic extracts outperform the monoterpenes found in essential oils, likely because they interact more effectively with target proteins.

However, although these results are promising and informative, further experimental validations are needed to confirm their effectiveness in real biological systems. This study paves the way for the therapeutic use of fennel, suggesting that the combination of phenolic extracts and essential oils in targeted formulations could lead to more effective therapeutic agents, a concept that aligns with the similar conclusions drawn by the research of Hosseini et al. (2021) [73].

The study of the antioxidant and antimicrobial activities of essential oils and fennel extracts reveals the richness of this plant in bioactive compounds that work synergistically to combat free radicals and pathogens. In vitro tests demonstrated that both the essential oil and extracts exhibit significant antioxidant capacities. Among the identified phenolic compounds, chlorogenic acid, quercetin-3-glucuronide, and dihydrocaffeic acid were found to be particularly effective in neutralizing free radicals, with IC50 values of 51.45 µg/mL for the essential oil and 93.71 µg/mL for the hydroethanolic extract. The antioxidant efficacy of the essential oil surpassed that of the aqueous extracts in the DPPH tests, confirming the superior potential of this extract form to combat oxidative stress, as noted in other similar studies [30,31].

Fennel extracts demonstrated notable antimicrobial activity, with particularly strong effectiveness against fungal pathogens, such as *Candida albicans* and *Aspergillus niger*, exhibiting minimum inhibitory concentrations (MIC) of 3.13 mg/mL and 6.25 mg/mL, respectively. Antimicrobial activity was also observed against bacteria, such as *Escherichia coli* and *Staphylococcus aureus*, highlighting a specificity of action depending on the type of microorganism. The in silico analyses further enriched these findings by identifying molecular mechanisms involving bioactive compounds, such as chlorogenic acid and quercetin-3-glucuronide. These compounds demonstrated strong affinities for antimicrobial target proteins, playing a central role in the observed activity. These findings open promising avenues for the development of natural therapeutic formulations and the exploration of synergies between active molecules.

Finally, although the main monoterpenes in the essential oil, such as fenchone (24.72%), trans-anethole (22.22%), and limonene (20.48%), play an important role, they contribute less to the antimicrobial activity than the phenolic extracts. These oxygenated monoterpenes, while displaying moderate antimicrobial action, enhance the overall effectiveness of the essential oil by complementing other components, in line with the findings of Alrub et al. (2023) [74]. This underscores the importance of combining different types of compounds found in fennel, where phenolic extracts dominate, but the monoterpenes also make a significant contribution to the overall biological activity.

3. Materials and Methods

3.1. Plant Material

Foeniculum vulgare, a member of the Apiaceae family, is a perennial herbaceous plant characterized by its yellow flowers and feathery, delicate foliage (Figure 8). The plant samples were collected from the Meknes region (Table 11). A summary of the species' key identification criteria is provided in Table 12. The seeds were carefully dried in the shade and ground into a fine powder, which was subsequently used for the preparation of various extracts.

Figure 8. Morphological aspects of *Foeniculum vulgare* Mill.

Table 11. Characteristics of the studied plant.

Plant Species	Vernacular	Harvest Site		Parts Used	Latitude (x)	Longitude (y)	Altitude (m)	Harvest Year
		Region	Locality					
F. vulgare Mill	Fenouil	Meknès	Ain jerry	Seeds	5°49′55″ O	33°85′86″ N	546	2024

Table 12. Key determination of the species.

Reign	Plantae
Kingdom	Fenouil
Class	Equisetopsida
Order	Apiales
Family	Apiaceae
Genus	*Foeniculum*
Species	*Foeniculum vulgare*

3.2. Quality Control of Plant Material

3.2.1. Moisture Content (MC)

The water content of the plant material was determined as previously assayed by Saidi et al. (2023) [75]. In this method, 5 g of the plant was maintained in the oven at 105 °C for 24 h. The humidity level is calculated by the following Equation (1):

$$MC\ (\%) = \frac{(m_1 - m_2)}{m_1} \times 100 \quad (1)$$

m_1: initial mass of the plant before drying in the oven (g),
m_2: final mass of the plant after drying in the oven (g).

3.2.2. Determination of pH

The pH is determined after adding 10 mL of hot distilled water to 2 g of the plant. The STPURE double electrode was immersed in the filtrate of the mixture to note the pH value on an Ohaus Starter 3100 pH meter (Remok et al., 2023) [76].

3.2.3. Determination of Titratable Acidity

The assay of titratable acidity, expressed as citric acid content per unit volume, is determined by titrimetry using a 0.01 N sodium hydroxide solution, in the presence of phenolphthalein as a colored indicator. Ten grams of plant powder was added to 50 mL of boiling distilled water, then the solution was stirred for 15 min. The mixture was adjusted to 100 mL with distilled water. After filtration, 10 mL of the filtrate to which 20 mL of distilled water were added, and were titrated with a solution of NaOH (0.01 N) with a few drops of phenolphthalein, until the color changed to pink. The noted titration volume was converted into citric acid equivalent following the Equation (2) below (Bergeron, 1995) [77]:

$$\text{Total Acidity} = \frac{\text{Dilution factor} \times \text{Weight of eq. Acide} \times \text{normality of NaOH} \times \text{titration vol. (mL)}}{Sample\ mass\ (g)} \quad (2)$$

3.2.4. Ash Content

The ash content was produced by incineration, according to the standard (1977). 2 g of the crushed plant material were placed in nickel crucibles, then puted into the muffle furnace at a temperature of 550 °C until all carbon particles were completely destroyed. The organic matter content is calculated by the following Formula (3):

$$OM\% = \frac{m_1 - m_2}{TS} \times 100 \quad (3)$$

Bulleted lists look like this:
OM%: Organic matter;
m_1: Pre-calcination capsule and sample mass;
m_2: Post-calcination capsule and sample mass;
TS: Test sample.
The ash content was calculated as follows (4):

$$\text{Ash\%} = 100 - \text{MO\%} \quad (4)$$

3.2.5. Dosage of Metallic Trace Elements (MTE) by ICP-AES

The trace metals analysis (As, Cr, Sb, Pb, Cd, Fe, Cu and Ti) were carried out using the standardized mineralization protocol (AFNOR, 1985), using aqua regia reagent (HNO_3 + 3HCl). One-tenth gram of plant material was mixed with 3 mL of aqua regia prepared from 1 mL of nitric acid HNO_3 (99%) and 2 mL of hydrochloric acid HCl (37%). This mixture was placed in a reflux assembly at 200 °C for 2 h. After decantation, the supernatant was filtered through a 0.45 µm membrane. The filtrate was made up to 15 mL with distilled water. The concentrations of heavy metals were determined through ICP-AES analysis (Ultima 2 Jobin Yvon) at the UATRS laboratory (Technical Support Unit for Scientific Research) of the CNRST in Rabat [78].

3.3. Phytochemical Screening

The characterization tests for the different chemical groups are based on coloring, complexation and precipitation reactions, according to the methods described by Dohou et al., Judith, Mezzoug, et al., Bekro et al., Bruneton, and N'Guessan et al. [79–84].

3.3.1. Primary Metabolites

The determination of primary metabolites in plants is crucial for understanding their physiological and biochemical processes. The extract's interaction with iodized water allowed for the identification of the polysaccharide's presence and type, while the Fehling method was applied to identify the reduced sugars. Two methods have been used to characterize the proteins: the reaction with biuret, which produces a colored complex (purple or mauve) when a few drops of copper sulfate are added to a base environment, and the reaction with xanthoproteins, which reveals the presence of specific aminated acids by heating nitric acid upon contact with the solution to be analyzed. The process of lipid detection involves adding an acid anhydride in an acidic environment to the extract to be analyzed, which turns red.

3.3.2. Secondary Metabolites

Determining the secondary metabolites of plants is essential to comprehending their ecological, medicinal, and industrial functions. Several analytical techniques are used to identify and quantify these compounds. The precipitation of salts following the use of Mayer and Dargendorff's reagent made it possible to highlight the presence of alkaloids. The tannins, or gallic tannins, have been identified using the Stiasny reaction, sodium acetate, and ferric chloride, while concentrated hydrochloric acid and isoamylic alcohol have been found to contain catechin tannins. The cyanidine reaction also revealed the leuco-anthocyanes, but without the addition of magnesium copeaux that were used to highlight the free flavonoids. Ten percent sulfuric acid and 25% NH_4OH were added, which allowed the anthocyanes to be detected. A 25% dilution of ammoniac was used to highlight the anthracene derivatives. By using the right chemical products and potassium hydroxyde, cardiac glycosides have been discovered. The addition of strong sulfuric acid revealed the presence of sterols and triterpenes. Analyzing each sample's mousse index revealed the presence of saponosides, which are distinguished by their ability to foam in aqueous solutions. One hundred percent ethanol was added to strengthen the aqueous extraction. After that, the watery extraction is enhanced with strong sulfuric acid and an ethanol solution infused with thymol to extract the oses and holosides.

3.4. *Preparation of Seed Extracts of F. vulgare*

3.4.1. Extraction by Soxhlet

A sample of 30 g of powdered fennel seeds was placed in a cartridge inside the soxhlet extraction chamber. The plant material is extracted with 350 mL of solvent consisting of either a mixture of ethanol/water (70/30) E (1) or water alone E (0) (Table 13). Several cycles are necessary to exhaust the plant material. After filtration, the solvent was evaporated to dryness under reduced pressure at 50 °C using a rotary evaporator. Then the extracts were stored at 4 °C until they are used for further analyses [85].

Table 13. Coding of extracts.

Extraction Method	Solvants	Codification
Decoction	Water	E (2)
Soxhlet	Ethanol/Water (70/30) Water	E (1) E (0)

3.4.2. Extraction by Decoction

The preparation of the aqueous extract (E (2)) was carried out by the seeds of the plant dried and pulverized into a fine powder (Table 13). Then, 30 g of the crushed plant were added to 350 mL of distilled water, then this mixture was introduced into a reflux assembly, and heated to 80 °C with stirring for one hour, then filtered. The extract was dried in an oven at 70 °C for 18 h then placed in amber glass vials [85].

3.5. Dosage of Phenolic Compounds

3.5.1. Determination of Total Polyphenols

The contents of total phenolic compounds were determined by the Folin–Ciocalteu colorimetric method described by Singleton and Rossi [86]. The reaction mixture is composed of 20 µL of plant extract studied, 1.5 mL of a sodium carbonate solution (75 g/L) and 1.5 mL of Folin–Ciocalteu reagent (phosphomolybdic $H_3PMO_{12}O_{40}$) at 10% (*V*/*V*). Then, the tubes were incubated at room temperature and protected from light for 2 h. The absorbance was measured at 760 nm and the results are expressed in milligrams of gallic acid equivalent per gram of dry matter (mg EAG/g of plant).

3.5.2. Dosage of Flavonoids

The determination of the total flavonoid content of plant extracts was carried out by the aluminum chloride ($AlCl_3$) method. Indeed, 2 mL of distilled water and 10 µL of aluminum chloride prepared in methanol (10% m/V) were added. The mixture was supplemented with absolute methanol up to a total volume of 5 mL. The solutions were homogenized and then left for 2 h in the dark. The absorbance was determined at 433 nm. Flavonoid contents were expressed as quercetin equivalents (QE) [87,88].

3.5.3. Dosage of Condensed Tannins

The determination of condensed tannins was carried out spectrophotometrically with vanillin in an acidic methanolic medium according to Price et al. (1978) [89]. The reaction mixture was composed of one volume of extract, 3 mL of methanolic vanillin solution (4%, m/V) and 1.5 mL of HCl (37%). This mixture was shaken then incubated at room temperature and in the dark for 20 min. The absorbance was measured at 499 nm and the concentration of condensed tannins was expressed as milligram equivalent to catechin per gram of extract (mg CE/g).

3.6. Identification of Chemical Composition by HPLC/UV-DAD

The analysis of the chemical composition of the extracts of the plants studied was carried out by HPLC-UV-DAD (Thermofisher Scientific Ultimate 3000, Sunnyvale, CA, USA). A volume of 10 µL of each extract was dissolved in distilled water to obtain a concentration of 100 µg/mL, filtered on PTFE syringe filters with 0.20 µm pore size (Interchim®, Montluçon, France), then injected into a C18 column 100 mm long, 2.1 mm in diameter and whose pores have a diameter of 1.7 µm. The temperature was set at 30 °C and the flow rate at 0.45 mL/min. The mobile phase was composed of two solvents: solvent A (Water + formic acid (0.1%), *v*/*v*) and solvent B (Acetonitrile + formic acid (0.1%), *v*/*v*). The elution gradient established was A + B [98:2] (0–19 min), A + B [70:30] (20–24 min), A + B [5:95] (25 min) and A + B [98:2] (26–30 min). Detection was carried out using a diode array detector at wavelengths of 280 nm, 320 nm and 360 nm. The standards used are chlorogenic acid, quinic acid, quercetin glucuronide, rosmarinate methyl, Syringic acid hexoside, luteolin, Kaempferol, diosmetin, Ascorbyl monomyristate pinoresinol, Galloylquinic acid, catechin obtained from Sigma-Aldrich®, Allentown, PA, USA, cinnamic acid (Rhône-Poulenc, Lyon, France), rosmarinic acid (Extrasynthesis, Genay, France), gallic acid (Prolabo, Paris, France), protocatechuic acid (Koch-Light Laboratories LTD, London, UK), apigenin (Carl Roth, Karlsruhe, Germany), coumarin (HPLC grade, Sigma-Aldrich, Behringer, Germany), luteolin, myricetin, caffeic acid and ferulic acid (Sigma, USA), apigetrin, vanillic acid, naringenin7-O-glucoside, and 3-Feruloylquinic acid (Merck, Darmstadt, Germany). The compounds were characterized solely by matching the retention times and UV spectra of the obtained peaks [90].

3.7. Extraction and Determination of Essential Oil Yield

The dried fennel seeds were subjected to hydrodistillation for 3 h using a Clevenger type apparatus. The essential oil obtained was dried over anhydrous sodium sulfate, filtered and stored at −4 °C until tested and analyzed. The essential oil (EO) yield is expressed in mL/100 g of dry matter [91].

3.7.1. Density

The density of an EO at 20 °C is the ratio between the density of this oil and the density of water at the same temperature. It was determined according to the following formula [92].

$$d_{20} = \frac{m_2 - m_0}{m_1 - m_0} \tag{5}$$

With:
m_0(g): Mass of the empty pycnometer;
m_1(g): Mass of the pycnometer filled with water;
m_2(g): Mass of the pycnometer filled with oil.

3.7.2. Analysis of Essential Oil by Gas Chromatography Mass Spectrometry

The EO were analyzed chromatographically using a Thermo Electron Trace GC Ultra gas chromatograph in conjunction with a Thermo Electron Trace MS mass spectrometer (Thermo Electron: Trace GC Ultra; Polaris Q MS, Rodano, Italy). Fragmentation occurred via electron impact at an intensity of 70 eV. The chromatograph utilized a DB-5 column (5% phenyl-methyl-siloxane, 30 m × 0.25 mm × 0.25 µm film thickness) in conjunction with a flame ionization detector (FID) operated by a H_2/Air gas mixture. The column temperature was set to rise at a rate of 4 °C per minute, beginning at 50 °C and reaching 200 °C, followed by a 5-min hold period. The injection mode was configured to split, with a split ratio of 1:70 and a flow rate of 1 mL/min. Nitrogen was utilized as the carrier gas, also at a flow rate of 1 mL/min. The essential oils' chemical composition was determined by comparing their calculated Kovats indices (IK) with those documented by Adams and other reference products in the literature [93–95]. Furthermore, retention indices and mass spectra were compared with data from the National Institute of Standards and Technology (NIST) mass spectral libraries. Experimental retention indices were compared with those found in the NIST online database (https://webbook.nist.gov/chemistry/name-ser/, accessed on 23 December 2024). The individual component proportions were automatically calculated from the total ion count identified by the GC-MS and presented as percentage compositions.

3.8. Antimicrobial Activity

3.8.1. Microbial Material

To determine the antimicrobial activity of the aqueous extract of the plants studied, we chose five bacterial strains common in human pathologies, belonging to Gram-positive and Gram-negative bacteria and five fungal strains (yeast and fungi) (Table 14).

Table 14. List of bacterial strains tested with their references.

Bacterial Strains	References	Fungal Strains	References
Enterobacter cloacae	02EV317	*Candida albicans*	Ca
Klebsiella pneumoniae	3DT1823	*Candida dubliniensis*	Cd
Escherichia coli sauvage	3DT1938	*Candida tropicalis*	Ct
Staphylococcus aureus BLACT	4IH2510	*Candida parapsilosis*	Cpa
Staphylococcus epidermidis	5994	*Aspergillus niger*	AspN

3.8.2. Determination of Minimum Inhibitory Concentration, Minimum Bactericidal Concentration, and Minimum Fungicidal Concentration

The Minimum Inhibitory Concentration (MIC) is defined as the lowest concentration of an essential oil (EO) or extract that effectively prevents the growth of a microorganism. The MIC was assessed utilizing the microdilution method [96]. A stock solution of the essential oil, formulated in 10% DMSO, was serially diluted to obtain concentrations from 5 to 0.93×10^{-2} mg/mL for the essential oil. A stock solution was prepared for the extracts and subsequently diluted to the required concentrations, expressed in mg/mL. Dilutions of essential oils and extracts were prepared in Mueller–Hinton broth for bacterial cultures and Sabouraud broth for fungal cultures, achieving a final volume of 100 µL per concentration. Subsequently, 100 µL of the microbial inoculum, calibrated to a final concentration of 10^6 CFU/mL for bacteria or 10^4 CFU/mL for fungi, was incorporated into the dilution series. Following 24 h of incubation at 37 °C, 10 µL of resazurin was introduced to each well to indicate microbial growth. After a further 2 h incubation at 37 °C, microbial growth was evidenced by a color transition from violet to pink. The MIC was identified as the minimal concentration that inhibited this color alteration. The 11th and 12th wells of each series functioned as the growth control and sterility control, respectively. The method was conducted twice for both the essential oil and the extracts. To ascertain the Minimum Bactericidal Concentration (MBC) or Minimum Fungicidal Concentration (MFC), 10 µL was extracted from each well exhibiting no apparent growth and put into Mueller–Hinton agar (MH) for bacterial assessment or Sabouraud agar for fungal evaluation. The plates were incubated for 24 h at 37 degrees Celsius. The MBC or MFC was established as the minimal concentration of the sample that achieved a 99.99% decrease in CFU/mL relative to the control. The MBC/MIC or MFC/MIC ratio was computed to assess the antibacterial efficacy. A ratio below 4 signifies a bactericidal/fungicidal impact, whilst a ratio beyond 4 indicates a bacteriostatic/fungistatic effect of the material.

3.9. Antioxidant Activity

3.9.1. DPPH* Trapping Free Radicals

The antioxidant activity of the essential oil and various seed extracts were evaluated to measure their scavenging capacities for stable 2,2′-diphenyl-1-picrylhydrazyl radicals. The DPPH test was performed as described [97]. The test is carried out by mixing a volume of 2.8 mL of the previous solution of DPPH˙ with 200 µL (0.004% methanolic solution of DPPH) of each sample to be tested or of standard antioxidant (ascorbic acid) at different concentrations (from 0 to 200 µg/mL) in methanol. At the same time, a negative control is prepared by mixing 200 µL of ethanol with 2.8 mL of the ethanol solution of DPPH*. After 30 min of incubation at room temperature in the dark. The absorbance reading is taken against a blank at 517 nm. The results obtained were expressed as percentage inhibition (PI%):

$$PI(\%) = \frac{A_{Control} - A_{test}}{A_{control}} * 100 \quad (6)$$

With:

$A_{control}$: Absorbance of the solution containing only the solution of the DPPH radical;

A_{sample}: Absorbance of the solution of the samples to be tested in the presence of DPPH.

IC_{50} or 50% Inhibitory Concentration is the concentration of the test sample necessary to reduce 50% of the DPPH* radical. The IC_{50} are calculated graphically by linear regressions.

3.9.2. FRAP Iron Reduction Power Test

The ability of phenolic extracts from C. sativum to reduce ferric iron (Fe^{3+}) to ferrous iron (Fe^{2+}) in potassium ferricyanide was assessed using the method outlined by Oyaizu [98]. In this assay, 1 mL of the plant extract was combined with 2.5 mL of phosphate buffer (0.2 M, pH 6.6) and 2.5 mL of a 1% potassium ferricyanide solution ($K_3Fe(CN)_6$). The mixture underwent incubation in a water bath at 50 °C for 20 min, followed by the addition of 2.5 mL of 10% trichloroacetic acid to terminate the reaction. The solution was centrifuged at 3000 rpm for a duration of 10 min. Following this, 2.5 mL of the supernatant was mixed with 2.5 mL of distilled water and 0.5 mL of a 0.1% aqueous $FeCl_3$ solution. The absorbance of the reaction medium was measured at 700 nm using a UV-Vis spectrophotometer, with a blank prepared by replacing the extract with distilled water for calibration. Butylated Hydroxyanisole (BHA) was used as the positive control, and its absorbance was measured under identical conditions. All experiments were performed in triplicate. The reducing power was plotted as a function of BHT concentration or the extract concentration, and the EC_{50} value, corresponding to an absorbance of 0.5, was determined from the graph.

3.9.3. Total Antioxidant Capacity (TAC)

The Total Antioxidant Capacity (TAC) of the plant extracts was assessed using the phosphomolybdenum method, as described by Khiya [99]. This technique is based on the reduction in molybdenum (VI), present as molybdate ions ($MoO_4{}^{2-}$), to molybdenum (V) ($MoO_2{}^{+}$) in the presence of antioxidant compounds, leading to the formation of a green phosphate/Mo(V) complex under acidic conditions. For the assay, 3 mL of the reagent solution was mixed with 0.3 mL of the extract, with the reagent solution composed of 0.6 M sulfuric acid, 28 mM sodium phosphate, and 4 mM ammonium molybdate. The reaction mixtures were then sealed and incubated at 95 °C for 90 min to ensure complete reduction. After cooling to room temperature, the absorbance of the resulting solutions was recorded at 695 nm using a UV-Vis spectrophotometer, with a blank sample prepared under identical conditions serving as the reference. To quantify the antioxidant capacity, several concentrations of ascorbic acid were used to construct a standard calibration curve. The results were expressed as milligrams of ascorbic acid equivalents per gram of crude extract (mg AAE/g), providing a comparative measure of the extracts' overall antioxidant potential.

3.10. Molecular Docking

The three-dimensional structures of the target proteins listed in Table 15 were obtained from the RCSB Protein Data Bank (https://www.rcsb.org/, accessed on 23 December 2024) and subsequently visualized using the UCSF Chimera software (Chimera-1.17). This preliminary step was crucial for facilitating the subsequent molecular docking analyses. The protein structures underwent a rigorous preprocessing protocol, leveraging the combined functionalities of Chimera and AutoDock Tools (version 1.5.6, The Scripps Research Institute, La Jolla, CA, USA). This protocol involved the removal of water molecules, heteroatoms, non-essential protein chains, and co-crystallized ligands, with the aim of optimizing the accessibility and specificity of the active site. To ensure an accurate representation of molecular interactions, polar hydrogen atoms were added, and Gasteiger partial charges were assigned. The processed protein structures were then converted into the pdbqt format, required for docking simulations using AutoDock Vina 1.2.0 (accessed on 23 December 2024). As for the ligands, three-dimensional models of the eighteen compounds under investigation were retrieved from the PubChem database and subjected to energy minimization to ensure conformational stability. These minimized structures were converted into the pdbqt format using OpenBabel (http://openbabel.org/; accessed on 23 December 2024), thereby ensuring compatibility with the docking engine. Molecular docking simulations were performed using AutoDock Vina, which employs an empirical scoring function to estimate the binding affinities between ligands and their target proteins. For each protein, a three-dimensional grid was defined around the active site, with meticulously adjusted coordinates and dimensions to encompass the targeted binding pocket. Post-docking analysis was carried out using PyMOL version 2.5.5 (accessed on 23 December 2024), enabling detailed visualization of molecular interactions (such as hydrogen bonds and hydrophobic contacts) and the assessment of associated binding free energies. To validate the reliability of the docking protocol, a re-docking procedure was implemented as follows: the co-crystallized ligand was removed and subsequently re-docked into the original binding region. The spatial overlap between the docked and crystallographic conformations was evaluated using the Root Mean Square Deviation (RMSD) value. An RMSD value below 2 Å was considered indicative of the structural and statistical validity of the docking methodology employed.

Table 15. Protein targets and molecular docking parameters.

Activities	Targets	PDB ID	Grid Box Center Coordinates	Grid Box Size
	Beta-lactamase	7TI1	center_x = −56.105 center_y = 21.017 center_z = 50.112	size_x = 30 size_y = 22 size_z = 24
	DNA topoisomerase 4 subunit A	3RAE	center_x = −54.023 center_y = 68.181 center_z = −18.042	size_x = 22 size_y = 36 size_z = 34
	DNA gyrase subunit B	4DUH	center_x = 21.304 center_y = 12.134 center_z = 25.205	size_x = 24 size_y = 24 size_z = 22
	DIHYDROFOLATE REDUCTASE TYPE 1 FROM TN4003	2W9S	center_x = 6.027 center_y = −1.060 center_z = 30.037	size_x = 24 size_y = 28 size_z = 30
Antibacterial activity	tyrosyl-tRNA synthetase	1JIJ	center_x = −9.074 center_y = 18.180 center_z = 93.030	size_x = 24 size_y = 28 size_z = 30
	Transcriptional regulator TcaR	3KP5	center_x = −27.301 center_y = −30.531 center_z = −1.040	size_x = 26 size_y = 28 size_z = 24
	Ubiquinol--cytochrome-c reductase subunit	7RJB	center_x = 148.091 center_y = 127.219 center_z = 147.014	size_x = 24 size_y = 24 size_z = 28
	Lanosterol 14-alpha demethylase	5V5Z	center_x = −44.181 center_y = −14.027 center_z = 22.108	size_x = 28 size_y = 30 size_z = 36
	Candidapepsin-2	4YBF	center_x = 9.025 center_y = 1.007 center_z = 15.741	size_x = 26 size_y = 32 size_z = 34
	Structure of A. niger Fdc1	4ZA5	center_x = 19.027 center_y = 1.012 center_z = 19.118	size_x = 24 size_y = 24 size_z = 22

Table 15. *Cont.*

Activities	Targets	PDB ID	Grid Box Center Coordinates	Grid Box Size
Antioxidant activity	Myeloperoxidase	5qj2	center_x = −49.012 center_y = 9.017 center_z = 29.115	size_x = 22 size_y = 28 size_z = 32
	Xanthine dehydrogenase/oxidase	3NRZ	center_x = 58.097 center_y = 3.009 center_z = 35.108	size_x = 22 size_y = 28 size_z = 32
	Cytochrome P450 2C9	1OG5	center_x = −38.207 center_y = 61.001 center_z = 27.024	size_x = 24 size_y = 22 size_z = 28
	Lipoxygenase-3	1N8Q	center_x = 26.027 center_y = 0.050 center_z = 16.130	size_x = 20 size_y = 28 size_z = 34
	NADPH oxidase	2CDU	center_x = 11.204 center_y = 1.035 center_z = 24.135	size_x = 20 size_y = 28 size_z = 34

3.11. Statistical Analysis

The data were expressed as mean ± standard error of the mean (SEM). The threshold for statistical significance was set at $p < 0.05$. Statistical analyses were carried out using a one-way analysis of variance (ANOVA), followed by a Tukey–Kramer multiple comparisons post hoc test, in order to determine significant differences between groups. All analyses were performed using GraphPad Prism version 8.0 for Windows (GraphPad Software Inc., San Diego, CA, USA).

4. Conclusions

This study demonstrates the remarkable therapeutic potential of *F. vulgare*, whose extracts and essential oil are characterized by a composition rich in bioactive compounds, notably, phenolic acids (chlorogenic acid, 14.79%), flavonoids (quercetin-3-glucuronide), and terpenes (fenchone 24.72%, trans-anethole 22.22%). The analyses revealed significant antioxidant activity, particularly pronounced in the essential oil (IC_{50} = 51.45 μg/mL) and the hydroethanolic extract, which is correlated with their high polyphenol content. Regarding antimicrobial activity, notable fungicidal action was observed against *Candida albicans* (MIC = 3.13 mg/mL) and *Aspergillus niger* (6.25 mg/mL), attributed to the predominant monoterpenes. In silico approaches confirmed the preferred interactions of these compounds with key enzymatic targets involved in oxidative stress and microbial proteases, thus validating the proposed mechanisms of action. These results position *F. vulgare* as a promising resource for the development of natural therapeutic agents, although preclinical and clinical studies remain necessary to optimize its applications in human health. The complementarity of phenolic extracts and essential oil particularly opens interesting perspectives for synergistic formulations in pharmacology and nutraceuticals. Thus, to ensure the safe application of *F. vulgare* extracts, it is imperative to elucidate their toxicological profile. In the absence of such data, in vitro cytotoxicity assays or comprehensive safety assessments should be considered as essential steps to guide future research directions.

Supplementary Materials: The following supporting information can be downloaded at: https://www.mdpi.com/article/10.3390/ijms26104499/s1.

Author Contributions: Conceptualization, B.E.M. and A.D.; methodology, B.E.M., A.D., A.A.S., O.A.K. and T.Z.; software, B.E.M., A.D. and M.E.; validation, A.B. (Amina Bari) and T.Z.; formal analysis, A.A.S., O.A.K., H.E.O. and K.F.; investigation, B.E.M., H.T., N.H., S.K. and A.C.; data curation, H.E.O.; writing—original draft preparation, B.E.M., A.D., A.B. (Amal Bouzoubaa) and T.Z.; writing—review and editing, B.E.M. and A.D.; visualization, A.B. (Amina Bari) and T.Z.; supervision, T.Z.; project administration, T.Z.; funding acquisition, O.A.K. and A.A.S. All authors have read and agreed to the published version of the manuscript.

Funding: This research was funded by Princess Nourah bint Abdulrahman University Researchers Supporting Project number (PNURSP2025R917), Princess Nourah bint Abdulrahman University, Riyadh, Saudi Arabia and was also funded by Researchers Supporting Project number (RSPD2025R1057), King Saud University, Riyadh, Saudi Arabia.

Institutional Review Board Statement: Not applicable.

Informed Consent Statement: Not applicable.

Data Availability Statement: Data are contained within this article.

Acknowledgments: The authors extend their appreciation to Princess Nourah bint Abdulrahman University Researchers Supporting Project number (PNURSP2025R917), Princess Nourah bint Abdulrahman University, Riyadh, Saudi Arabia, and are also grateful for Researchers Supporting Project number (RSPD2025R1057), King Saud University, Riyadh, Saudi Arabia.

Conflicts of Interest: The authors declare that they have no known competing financial interests or personal relationships that could have appeared to influence the work reported in this paper.

References

1. Badgujar, S.B.; Patel, V.V.; Bandivdekar, A.H. *Foeniculum vulgare* Mill: A Review of Its Botany, Phytochemistry, Pharmacology, Contemporary Application, and Toxicology. *BioMed Res. Int.* **2014**, *2014*, 842674. [CrossRef] [PubMed]
2. Noreen, S.; Tufail, T.; Badar Ul Ain, H.; Awuchi, C.G. Pharmacological, Nutraceutical, Functional and Therapeutic Properties of Fennel (*Foeniculum vulgare*). *Int. J. Food Prop.* **2023**, *26*, 915–927. [CrossRef]
3. Rahimi, R.; Ardekani, M.R.S. Medicinal Properties of *Foeniculum vulgare* Mill. in Traditional Iranian Medicine and Modern Phytotherapy. *Chin. J. Integr. Med.* **2013**, *19*, 73–79. [CrossRef] [PubMed]
4. Saddiqi, H.A.; Iqbal, Z. Usage and Significance of Fennel (*Foeniculum vulgare* Mill.) Seeds in Eastern Medicine. In *Nuts and Seeds in Health and Disease Prevention*; Preedy, V.R., Watson, R.R., Patel, V.B., Eds.; Academic Press: San Diego, CA, USA, 2011; pp. 461–467, ISBN 978-0-12-375688-6.
5. Grover, S.; Malik, C.; Hora, A.; Kushwaha, H. Botany, Cultivation, Chemical Constituents and Genetic Diversity in Fennel (*Foeniculum vulgare* Mill): A Review. *LS Int. J. Life Sci.* **2013**, *2*, 128. [CrossRef]
6. Rani, S.N.; Aslam, Z.N.; Babar, S.N. Foeniculum vulgare (Fennel): A Versatile Herb With Culinary, Medicinal, and Environmental Significance. In *Therapeutic and Pharmacological Applications of Ethnobotany*; IGI Global Scientific Publishing: Hershey, PA, USA, 2024; pp. 155–188, ISBN 979-8-3693-1986-4.
7. Magagula, T.I.P. *Assessment of the Invasive Potential of Foeniculum vulgare (Fennel) in the Sarah Baartman District and Nelson Mandela Bay Municipalities in South Africas Eastern Cape Province*; University of the Witwatersrand: Johannesburg, South Africa, 2018.
8. Debnath, S.; Kumar, H.; Sharma, A. *Foeniculum vulgare* from Spice to Pharma: Recent Advances in Its Medicinal Value, Bioactivities and Perspectives. *Tradit. Integr. Med.* **2023**, 217–229. [CrossRef]
9. Motti, R. Wild Plants Used as Herbs and Spices in Italy: An Ethnobotanical Review. *Plants* **2021**, *10*, 563. [CrossRef]
10. Authority (EFSA), E.F.S. Protocol for the Scientific Opinion on the Evaluation of the Safety in Use of Preparations from the Fruits of Sweet and Bitter Fennel (*Foeniculum vulgare* Mill. and Foeniculum Piperitum (Ucria) C.Presl). *EFSA Support. Publ.* **2023**, *20*, 8245E. [CrossRef]
11. Guidi, L.; Landi, M. Aromatic Plants: Use and Nutraceutical Properties. In *Novel Plant Bioresources*; John Wiley & Sons, Ltd.: Hoboken, NJ, USA, 2014; pp. 303–345, ISBN 978-1-118-46056-6.
12. Zafar, S.; Khan, M.K.; Perveen, S.; Iqbal, M.; AL-Huqail, A.A. Fennel. In *Essentials of Medicinal and Aromatic Crops*; Zia-Ul-Haq, M., Abdulkreem AL-Huqail, A., Riaz, M., Farooq Gohar, U., Eds.; Springer International Publishing: Cham, Switzerland, 2023; pp. 483–514, ISBN 978-3-031-35403-8.
13. Sayed Ahmad, B.; Talou, T.; Saad, Z.; Hijazi, A.; Cerny, M.; Kanaan, H.; Chokr, A.; Merah, O. Fennel Oil and By-Products Seed Characterization and Their Potential Applications. *Ind. Crops Prod.* **2018**, *111*, 92–98. [CrossRef]
14. Rasul, A.; Akhtar, N.; Khan, B.A.; Mahmood, T.; Zaman, S.U.; Khan, H.M.S. Formulation Development of a Cream Containing Fennel Extract: In Vivo Evaluation for Anti-Aging Effects. *Pharm. Int. J. Pharm. Sci.* **2012**, *67*, 54–58. [CrossRef]

15. Milenković, A.; Ilić, Z.; Stanojević, L.; Milenković, L.; Šunić, L.; Lalević, D.; Stanojević, J.; Danilović, B.; Cvetković, D. Essential Oil Yield, Composition, Antioxidant and Microbial Activity of Wild Fennel (*Foeniculum vulgare* Mill.) from Monte Negro Coast. *Horticulturae* **2022**, *8*, 1015. [CrossRef]
16. Zeng, H.; Chen, X.; Liang, J. In Vitro Antifungal Activity and Mechanism of Essential Oil from Fennel (*Foeniculum vulgare* L.) on Dermatophyte Species. *J. Med. Microbiol.* **2015**, *64*, 93–103. [CrossRef]
17. Syed, F.Q.; Mirza, M.B.; Elkady, A.I.; Hakeem, K.R.; Alkarim, S. An Insight of Multitudinous and Inveterate Pharmacological Applications of *Foeniculum vulgare* (Fennel). In *Plant and Human Health, Volume 3: Pharmacology and Therapeutic Uses*; Ozturk, M., Hakeem, K.R., Eds.; Springer International Publishing: Cham, Switzerland, 2019; pp. 231–254, ISBN 978-3-030-04408-4.
18. Lim, T.K. Foeniculum vulgare. In *Edible Medicinal And Non-Medicinal Plants: Volume 5, Fruits*; Lim, T.K., Ed.; Springer: Dordrecht, The Netherlands, 2013; pp. 36–59, ISBN 978-94-007-5653-3.
19. Singh, S.; Datta, R.; Johri, P.; Trivedi, M. *Phytopharmaceuticals and Biotechnology of Herbal Plants*; CRC Press: Boca Raton, FL, USA, 2022; ISBN 978-1-000-81129-2.
20. Choi, E.-M.; Hwang, J.-K. Antiinflammatory, Analgesic and Antioxidant Activities of the Fruit of *Foeniculum vulgare*. *Fitoterapia* **2004**, *75*, 557–565. [CrossRef] [PubMed]
21. Syed, R.; Saleh, A.; Mohammed, A.-D.; Mansour, A.-S.; Mohammed, A.-Y.; Ibrahim, A.-M. Gastroprotective Effect of Fennel "*Foeniculum vulgare*" a Commonly Used Spice in Arab Traditional Medicine. Обзоры по клинической фармакологии и лекарственной терапии **2012**, *10*, 91.
22. Cioanca, O.; Hancianu, M.; Mircea, C.; Trifan, A.; Hritcu, L. Essential Oils from Apiaceae as Valuable Resources in Neurological Disorders: *Foeniculi Vulgare Aetheroleum*. *Ind. Crops Prod.* **2016**, *88*, 51–57. [CrossRef]
23. Diao, W.-R.; Hu, Q.-P.; Zhang, H.; Xu, J.-G. Chemical Composition, Antibacterial Activity and Mechanism of Action of Essential Oil from Seeds of Fennel (*Foeniculum vulgare* Mill.). *Food Control* **2014**, *35*, 109–116. [CrossRef]
24. Mandras, N.; Nostro, A.; Roana, J.; Scalas, D.; Banche, G.; Ghisetti, V.; Del Re, S.; Fucale, G.; Cuffini, A.M.; Tullio, V. Liquid and Vapour-Phase Antifungal Activities of Essential Oils against Candida Albicans and Non-Albicans Candida. *BMC Complement. Altern. Med.* **2016**, *16*, 330. [CrossRef]
25. Barakat, H.; Alkabeer, I.A.; Aljutaily, T.; Almujaydil, M.S.; Algheshairy, R.M.; Alhomaid, R.M.; Almutairi, A.S.; Mohamed, A. Phenolics and Volatile Compounds of Fennel (*Foeniculum vulgare*) Seeds and Their Sprouts Prevent Oxidative DNA Damage and Ameliorates CCl4-Induced Hepatotoxicity and Oxidative Stress in Rats. *Antioxidants* **2022**, *11*, 2318. [CrossRef]
26. Mohamad, R.H.; El-Bastawesy, A.M.; Abdel-Monem, M.G.; Noor, A.M.; Al-Mehdar, H.A.R.; Sharawy, S.M.; El-Merzabani, M.M. Antioxidant and Anticarcinogenic Effects of Methanolic Extract and Volatile Oil of Fennel Seeds (*Foeniculum vulgare*). *J. Med. Food* **2011**, *14*, 986–1001. [CrossRef]
27. Abou, N.; Abou El-Soud, N.; El-Laithy, N.; El-Saeed, G.; Wahby, M.S.; Khalil, M.; Morsy, F.; Shaffie, N.; Abou, C.; Wahby, S. Antidiabetic Activities of *Foeniculum vulgare* Mill. Essential Oil in Streptozotocin-Induced Diabetic Rats. *Maced. J. Med. Sci.* **2011**, *4*, 139–146.
28. Koppula, S.; Kumar, H. *Foeniculum vulgare* Mill (Umbelliferae) Attenuates Stress and Improves Memory in Wister Rats. *Trop. J. Pharm. Res.* **2013**, *12*, 553–558. [CrossRef]
29. Kooti, W.; Moradi, M.; Ali-Akbari, S.; Sharafi-Ahvazi, N.; Asadi-Samani, M.; Ashtary-Larky, D. Therapeutic and Pharmacological Potential of *Foeniculum vulgare* Mill: A Review. *J. Herb. Med. Pharmacol.* **2015**, *4*, 1–9.
30. Anwar, F.; Ali, M.; Hussain, A.I.; Shahid, M. Antioxidant and Antimicrobial Activities of Essential Oil and Extracts of Fennel (*Foeniculum vulgare* Mill.) Seeds from Pakistan. *Flavour. Fragr. J.* **2009**, *24*, 170–176. [CrossRef]
31. Roby, M.H.H.; Sarhan, M.A.; Selim, K.A.-H.; Khalel, K.I. Antioxidant and Antimicrobial Activities of Essential Oil and Extracts of Fennel (*Foeniculum vulgare* L.) and Chamomile (*Matricaria chamomilla* L.). *Ind. Crops Prod.* **2013**, *44*, 437–445. [CrossRef]
32. Vella, F.M.; Pignone, D.; Laratta, B. The Mediterranean Species Calendula Officinalis and *Foeniculum vulgare* as Valuable Source of Bioactive Compounds. *Molecules* **2024**, *29*, 3594. [CrossRef]
33. He, W.; Huang, B. A Review of Chemistry and Bioactivities of a Medicinal Spice: *Foeniculum vulgare*. *J. Med. Plants Res.* **2011**, *5*, 3595–3600.
34. Anka, Z.M.; Gimba, S.; Nanda, A.; Salisu, L. Phytochemistry and Pharmacological Activities of *Foeniculum vulgare*. *IOSR J. Pharm.* **2020**, *10*, 1–10.
35. Tomou, E.-M.; Papakyriakopoulou, P.; Skaltsa, H.; Valsami, G.; Kadoglou, N.P.E. Bio-Actives from Natural Products with Potential Cardioprotective Properties: Isolation, Identification, and Pharmacological Actions of Apigenin, Quercetin, and Silibinin. *Molecules* **2023**, *28*, 2387. [CrossRef]
36. Jamal, K.; Al-Taweel, A.; Bukhari, S.I.; Orfali, R.; Moubayed, N.M.S.; Al-Qahtani, J.; Aati, H.; Taglialatela-Scafati, O.; Peng, J.; Perveen, S. Isochlorogenic Acid Glucosides from the Arabian Medicinal Plant Artemisia Sieberi and Their Antimicrobial Activities. *Molecules* **2023**, *28*, 7460. [CrossRef]

37. Miao, M.; Xiang, L. Chapter Three-Pharmacological Action and Potential Targets of Chlorogenic Acid. In *Advances in Pharmacology*; Du, G., Ed.; Pharmacological Advances in Natural Product Drug Discovery; Academic Press: Cambridge, MA, USA, 2020; Volume 87, pp. 71–88.
38. Raal, A.; Orav, A.; Arak, E. Essential Oil Composition of *Foeniculum vulgare* Mill. Fruits from Pharmacies in Different Countries. *Nat. Prod. Res.* **2012**, *26*, 1173–1178. [CrossRef]
39. Božović, M.; Garzoli, S.; Vujović, S.; Sapienza, F.; Ragno, R. *Foeniculum vulgare* Miller, a New Chemotype from Montenegro. *Plants* **2022**, *11*, 42. [CrossRef]
40. Akarca, G.; Şevik, R.; Denzikara, A.; Kilinç, M.; Aşçioğlu, Ç. Bioactive Properties and Volatile Compounds of the Essential Oil of the Leaves of Giant Fennel (Çaksır) Growing in Türkiye. *J. Essent. Oil Bear. Plants* **2023**, *26*, 1338–1349. [CrossRef]
41. Afifi, S.M.; El-Mahis, A.; Heiss, A.G.; Farag, M.A. Gas Chromatography–Mass Spectrometry-Based Classification of 12 Fennel (*Foeniculum vulgare* Miller) Varieties Based on Their Aroma Profiles and Estragole Levels as Analyzed Using Chemometric Tools. *ACS Omega* **2021**, *6*, 5775–5785. [CrossRef] [PubMed]
42. Garga, C.; Khan, S.; Ansari, S.; Suman, A.; Garg, M. Chemical Composition, Therapeutic Potential and Perspectives of *Foeniculum vulgare*. *Pharmacogn. Rev.* **2009**, *3*, 346.
43. Ramteke, A.M. Naturally Isolated Compounds from Spices and Herbs and Their Medicinal Uses. In *The Chemistry Inside Spices & Herbs: Research and Development*; Bentham Science Publishers: Sharjah, United Arab Emirates, 2022; pp. 138–153.
44. Senatore, F.; Oliviero, F.; Scandolera, E.; Taglialatela-Scafati, O.; Roscigno, G.; Zaccardelli, M.; De Falco, E. Chemical Composition, Antimicrobial and Antioxidant Activities of Anethole-Rich Oil from Leaves of Selected Varieties of Fennel [*Foeniculum vulgare* Mill. Ssp. Vulgare Var. Azoricum (Mill.) Thell]. *Fitoterapia* **2013**, *90*, 214–219. [CrossRef]
45. Lei, J.-D.; Li, Q.; Zhang, S.-B.; Lv, Y.-Y.; Zhai, H.-C.; Wei, S.; Ma, P.-A.; Hu, Y.-S. Transcriptomic and Biochemical Analyses Revealed Antifungal Mechanism of Trans-Anethole on Aspergillus Flavus Growth. *Appl. Microbiol. Biotechnol.* **2023**, *107*, 7213–7230. [CrossRef]
46. Šunić, L.; Ilić, Z.S.; Stanojević, L.; Milenković, L.; Stanojević, J.; Kovač, R.; Milenković, A.; Cvetković, D. Comparison of the Essential Oil Content, Constituents and Antioxidant Activity from Different Plant Parts during Development Stages of Wild Fennel (*Foeniculum vulgare* Mill.). *Horticulturae* **2023**, *9*, 364. [CrossRef]
47. Cvetkovic, D.; Stanojevic, J.; Djordjevic, N.; Karabegović, I.; Stanojevic, L.; Pavlic, B.; Danilović, B. Effect of Different Extraction Techniques on the Composition of Essential Oil Isolated from Fennel (*Foeniculum vulgare* Mill.) Rhizome. *J. Essent. Oil Res.* **2023**, *35*, 24–34. [CrossRef]
48. Pessoa, M.L.d.S.; Silva, L.M.O.; Araruna, M.E.C.; Serafim, C.A.d.L.; Alves Júnior, E.B.; Silva, A.O.; Pessoa, M.M.B.; Diniz Neto, H.; de Oliveira Lima, E.; Batista, L.M. Antifungal Activity and Antidiarrheal Activity via Antimotility Mechanisms of (-)-Fenchone in Experimental Models. *World J. Gastroenterol.* **2020**, *26*, 6795–6809. [CrossRef]
49. Araruna, M.E.C.; Alves Júnior, E.B.; de Lima Serafim, C.A.; Pessoa, M.M.B.; de Souza Pessôa, M.L.; Alves, V.P.; Sobral, M.V.; da Silva, M.S.; Alves, A.F.; de Paiva Sousa, M.C.; et al. (-)-Fenchone Ameliorates TNBS-Induced Colitis in Rats via Antioxidant, Immunomodulatory, and Cytoprotective Mechanisms. *Pharmaceuticals* **2025**, *18*, 18. [CrossRef]
50. Aćimović, M.; Jeremić, J.S.; Cvetković, M.; Lončar, B.; Pezo, L. Comparative Analysis of Essential Oils and Hydrolates from Aniseed, Coriander, and Fennel Fruits. *Chem. Biodivers.* **2025**, *n/a*, e202500022. [CrossRef]
51. Akbar, S. Foeniculum vulgare Mill. (Apiaceae/Umbelliferae). In *Handbook of 200 Medicinal Plants: A Comprehensive Review of Their Traditional Medical Uses and Scientific Justifications*; Akbar, S., Ed.; Springer International Publishing: Cham, Switzerland, 2020; pp. 931–945, ISBN 978-3-030-16807-0.
52. Anandakumar, P.; Kamaraj, S.; Vanitha, M.K. D-Limonene: A Multifunctional Compound with Potent Therapeutic Effects. *J. Food Biochem.* **2021**, *45*, e13566. [CrossRef] [PubMed]
53. Goswami, N.; Chatterjee, S. Assessment of Free Radical Scavenging Potential and Oxidative DNA Damage Preventive Activity of *Trachyspermum ammi* L. (Carom) and *Foeniculum vulgare* Mill. (Fennel) Seed Extracts. *BioMed Res. Int.* **2014**, *2014*, 582767. [CrossRef] [PubMed]
54. Lu, Y.; Foo, L.Y. Antioxidant Activities of Polyphenols from Sage (Salvia Officinalis). *Food Chem.* **2001**, *75*, 197–202. [CrossRef]
55. Razzaghi-Asl, N.; Garrido, J.; Khazraei, H.; Borges, F.; Firuzi, O. Antioxidant Properties of Hydroxycinnamic Acids: A Review of Structure- Activity Relationships. *Curr. Med. Chem.* **2013**, *20*, 4436–4450. [CrossRef]
56. Akinsemolu, A.A.; Onyeaka, H.N. Microorganisms Associated with Food Spoilage and Foodborne Diseases. In *Food Safety and Quality in the Global South*; Ogwu, M.C., Izah, S.C., Ntuli, N.R., Eds.; Springer Nature: Singapore, 2024; pp. 489–531, ISBN 978-981-97-2428-4.
57. Hashemi, S.M.B.; Khaneghah, A.M.; Sant'Ana, A.d.S. *Essential Oils in Food Processing: Chemistry, Safety and Applications*; John Wiley & Sons: Hoboken, NJ, USA, 2017; ISBN 978-1-119-14934-7.
58. Khammassi, M.; Habiba, K.; Mighri, H.; Mouna, S.; Oumayma, K.; Seçer, E.; Ismail, A.; Jamoussi, B.; Yassine, M. Phytochemical Screening of Essential Oils and Methanol Extract Constituents of Wild *Foeniculum vulgare* Mill.: A Potential Natural Source for Bioactive Molecules. *Chem. Afr.* **2023**, *6*, 1227–1240. [CrossRef]

59. Santoro, V.; Rosa, E.; Donadio, G.; Polito, F.; De Feo, V.; De Tommasi, N. *Foeniculum vulgare* Miller Bracts, Revalorization of a Local Food Waste. *Sci. Rep.* **2024**, *14*, 31287. [CrossRef]
60. Adorjan, B.; Buchbauer, G. Biological Properties of Essential Oils: An Updated Review. *Flavour. Fragr. J.* **2010**, *25*, 407–426. [CrossRef]
61. Naaz, S.; Ahmad, N.; Qureshi, M.I.; Hashmi, N.; Akhtar, M.S.; A. Khan, M.M. Antimicrobial and Antioxidant Activities of Fennel Oil. *Bioinformation* **2022**, *18*, 795–800. [CrossRef]
62. Jadid, N.; Widodo, A.F.; Ermavitalini, D.; Sa'adah, N.N.; Gunawan, S.; Nisa, C. The Medicinal Umbelliferae Plant Fennel (*Foeniculum vulgare* Mill.): Cultivation, Traditional Uses, Phytopharmacological Properties, and Application in Animal Husbandry. *Arab. J. Chem.* **2023**, *16*, 104541. [CrossRef]
63. Rafieian, F.; Amani, R.; Rezaei, A.; Karaça, A.C.; Jafari, S.M. Exploring Fennel (*Foeniculum vulgare*): Composition, Functional Properties, Potential Health Benefits, and Safety. *Crit. Rev. Food Sci. Nutr.* **2024**, *64*, 6924–6941. [CrossRef]
64. Elbaz, M.; Abdesslem, S.B.; St-Gelais, A.; Boulares, M.; Moussa, O.B.; Timoumi, M.; Hassouna, M.; Aider, M. Essential Oils Profile, Antioxidant and Antibacterial Potency of Tunisian Fennel (*Foeniculum vulgare* Mill.) Leaves Grown under Conventional and Organic Conditions. *Food Chem. Adv.* **2024**, *4*, 100734. [CrossRef]
65. Miguel, M.G.; Cruz, C.; Faleiro, L.; Simões, M.T.F.; Figueiredo, A.C.; Barroso, J.G.; Pedro, L.G. *Foeniculum vulgare* Essential Oils: Chemical Composition, Antioxidant and Antimicrobial Activities. *Nat. Prod. Commun.* **2010**, *5*, 1934578X1000500231. [CrossRef]
66. Peixoto, J.A.B.; Álvarez-Rivera, G.; Alves, R.C.; Costa, A.S.G.; Machado, S.; Cifuentes, A.; Ibáñez, E.; Oliveira, M.B.P.P. Comprehensive Phenolic and Free Amino Acid Analysis of Rosemary Infusions: Influence on the Antioxidant Potential. *Antioxidants* **2021**, *10*, 500. [CrossRef]
67. Chaowuttikul, C.; Palanuvej, C.; Ruangrungsi, N. Quantification of Chlorogenic Acid, Rosmarinic Acid, and Caffeic Acid Contents in Selected Thai Medicinal Plants Using RP-HPLC-DAD. *Braz. J. Pharm. Sci.* **2020**, *56*, e17547. [CrossRef]
68. Panda, P.K.; Yang, J.-M.; Chang, Y.-H.; Su, W.-W. Modification of Different Molecular Weights of Chitosan by P-Coumaric Acid: Preparation, Characterization and Effect of Molecular Weight on Its Water Solubility and Antioxidant Property. *Int. J. Biol. Macromol.* **2019**, *136*, 661–667. [CrossRef]
69. Zhou, J.; Lin, Z.; Ju, Y.; Rahim, M.A.; Richardson, J.J.; Caruso, F. Polyphenol-Mediated Assembly for Particle Engineering. *Acc. Chem. Res.* **2020**, *53*, 1269–1278. [CrossRef]
70. Chen, Z.; Świsłocka, R.; Choińska, R.; Marszałek, K.; Dąbrowska, A.; Lewandowski, W.; Lewandowska, H. Exploring the Correlation Between the Molecular Structure and Biological Activities of Metal–Phenolic Compound Complexes: Research and Description of the Role of Metal Ions in Improving the Antioxidant Activities of Phenolic Compounds. *Int. J. Mol. Sci.* **2024**, *25*, 11775. [CrossRef]
71. Razaviamri, S.; Wang, K.; Liu, B.; Lee, B.P. Catechol-Based Antimicrobial Polymers. *Molecules* **2021**, *26*, 559. [CrossRef]
72. Dangles, O. Antioxidant Activity of Plant Phenols: Chemical Mechanisms and Biological Significance. *Curr. Org. Chem.* **2012**, *16*, 692–714. [CrossRef]
73. Hosseini, E.; Majidi, M.M.; Ehtemam, M.H.; Ghanadian, M.; Huyghe, C. Variation in a Worldwide Collection of Fennel (*Foeniculum vulgare* Var. Vulgare). *Crop Pasture Sci.* **2021**, *72*, 1008–1021. [CrossRef]
74. Alrub, M.A. Unveiling the Chemical Profiling, Antioxidant, Anticancer and Antibacterial Activities of Essential Oils Derived from Fennel (*Foeniculum vulgare*), M. Fruticose and Alyosia Citriodora. Ph.D. Thesis, Faculty of Graduate Studies, An-Najah National University, Nablus, Palestine, 2023.
75. Saidi, S.; Remok, F.; Handaq, N.; Drioiche, A.; Gourich, A.A.; Menyiy, N.E.; Amalich, S.; Elouardi, M.; Touijer, H.; Bouhrim, M.; et al. Phytochemical Profile, Antioxidant, Antimicrobial, and Antidiabetic Activities of Ajuga Iva (L.). *Life* **2023**, *13*, 1165. [CrossRef] [PubMed]
76. Remok, F.; Saidi, S.; Gourich, A.A.; Zibouh, K.; Maouloua, M.; Makhoukhi, F.E.; Menyiy, N.E.; Touijer, H.; Bouhrim, M.; Sahpaz, S.; et al. Phenolic Content, Antioxidant, Antibacterial, Antihyperglycemic, and α-Amylase Inhibitory Activities of Aqueous Extract of Salvia Lavandulifolia Vahl. *Pharmaceuticals* **2023**, *16*, 395. [CrossRef] [PubMed]
77. Bergeron, L. *Effet de La Teneur En Eau Du Sol Sur Le Rendement et La Qualité Des Fruits Du Bleuet Nain*; Université du Québec à Chicoutimi: Chicoutimi, QC, Canada, 1995.
78. Skujins, S. *Handbook for ICP-AES (Varian-Vista)*; A short guide to vista series ICP-AES operation. Version 1; Varian Int. AG: Zug, Switzerland, 1998.
79. Dohou, R.; Yamni, K.; Tahrouch, S.; Hassani, L.I.; Badoc, A.; Gmira, N. Screening Phytochimique d'une Endémique Iberomarocaine, Thymelaea Lythroides. *Bull. Bull.-Société De Pharm. De Bordx.* **2003**, *142*, 61–78.
80. Judith, M.D. Étude Phytochimique et Pharmacologique de Cassia Nigricans Vhal (Caeslpiniaceae) Utilisée Dans Le Traitement Des Dermatoses Au Tchad. Ph.D. Thesis, Université de Bamako, Bamako, Mali, 2005.
81. Mezzoug, N.; Elhadri, A.; Dallouh, A.; Amkiss, S.; Skali, N.S.; Abrini, J.; Zhiri, A.; Baudoux, D.; Diallo, B.; El Jaziri, M. Investigation of the Mutagenic and Antimutagenic Effects of Origanum Compactum Essential Oil and Some of Its Constituents. *Mutat. Res./Genet. Toxicol. Environ. Mutagen.* **2007**, *629*, 100–110. [CrossRef]

82. Bekro, Y.A.; Jana, A.; Mamyrbekova, B.; Bousa, B.; Fézan, H.; Tra, B.; Ehile, E.E. Etude Ethnobotanique et Sceening Phtochimque de Caesalpina Benthamiana (Ball). Herend et Zarucchi (Caesalpiniaceae). *Sci. Nat.* **2007**, *4*, 217–225.
83. Bruneton, J. *Pharmacognosie: Phytochimie, Plantes Médicinales*, 4th ed.; Tec & Doc Lavoisier: Paris, France, 2009; ISBN 978-2-7430-1188-8.
84. N'Guessan, K.; Kadja, B.; Zirihi, G.; Traoré, D.; Aké-Assi, L. Screening Phytochimique de Quelques Plantes Médicinales Ivoiriennes Utilisées En Pays Krobou (Agboville, Côte-d'Ivoire). *Sci. Nat.* **2009**, *6*, 1–15. [CrossRef]
85. El Karkouri, J.; Kchibale, A.; Chroho, M.; Eddamsyry, B.; Touijer, H.; El Makhoukhi, F.; Handaq, N.; Eto, B.; Salamatullah, A.M.; Bourhia, M.; et al. Phytochemical Profile, Antioxidant Activity, Anti-Hyperglycemic Effect and Toxicity Assessment of Ridolfia Segetum (L.) Moris Extract. *Life* **2023**, *13*, 44. [CrossRef]
86. Singleton, V.L.; Rossi, J.A. Colorimetry of Total Phenolics with Phosphomolybdic-Phosphotungstic Acid Reagents. *Am. J. Enol. Vitic.* **1965**, *16*, 144–158. [CrossRef]
87. Hung, P.V.; Maeda, T.; Miyatake, K.; Morita, N. Total Phenolic Compounds and Antioxidant Capacity of Wheat Graded Flours by Polishing Method. *Food Res. Int.* **2009**, *42*, 185–190. [CrossRef]
88. Djeridane, A.; Yousfi, M.; Nadjemi, B.; Boutassouna, D.; Stocker, P.; Vidal, N. Antioxidant Activity of Some Algerian Medicinal Plants Extracts Containing Phenolic Compounds. *Food Chem.* **2006**, *97*, 654–660. [CrossRef]
89. Price, M.L.; Van Scoyoc, S.; Butler, L.G. A Critical Evaluation of the Vanillin Reaction as an Assay for Tannin in Sorghum Grain. *J. Agric. Food Chem.* **1978**, *26*, 1214–1218. [CrossRef]
90. Gourich, A.A.; Touijer, H.; Drioiche, A.; Asbabou, A.; Remok, F.; Saidi, S.; Siddique, F.; Ailli, A.; Bourhia, M.; Salamatullah, A.M.; et al. Insight into Biological Activities of Chemically Characterized Extract from *Marrubium vulgare* L. in Vitro, in Vivo and in Silico Approaches. *Front. Chem.* **2023**, *11*, 1238346. [CrossRef] [PubMed]
91. Tagnaout, I.; Zerkani, H.; Bencheikh, N.; Amalich, S.; Bouhrim, M.; Mothana, R.A.; Alhuzani, M.R.; Bouharroud, R.; Hano, C.; Zair, T. Chemical Composition, Antioxidants, Antibacterial, and Insecticidal Activities of Origanum Elongatum (Bonnet) Emberger & Maire Aerial Part Essential Oil from Morocco. *Antibiotics* **2023**, *12*, 174. [CrossRef] [PubMed]
92. Ailli, A.; Handaq, N.; Touijer, H.; Gourich, A.A.; Drioiche, A.; Zibouh, K.; Eddamsyry, B.; El Makhoukhi, F.; Mouradi, A.; Bin Jardan, Y.A. Phytochemistry and Biological Activities of Essential Oils from Six Aromatic Medicinal Plants with Cosmetic Properties. *Antibiotics* **2023**, *12*, 721. [CrossRef]
93. Adams, R.P. *Identification of Essential Oil Components by Gas Chromatography/Mass Spectrometry*; Allured Publishing Corporation: Carol Stream, IL, USA, 2007; Volume 456.
94. Babushok, V.I.; Linstrom, P.J.; Zenkevich, I.G. Retention Indices for Frequently Reported Compounds of Plant Essential Oils. *J. Phys. Chem. Ref. Data* **2011**, *40*, 043101. [CrossRef]
95. Kovats, E.S. Gas Chromatographic Characterization of Organic Substances in the Retention Index System. *Adv. Chromatogr.* **1965**, *1*, 229–247.
96. Drioiche, A.; Baammi, S.; Zibouh, K.; Al Kamaly, O.; Alnakhli, A.M.; Remok, F.; Saidi, S.; Amaiach, R.; El Makhoukhi, F.; Elomri, A. A Study of the Synergistic Effects of Essential Oils from Origanum Compactum and Origanum Elongatum with Commercial Antibiotics against Highly Prioritized Multidrug-Resistant Bacteria for the World Health Organization. *Metabolites* **2024**, *14*, 210. [CrossRef]
97. Liu, M.H.; Otsuka, N.; Noyori, K.; Shiota, S.; Ogawa, W.; Kuroda, T.; Hatano, T.; Tsuchiya, T. Synergistic Effect of Kaempferol Glycosides Purified from Laurus Nobilis and Fluoroquinolones on Methicillin-Resistant Staphylococcus Aureus. *Biol. Pharm. Bull.* **2009**, *32*, 489–492. [CrossRef]
98. Oyaizu, M. Studies on Products of Browning Reaction. Antioxidative Activities of Products of Browning Reaction Prepared from Glucosamine. *Jpn. J. Nutr. Diet.* **1986**, *44*, 307–315. [CrossRef]
99. Khiya, Z.; Oualcadi, Y.; Gamar, A.; Berrekhis, F.; Zair, T.; Hilali, F.E. Correlation of Total Polyphenolic Content with Antioxidant Activity of Hydromethanolic Extract and Their Fractions of the Salvia Officinalis Leaves from Different Regions of Morocco. *J. Chem.* **2021**, *2021*, e8585313. [CrossRef]

International Journal of
Molecular Sciences

MDPI

Article

Effect of Fenugreek Extract on Testosterone Propionate-Induced Benign Prostatic Hyperplasia

Jeong Yoon Lee [1], Jiyoung Bang [1], Jinhak Kim [2], Kwang-Soo Baek [2], Dongchan Oh [2] and Yoo-Hyun Lee [1,*]

1 Department of Food and Nutrition, The University of Suwon, Hwasung 445743, Republic of Korea; 109920@naver.com (J.Y.L.); jybbb1204@naver.com (J.B.)

2 R&D Division, Daehan Chemtech Co., Ltd., Gwacheon-si 13840, Republic of Korea; jhkim@dhchemtech.com (J.K.); rnd@dhchemtech.com (K.-S.B.); fpaamgi@naver.com (D.O.)

* Correspondence: creamut@suwon.ac.kr; Tel.: +82-31-229-8194

Abstract: Benign prostatic hyperplasia (BPH) is a noncancerous urinary disorder that is common in older adult men; however, its underlying mechanisms remain unclear. Fenugreek has some biological effects, including hyperglycemia regulation, immune response modulation, and anti-cancer properties; In this study, we investigated the ameliorative effects of fenugreek seed extract (Forceterone® [FCT]) in a testosterone propionate (TP)-induced BPH animal model and its mechanisms in BPH-1 human prostate epithelial cells. Sprague Dawley (SD) rats were injected subcutaneously with TP (3 mg/kg) for 8 weeks to induce BPH while FCT was administered orally at 25, 50, and 100 mg/kg. In addition, BPH-1 cells were used to evaluate the inhibitory effects on cell proliferation and examine inflammatory cytokine expression. Treating rats with FCT decreased prostate weight, dihydrotestosterone (DHT) level, and proliferating cell nuclear antigen (PCNA) expression in the prostate. Furthermore, it decreased androgen receptor (AR), 5α-reductase 2, B-cell lymphoma 2 (Bcl-2), interleukin-6 (IL-6), tumor necrosis factor-alpha (TNF-α), and NF-κB expression in vitro and in vivo and increased Bcl-2-associated X protein (Bax) expression. FCT also inhibited cell proliferation dose dependently in BPH-1 cells. These findings showed the potential use of FCT as an alternative treatment for BPH.

Keywords: fenugreek seed extract; benign prostatic hyperplasia; androgen receptor; dihydrotestosterone

1. Introduction

The prostate is an essential accessory organ involved in seminal fluid formation, which is crucial for male reproduction [1]. It can be a target for conditions such as benign prostatic hyperplasia (BPH) with aging. BPH is an obstructive disease characterized by abnormal prostate enlargement surrounding the proximal urethra, anterior to the rectum, and below the bladder. In addition, BPH leads to urinary symptoms. The prevalence of BPH is approximately 20%, exceeds 50%, and reaches 80% in 40, 50, and 70-year-old men, respectively [2]. Lower urinary tract symptoms (LUTSs) include a weak stream, dribbling, hesitancy, and nocturia due to bladder outlet obstruction, reducing quality of life [3,4].

Surgical treatment may be considered in severe BPH cases; however, most mild-to-moderate patients use non-invasive treatments, comprising a combination of medications such as alpha-1-adrenoceptor antagonists (alpha-1 blockers) and 5 alpha-reductase inhibitors [5]. Alpha-1 blockers, including terazosin and alfuzosin, which relax the prostate muscles, and phosphodiesterase type 5 inhibitors, which are vasodilators, rapidly improve LUTSs; however, their application is limited, and they are unable to provide fundamental treatment [4]. In addition, 5 alpha-reductase inhibitors, including finasteride and dutasteride, which reduce prostate size, must be taken for a minimum of six months and popularly cause adverse side effects, such as decreased libido, erectile dysfunction, and delayed ejaculation [6,7].

However, the etiology of BPH is uncertain. Androgens, including testosterone and dihydrotestosterone (DHT), play key roles in prostate growth [8]. DHT is a potent androgen with approximately 10 times greater affinity for the androgen receptor (AR) than testosterone [9]. Testosterone is converted to DHT by 5 alpha-reductase. The use of 5 alpha-reductase inhibitors to improve BPH symptoms highlights the importance of managing DHT levels during treatment. Several studies have reported elevated DHT levels and AR expression in BPH tissues [10]. Recently, studies have suggested that reduced apoptosis [11,12] and chronic inflammation [13] are associated with BPH development.

Trigonella foenum-graecum (Fenugreek) is an annual herb belonging to the family Fabaceae that is widely cultivated in various regions, including North Africa [14,15]. Fenugreek has long been used as a medicinal plant, as a lactation stimulant [16], and tonic agent [17]. Recent clinical studies have reported its benefit on blood glucose control in patients with type 2 diabetes [18]. Animal studies have demonstrated several effects, including cholesterol-lowering properties [18], immune response modulation [19], and anti-cancer [20] and neuroprotective effects [19]. Fenugreek seeds contain chemical constituents including alkaloids, amino acids, and saponins [14]. In this study, we investigated the effects of fenugreek seed ethanol extract (Forceterone® [FCT]) on BPH through in vitro and in vivo experiments. The fenugreek seed extract used in Rao's studies differs from the fenugreek seed extract we used in terms of the extraction method and marker compound content [21]. We conducted this study to evaluate the efficacy of FCT on BPH and to investigate its underlying mechanisms.

2. Results

2.1. Identification of Phenolic Compound in FCT

The components present in FCT were identified and represented as peaks (Figure 1). HPLC characterization showed that compounds a, b, c, and d were identified based on their respective retention times. Compound a was identified as iso-orientin, with a retention time of 26.2 min, while compound b was identified as orientin, with a retention time of 27.2 min. Compounds c and d were identified as vitexin and iso-vitexin, with retention times of 32.1 min and 33.1 min, respectively.

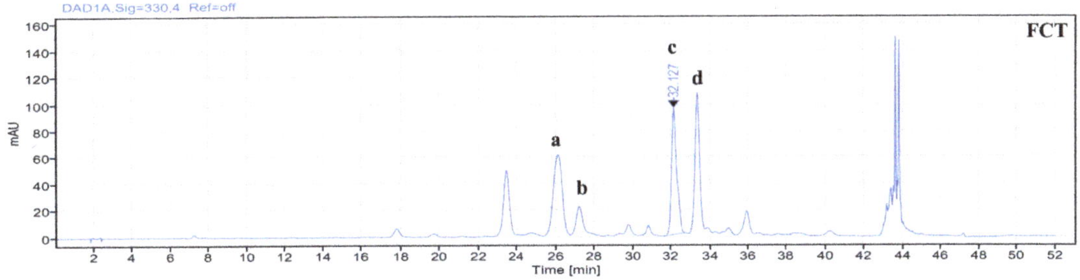

Figure 1. High-performance liquid chromatography analysis of standard FCT. a: iso-orientin (retention time: 26.2 min); b: orientin (retention time: 27.2); c: vitexin (retention time: 32.1 min); d: iso-vitexin (retention time: 33.1 min).

2.2. FCT Decreased Prostate Weight, DHT Levels, and 5a-Reductase 2 Expression in Rats Induced with BPH

There was no difference in body weight among the experimental animals (Supplemental Table S2). The BPH-induced group showed an 86.30% increase ($p < 0.001$) in prostate weight compared with the control group (Table 1). Administration of FCT-M (50 mg/kg) and FCT-H (100 mg/kg) resulted in significant reductions in prostate weight by 22.08 ($p < 0.05$) and 25.82% ($p < 0.01$), respectively, compared with the BPH group (Table 1). The positive control, the SAW group, had an 18.28% reduction ($p < 0.05$). The prostate ratio (prostate weight/body weight x100) in the FCT-M and FCT-H groups decreased by 22.81 and 26.91%, respectively, compared with that in the BPH group (Table 1). In contrast, the SAW group experienced a 27.76% reduction ($p < 0.01$).

Table 1. The effects of FCT on prostate in TP-induced BPH rats.

Group	Prostate Weight (g)	Prostate Ratio [1]
CON	0.66 ± 0.15	0.16 ± 0.03
BPH	1.23 ± 0.18 ###	0.33 ± 0.05 ###
SAW	1.00 ± 0.15 *	0.24 ± 0.05 **
FCT-L	1.05 ± 0.13	0.28 ± 0.05
FCT-M	0.96 ± 0.18 *	0.26 ± 0.06 *
FCT-H	0.91 ± 0.22 **	0.24 ± 0.06 **

[1] Prostate ratio: prostate weight (mg)/body weight (g) × 100. CON: normal control group; BPH: 3 mg/kg TP injection; SAW: saw palmetto 100 mg/kg + 3 mg/kg TP injection; FCT-L: FCT 25 mg/kg + 3 mg/kg TP injection; FCT-M: FCT 50 mg/kg + 3 mg/kg TP injection; FCT-H: FCT 100 mg/kg + 3 mg/kg TP injection. Values are presented as means ± SD. Significant differences between CON and BPH group were denoted by ### $p < 0.001$, and those versus the BPH group were denoted by * $p < 0.05$ and ** $p < 0.01$, respectively.

The serum DHT level in the BPH group increased by approximately 2.7 times compared to the CON group (Figure 2B). The FCT-H groups had significant reductions in serum DHT levels, by 32.78%, which was higher than that of the SAW group ($p < 0.01$). Prostate DHT levels in the BPH group were 2.5 times higher than those in the CON group; however, FCT significantly reduced prostate DHT levels in all groups. (Figure 2C). In addition, the prostate DHT levels in the SAW group significantly decreased compared to the BPH group ($p < 0.01$).

Prostate 5α-reductase 2 levels (Figure 2D), increased by approximately 29.43% in the BPH group compared to the CON group (no significant difference). All FCT-treated groups experienced significant reductions in 5α-reductase 2 levels compared to the BPH group ($p < 0.001$), with reductions greater than those observed in the SAW group.

Figure 2E shows the results of H&E staining of prostate tissue to examine the histological changes induced by FCT. In the BPH group, compared to the CON group, epithelial thickness was significantly increased due to epithelial cells protruding into the glandular lumen and glands surrounded by double-layered epithelium, while the glandular lumen size was reduced ($p < 0.05$). In contrast, the high-dose FCT-treated group showed a significant reduction in epithelial thickness, by approximately 21.03%, compared to the BPH group and exhibited a structure with thin-walled tubular acini, similar to that of the CON group ($p < 0.05$). The SAW group also showed a reduction in epithelial thickness by approximately 20.26% compared to the BPH group, reaching a level similar to that of the FCT-L group. Furthermore, the lumen area in the FCT-M and FCT-L groups significantly increased by 77.90% and 65.20%, respectively, compared to the BPH group, demonstrating a marked increase in glandular lumen area and a reduction in the height of the glandular epithelium ($p < 0.05$). The SAW group also showed an increase in lumen area compared to the BPH group ($p < 0.05$).

A

No.	Group	Castration	n	Testosterone propionate	Administration
1	CON	N	8	Corn oil	Saline
2	BPH	Y	8	3 mg/kg of Body weight /day	Saline
3	SAW	Y	8		Saw palmetto 100mg/kg of Body weight /day
4	FCT-L	Y	8		FCT 25 mg/kg of Body weight /day
5	FCT-M	Y	8		FCT 50 mg/kg of Body weight /day
6	FCT-H	Y	8		FCT 100 mg/kg of Body weight /day

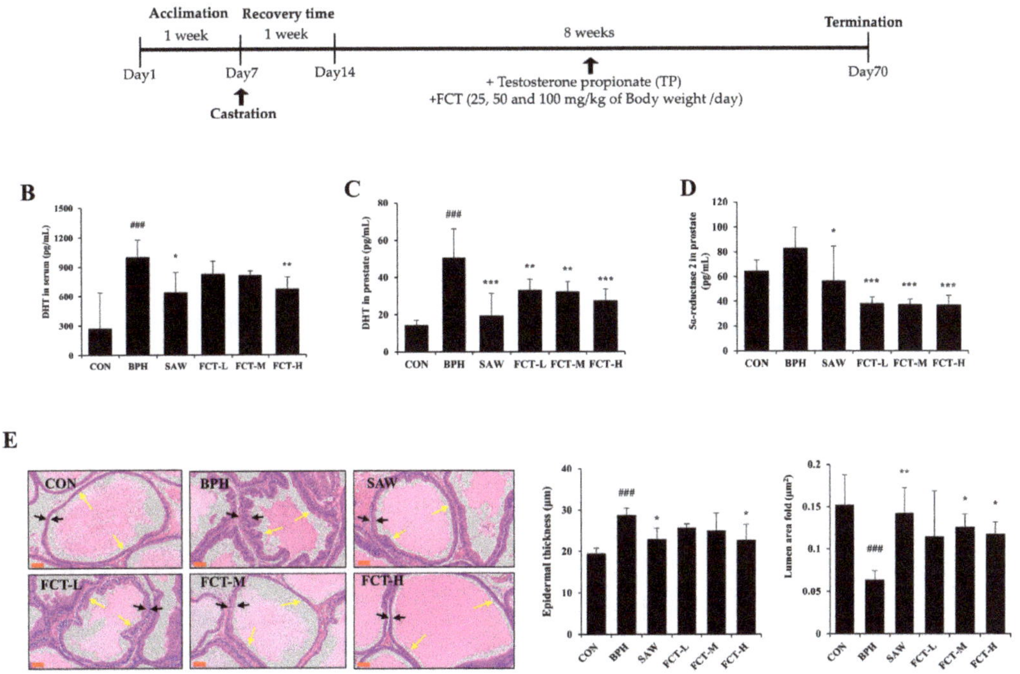

Figure 2. Effects of FCT on prostate weight, DHT levels, 5α-reductase 2 expression, and prostate H&E staining in BPH-induced rats. (**A**) Description of experimental animal groups and study design. (**B**) Serum DHT levels, (**C**) prostate DHT levels and (**D**) prostate 5α-reductase 2 were measured using an ELISA assay. (**E**) H&E staining analyses were performed on prostate tissue (magnification 20×, scale bar 60 μm). Black arrows indicate the epithelial layer and yellow arrows mark the lumen area. CON: normal control group; BPH: 3 mg/kg TP injection; SAW: saw palmetto 100 mg/kg + 3 mg/kg TP injection; FCT-L: FCT 25 mg/kg + 3 mg/kg TP injection; FCT-M: FCT 50 mg/kg + 3 mg/kg TP injection; FCT-H: FCT 100 mg/kg + 3 mg/kg TP injection. Data are presented as mean ± standard deviation (SD). Significant differences between CON and BPH group were denoted by ### $p < 0.001$, and versus the BPH group was denoted by * $p < 0.05$, ** $p < 0.01$, and *** $p < 0.001$, respectively.

2.3. FCT Regulated Apoptosis Signaling via Androgen Receptor in BPH-Induced Rats

Bcl-2 mRNA expression in the prostate tissue increased by 87.98% in the BPH group compared to that in the control group (Figure 3A). However, it was significantly decreased in the FCT-treated groups, similar to that in the SAW group. In addition, the expression of Bcl-2 protein levels was significantly reduced in the FCT group compared to that in the BPH group (Figure 3B). Bax mRNA expression decreased by 53.67% in the BPH group but increased by 55.10% in the FCT-H group compared to the BPH group, exceeding the increase observed in the SAW group. The Bax/Bcl-2 ratio in the BPH group decreased by 74.49% compared to the CON group. However, it significantly increased in FCT-L and

FCT-H groups, with the FCT-H group showing an approximate 2.7-fold increase, 1.5 times higher than that of the SAW group ($p < 0.001$). Next, AR protein levels increased more in the BPH group compared to those in the CON group; however, they decreased in the FCT-M, FCT-H, and SAW groups (Figure 3B).

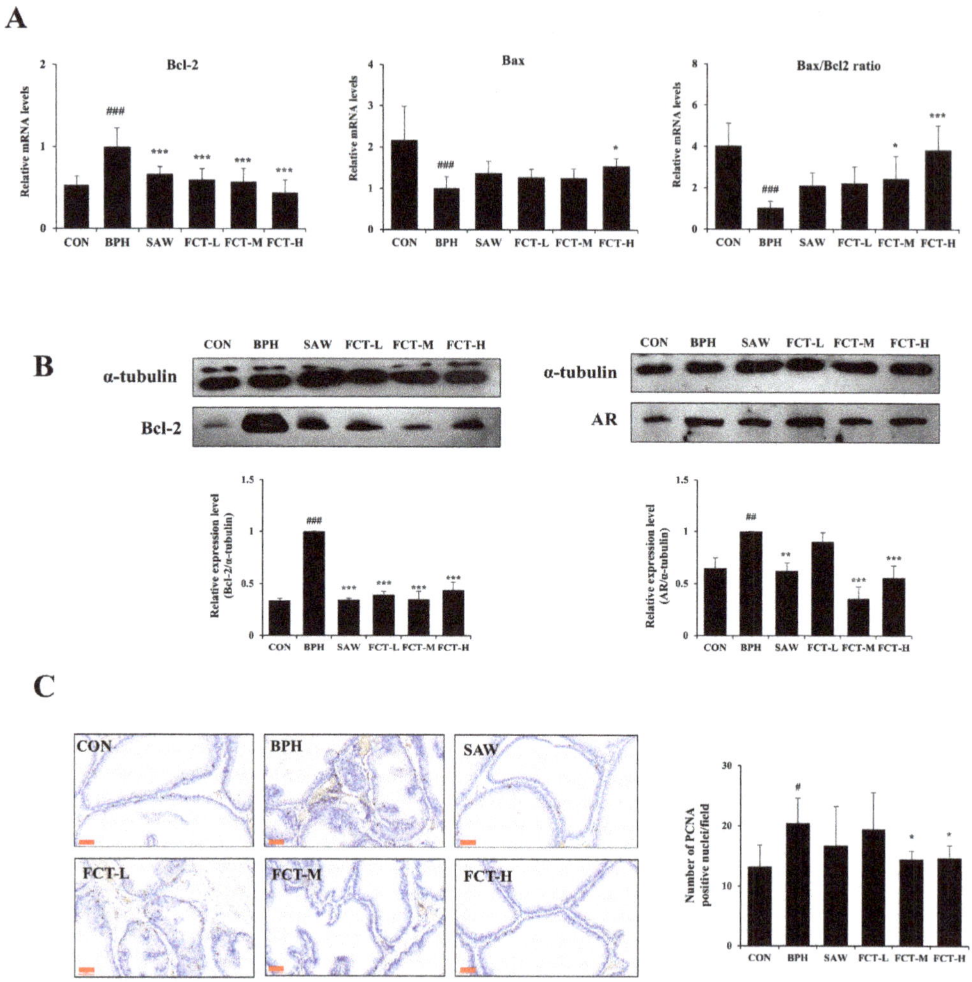

Figure 3. Effects of FCT on Bcl-2, Bax, AR, and PCNA expression in BPH-induced rats. (**A**) The mRNA expression levels of the Bcl-2, Bax and Bax/Bcl-2 ratio were measured by qRT-PCR with GAPDH as the internal control. (**B**) Bcl-2 and AR protein expression were detected by Western blot analysis, using α-tubulin as the loading control. (**C**) PCNA expression was identified using immunohistochemical (IHC) staining of prostate tissue (magnification 20×, scale bar 60 μm). CON: normal control group; BPH: 3 mg/kg TP injection; SAW: saw palmetto 100 mg/kg + 3 mg/kg TP injection; FCT-L: FCT 25 mg/kg + 3 mg/kg TP injection; FCT-M: FCT 50 mg/kg + 3 mg/kg TP injection; FCT-H: FCT 100 mg/kg + 3 mg/kg TP injection. Data are presented as mean ± standard deviation (SD). Significant differences between CON and BPH group are denoted by # $p < 0.05$, ## $p < 0.01$, and ### $p < 0.001$, and, compared to the BPH group, denoted by * $p < 0.05$, ** $p < 0.01$, and *** $p < 0.001$, respectively.

PCNA-positive nuclei in the prostate tissue were markedly increased in the BPH group compared to those in the CON group ($p < 0.05$) (Figure 3C). In addition, PCNA-positive cells significantly decreased in the FCT-M and FCT-H groups (29.21 and 28.10%, respectively) ($p < 0.05$).

2.4. FCT Reduced the mRNA Expression of NF-κB and Pro-Inflammatory Cytokines in BPH-Induced Rats

The mRNA levels of pro-inflammatory cytokines were examined to assess changes in the inflammatory response in rats with TP-induced BPH (Figure 4). The mRNA expressions of NF-κB, COX-2, TNF-α, IL-6, and IL-8 were significantly increased in the BPH group compared to the normal group; however, FCT administration reduced their expression. The NF-κB mRNA expression levels were reduced above 50% compared to the BPH group following FCT treatment ($p < 0.001$) (Figure 4A). COX-2 expression decreased by more than 60% in the FCT-treated groups ($p < 0.001$) (Figure 4B). In addition, FCT treatment led to a decrease in the expression levels of TNF-α, IL-6, and IL-8 in all groups ($p < 0.001$), with significant reductions observed in the SAW group. (Figure 4C–E). Notably, IL-8 expression decreased by more than 80% in the FCT-H group ($p < 0.001$).

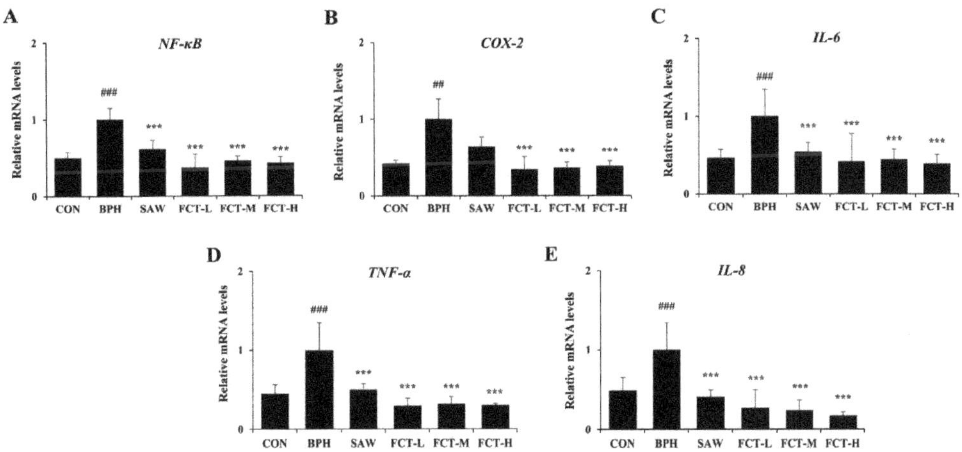

Figure 4. Effects of FCT on pro-inflammatory cytokines in BPH-induced rats. mRNA expression levels of (**A**) NF-κB, (**B**) COX-2, (**C**) IL-6, (**D**) TNF-α, and (**E**) IL-8 were quantified by qRT-PCR. GAPDH was used for data normalization. CON: normal control group, BPH: 3 mg/kg TP injection, SAW: saw palmetto 100 mg/kg + 3 mg/kg TP injection, FCT-L: FCT 25 mg/kg + 3 mg/kg TP injection, FCT-M: FCT 50 mg/kg + 3 mg/kg TP injection, FCT-H: FCT 100 mg/kg + 3 mg/kg TP injection. Data are presented as mean ± standard deviation (SD). Significant differences between CON and BPH group were denoted by ## $p < 0.01$, and ### $p < 0.001$, and versus the BPH group was denoted by *** $p < 0.001$, respectively.

2.5. FCT Inhibited Cell Proliferation and Decreased the mRNA Expression Levels of AR, Steroid 5 Alpha-Reductase 2 (SRD5A2), and Bcl-2 in BPH-1 Cells

We examined the changes in cell proliferation in BPH-1 epithelial cells following FCT treatment (Figure 5A). A significant reduction in cell viability was observed at a concentration of 5 μg/mL of the extract, with decreases of 6.37, 39.03, and 78.79% at concentrations of 5, 10, and 25 μg/mL, respectively, showing a dose-dependent decrease. Cell viability decreased by over 80% at concentrations of 50 μg/mL or higher, indicating a strong inhibitory effect on cell proliferation due to FCT ($p < 0.001$).

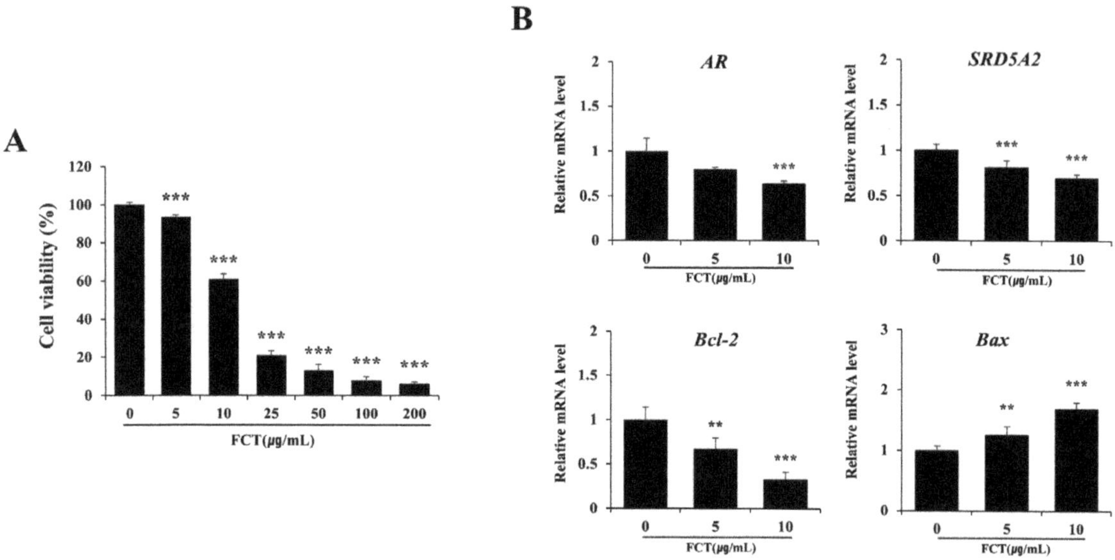

Figure 5. Effects of FCT on cell proliferation and mRNA expression levels of AR, SRD5A2, Bcl-2, and Bax in BPH-1 cells. (**A**) Cell viability was measured in BPH-1 cells after FCT treatment (0 to 200 μg/mL) for 24 h using the MTT assay. (**B**) mRNA expression levels of AR, SRD5A2, Bcl-2, and Bax were analyzed after FCT treatment (0 to 10 μg/mL) for 24 h by qRT-PCR, using GAPDH as the control. Data are presented as mean ± standard deviation (SD). Significant differences relative to the control were denoted by ** $p < 0.01$ and *** $p < 0.001$, respectively.

In addition to the inhibition of cell proliferation, we investigated changes in the mRNA expression levels of AR, SRD5A2, Bcl-2, and Bax, which are involved in the regulation of apoptosis, in the FCT-treated groups (Figure 5B). AR expression significantly decreased by 35.52% following treatment with FCT (10 μg/mL), whereas SRD5A2 showed a dose-dependent decrease starting from 5 μg/mL. Bcl-2 expression significantly decreased, by 66.80%, at 10 μg/mL. In contrast, Bax expression increased in a dose-dependent manner from 5 to 10 μg/mL.

2.6. FCT Suppressed the Protein Expression of NF-κB p65, COX-2, TNF-A, IL-6, and IL-1β in Cells

Treatment of BPH-1 cells with FCT resulted in a dose-dependent reduction in NF-κB p65 expression (Figure 6A). FCT at 25 μg/mL reduced nuclear translocation of NF-κB p65 and decreased phosphorylated NF-κB p65 levels within the nucleus. Consequently, FCT suppressed NF-κB p65 activation, alongside a decrease in pro-inflammatory cytokine expression.

We examined the protein expression changes in COX-2, TNF-α, IL-6, and IL-1β to investigate the relationship between FCT and inflammatory responses in BPH-1 cells (Figure 6B). FCT at 5 μg/mL did not affect the expression of TNF-α and IL-6. However, we observed a decrease in protein expression at 10 μg/mL. Notably, at 25 μg/mL, the expression of COX-2, TNF-α, IL-6, and IL-1β were significantly reduced.

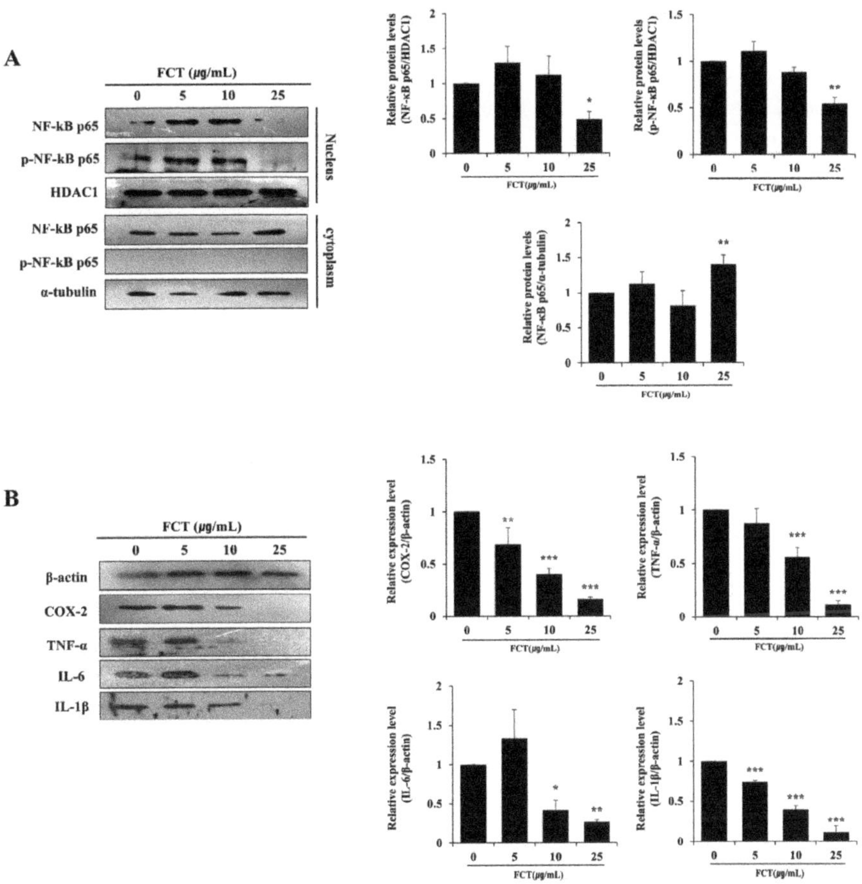

Figure 6. Effects of FCT on NF-kB p65 and pro-inflammatory cytokines' protein expression in BPH-1 cells. (**A**) Western blot analysis was performed to detect nucleus/cytoplasm NF-kB p65 protein levels in BPH-1 cells. HDAC1 in the nucleus and α-tubulin in the cytoplasm were used as loading controls. (**B**) Pro-inflammatory cytokine protein levels were measured by Western blot analysis, with β-actin as the internal control. Data are presented as mean ± standard deviation (SD). Significant differences versus control were denoted by * $p < 0.05$, ** $p < 0.01$, and *** $p < 0.001$, respectively.

3. Discussion

BPH has become a major urological health problem that reduces the quality of life of men increasingly with the age of the population [22]. The pathophysiology of BPH is poorly understood; however, hormones, imbalances between prostate proliferation and apoptosis, and inflammation are reportedly associated with the development and progression of the disease [23,24].

In the prostate gland, the more potent androgen, DHT, is synthesized from testosterone and catalyzed by 5AR. DHT has a three-fold higher affinity for AR compared to testosterone, mediating BPH development [25]. In addition, AR expression was significantly higher in patients with BPH than in normal prostate tissue, as observed in both epithelial and stromal cells [10]. In this study, FCT-treated groups (FCT-M and -H) showed reduced prostate weight, DHT levels in both serum and prostate tissue, and reduced 5AR and AR expression levels in a testosterone-propionate induced BPH animal model.

An imbalance between cell proliferation and death is a prominent factor in BPH. The anti-apoptotic gene Bcl-2 was significantly upregulated, whereas Bax, p53, and caspase-3 were significantly downregulated in the BPH-induced rat model [10,24]. Various studies reported that the expression of NF-κB was enhanced in the BPH model, and it affects the cell proliferation/apoptosis balance, including Bcl-2 upregulation and inflammation. Specifically, NF-κB was activated in the BPH-induced group, and the expression of IL-8, TNF-α, and COX-2 increased significantly [24,26]. In this study, the FCT-treated group experienced downregulated Bcl-2 expression, upregulated relative expression of Bax/Bcl-2, and reduced expression of PCNA (the proliferation marker). In addition, the FCT-treated group had reduced NF-κB and inflammatory factors, comprising COX-2, TNF-α, IL-6, and IL-8, compared to the BPH group.

NF-κB is a transcription factor that regulates genes involved in cell proliferation, apoptosis, inflammation, and other biological processes [27,28]. The transcription factor NF-κB, activated by various stimuli including cytokines and growth factors, translocates from the cytoplasm to the nucleus to initiate transcriptional responses, upregulating genes mediating cell proliferation such as TNF-α, IL-6, and IL-1β upon appropriate stimulation [29,30]. Inflammation is important in BPH, and TNF-α, IL-6, and IL-8 are major pro-inflammatory cytokines involved in the occurrence and development of BPH [23,26]. IL-1β is a key cytokine in regulating the inflammatory process and inducing COX-2 expression [27]. COX-2 is upregulated in the prostate during significant inflammation and is associated with Bcl-2 overexpression in BPH [31]. In this study, NF-κB was activated by its translocation from the cytosol to the nucleus of BPH-1 cells. However, following FCT treatment of BPH-1 cells, nuclear translocation of NF-κB was suppressed, resulting in the downregulation of genes associated with cell proliferation and inflammation. Moreover, FCT treatment of BPH cells suppressed pro-inflammatory cytokines, which is consistent with the in vivo results.

Fenugreek seed contains alkaloids, flavonoids, and oils, primarily linolenic acid, linoleic acid, and oleic acid [14]. Vitexin was used as the standard compound in high-performance liquid chromatography (HPLC) analysis for FCT. The key distinguishing factor of the fenugreek extract used in this study compared to conventional fenugreek extracts (Testofen) lies in the composition and content of polyphenolic compounds. The vitexin content of the fenugreek extract used in this experiment is approximately 3–4 times higher than that of the extract in previous studies. Structural differences in polyphenolic compounds result in changes in physical and chemical properties, and bioactivity, which may lead to different effects between the two extracts [32]. Vitexin is a polyphenolic compound known for its prominent anti-inflammatory properties by modulating the NF-κB pathway and pro-inflammatory cytokines [33,34]. A previous study reported that the fenugreek seed extract reduced IL-6 in an ovariectomized rat model [35] and downregulated NF-κB, TNF-α, IL-6, IL-8, and IL-1β in a pulmonary fibrosis animal model [36]. In the current study, the effect of FCT on BPH is attributed to its various bioactive compounds, which modulate the NF-κB pathway and pro-inflammatory cytokines. We hypothesize that the effects of FCT on BPH are attributed to vitexin and its anti-inflammatory and anti-proliferative properties.

In this study, we showed that FCT includes flavone-c-glycoside such as vitexin, iso-vitexin, orientin, and iso-orientin through HPLC analysis. The flavone-c-glycoside generates aglycones, such as deglycosylated and acetylated metabolites, in the intestine through the metabolic pathways like glucuronidation and sulfation [37]. Dong's study reported that vitexin, when absorbed and fragmented in vivo, generates intermediates such as apigeni, and fragment ions with high abundance [38]. Additionally, Shi's study found that orientin is metabolized in the human intestine, releasing aglycones, and specific strains generate luteolin [37].

The structural diversities of flavonoid compounds largely contribute to their physical, chemical properties, and antioxidant activity [32]. Therefore, it is important to note that the effective forms of C-glycoside flavonoids such as vitexin and orientin may be the metabolites, like flavonols or aglycones, arising from in vivo biotransformation. In contrast, flavonoids exhibit high antioxidant activity in vitro, which may not be the case in vivo [32]. Thus, our further study will focus on examining the changes in physiological activity when the metabolites of FCT, including vitexin, are absorbed in vivo.

We observed a discrepancy between the in vitro and in vivo results with FCT treatment. For example, in the in vitro study, the apoptosis-related genes Bax and Bcl-2, as well as anti-inflammatory cytokines, were dose-dependently downregulated following FCT treatment. However, in the in vivo study, their downregulation did not exhibit a dose-dependent pattern. In a previous report [39,40], flavone C-glycosides were found to have poor absorption in the rat intestine and to potentially resist degradation by gut bacteria in rats, as less than 1% was excreted in urine and 10–88% was recovered from feces. Although the efficacy of flavonoids, including flavone C-glycoside, in animal studies is not easy to predicted based on in vitro results, the activity of flavone could be explained in the intestinal lumen rather than in circulation [39].

This discrepancy may be attributed to the bioavailability of flavonoids and the complexity of their metabolic pathways. In future studies, the analysis of flavonoids in blood, urine, and feces will be necessary to clarify the underlying causes of these inconsistencies between in vitro and in vivo results. We plan to investigate the effects of vitexin and orientin, including intermediates like apigenin, luteolin, and fragment ions on BPH and clarify their mechanisms of action.

4. Materials and Methods

4.1. Preparation of Fenugreek Seed Extracts

We obtained the FCT powder (Fenugreek seed ethanol extract) from Daehan Chemtech Co., Ltd. (Gwacheon, Republic of Korea). The fenugreek seed was prepared through hydroalcoholic extraction. It was first extracted with a solvent mixture of ethanol and water (80:20 v/v) for 8 h. The mixture was then filtered and concentrated. The concentrated product was dried, sieved, and blended. FCT was standardized to have a vitexin content of 3–4% and was manufactured in a GMP-compliant facility.

4.2. Liquid Chromatography

FCT powder was accurately weighed in a volumetric flask for HPLC analysis. Subsequently, ethanol (80%) was added to the flask, and the contents were sonicated for 10 min. We performed the analysis at 30 °C using a LiChrosorb® RP18 column (4.6 × 250 mm, 5 μm particle size) and at a wavelength of 330 nm. The mobile phases were solvent A (0.1% orthophosphoric acid in water) and B (acetonitrile). Afterward, the analysis was performed at a flow rate of 1.5 mL/min under the following linear gradient conditions: 90% A to 10% B for 8 min; 87% A to 13% B for 17 min; 80% A to 20% B for 5 min; 80% A to 20% B for 10 min; 0% A to 100% B for 5 min; and 90% A to 10% B for 2 min.

4.3. Cell Proliferation

We cultured BPH-1, a human benign prostate hyperplastic cell line, in RPMI 1640 medium containing 20% fetal bovine serum, 1% antibiotics, 10 ng/mL insulin–transferrin–sodium selenite, and 20 ng/mL testosterone. The method for cell viability measurement followed a previous study [41]. Subsequently, we seeded a 96-well plate with 10,000 cells/well, added 0 to 200 μg/mL of FCT, and cultured for 24 h. Afterwards, 5 mg/mL of 3-(4,5-Dimethylthiazol-2-yl)-2,5-diphenyltetrazolium bromide was added. The absorbance was measured after 2 h by subtracting the absorbance at 630 nm from that at 570 nm. Cell proliferation was calculated considering the control as 100%.

4.4. Separation of the Nuclear and the Cytoplasmic Fractions in BPH-1 Cells

Cells were seeded at a density of 10^6 cells/mL in a 100 mm dish and incubated for 24 h. After treating the cells with FCT at concentrations of 0, 5, 10, and 25 μg/mL for 24 h, the cells were centrifuged at 500× *g* for 2–3 min to collect the pellet. Nuclear and cytoplasmic fractions were separated from the cells using NE-PER™ Nuclear and Cytoplasmic Extraction Reagents (Thermo Fisher Scientific, Waltham, MA, USA) according to the manufacturer's instructions. The separated fractions were then subjected to lysis for subsequent Western blot analysis.

4.5. Quantitative Reverse Transcription Polymerase Chain Reaction (qRT-PCR)

We used RNA-iso Plus (Takara, Shiga, Japan) for RNA extraction according to the manufacturer's protocol. Complementary DNA was synthesized from RNA using reverse transcriptase reagent (Takara), a T100 Thermal Cycler (Bio-Rad, Hercules, CA, USA). qRT-PCR was performed using a Light Cycler 96 (Roche, Basel, Switzerland) and FastStart Essential DNAS Green Master Mix (Roche, Basel, Switzerland). The samples were initially preincubated at 95 °C for 10 min, followed by 40 amplification cycles under the following conditions: denaturation at 95 °C for 10 s, annealing at 55 °C for 10 s, and extension at 72 °C for 10 s. This was succeeded by a final step of 95 °C for 5 s and 65 °C for 60 s. The cooling step was carried out at 37 °C for 30 s. All experiments were performed in triplicate. The primer sequences are listed in Supplemental Table S1.

4.6. Animals

We acclimatized seven-week-old male Sprague Dawley rats (n = 48; Saeron Bio, Uiwang, Republic of Korea) for one week under controlled conditions at 23 ± 2 °C, 55 ± 5% relative humidity, and a 12 h light–dark cycle. The study protocol was approved by the Institutional Animal Care and Utilization Committee of Suwon University (approval code and date: USW-IACUC-2023-001 and 13 April 2023). Subsequently, castration was performed in all groups, excluding the control group (n = 8). Testosterone propionate (TP) (3 mg/kg body weight/day) was injected subcutaneously to induce prostate hypertrophy after stabilization for a week. The above experiment was conducted for each animal group as follows: Water and food were freely consumed throughout the experiment, and the samples were administered orally daily for eight weeks. The rats were used and randomly divided into the following 6 groups: (A) injected with corn oil subcutaneously and treated with saline by oral gavage (CON; n = 8); (B) injected with TP (3 mg/kg B.W/day) and treated with saline (BPH; n = 8); (C) injected with TP (3 mg/kg B.W/day) and treated with FCT (25 mg/kg B.W/day) (FCT-L; n = 8); (D) injected with TP (3 mg/kg B.W/day) and treated with FCT (50 mg/kg B.W/day) (FCT-M; n = 8), (E) injected with TP (3 mg/kg B.W/day) and treated with FCT (100 mg/kg B.W/day) (FCT-H; n = 8); (F) injected with TP (3 mg/kg B.W/day) and treated with saw palmetto (100 mg/kg B.W/day) (SAW; n = 8). After eight weeks, all animals were sacrificed using Carbon dioxide (CO_2) anesthesia. Blood was collected via cardiac puncture and centrifuged at 1500× *g* for 20 min to obtain serum. To isolate the prostate from the abdomen, the seminal vesicles, bladder, and prostate were separated sequentially. Using fine scissors, the bladder was removed first, followed by the seminal vesicles, leaving only the prostate surrounding the urethra. Prostate tissues were weighed, and organs were stored at −80 °C.

4.7. Castration

Castration was performed based on the OECD Hershberger Bioassay [42]. The scrotal skin and peritoneum of the animal are incised to expose the testes, and the blood vessels connecting to the vas deferens are ligated with surgical sutures. The testes are removed, and the open scrotal skin is closed by applying automatic clips to the incision edges. After a 7-day recovery period following the surgery, the administration of FCT and TP injection begins.

4.8. H&E Staining

The ventral and lateral prostate tissues were fixed in 4% formaldehyde and embedded in melting wax to prepare the paraffin blocks. The prostate tissue staining method followed a previous study [41]. The paraffin sections were dewaxed in xylene solution and dehydrated in ethanol for staining. The tissue was dehydrated, followed by eosin staining, and finally covered with coverslips after staining with hematoxylin.

4.9. Immunohistochemistry (IHC)

Prostate tissues were embedded in paraffin blocks for proliferating cell nuclear antigen antibody staining after fixation with 4% formaldehyde. The overall method for IHC staining followed a previous study [43]. Each section was dehydrated by washing with xylene and ethanol. Subsequently, the sections were incubated with 3% hydrogen peroxide and washed for blocking. The primary antibody was incubated overnight, and after washing with the wash buffer, the secondary antibody was incubated for 30 min. The sections were incubated using ABC (Avidin/Biotin Complex), and the staining results were visualized using a DAB kit as a substrate. Afterward, the sections were dehydrated again and mounted on coverslips.

4.10. Western Blot Analysis

The cells were lysed in Radioimmunoprecipitation Assay (RIPA) buffer and centrifuged at 13,000× *g* for 15 min to obtain the supernatant. Similarly, prostate tissue was homogenized in RIPA buffer and centrifuged for protein quantification. The method and procedure for Western blot analysis followed a previous study [41]. The protein samples (15 μg) were separated by sodium dodecyl sulfate–polyacrylamide gel electrophoresis (SDS-PAGE) gels and transferred onto nitrocellulose membranes. The membranes were blocked with 5% skim milk and incubated overnight at 4 °C with primary antibodies. Furthermore, the membranes were incubated with horseradish peroxidase (HRP)-conjugated secondary antibodies. Signals were detected using an enhanced chemiluminescence (ECL) substrate and an X-ray film. The primary and secondary antibodies used are as follows: AR (Thermo Fisher Scientific, Waltham, MA, USA), Nuclear factor-kappa B, tumor necrosis factor-alpha, Histone Deacetylase, interleukin-6, and interleukin-1β (Santa Cruz Biotechnology, Dallas, TX, USA, respectively), phosphorylated NF-κB (Cell Signaling Technology, Danvers, MA, USA), α-tubulin (Merk Millipore, Burlington, MA, USA), Cyclooxygenase-2 (Cell Signaling Technology). Anti-rabbit IgG and anti-mouse IgG (Cell Signaling Technology) were used. All antibodies were diluted at a ratio of 1:1000 in 5% skim milk.

4.11. Chemical Analysis of Serum

The NX500i (Fujifilm, Tokyo, Japan) was used to analyze serum samples obtained by centrifuging blood from experimental animals at 1500× *g* for 30 min. Alanine aminotransferase (ALT) levels and aspartate aminotransferase (AST) levels were measured. The detailed protocol for serum analysis followed a previous study [41].

4.12. ELISA in Serum and Prostate

DHT levels were assessed using a DHT ELISA assay kit (Cusabio, Houston, TX, USA) in both the serum and prostate. ELISA analysis was performed following the manufacturer's instructions, and the detailed methods were referenced from a previous study [41].

Prostate tissues were homogenized in phosphate-buffered saline and incubated at −20 °C overnight. The samples were centrifuged at 4 °C and 5000 rpm for 5 min, and the supernatant was collected for analysis after two freeze–thaw cycles were performed to damage the cell membrane. The supernatant and serum were incubated with HRP-conjugated DHT and DHT-specific antibody in a 96-well microplate pre-coated with goat-anti-rabbit antibody at 37 °C for 1 h. Finally, 5-alpha reductase 2 activity in the prostate samples was evaluated using the 5-alpha reductase 2 ELISA kit (Cusabio, Houston, TX, USA).

4.13. Statistical Analysis

The experimental groups in this experiment were expressed as mean ± SD. Statistical significance was assessed using one-way analysis of variance (ANOVA), followed by Dunnett's post hoc test after confirming normality assumptions. The Kruskal–Wallis H test was performed for nonparametric analysis, followed by Dunn's post hoc test. Statistical significance was set at $p < 0.05$. The Statistical Package for Social Science, version 22.0 (IBM Co., Armonk, NY, USA) was used for statistical analysis.

5. Conclusions

In summary, our results showed that FCT treatment reduced DHT and 5-alpha reductase levels, prostate weight in the BPH-rat model, and regulated cell proliferation and apoptosis through modulating NF-kB and pro-inflammatory cytokines, including TNF-α, IL-6, IL-8, and IL-1β. These suggest that FCT could ameliorate BPH.

Supplementary Materials: The following supporting information can be downloaded at https://www.mdpi.com/article/10.3390/ijms26031261/s1.

Author Contributions: Conceptualization, J.K., K.-S.B., D.O. and Y.-H.L.; data curation, J.Y.L., J.B. and Y.-H.L.; formal analysis, J.K., K.-S.B., D.O., J.Y.L. and Y.-H.L.; funding acquisition, Y.-H.L.; investigation, J.Y.L., J.B. and Y.-H.L.; methodology, J.Y.L., J.B. and Y.-H.L.; project administration, Y.-H.L.; resources, J.K., K.-S.B., D.O. and Y.-H.L.; supervision, Y.-H.L.; validation, Y.-H.L.; visualization, J.K., K.-S.B., D.O. and Y.-H.L.; writing—original draft preparation, J.Y.L., J.B. and Y.-H.L.; writing—review and editing, J.K., K.-S.B., D.O. and Y.-H.L. All authors have read and agreed to the published version of the manuscript.

Funding: This research was funded by Daehan Chemtech Co., Ltd., grant number 2023-0148.

Institutional Review Board Statement: The study protocol was approved by the Institutional Animal Care and Utilization Committee of Suwon University (approval code and date: USW-IACUC-2023-001 and 13 April 2023).

Informed Consent Statement: Not applicable.

Data Availability Statement: The original contributions presented in this study are included in the article/Supplementary Materials. Further inquiries can be directed to the corresponding author.

Conflicts of Interest: The authors declare that this study received funding from Daehan Chemtech Co., Ltd. Three employees of Daehan Chemtech Co., Ltd. are co-authors of this study. These co-authors contributed to the conceptualization of the study together with other authors, provided resources by supplying the material (Fenugreek seed extract) used in the research, and generated a graphical representation of the HPLC analysis. Additionally, they reviewed the manuscript and provided comments. However, all experimental procedures, data analysis, and interpretation of results were conducted independently by the remaining authors. The final decisions regarding the manuscript's content and conclusions were made solely by the independent authors, without any undue influence from the funding organization.

Abbreviations

The following abbreviations are used in this manuscript:

BPH	Benign Prostatic Hyperplasia
DHT	Dihydrotestosterone
T	Testosterone
TP	Testosterone Propionate
SD	Sprague–Dawley
FCT	Fenugreek seed extract (Forceterone®)
COX-2	Cyclooxygenase-2
TNF-α	Tumor Necrosis Factor-α
IL-1β	Interleukin-1β
IL-6	Interleukin-6
IL-8	Interleukin-8
Bcl-2	B-cell lymphoma 2
Bax	Bcl-2 associated protein X
NF-κB	Nuclear factor κB
pNF-κB	Phosphorylated NF-κB
PCNA	Proliferating Cell Nuclear Antigen
AR	Androgen Receptor

References

1. Passos, G.R.; Ghezzi, A.C.; Antunes, E.; de Oliveira, M.G.; Mónica, F.Z. The Role of Periprostatic Adipose Tissue on Prostate Function in Vascular-Related Disorders. *Front. Pharmacol.* **2021**, *12*, 626155. [CrossRef] [PubMed]
2. Xie, Z.; Liu, X.; Hu, Z.; Dong, C. Longitudinal Association between the Dynamic Nature of Depression with Lower Urinary Tract Symptoms Suggestive of Benign Prostatic Hyperplasia (LUTS/BPH). *BMC Public Health* **2024**, *24*, 1121. [CrossRef]
3. Chess-Williams, R.; McDermott, C.; Sellers, D.J.; West, E.G.; Mills, K.A. Chronic Psychological Stress and Lower Urinary Tract Symptoms. *Low. Urin. Tract Symptoms* **2021**, *13*, 414–424. [CrossRef] [PubMed]
4. Csikós, E.; Horváth, A.; Ács, K.; Papp, N.; Balázs, V.L.; Dolenc, M.S.; Kenda, M.; Glavač, N.K.; Nagy, M.; Protti, M.; et al. Treatment of Benign Prostatic Hyperplasia by Natural Drugs. *Molecules* **2021**, *26*, 7141. [CrossRef] [PubMed]
5. Miernik, A.; Gratzke, C. Current Treatment for Benign Prostatic Hyperplasia. *Dtsch. Arztebl. Int.* **2020**, *117*, 843–854. [CrossRef]
6. Shin, Y.S.; Karna, K.K.; Choi, B.R.; Park, J.K. Finasteride and Erectile Dysfunction in Patients with Benign Prostatic Hyperplasia or Male Androgenetic Alopecia. *World J. Men's Health* **2019**, *37*, 157–165. [CrossRef]
7. Li, Y.; Ma, J.; Qin, X.H.; Hu, C.Y. The Efficacy and Safety of Dutasteride and Finasteride in Patients with Benign Prostatic Hyperplasia: A Systematic Review and Meta-Analysis. *Transl. Androl. Urol.* **2022**, *11*, 313–324. [CrossRef]
8. Cannarella, R.; Condorelli, R.A.; Barbagallo, F.; La Vignera, S.; Calogero, A.E. Endocrinology of the Aging Prostate: Current Concepts. *Front. Endocrinol.* **2021**, *12*, 554078. [CrossRef]
9. Deslypere, J.P.; Young, M.; Wilson, J.D.; McPhaul, M.J. Testosterone and 5 Alpha-Dihydrotestosterone Interact Differently with the Androgen Receptor to Enhance Transcription of the MMTV-CAT Reporter Gene. *Mol. Cell. Endocrinol.* **1992**, *88*, 15–22. [CrossRef]
10. Tong, Y.; Zhou, R.Y. Review of the Roles and Interaction of Androgen and Inflammation in Benign Prostatic Hyperplasia. *Mediat. Inflamm.* **2020**, *2020*, 7958316. [CrossRef]
11. Li, Y.; Shi, B.; Dong, F.; Zhu, X.; Liu, B.; Liu, Y. Effects of Inflammatory Responses, Apoptosis, and STAT3/NF-KB- and Nrf2-Mediated Oxidative Stress on Benign Prostatic Hyperplasia Induced by a High-Fat Diet. *Aging* **2019**, *11*, 5570–5578. [CrossRef]
12. Quiles, M.T.; Arbós, M.A.; Fraga, A.; De Torres, I.M.; Reventós, J.; Morote, J. Antiproliferative and Apoptotic Effects of the Herbal Agent *Pygeum Africanum* on Cultured Prostate Stromal Cells from Patients with Benign Prostatic Hyperplasia (BPH). *Prostate* **2010**, *70*, 1044–1053. [CrossRef] [PubMed]
13. Madersbacher, S.; Sampson, N.; Culig, Z. Pathophysiology of Benign Prostatic Hyperplasia and Benign Prostatic Enlargement: A Mini-Review. *Gerontology* **2019**, *65*, 458–464. [CrossRef] [PubMed]
14. Visuvanathan, T.; Than, L.T.L.; Stanslas, J.; Chew, S.Y.; Vellasamy, S. Revisiting *Trigonella foenum-graecum* L.: Pharmacology and Therapeutic Potentialities. *Plants* **2022**, *11*, 1450. [CrossRef] [PubMed]
15. Yao, D.; Zhang, B.; Zhu, J.; Zhang, Q.; Hu, Y.; Wang, S.; Wang, Y.; Cao, H.; Xiao, J. Advances on Application of Fenugreek Seeds as Functional Foods: Pharmacology, Clinical Application, Products, Patents and Market. *Crit. Rev. Food Sci. Nutr.* **2020**, *60*, 2342–2352. [CrossRef]

16. Sevrin, T.; Boquien, C.Y.; Gandon, A.; Grit, I.; de Coppet, P.; Darmaun, D.; Alexandre-Gouabau, M.C. Fenugreek Stimulates the Expression of Genes Involved in Milk Synthesis and Milk Flow through Modulation of Insulin/GH/IGF-1 Axis and Oxytocin Secretion. *Genes* **2020**, *11*, 1208. [CrossRef]
17. Sun, W.; Shahrajabian, M.H.; Cheng, Q. Fenugreek Cultivation with Emphasis on Historical Aspects and Its Uses in Traditional Medicine and Modern Pharmaceutical Science. Mini-Rev. *Med. Chem.* **2021**, *21*, 724–730.
18. Geberemeskel, G.A.; Debebe, Y.G.; Nguse, N.A. Antidiabetic Effect of Fenugreek Seed Powder Solution (*Trigonella foenum-graecum* L.) on Hyperlipidemia in Diabetic Patients. *J. Diabetes Res.* **2019**, *2019*, 8507453. [CrossRef]
19. Paneru, D.; Tellez-Isaias, G.; Bottje, W.G.; Asiamah, E.; Abdel-Wareth, A.A.; Salahuddin, M.; Lohakare, J. Modulation of Immune Response and Cecal Microbiota by Dietary Fenugreek Seeds in Broilers. *Vet. Sci.* **2024**, *11*, 57. [CrossRef]
20. Al-Timimi, L.A.N. Antibacterial and Anticancer Activities of Fenugreek Seed Extract. *Asian Pac. J. Cancer Prev.* **2019**, *20*, 3771–3776. [CrossRef]
21. Rao, A.; Grant, R. The Effect of Trigonella foenum-graecum Extract on Prostate-Specific Antigen and Prostate Function in Otherwise Healthy Men with Benign Prostate Hyperplasia. *Phytother. Res.* **2020**, *34*, 634–639. [CrossRef]
22. Ye, Z.; Wang, J.; Xiao, Y.; Luo, J.; Xu, L.; Chen, Z. Global Burden of Benign Prostatic Hyperplasia in Males Aged 60-90 Years from 1990 to 2019: Results from the Global Burden of Disease Study 2019. *BMC Urol.* **2024**, *24*, 193. [CrossRef] [PubMed]
23. Chughtai, B.; Lee, R.; Te, A.; Kaplan, S. Role of Inflammation in Benign Prostatic Hyperplasia. *Rev. Urol.* **2011**, *13*, 147. [PubMed]
24. Sudeep, H.V.; Venkatakrishna, K.; Amrutharaj, B.; Anitha Shyamprasad, K. A Phytosterol-Enriched Saw Palmetto Supercritical CO_2 Extract Ameliorates Testosterone-Induced Benign Prostatic Hyperplasia by Regulating the Inflammatory and Apoptotic Proteins in a Rat Model. *BMC Complement. Altern. Med.* **2019**, *19*, 270. [CrossRef] [PubMed]
25. Liu, J.; Fang, T.; Li, M.; Song, Y.; Li, J.; Xue, Z.; Li, J.; Bu, D.; Liu, W.; Zeng, Q.; et al. Pao Pereira Extract Attenuates Testosterone-Induced Benign Prostatic Hyperplasia in Rats by Inhibiting 5α-Reductase. *Sci. Rep.* **2019**, *9*, 19703. [CrossRef]
26. Mostafa, F.; Mantawy, E.M.; Azab, S.S.; El-Demerdash, E. The Angiotensin Converting Enzyme Inhibitor Captopril Attenuates Testosterone-Induced Benign Prostatic Hyperplasia in Rats; a Mechanistic Approach. *Eur. J. Pharmacol.* **2019**, *865*, 172729. [CrossRef]
27. Wang, S.; Li, Y.; Li, W.; Zhang, K.; Yuan, Z.; Cai, Y.; Xu, K.; Zhou, J.; Du, Z. Curcuma Oil Ameliorates Benign Prostatic Hyperplasia through Suppression of the Nuclear Factor-Kappa B Signaling Pathway in Rats. *J. Ethnopharmacol.* **2021**, *279*, 113703. [CrossRef]
28. Bhatt, D.; Ghosh, S. Regulation of the NF-KB-Mediated Transcription of Inflammatory Genes. *Front. Immunol.* **2014**, *5*, 71. [CrossRef]
29. Hayden, M.S.; Ghosh, S. Shared Principles in NF-KappaB Signaling. *Cell* **2008**, *132*, 344–362. [CrossRef]
30. Sethi, G.; Sung, B.; Aggarwal, B.B. Nuclear Factor-KappaB Activation: From Bench to Bedside. *Exp. Biol. Med.* **2008**, *233*, 21–31. [CrossRef]
31. Wang, W.; Bergh, A.; Damber, J.E. Chronic Inflammation in Benign Prostate Hyperplasia Is Associated with Focal Upregulation of Cyclooxygenase-2, Bcl-2, and Cell Proliferation in the Glandular Epithelium. *Prostate* **2004**, *61*, 60–72. [CrossRef]
32. Xie, L.; Deng, Z.; Zhang, J.; Dong, H.; Wang, W.; Xing, B.; Liu, X. Comparison of Flavonoid O-Glycoside, C-Glycoside and Their Aglycones on Antioxidant Capacity and Metabolism during In Vitro Digestion and In Vivo. *Foods* **2022**, *11*, 882. [CrossRef] [PubMed]
33. Zhou, G.; Cui, J.; Xie, S.; Wan, H.; Luo, Y.; Guo, G. Vitexin, a Fenugreek Glycoside, Ameliorated Obesity-Induced Diabetic Nephropathy via Modulation of NF-KB/IkBα and AMPK/ACC Pathways in Mice. *Biosci. Biotechnol. Biochem.* **2021**, *85*, 1183–1193. [CrossRef] [PubMed]
34. Dong, L.Y.; Li, S.; Zhen, Y.L.; Wang, Y.N.; Shao, X.; Luo, Z.G. Cardioprotection of Vitexin on Myocardial Ische-mia/Reperfusion Injury in Rat via Regulating Inflammatory Cytokines and MAPK Pathway. *Am. J. Chin. Med.* **2013**, *41*, 1251–1266. [CrossRef] [PubMed]
35. Lodra, E.H.; Effendi, M.C.; Pematasari, N.; Dradjat, R.S. Fenugreek Seed Ethanolic Extract Improves Alveolar Bone Parameters by Attenuating Inflammation in Ovariectomized Rats. *J. Inflamm. Res.* **2023**, *16*, 4933–4940. [CrossRef]
36. Kandhare, A.D.; Bodhankar, S.L.; Mohan, V.; Thakurdesai, P.A. Effect of Glycosides Based Standardized Fenugreek Seed Extract in Bleomycin-Induced Pulmonary Fibrosis in Rats: Decisive Role of Bax, Nrf2, NF-KB, Muc5ac, TNF-α and IL-1β. *Chem. Biol. Interact.* **2015**, *237*, 151–165. [CrossRef]
37. Xu, J.; Qian, D.; Jiang, S.; Zhang, X.; Shang, E.; Guo, J.; Duan, J. Application of Ultra-Performance Liquid Chromatography Coupled with Quadrupole Time-of-Flight Mass Spectrometry to Determine the Metabolites of Orientin Produced by Human Intestinal Bacteria. *J. Chromatogr. B* **2014**, *944*, 123–127. [CrossRef]
38. Dong, P.; Shi, L.; Wang, S.; Jiang, S.; Li, H.; Dong, F.; Xu, J.; Dai, L.; Zhang, J. Rapid Profiling and Identification of Vitexin Metabolites in Rat Urine, Plasma, and Faeces after Oral Administration Using a UHPLC-Q-Exactive Orbitrap Mass Spectrometer Coupled with Multiple Data-Mining Methods. *Curr. Drug Metab.* **2021**, *22*, 185–197.
39. Hostetler, G.L.; Ralston, R.A.; Schwartz, S.J. Flavones: Food Sources, Bioavailability, Metabolism, and Bioactivity. *Adv. Nutr.* **2017**, *8*, 423–435. [CrossRef]

40. Zheng, S.; Geng, D.; Liu, S.; Wang, Q.; Liu, S.; Wang, R. A Newly Isolated Human Intestinal Bacterium Strain Capable of Deglycosylating Flavone C-Glycosides and Its Functional Properties. *Microb. Cell Fact.* **2019**, *18*, 1–10. [CrossRef]
41. Lee, J.Y.; Kim, S.; Kim, S.; Kim, J.H.; Bae, B.S.; Koo, G.B.; So, S.H.; Lee, J.; Lee, Y.H. Effects of Red Ginseng Oil (KGC11o) onTestosterone-Propionate-Induced Benign Prostatic Hyperplasia. *J. Ginseng Res.* **2022**, *46*, 473–480. [CrossRef]
42. OECD. *Test No. 441: Hershberger Bioassay in Rats: A Short-Term Screening Assay for (Anti)Androgenic Properties;* OECD Guidelines for the Testing of Chemicals, Section 4; OECD Publishing: Paris, France, 2009.
43. Lee, J.Y.; Kim, S.; Kim, H.; Yeon, S.H.; Kim, S.Y.; Son, R.H.; Park, C.L.; Lee, Y.H. Improvement in Testosterone Production by *Acorus gramineus* for the Alleviation of Andropause Symptoms. *J. Med. Food* **2024**, *27*, 740–748. [CrossRef]

International Journal of
Molecular Sciences

MDPI

Article

Improved Skin Barrier Function Along with Hydration Benefits of *Viola yedoensis* Extract, Aesculin, and Schaftoside and LC-HRMS/MS Dereplication of Its Bio-Active Components

Sreelatha Thonthula [1], Sandra De Sousa [2], Alexis Dubuis [2,*], Samia Boudah [2], Richa Mehta [1], Akanksha Singh [1], Joan Eilstein [2], Jean-Claude Tabet [3,4], Sherluck John [1], Dhimoy Roy [5] and Steve Thomas Pannakal [1,*]

1 L'Oréal Research and Innovation, Bangalore 560067, India; sreelatha.thonthula@loreal.com (S.T.); richa.mehta@loreal.com (R.M.); akanksha.singh@loreal.com (A.S.); sherluck.john@loreal.com (S.J.)
2 L'Oréal Research and Innovation, 93600 Aulnay-Sous-Bois, France; sandra.desousa@loreal.com (S.D.S.); samia.boudah@loreal.com (S.B.); joan.eilstein@loreal.com (J.E.)
3 Faculty of Sciences and Engineering, Institut Parisien de Chimie Moléculaire, Sorbonne University, 75005 Paris, France; jean-claude.tabet@sorbonne-universite.fr
4 Medicines and Health Technologies Department (DMTS), CEA, INRAE, MetaboHUB, Paris-Saclay University, 91190 Gif sur Yvette, France
5 L'Oréal Research and Innovation, Mumbai Maharashtra 410210, India; dhimoy.roy@loreal.com
* Correspondence: alexis.dubuis@loreal.com (A.D.); steve.pannakal@loreal.com (S.T.P.); Tel.: +91-80-67618610 (S.T.P.)

Abstract: The skin hydration level is a key factor that influences the physical and mechanical properties of the skin. The stratum corneum (SC), the outermost layer of the epidermis, is responsible for the skin's barrier function. In this study, we investigated the role of a unique composition of *Viola yedoensis* extract for its ability to activate CD44, a cell-surface receptor of hyaluronic acid, and aquaporin-3, a water-transporting protein, in human keratinocytes (HaCaT). An ELISA assay evaluating the protein expression levels of CD44, aquaporin-3 (AQP3), filaggrin, and keratin-10 revealed that *V. yedoensis* extract upregulated the levels of CD44 and AQP3 by 15% and 78%, respectively. Additionally, *V. yedoensis* extract demonstrated a comparative effect on water vapor flux in TEWL and lipid perturbation in DSC versus the reference, glycerin. In light of this new biological efficacy, a detailed phytochemical characterization was undertaken using an integrated LC-HRMS/MS-based metabolomics approach, which provided further insights on the chemistry of *V. yedoensis*. This led to the identification of 29 secondary metabolites, 14 of which are reported here for the first time, including esculetin, aesculin, apigenin and kaempferol C-glycosides, megastigmane glycosides, roseoside, platanionoside B, and an eriojaposide B isomer, along with the rare, calenduloside F and esculetin diglucoside, which are reported for the first time from the genus, *Viola*. Notably, two active components identified in the *V. yedoensis* extract, namely, aesculin and schaftoside, showed an upregulation of the protein expression of CD44 in HaCaT cells by 123% and 193% within 24 h of treatment, respectively, while aesculin increased AQP3 levels by 46%. Aesculin and schaftoside also significantly upregulated the expression of K-10 levels by 299% and 116%, which was considerably higher than sodium hyaluronate, the positive control. The rationale used to characterize the new structures is outlined along with the related biosynthetic pathways envisioned to generate roseoside and Eriojaposide B. These findings provide new molecular insights to deepen the understanding of how *V. yedoensis* extract, along with the biomarkers aesculin and schaftoside, restores the skin barrier and skin hydration benefits.

Keywords: *Viola yedoensis*; roseoside; platanionoside B; eriojaposide B isomer; calenduloside F; skin barrier; hydration; structural elucidation; LC-HRMS

1. Introduction

The skin, our body's largest interfacial film, acts as a crucial barrier separating the human body from the external environment [1,2]. Primarily composed of stratum corneum, it consists predominantly of keratin and lipids, forming a physical layer that serves as a barrier against trans-epidermal water loss (TEWL) and pathogens. Meanwhile, the lipids and natural moisturizing factors within this layer contribute to its chemical barrier properties [3,4]. Furthermore, the stratum corneum is structured with multiple levels of corneocytes embedded in a lipid matrix comprising ceramides, cholesterol, and fatty acids. These lipids are distributed across two layers, known as the stratum disjunctum and stratum compactum. The cornified cell envelope (CE) present on the outer surface of the epidermis is a part of the skin barrier. The primary constituents of the CE are CE keratins, making up 85% of differentiated keratinocytes. The cornified cell envelope is made of proteins such as involucrin, loricrin, and proline-rich proteins such as SPRR [5]. The upregulation of these proteins help in providing an adaptive response of the epidermis to UV radiation by maintaining the CE barrier function. Loricrin and involucrin are structural proteins crucial for skin barrier function and integrity [6]. Filaggrin, a filament aggregating protein is an important component of the cornified cell envelope. Produced as the precursor proprotein, profilaggrin, it is found in the keratohyalin granules of the granular layer. As the granular cells differentiate into cornified cells, profilaggrin is dephosphorylated and cleaved by several endoproteases to generate filaggrin monomers–amino acids and their derivatives. The filaggrin breakdown products form the natural moisturizing factors (NMFs), which contribute to epidermal hydration and barrier function [7]. This intricate composition reinforces the skin's protective function [8]. The retention of water in the SC is dependent on the presence of natural hygroscopic agents within the corneocytes, collectively referred to as natural moisturizing factors or NMFs [9,10]. Glycerol, an endogenous humectant derived from sebaceous glands, along with humectants like glycerol, draws water from the dermis to the SC and upregulates the expression of aquaporin channels, claudins, and filaggrin, thereby increasing hydration and preventing water loss [11]. Aquaporins constitute a family of transmembrane channels responsible for transporting water and, in certain instances, small solutes across the plasma membrane through osmotic gradients. Among these, aquaporin-3 (AQP3) stands out as the predominant and extensively studied aquaporin functioning as an aquaglyceroporin, demonstrating its capability to transport water, glycerol, urea, and hydrogen peroxide [12].

Another important component of our skin is hyaluronic acid (HA), a natural, non-sulfated, non-branched glycosaminoglycan that is a major component of the dermis. HA is also reported to be present in the epidermis and interacts with CD44, which functions as a coupling site on keratinocyte cell surface, thereby regulating keratinocyte differentiation and SC-extracellular lipid formation [13]. CD44 is associated with the regulation of keratinocyte proliferation in response to extracellular stimuli and the maintenance of local HA homeostasis [14]. Concurrently, aquaporins, filaggrin, and CD44 contribute to the maintenance of moisture and lipid content in the skin. The differentiation of keratinocytes from the basal to the spinous layer is characterized by a shift in keratin production from keratins 5 and 14 to keratins 1 and 10. In addition to keratins 1 and 10, the cells in the spinous layer also synthesize involucrin. As cells mature, they transition to the granular layer, expressing profilaggrin and loricrin [15]. These proteins, along with the lipids synthesized by these cells, form the cornified layer, which is an essential part of the epidermal barrier, preventing fluid loss and protecting the organism against invasion by pathogens.

Over the past decade, there has been a notable surge in interest among researchers, the cosmetic industry, and consumers worldwide for botanical ingredients and their secondary metabolites sourced from natural origins. This increased attention is driven by the perception that these natural compounds are not only effective but also safer, non-toxic, and free from adverse side effects. The market for natural cosmetic products is experiencing rapid growth and is considered one of the fastest-growing sectors globally. Projections indicate that it is poised to achieve a substantial value, reaching approximately

USD 30.1 billion by the year 2026 [16]. This reflects the increasing consumer preference for natural and sustainable beauty solutions, driving the expansion of the natural cosmetics market. Cosmetics enriched with bioactive compounds are considered well suited to meet the specific needs of the skin, and they are often viewed as more environmentally friendly compared to conventional cosmetics. In recent years, there has been a growing interest in dermocosmetics and cosmeceuticals derived from plant materials. This heightened attention reflects an increased focus on plant-based products known for their beneficial properties in skincare [17]. The trend reflects a broader consumer interest in natural and botanical solutions within the realm of skincare and cosmetic formulations.

In recent years, there has been renewed focus on investigating plants for their potential benefits in skin hydration. Literature insights reveal *Aloe barbadensis*, commonly known as *Aloe vera*, yields a gel extracted from its succulent leaves. The gel acts as a humectant and attracts water from the dermis below, aiding in maintaining hydration by binding this water within the stratum corneum [18]. *Saccharum officinarum* (sugarcane) contains starches and sugars rich in potassium and dimethyl sulfoniopropionate (DMSP) that have skin hydrating properties [19]. However, despite this potential, a limited number of natural products have been thoroughly investigated for their benefits regarding skin hydration and epidermal barrier restoration. This gap in research prompted our exploration of plants within the genus *Viola* for their potential hydration and skin restoration benefits. To date, there have been several prior investigations on two related *Viola* species, *Viola tricolor* and *Viola*, which have been an occasional source of flavone glycosides and coumarins [20,21]. Our attention was drawn to *Viola yedoensis* Makino or Chinese violet, a well-known traditional Chinese medicinal plant because its extract was potently active in a primary screen for hydration benefits. *V. yedoensis* is reported to be used to treat various skin disorders such as allergic dermatitis [22], hyperplasia, and edema. The plant is a rich source of flavone glycosides [23], coumarins [24], and phenolic acids, which are known for their diverse bioactivities. Building on this foundation, a flavonoid-enriched extract of *V. yedoensis* was developed and evaluated for its effects on skin hydration and barrier function. Thus, the whole plant of *V. yedoensis* was subjected to a one step hydroalcoholic extraction, and the extract enriched with flavonoid glycosides was evaluated in the enzyme-linked immunosorbent assay (ELISA) against the keratinocyte differentiation markers CD44, involucrin, filaggrin, keratin-10, and aquaporin-3 with sodium hyaluronate as the positive reference in HaCaT cells. The extract enriched with the flavonoid glycosides was fully characterized using liquid chromatography coupled to high-resolution tandem mass spectrometry (LC-HRMS/MS). Dereplication with product ion spectrum databases, divulging the fragmentation pathways of deprotonated glycoconjugates and the interpretation of dissociation mechanisms, was proposed to annotate 29 secondary metabolites, 14 of them being identified for the first time for the genus *Viola*. Additionally, an evaluation of the trans-epidermal water loss (TEWL, directly proportional to skin hydration in healthy skin) and differential scanning calorimetry (DSC, reference method for characterizing free water molecule inside SC) analysis of the extract was performed and compared with the reference glycerol and water. Finally, the major metabolites, aesculin and schaftoside, were evaluated for their in vitro properties on epidermal markers and on improving the skin hydration properties as well.

2. Results

2.1. Epidermal Barrier Function and Skin Hydration Evaluation of V. yedoensis Extract

The *V. yedoensis* extract was evaluated for the protein expression levels of CD44, aquaporin-3 (AQP3), keratin-10 (K-10), filaggrin (FLG), and involucrin (INVL) in HaCaT cells by sandwich-ELISA assay. A cytotoxicity assessment of *V. yedoensis* extract was performed, and the non-cytotoxic dose was evaluated for the modulation of AQP3 and CD44 markers.

2.2. *Cell Viability Assay*

To investigate the cytotoxicity of *V. yedoensis* extract in HaCaT cells, we conducted viability assays using a cell-counting kit. The immortalized human keratinocyte cell line HaCaT was used as a model to study of keratinocyte cytotoxicity. HaCaT cells were seeded in 24-well plates (Thermo Fisher Scientific, Carlsbad, CA, USA) at a density of 10,000 cells/cm^2. Cell culture monolayers in 24-well plates were treated with *V. yedoensis* extract, ranging in concentration between 0.02 and 2.0%, which were applied after 14 days. All solutions were diluted in DPBS. According to our results, the IC_{50} (cytotoxic concentration) of *V. yedoensis* extract was found to be 0.07%. In contrast, treatments with 0.03% did not significantly affect cell viability, which remained close to the untreated control in the HaCaT cell type; therefore, we chose a 0.03% concentration as our test concentration for the study on the epidermal differentiation of markers. Overall, the cells looked visually healthy, and no cell peeling was observed. Sodium hyaluronate (5 kDa, 0.1%) was used as the positive control (Figure 1).

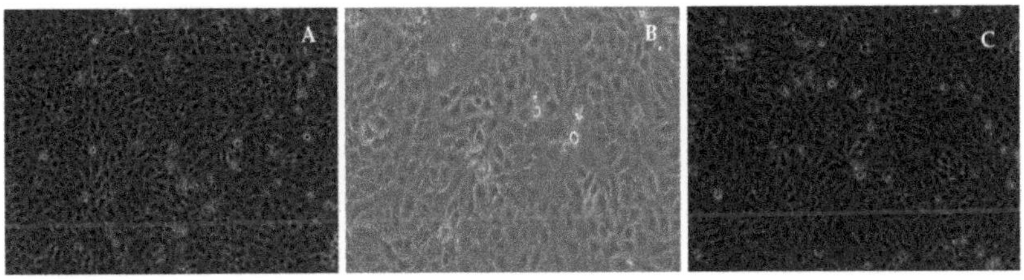

Figure 1. Representative images (10× magnification) of HaCaT cell line. (**A**) Untreated, (**B**) treated with *V. yedoensis* extract (0.03%), and (**C**) positive control (sodium hyaluronate 0.1%).

In-Vitro Study of Epidermal Differentiation Markers

The results of the treatment of HaCaT cells with *V. yedoensis* extract and the positive control, sodium hyaluronate, at 0.1% displayed an expression of epidermal differentiation markers, CD44, AQP3, and K-10, and are shown in Figure 2. The results of the ELISA assay indicate that our *V. yedoensis* extract and the low molecular weight hyaluronic acid sodium salt (5 kDa) were able to stimulate the protein expression levels of CD44 expression in HaCaT cell line by 15% and 86%, respectively. The CD44 measurement was normalized to total protein, so it shows that the increase in CD44 was due to an increase in expression (more CD44 per μg protein) and not due to an increase in the number of cells (no change in CD44 per μg protein). This information is important as it suggests that the increase in CD44 protein is not related to keratinocyte cellular proliferation but to actual protein expression increase. Once again, human keratinocytes (HaCaTs) were used in this study to examine the effect of *V. yedoensis* extract and the positive control on AQP3 expression because the expressions of cytokine/chemokine receptors are more stable in HaCaT than in NHKs. Our *V. yedoensis* extract upregulated the protein expression of AQP3 in HaCaT keratinocytes cells by 78%, which was significantly higher than the positive reference at 54% within 24 h treatment, respectively (Figure 2), suggesting that *V. yedoensis* extract may facilitate the transport and metabolism of water and glycerol in the skin epidermis and the function of the epidermal water permeability barrier. Previous reports suggested that the overexpression of Akt could at least partially contribute to the increase in AQP3 and consequently lead to autophagic responses, which means P13K/AKT could mainly control the upregulation of AQP3. However, we did not examine the P13/AKT pathway and confirm the same in our study. There are possibilities that Ras, ERK, and AMPK could be mediated to induce AQP-3, which we plan to examine in the future. To conclude, *V. yedoensis* extract did not upregulate the expression of FLG, INVL, and K-10 in HaCaT cells. All values were normalized with total protein yield.

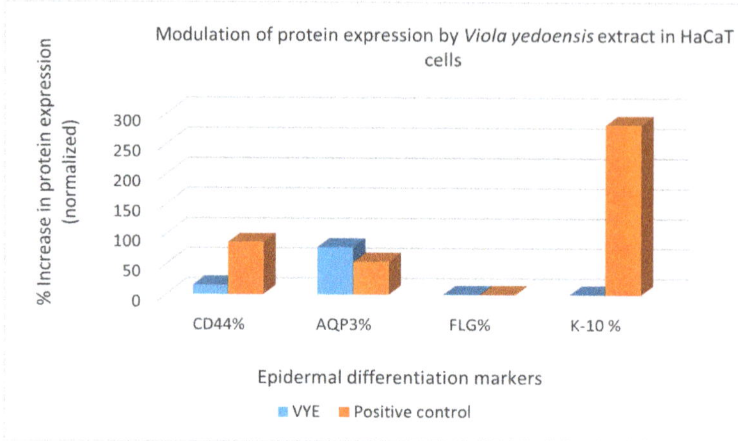

Figure 2. Regulation of epidermal differentiation markers—CD44, AQP3, and K-10—in HaCaT cells treated by *V. yedoensis* extract and positive control (sodium hyaluronate, 0.1%). Values are average of experiments in duplicate.

2.3. Skin Hydration Evaluation

TEWL analysis: In normal conditions, with an intact skin barrier, TEWL on human skin is on the order of 5–10 $g/m^2/h$. The TEWL measurements with isolated stratum corneum treated by *V. yedoensis* extract were comparable to water and 5% glycerin in water. It has a similar hydration effect as 5% glycerin (Table 1).

Table 1. TEWL and DSC analysis of *V. yedoensis* extract.

Analysis Parameters	*V. yedoensis* Extract	Water	5% Glycerine in Water
TEWL (water flux ΔTEWL, %)	±5.2%	±5%	±5.5%
DSC (lipid perturbation ΔTm °C)	±0.30 °C	±2 °C	±0.11 °C

DSC analysis: The thermogram of the untreated stratum corneum is characterized by four phase transitions—T1 (40 °C), T2 (72 °C), T3 (85 °C), T4 (>110 °C). The endothermic peaks corresponding to transitions T1, T2, and T3 for a typical stratum corneum sample are shown in Figure 3. *V. yedoensis* extract shows a lipid perturbation of ±0.30 °C in DSC analysis (Table 1).

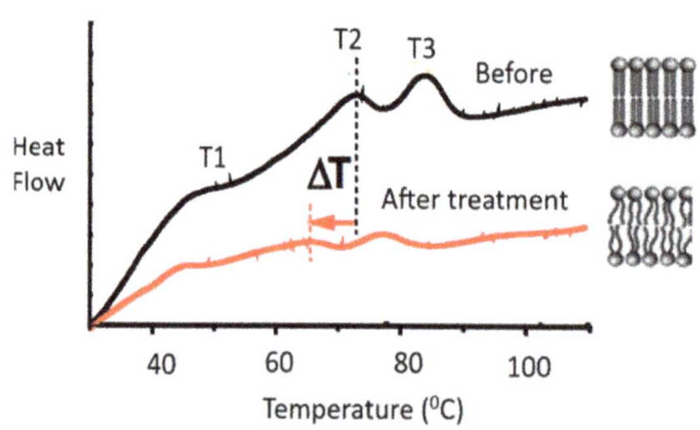

Figure 3. Endothermic peaks in DSC with *V. yedoensis* extract.

2.4. Chemical Characterization of the Viola yedoensis Extract

2.4.1. Methodology

The *V. yedoensis* extract was fractionated and chemically characterized by LC-V/HRMS. This fractionation aimed to enrich minor secondary metabolites within the extract. Spectral comparison with databases and the interpretation of dissociation mechanisms led to the identification of 29 secondary metabolites, 14 of which are reported here for the first time in *V. yedoensis*. Figure 4 presents the methodology and data used to annotate both known and novel compounds from *V. yedoensis* extract.

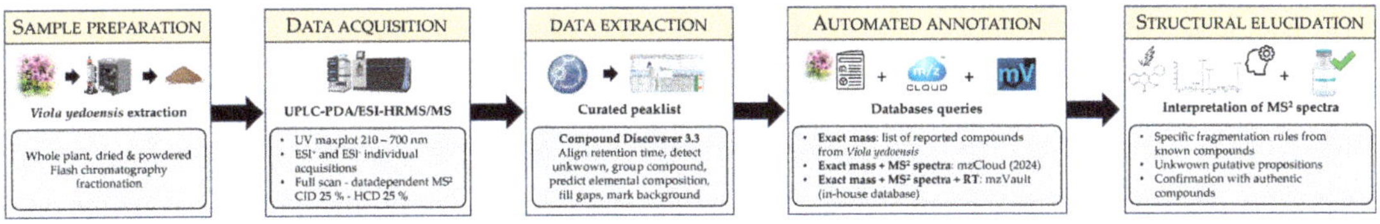

Figure 4. Methodology used to annotate secondary metabolites of *V. yedoensis* extract, according to recommendations from the Chemical Analysis Working Group (CAWG) [25].

Of the 29 compounds identified, 12 were annotated with the highest level of confidence by comparing raw data, collisional spectra (conditions in experimental part), and retention time with authentic compounds (10 were already reported in *V. yedoensis*). Putative identifications were proposed for 6 compounds based on MS/MS spectral matching with databases (without RT confirmation), and 11 compounds were tentatively identified through MS/MS spectra interpretation (of which, 5 were already reported in *V. yedoensis*).

To do so, stable ionized molecular species are essential to provide specific structural information that can be used for the elucidation of unknown metabolites or to propose putative structures. As polar molecules, glycoconjugates require electrospray ionization (ESI) conditions to be desorbed as stable even-electron molecular species [26]. Due to the presence of OH groups, numerous acidic sites are located on both aglycones and carbohydrates. Consequently, the negative ionization mode was chosen to generate deprotonated molecules. Under these conditions, a mixture of competitive deprotomers arises from desolvation of charged aggregates during the ESI process. The negative charge of these activated deprotomers probably favors different fragmentation orientations, providing complementary structural information using high-resolution tandem mass spectrometry. Notably, unlike protonated molecules where C-C and C-O aglycon linkages lead to cross-ring dissociations of hexose/pentose moiety, only deprotonated molecules with a C-C bond to the aglycone lead to such cleavages, whereas those with a C-O bond led to the ether cleavage yielding a dehydrated carbohydrate release. This distinction provides a reliable way to differentiate between O-aglycon and C-aglycon linkage with sugars. Furthermore, to obtain diverse structural information such as carbohydrate location on the aglycon and sugar linkage patterns, non-resonant ion excitation was employed to enable the construction of informative MS/MS spectra databases, consisting of known compounds. These databases facilitated the interpretation of deprotonated molecule fragmentations, leading to the establishment of specific decomposition rules. These observations were then applied to propose putative structures for unknown compounds. To distinguish the location of different hexose and pentose moieties on the aglycone skeleton, product ion spectra obtained in resonant excitation mode were used. This mode mainly yields competitive fragmentations directly from the selected precursor ion, which results in the first generation of product ions [27]. To illustrate this approach, flavonoid and terpenoid glycoconjugates within *V. yedoensis* extract were selected to propose putative structures and highlight fragmentation rules applicable to the structural elucidation of other natural products.

2.4.2. Identification of Known Compounds from *Viola yedoensis*

Figure 5 shows LC-UV/HRMS chromatograms of *V. yedoensis* extract, including (a) the UV maxplot and (b, c) total ion current (TIC) acquired in positive and negative ion modes, respectively.

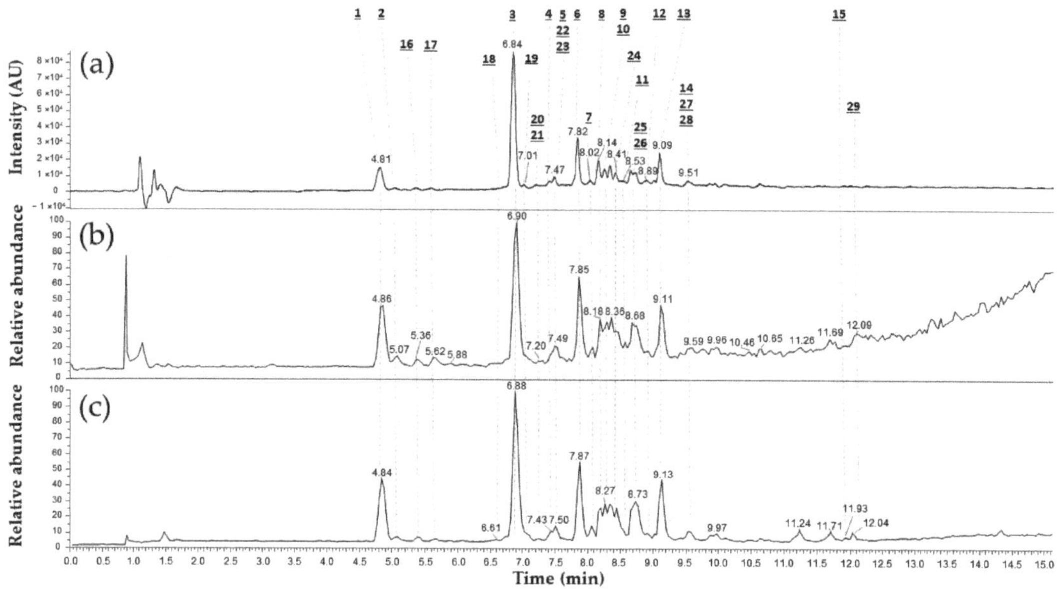

Figure 5. LC-UV/HRMS chromatograms of *V. yedoensis* extract. (**a**) UV maxplot 210–700 nm, TIC with (**b**) ESI positive and (**c**) ESI negative, *m/z* 100–1100.

UV maxplot and TIC profiles exhibit a high degree of superposition, indicating that major UV-absorbing components were also detectable by mass spectrometry. This observation confirms the suitability of HRMS/MS for their structural characterization. Notably, positive and negative ionization modes demonstrated comparable efficiency for these compounds. Many of the observed signals were compounds previously reported from *V. yedoensis*. Their spectral characteristics are summarized in Table 2.

In Table 2, identified compounds from *V. yedoensis* extract already known from the plant belongs to coumarins (e.g., **3** and **10**), coumarin glycosides (e.g., **1**, **2**, and **7**), flavonoids (e.g., **15**), and flavonoid mono- and di-glycosides (e.g., **4**, **5**, **6**, **8**, **9**, **11**, **12**, **13**, and **14**). Most of the flavonoids are O-glycoside and C-glycoside forms of apigenin (flavone), luteolin (flavone), and kaempferol (flavonol). Among the coumarins, aesculin (**2**) was identified as the main compound, while schaftoside (**6**) was found to be the major C-glycoside flavone in *V. yedoensis* extract, considering all the modes of detection (see Figure 4). Because of their abundance in *V. yedoensis* extract, aesculin (**2**) and schaftoside (**6**) were later studied in vitro on epidermal differentiation markers and for their skin hydration benefits.

Table 2. Reported compounds in *V. yedoensis* identified in extract with *m/z* of respective precursor $[M - H]^-$ ions and their product ions using LC/ESI-HRMS/MS under non-resonant collisional excitation conditions (HCD 20%).

Index	Rt (min)	Name/CAS Number (Identified with [a] MS² Spectra, [b] RT, [c] De Novo)	Glycosyl Linkage to Aglycon Backbone	Precursor Ion *m/z*	δ ppm	Elemental Composition	Product Ion *m/z* (‡ Oxidation)	References
1	4.75	Cichoriin [c]/531-58-8	O-coumarin	339.0726	3.1	$C_{15}H_{16}O_9$	177.0188, 133.0294, 105.0350	[28]
2	5.17	Aesculin (Esculin) [a,b]/531-75-9	O-coumarin	339.0725	2.9	$C_{15}H_{16}O_9$	177.0189, 133.0294, 105.0347	[29]
3	6.85	Esculetin [a,b]/305-01-1	Coumarin	177.0188	0.1	$C_9H_6O_4$	133.0295, 105.0348	[29]
4	7.40	Vicenin 2 [a,b]/23666-13-9	Di-C-flavone	593.1519	2.1	$C_{27}H_{30}O_{15}$	503.1203, 473.1092, 383.0778, 353.0672	[30]
5	7.61	Isocarlinoside [c]/83151-90-0	Di-C-flavone	579.1361	1.9	$C_{26}H_{28}O_{15}$	489.1044, 459.0936, 429.0830, 399.0725, 369.0620	[31]
6	7.84	Schaftoside [a,b]/51938-32-0	Di-C-flavone	563.1412	2.0	$C_{26}H_{28}O_{14}$	473.1094, 443.0990, 383.0778, 353.0672	[31]
7	8.02	Prionanthoside [c]/161842-81-5	O-coumarin	381.0831	2.4	$C_{17}H_{18}O_{10}$	177.0189	[23]
8	8.13	Isoorientin [a,b]/4261-42-1	C-flavone	447.0937	2.1	$C_{21}H_{20}O_{11}$	4290829, 357.0621, 327.0515	[31]
9	8.21	Apigenin 6-C-α-L-arabinopyranosyl-8-C-β-D-xylopyranoside [c]/677021-30-6	Di-C-flavone	533.1307	2.2	$C_{25}H_{26}O_{13}$	515.1203, 473.1094, 443.0987, 383.0782, 353.0673	[31]
10	8.24	Euphorbetin [c]/35897-99-5	Coumarin dimer	353.0309	3.3	$C_{18}H_{10}O_8$	335.0204, 309.0409, 177.0188	[32]
11	8.49	Isovitexin [a,b]/38953-85-4	C-flavone	431.0988	2.3	$C_{21}H_{20}O_{10}$	3410669, 311.0566, 283.0617	[33]
12	8.85	Cynaroside (luteolin 7-O-glucoside) [a,b]/5373-11-5	O-flavone	447.0938	2.4	$C_{21}H_{20}O_{11}$	327.0525, 285.0407 ‡	[32]
13	9.14	Quercetin 3-glucoside [a,b]/482-35-9	O-flavonol	463.0887	2.3	$C_{21}H_{20}O_{12}$	300.0278 ‡, 271.0254, 255.0299, 151.0033	[29]
14	9.51	Astragalin (kaempferol 3-O glucoside) [a,b]/480-10-4	O-flavonol	447.0938	2.4	$C_{21}H_{20}O_{11}$	357.0630, 327.0515, 284.0330 ‡, 255.0300, 227.0354, 151.0037	[30]
15	11.86	Luteolin [a,b]/491-70-3	Flavone	285.0408	3.1	$C_{15}H_{10}O_6$	175.0397, 151.0034	[34]

Identified with authentic compound: [a] MS² spectrum, [b] RT, and [c] de novo structural elucidation. ‡ Oxidation of aglycone to quinone-like structure characterized by high electron affinity yielding preferentially radical anion by electron transfer from deprotonated carbohydrate in ion/dipole complex (vide infra).

2.4.3. Confirming Known and Elucidating New Bioactive Structures in *Viola yedoensis* Flavonoid Glycoconjugates

Fragmentation pathways of deprotonated glycoconjugates are elucidated by considering the mobile proton theory or charge delocalization through overlapping orbitals. When these processes are hindered by steric constraints or significant spatial separation between the charge site and the bond involved for cleavage, deprotonated molecule isomerization into an ion/dipole complex must be considered [35]. This facilitates negative charge transfer between distant sites, driven by the relative gas-phase acidities of the participating functional groups (hydroxyl groups).

For instance, the estimated difference of gas-phase acidity between an OH group of aglycone and a dehydrated hexose is larger than 70 kJ/mole (details in experimental part). Consequently, proton transfer from the phenolic group can neutralize the deprotonated hexose (e.g., alkoxide site) that occurs since it is an exothermic pathway. Conversely, without a mobile proton on the aglycon, proton transfer is precluded. Consequently, negative charge transfer to the aglycon requires hydride transfer from the deprotonated carbohydrate within the ion/dipole complex, effectively reducing the quinone form of oxidized aglycon. This proposed mechanism is illustrated in Figure 6 with quercetin-3-sophoroside-7-O-rhamnoside (**21**). Under these conditions, fragmentations of deprotonated luteolin (15) are essentially limited to two competitive processes initiated by the heterocycle opening at the C3-C4 bond induced by a phenoxy site, followed by the cleavage of either the C2-C1′ linkage (the formation of *m/z* 175) or the C2-O bond (the formation of *m/z* 151) (Figure S1). Conversely, deprotonated glycoconjugates listed in Table 1 decompose primarily through the loss of a carbohydrate moiety (either a complete carbohydrate unit or a fragment thereof) or a glucuronic acid unit. This fragmentation pathway consistently yields the deprotonated aglycone. The specific fragmentation patterns observed are contingent upon the nature of the glycosidic linkage connecting the aglycone and the carbohydrate, either O-glycoside or C-glycoside.

In all the studied cases, the release of saccharides linked to aglycone involves benzylic C-O bond cleavage to regenerate intact deprotonated aglycone, as opposed to the significantly less favorable cleavage of phenylic bond. Furthermore, the benzylic cleavage must proceed through a stepwise mechanism involving the formation of an ion/dipole complex intermediate [36,37]. Such a mechanism would enable the migration of the negative charge if required for consecutive/competitive possible pathway(s).

This fragmentation behavior is exemplified by the deprotonated glycosidic linkage positional isomers as cichoriin $[\mathbf{1} - H]^-$ and aesculin $[\mathbf{2} - H]^-$, both containing a single hexose moiety. Under resonant excitation conditions (i.e., collision-induced dissociation, CID), the product ion spectra of these anions, acquired at a normalized collision energy (NCE) of 25% (E_{lab}), consistently show a product ion at *m/z* 177, attributed to the dehydrated glucose ($C_6H_{10}O_5$, 162 Da) loss (Figure S2). Notably, this fragmentation pathway is more prominent for deprotonated cichoriin compared to aesculin. This difference can be rationalized by considering the lower gas-phase acidity of the phenolic group in (**1**) relative to (**2**). As a result, the phenoxide anion in deprotonated (**2**) is stabilized by greater charge delocalization compared to its analog in deprotonated (**1**) illustrated in Figure S2a. A plausible stepwise dissociation mechanism, illustrated in Figure S2b, involves a common ion/dipole intermediate, subsequently losing $C_6H_{10}O_5$ to form the common product ion at *m/z* 177. This rationalizes why an abundance of the survivor $[\mathbf{1} - H]^-$ precursor ion is lower than that of deprotonated aesculin, $[\mathbf{2} - H]^-$. A possible reactional pathway rationalizes the favorable dissociation of $[\mathbf{1} - H]^-$ compared to that of $[\mathbf{2} - H]^-$ where the charge is dispersed on two oxygen atoms (Figure S3).

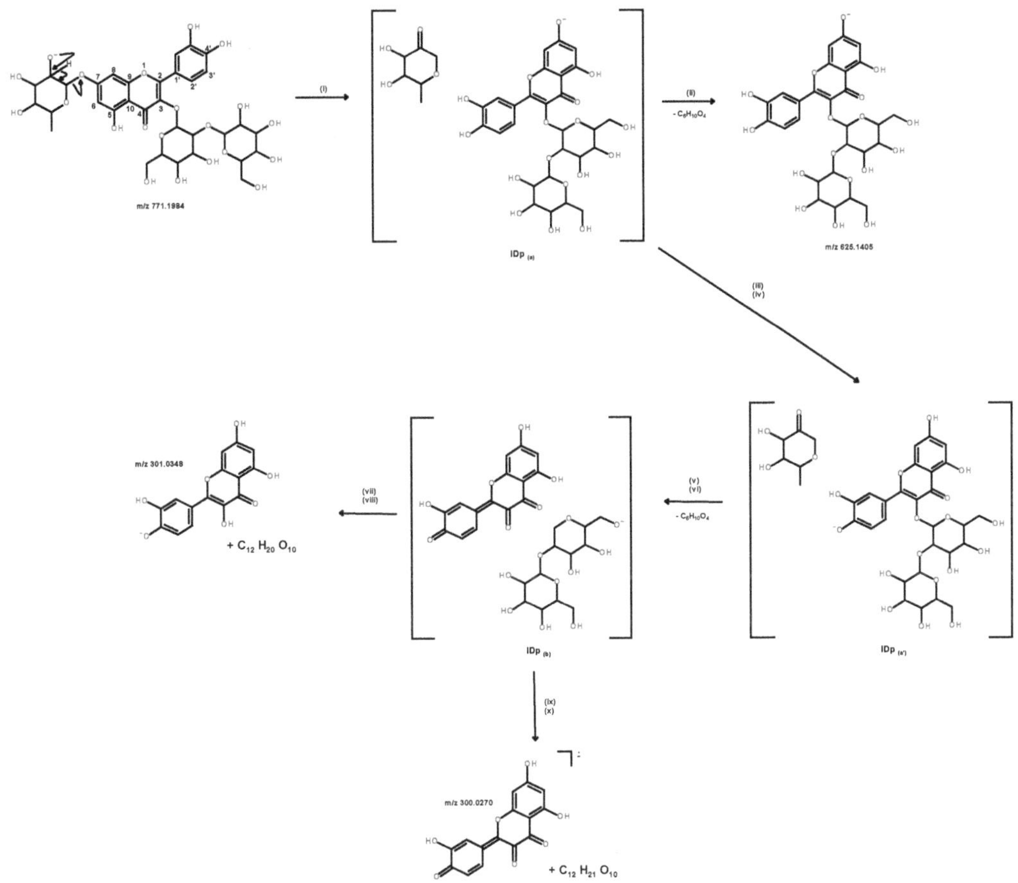

Figure 6. Proposed mechanisms of stepwise gas-phase fragmentations of collisionally activated deprotonated glycoconjugates, desorbed as a various deprotomers provided in LC-ESI/HRMS/MS from *V. yedoensis* extract. Main product ions displayed in collision spectrum of quercetin-3-sophoroside-7-O-rhamnoside deprotomer mixture are used as model for description of possible interpretation of various product ion: (i) stepwise mechanisms via molecular anion isomerization to ion/dipole ($IDp_{(a)}$) based on 1,2-hydride transfer from a secondary hydroxyl (or, eventually, the primary hydroxyl) at the rhamnose moiety promoting the C-O linkage cleavage at C(7), (ii) direct splitting of IDp(a) yielding *m/z* 625 by the dehydrated rhamnose neutral loss ($C_6H_{10}O_4$, 146 Da), or (iii) internally to $IDp_{(a)}$, deprotonation of dehydrated rhamnose by proton transfer to phenoxy at C(7), (iv) intermediate used for long-distance exothermic negative charge transfer from $C_6H_9O_4^-$ to phenol at C(4′) (intermediate $IDp_{(a')}$) as well as at other phenolic C(5) and C(3′) positions (but ineffective for the further reactions), (v) followed by $C_6H_{10}O_4$ loss, and (vi) resulting intermediate isomerization into $IDp_{(b)}$ induced by the charge migration from phenoxy at C(4′) to dehydrated sophoroside ($C_{12}H_{21}O_{10}$) by cleavage of C-O linkage at C(3), and (vii) internal $IDp_{(b)}$ hydride transfer from $C_{12}H_{21}O_{10}^-$ to aglycon followed by (viii) release of dehydrated $C_{12}H_{20}O_{10}$ sophoroside (324 Da) to yield deprotonated aglycon m/z 301; alternatively from $IDp_{(b)}$ (ix) internal electron transfer from $C_{12}H_{21}O_{10}^-$ to aglycon and (x) formation of oxidized aglycon radical anion *m/z* 300 by the $C_{12}H_{21}O_{10}^{\bullet}$ radical release. Note the sequences from (v) to (viii) and (v) to (x) can occur initially from *m/z* 771 to give rise to formation of *m/z* 447 and *m/z* 446 product ions, respectively, possibly yielding formation of *m/z* 301 and *m/z* 300 by release of dehydrated rhamnose and its radical form.

Glycoconjugates **12** (Table 2), **18**, **27**, and **28** (Table 3) appeared similar to compounds **1** and **2** as they all contain a monosaccharide linked via a benzylic C–O bond and exhibit a loss of dehydrated glucose (162 Da) upon CID. This fragmentation follows a pathway analogous to that depicted in Figure S2. Similarly, compound **7**, containing an acetylated glucose moiety, undergoes analogous fragmentation with a neutral loss of 204 Da. The deprotonated $[\mathbf{23} - H]^-$ ion (*m/z* 433), featuring a rhamnose-O-aryl linkage, loses dehydrated rhamnose (146 Da). The loss of 162 Da is also observed in disaccharide conjugates, such as the $[\mathbf{20} - H]^-$ precursor ion (*m/z* 597), which contains one O-phenyl linkage. However, the fragmentation pattern of disaccharide conjugates is influenced by the aglycone structure. For instance, the $[\mathbf{17} - H]^-$ ion (*m/z* 471) can directly lose both linked carbohydrates (294 Da), yielding the deprotonated aglycone at *m/z* 177. This process likely proceeds through a stepwise mechanism involving isomerization to an ion/dipole complex prior to dissociation, similar to the mechanism proposed for the $[\mathbf{1} - H]^-$ and $[\mathbf{2} - H]^-$ isomer precursor ions in Figure S2b. The presence of a phenol group ortho to the O-glycoside linkage appears to promote the direct loss of the disaccharide unit, whereas its absence allows for the consecutive losses of individual sugar.

Interestingly, a competing fragmentation pathway involving the loss of a radical species, formally $C_6H_{11}O_5^{\bullet}$ (163 Da), is observed for monosaccharide compounds **13** and **14**. This radical loss, potentially corresponding to a 1-deoxyglucose or 1,5-anhydroglucitol radical, occurs in competition with the loss of 162 Da that corresponds to dehydrated glucose. The observation of this radical loss is noteworthy, as it seemingly breaks the even-electron rule [38]. While the exact mechanism driving this radical loss remains to be fully elucidated, it may involve either a direct loss of $C_6H_{11}O_5^{\bullet}$ or a consecutive loss of $C_6H_{10}O_5$ and $H^{\bullet}$. Notably, the position of the saccharide moiety on the aglycone skeleton and the location of phenolic groups appear to influence the prevalence of this radical anion formation. For instance, deprotonated quercetin-3-glucoside (**13**, 5,7,3′,4′-tetrahydroxyflavonol backbone) and astragalin (**14**, 5,7,4′-trihydroxyflavonol) preferentially undergo the 163 Da loss. In contrast, deprotonated isorhamnetin 3-O-glucoside (**27**, 5,7,4′-trihydroxy, 3′-O-methoxyflavonol) and cynaroside (**12**, 5,7,3′,4′-tetrahydroxluteolin), with the glycosidic linkage at C(7), favor the neutral loss of dehydrated glucose (162 Da). Further investigations are currently underway to elucidate the factors governing the competition between these two fragmentation pathways and will be reported elsewhere.

In the case of non-linked mono- and di-saccharides with a C-phenyl linkage, a strong hindrance of the dehydrated saccharide loss is expected due to charge-driven fragmentations. O-ring opening generates an alkoxide intermediate. This alkoxide facilitates 1,2 or 1,3 hydride transfers, leading to the release of $C_2H_4O_2$ (60 Da loss), $C_3H_6O_3$ (90 Da loss), or $C_4H_8O_4$ (120 Da loss). Subsequent proton transfer from the neighboring phenolic site to the resulting allylic alkoxide can promote a second glycosidic ring opening, leading to further losses of 90 Da or 120 Da. However, in the case of linked disaccharides, such as iso-orientin 6′-O-glucopyranoside (**19**), the expected 162 Da loss is observed. This loss results from the cleavage of the glycosidic linkage between the hexose (or/and pentose) units without affecting the aglycone–hexose interaction. This cleavage pathway competes with the release of $(CH_2O)_n$ units ($n \geq 2$ from pentose linkages, as observed in compounds **9**, **24** and $n \geq 3$ from hexose linkages, as in compounds **4**, **5**, **6**, **8**, **11**, and **19**) through consecutive cleavages within each pentose/hexose ring.

Table 3. Unreported compounds in *V. yedoensis* identified in extract with *m/z* of precursor ions $[M-H]^-$ and $[M+H]^+$ and their product ions using LC/ESI-HRMS/MS under resonant (CID 25 %) and non-resonant collisional (HCD 20%) excitation conditions.

Index	Rt (min)	Name/CAS Number (Identified with: [a] MS² spectra, [b] RT, [c] De Novo)	Glycosyl Linkage to Aglycon Backbone	Precursor Ion	*m/z* of Precursor Ion	δ ppm	Molecular Formula	Excitation Conditions	Product Ion *m/z* (‡ Oxidation)	References
16	5.39	Esculetin diglucoside [c]	O-coumarin	$[M-H]^-$	501.1239	1.1	$C_{21}H_{26}O_{14}$	CID 25%	177.0192, 133.0289, 105.0339, 89.0390, 81.0339	[39]
17	5.58	5-hydroxy apiosylskimmin [c]/1149372-95-1	O-coumarin	$[M-H]^-$	471.1131	1.6	$C_{20}H_{24}O_{13}$	HCD 20%	177.0188	[40]
18	6.59	Roseoside [a,b]/54835-70-0	O-megastigmane	$[M-H]^-$	385.1856	1.7	$C_{19}H_{30}O_8$	CID 25%	223.1344, 205.1231, 161.0448, 153.0920	[41]
19	7.00	Meloside L [a]/55196-48-0	C-flavone	$[M-H]^-$	609.1485	4.8	$C_{27}H_{30}O_{16}$	HCD 20%	489.1040, 447.0924, 429.0815, 399.0722, 369.0635, 357.0615	[42]
20	7.24	Quercetin 3-arabinofuranoside 7-glucoside [c]/23394-51-6	Di-O-flavonol	$[M-H]^-$ $[M+H]^+$	595.1311 597.1452	0.6	$C_{26}H_{28}O_{16}$	HCD 20%	463.0888 ‡, 433.0787, 301.0355 ‡, 465.1029, 435.0934, 303.0504 ‡	[43]
21	7.25	Quercetin-3-sophoroside-7-O-rhamnoside [a,b]/64828-40-6	Di-O-flavonol	$[M-H]^-$	771.1978	0.8	$C_{33}H_{40}O_{21}$	CID 25%	625.1396, 446.0841 ‡, 301.0356 ‡	[44]
22	7.44	Kaempferol-3-sophoroside-7-rhamnoside [a]/93098-79-4	Di-O-flavonol	$[M-H]^-$ $[M+H]^+$	755.1470 757.2203	1.6	$C_{33}H_{40}O_{20}$	HCD 20%	609.1470, 431.0970 ‡, 285.0408 ‡, 433.1139, 287.0555	[45]
23	7.46	Kaempferol-7-O-rhamnoside [a]/20196-89-8	O-flavonol	$[M+H]^+$	433.1138	0.8	$C_{21}H_{20}O_{10}$	HCD 20%	287.0554	[46]
24	8.33	Vicenin-1 6″-O-acetate [c]/163345-99-1	Di-C-flavone	$[M-H]^-$	605.1518	1.9	$C_{28}H_{30}O_{15}$	HCD 20%	545.1312, 455.0990, 335.0570	[47]
25	8.64	Kaempferol 3-O-sophoroside [c]/19895-95-5	O-flavonol	$[M-H]^-$	609.1470	2.4	$C_{27}H_{30}O_{16}$	HCD 20%	429.0834, 327.0512, 285.0407 ‡, 255.0300, 227.0352, 151.0036	[48]
26	8.66	Platanionoside B [c]/23601-22-8	Di-O-megastigmane	$[M-H]^-$	533.2608	1.9	$C_{25}H_{42}O_{12}$	CID 25%	371.2072, 209.1534, 161.0448	[49]
27	9.54	Isorhamnetin 3-O-glucoside [a]/5041-82-7	O-flavonol	$[M-H]^-$	477.1045	2.5	$C_{22}H_{22}O_{12}$	HCD 20%	357.0609, 315.0507 ‡, 285.0406, 271.0257, 243.0301	[30]
28	9.55	(4R)-4-[(3R)-3-[[3-O-(4-carboxy-3-hydroxy-3-methyl-1-oxobutyl)-β-D-glucopyranosyl]oxy]butyl]-3,5,5-trimethyl-2-cyclohexen-1-one [a]/1039756-96-1	O-apocarotenoid	$[M-H]^-$ $[M+H]^+$	515.2484 517.2642	2.8 1.3	$C_{25}H_{40}O_{11}$	CID 50% CID 25%	471.2578, 453.2473, 413.2162, 371.2059, 353.1954, 209.1539, 161.0438, 125.0235, 99.0443, 211.1691	[50]
29	12.04	Calenduloside F [a]/51415-02-2	Di-O-triterpene	$[M-H]^-$ $[M+H]^+$	793.4389 795.4531	1.9 0.0	$C_{42}H_{66}O_{14}$	CID 35% CID 20%	631.3845, 613.3723, 509.3618, 455.3509, 439.3578, 249.1848, 217.1950, 203.1792, 191.1794	[51]

Identified with authentic compound: [a] MS² spectrum, [b] RT, and [c] de novo structural elucidation. ‡ Oxidation of aglycone to quinone-like structure characterized by high electron affinity yielding preferentially radical anion by electron transfer from deprotonated carbohydrate in ion/dipole complex (vide infra).

2.5. Identification and Characterization of Terpenoid Glycoconjugates

The fragmentation of both positively and negatively charged molecular species must be considered for the study of terpenoid glycoconjugates, as the dissociation provides complementary structural information. When a molecular species dissociates into two fragments, the charge distribution is determined by the gas-phase thermochemical properties of the corresponding neutrals of each fragment. Consequently, the charge-carrying fragment corresponds to the one who's neutral is the strongest base in positive mode and the strongest acid in negative mode.

The protonated [**28** + H]$^+$ molecule dissociates mainly to form *m/z* 211.1691 ion ($C_{13}H_{23}O_2$, 2 ppm), which suggests that the protonated aglycon releases H_2O (or 2 H_2O) independently of the collision energy used for its activation. On the other hand, the collisional spectra of the deprotonated molecule [**28** − H]$^-$ (*m/z* 515.2484, $C_{25}H_{39}O_{11}$) recorded under several collision energy conditions display more product ions than observed from other deprotonated compounds, as recorded for deprotonated platanionoside B (**26**). The [**28** − H]$^-$ and [**26** − H]$^-$ precursor ions dissociate into a common fragment *m/z* 209.1539 ion ($C_{13}H_{21}O_2$, 4 ppm), which is consistent with the *m/z* 211.1690 positive product ion ($C_{13}H_{23}O_2$, 1 ppm). Notably, very weak abundant *m/z* 161.0457 corresponds to a dehydrated hexose $C_6H_9O_5$ (4 ppm) formed with *m/z* 209.1539 from the competitive, stepwise dissociations of the *m/z* 371.2061 ion (a similar process is described in Figure S4). This further confirms that the aglycon of these compounds are similar, like β-ionone/β-ionol backbone. Two possible isomeric candidates can be considered initially, such as eriojaposide B and its apocarotenoid isomer. However, the former must be ruled out since the CO_2 and (CO_2 + C_3H_6O) losses cannot occur from hexoses or pentoses, unlike the apocarotenoid isomer with a 3-hydroxy-3-methylglutaryl end group and an ester linkage with a hexose. Indeed, from the latter, these losses can be explained by using processes promoted by the negative charge from the carboxylate end group. This CO_2 release is assisted by proton migration from the neighboring hydroxy group in the 3-hydroxy-3-methylglutaryl end group, which allows consecutively the C_3H_6O release (1 ppm) as acetone neutral (*m/z* 413.2162, Figure S4a). Alternatively, the complementary ion pair *m/z* 371.2061 ($C_{19}H_{31}O_4$, 4 ppm) and *m/z* 143.0344 ($C_6H_7O_4$, 3 ppm) can be rationalized by a stepwise dissociation of the deprotonated apocarotenoid isomer through isomerization into ion/dipole and internal proton transfer followed by the splitting of both partners, allowing the competitive formation of both fragment ions (Figure S4b). This latter ion can (i) directly lose CO_2 through the formation of a 5-membered lactone *m/z* 99 (Figure S4c) or (ii) be isomerized into a hydroxylated cyclic anhydride with a negative charge carried by an enolate site of the intermediate (Figure S4d). All these processes are consistent with a 3-hydroxy-3-methylglutaryl end group.

The positive mass spectrum of the compound **29** is characterized by a major ion at *m/z* 439.3578 ($C_{30}H_{47}O_2$, 1 ppm), likely attributable to prompt dissociations occurring during charged aggregate desolvation. In contrast, the negative ion mode spectrum is dominated by a highly abundant signal at a higher *m/z* value of 793.4363, corresponding to an elemental composition of $C_{42}H_{65}O_{14}$ (2 ppm error). Interestingly, a similar behavior is observed in tandem mass spectra of calenduloside E (retrieved from the mzCloud database) in positive mode at *m/z* 439.3567 ($C_{30}H_{47}O_2$, 3 ppm) and in negative mode at *m/z* 631.3830 ($C_{36}H_{55}O_9$, 2 ppm). These results suggest that both glycoconjugates share a common aglycone moiety, likely oleanolic acid, with calenduloside F (compound **29**) bearing an additional carbohydrate unit. Notably, product ions detected at *m/z* 249.1848 ($C_{16}H_{25}O_2$, 0.5 ppm) and 191.1792 ($C_{14}H_{23}$, 2 ppm) generated from the selected ion *m/z* 439.3567 ($C_{30}H_{47}O_2$, corresponding to dehydrated oleanolic acid) of compound **29** exhibit complementary elemental compositions relative to this precursor ion. These product ions are generated via molecular isomerization into ion/dipole intermediates that competitively undergo competitive fragmentations (Figures S5 and S6). Furthermore, a series of product ions in positive mode *m/z* 421, 393, 315, 217, and 203 are observed in collisional spectra of the prompt *m/z* 439.3567 fragment ion during MS/MS experiments of both protonated molecules. The formation of these product ions can be attributed to competitive, stepwise fragmentation processes.

2.6. Effect of Aesculin and Schaftoside on Epidermal Differentiation Markers In Vitro and Skin Hydration Benefits

Based on the ELISA assay results of *V. yedoensis* extract on epidermal differentiation markers, we were interested in decoding the active components responsible for the skin hydration benefits of this extract. Aesculin (coumarin glycoside) and schaftoside (apigenin-C-glycoside) were shortlisted to be evaluated for their effect on epidermal differentiation markers and hydration benefits based on their abundance in the extract, evident from the LC-HRMS/MS-based metabolomics approach under resonant (CID 25%) and non-resonant collisional (HCD 20%) excitation conditions as well as their wide range of biological activities previously reported, such as anti-inflammatory [52], wound healing [53], and protection against UVB-induced damage [3,54]. The two secondary metabolites were evaluated for the protein expression levels of CD44, AQP3, K-10, and FLG in HaCaT cells by sandwich-ELISA assay. A cytotoxicity assessment of aesculin and schaftoside was performed, and the non-cytotoxic dose was evaluated for the modulation of CD44, AQP-3, FLG, and K-10.

2.7. Cell Viability Assay

The half maximal inhibitory concentration, IC_{50} (cytotoxic concentration), of aesculin (0.5–2%) and schaftoside (0.05–0.2%) was performed to determine the non-cytotoxic concentrations to be used for the in vitro assay. The cell culture was maintained at 37 °C in a humidified incubator with 5% CO_2 atmosphere and grown to 70% to 80% confluence. Then, 1% aesculin and 0.1% schaftoside were tested on HaCaT keratinocyte cells, and the percentage of HaCaT cell death observed was found to be <38% after 24 h treatment. Visually, the cells looked healthy, and no cell peeling was observed (Figure 7). Thus, 1% aesculin and 0.1% schaftoside were considered as non-cytotoxic to cells and tested in the ELISA assay at the same concentration. Overall, the cells looked visually healthy, and no cell peeling was observed. Sodium hyaluronate (5 kDa, 0.1%) was used as the positive control (Figure 7).

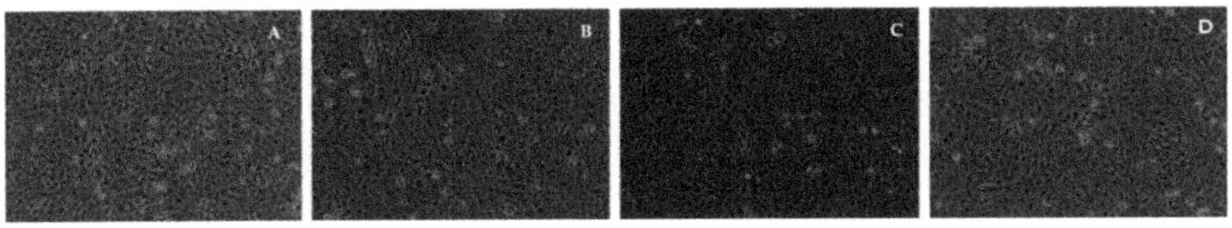

Figure 7. Representative images (10× magnification) of HaCaT cell line. (**A**) Untreated, (**B**) aesculin (1%), (**C**) schaftoside (0.1%), and (**D**) positive control (sod. hyaluronate 0.1%).

2.8. Epidermal Differentiation Markers In Vitro ELISA Assay

The two most abundant secondary metabolites of interest (aesculin and schaftoside) were added to HaCaT cells, and the AQP3 mRNA expression level was analyzed 24 h later. No morphological changes in cells were observed at a concentration of 1% and 0.1% for both the secondary metabolites. The coumarin glycoside, aesculin, significantly increased the mRNA expression level of AQP3 in the HaCaT cells compared to the control cells. Among them, the rate of increase in AQP3 levels induced by aesculin was significantly higher at 46%, closely similar to that of the positive control at a lower concentration (54% expression at 0.1%) and approximately 2-fold lower to that of the *V. yedoensis* extract (78% expression at 0.03%), suggesting that aesculin could be one of the main bioactive components responsible for the upregulation AQP3 levels in the *V. yedoensis* extract (Figure 1). These results suggest that aesculin contributes to an increase in AQP3 levels, which is important for the migration and proliferation of keratinocytes and the function of the epidermal water permeability barrier, by activating the P13K/Akt/mTOR signaling pathway, which we did not confirm in this study. Further, we observed that both aesculin and schaftoside upregulated the protein expression of CD44 in HaCaT cells by 123% and 193%, which was significantly superior to the positive control (86% at 0.1%) within 24 h treatment, respectively, suggesting that the topical application of the flavonoid, schaftoside, could provide both the biological benefits of CD44 stimulation and the physiological benefits of improved stratum corneum hydration that hyaluronic acid is popularly known to provide. In particular, the interaction between CD44 and HA is involved in keratinocyte differentiation and lipid synthesis. Additionally, the treatment of HaCaT keratinocytes with 1% aesculin, 0.1% schaftoside, and 0.1% sodium hyaluronate (positive control) upregulated the expression of K-10 levels by 299%, 116%, and 282%, respectively, suggesting that the two bioactive components could be responsible for maintaining epidermal integrity by increasing terminally differentiating epidermal keratinocyte differentiation. Interestingly, only schaftoside significantly upregulated the expression of filaggrin (FLG) by 93% (Figure 8), a structural protein that plays an important role in the skin's barrier function and is implicated in some of the most common dermatological diseases, such as atopic dermatitis (AD) and ichthyosis vulgaris. Overall, we also observed that there was no upregulation in the involucrin levels on treatment with aesculin, schaftoside, or sodium hyaluronate.

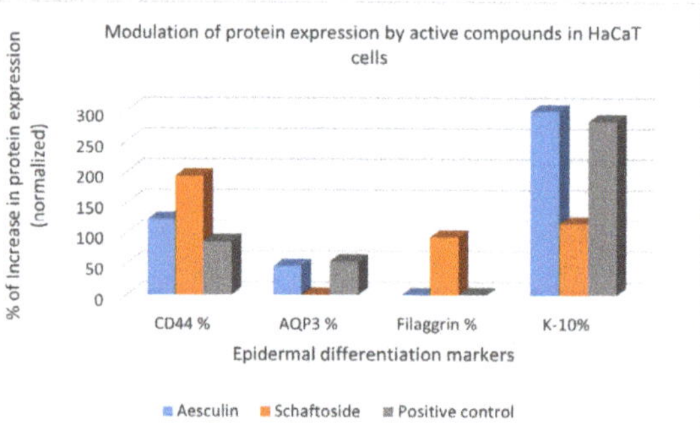

Figure 8. Regulation of epidermal differentiation markers CD44, AQP3, FLG, and K-10 in HaCaT cells treated by aesculin (1%), schaftoside (0.1%), and positive control (sod. hyaluronate 0.1%). Values are average of experiments in duplicates.

3. Discussion

Skin hydration is a critical aspect of maintaining healthy skin, and water is generally considered essential for the normal functions of the skin [55]. Today, several cosmetic ingredients aim to restore physiological hydration while helping to decrease skin evaporation and improve the appearance and tactile properties of the skin [56]. Hyaluronic acid (HA), a glycosaminoglycan (GAG), provides hydration and structural integrity to the dermis. It has been reported that HA is also naturally present in the epidermis, binds to the extracellular space via CD44, and may play a role in epidermal barrier function and SC hydration [57]. In normal skin, HA is found in relatively high concentrations in the basal layer of the epidermis where proliferating keratinocytes are found [58]. Generally, CD44 is the main receptor for hyaluronic acid in the skin. Hyaluronic acid, the main ligand for CD44, binds to and activates CD44, resulting in the activation of cell signaling pathways that induce cell proliferation. A deficiency of CD44 leads to a reduction in HA, which in turn alters the proliferation and differentiation of keratinocytes, lipid synthesis, and ultimately the skin barrier function [13]. In our investigation, *V. yedoensis* extract stimulated the synthesis of hyaluronic acid in the skin endogenously by increasing the expression of CD44, the main receptor for hyaluronic acid in keratinocytes. Thus, it acts as a natural humectant, enabling skin hydration by binding and retaining water molecules in the epidermis. AQP3 is a member of the family of aquaglyceroporins, which are membrane proteins that form water channels across the cell membrane. They facilitate the transport of water and solutes like glycerol or urea [59]. AQP3 expression in SC is responsible for the function of water loss prevention. Both *V. yedoensis* extract and aesculin treatments resulted in the increased expression of AQP3 protein, suggesting enhanced AQP3 activity. Compounds that increase AQP3 in the skin are reported as skin moisturizers. Recent studies have demonstrated that ligands of either PPARγ or LXR increase the expression level of AQP3 in epidermal keratinocytes [60,61]. Furthermore, aesculin has been reported to suppress LPS-induced inflammation by activating PPAR-γ in RAW 264.7 cells (murine macrophage cell line) [62]. These results possibly suggest that aesculin may upregulate the expression of AQP3 via its PPARγ agonist action.

Filaggrin (FLG), a large protein localized in the corneocyte layer in the SC, is converted to NMF in the more superficial layers of the SC. Proteolysis of FLG is dependent on water activity within the corneocytes [63]. Interestingly, a previous study reported that apigenin significantly increases the protein expression of FLG and AQP3 in HaCaT cells [64]. In our study, schaftoside, an apigenin-di-C-glycoside, upregulated the expression of AQP3 and FLG. FLG and involucrin are two epidermal proteins that are downregulated in response to damage to the skin barrier. Additional insights from the literature reveal that apigenin and vicenin-1 (6,8-diglycosylflavone) increased the expression of FLG and claudin-1 upon exposure to UV radiation in a 3D skin model study [53]. Authors offer an explanation that FLG expression in normal human epidermal keratinocytes (NHEKs) was upregulated by activation of the aryl hydrocarbon receptor (AhR) [65]. In addition, apigenin exhibited aryl hydrocarbon receptor (AhR) antagonist activity in Caco2 colon cancer cells. Therefore, one explanation that can be inferred is that schaftoside may possibly upregulate the FLG expression in keratinocytes by modulating AhR. However, this hypothesis must be validated by additional studies in the future. Aesculin and schaftoside upregulated the epidermal differentiation marker, keratin-10 (K-10). K-1 and K-10 are structural keratin proteins responsible for hyperdifferentiation and differentiation of the epidermis, thereby regulating skin barrier function. A downregulation of these proteins normally occurs in conditions such as atopic dermatitis and epidermolytic hyperkeratosis [66,67].

There are many biophysical methods available for measuring skin hydration. The measurement of trans-epidermal water loss (TEWL) rates through human skin is the most used method in the cosmetics and skincare industry because TEWL directly correlates with skin barrier dysfunction. Ideally, TEWL should be as low as possible in healthy skin. Lower numbers correlate with less water loss, and high numbers correlate with increased water loss and poor barrier function [68]. *V. yedoensis* extract showed a hydration effect comparable to 5% glycerin in water. Further, *V. yedoensis* extract decreases the trans-epidermal water loss. In normal conditions, with an intact skin barrier, TEWL on human skin is on the order of 5–10 g/m^2. Trans-epidermal water loss gives us the measure of the barrier function of the skin. The flux of water vapor that crosses the skin surface gives a measure of trans-epidermal water loss. As the water loss across the SC increases, the humidity next to the skin surface rises above ambient humidity [69]. This creates a humidity gradient above the skin surface that is proportional to the SC water loss. It characterizes the barrier function of the stratum corneum and thus allows the studying of the influence of physical or chemical factors on the integrity of the skin barrier. The more the skin barrier is perturbed, the higher the TEWL [70].

Primary intercellular barrier lipids, including cholesterol, essential fatty acids (EFAs), and ceramides, work together to form a barrier to prevent TEWL. Barrier lipids help maintain the skin's natural collagen, elastin, and proteins by minimizing TEWL, increasing hydration and reinforcing the protective barrier [71]. In the DSC experiment, it can be observed that the intracellular lamellar lipids in the SC provide a tight and effective barrier to the passage of water through the tissue. Lipid disordering on the molecular level is translated into a deterioration of the barrier properties of the SC at the tissue level. A decrease of $\geq$2 °C is considered significant. *V. yedoensis* extract showed an equivalent effect on water vapor flux in lipid perturbation compared to 5% glycerin in water. Thus, this extract maintained the intact skin barrier properties.

To summarize, the phytochemical analysis and structural characterization of secondary metabolites within the *Viola yedoensis* extract led to the discovery of 14 new bioactive components, notably belonging to megastigmane glycosides and triterpene saponins. This marks the first report of their occurrence and distribution within *Viola yedoensis*. A comprehensive understanding of their presence necessitates an examination of their biosynthetic pathways within the plant. Therefore, to establish a biosynthetic relationship between the newly identified megastigmane glycosides, roseoside and eriojaposide B, and the corresponding calenduloside, we decided to elaborate on the related biosynthetic pathways envisioned to generate the apocarotenoids. Roseoside and eriojaposide B are megastigmane glycosides with a 3-oxo-α-ionol skeleton, while platanionoside B has a 3-hydroxy-β-ionol skeleton. Megastigmanes belong to the family of apocarotenoids. The C13 norisoprenoids possess a structure representative of carotenoid cleavage products. Both sesquiterpenes and carotenoids share a common biosynthetic pathway originating from β-Ionone (Figure 9) [72]. Calendulosides are glycosylated triterpene saponins reported mainly from the genus Calendula. The biosynthetic precursor for these oleanane-type pentacyclic triterpenoids is β-Amyrin (Figure 10). The first step in the biosynthesis of triterpenoid saponins involves the cyclization of 2,3-oxidosqualene to cycloartenol. This undergoes further cyclization to dammarenyl cation, which undergoes rearrangement to oleanane-β-Amyrin and finally to oleanolic acid [73].

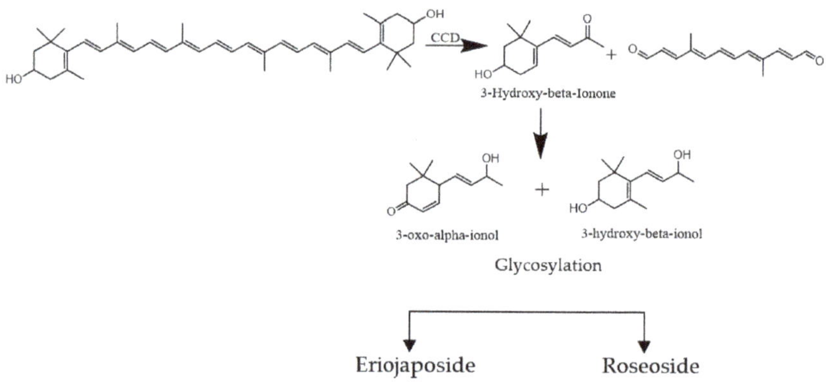

Figure 9. Biosynthetic pathway of roseoside and eriojaposide B.

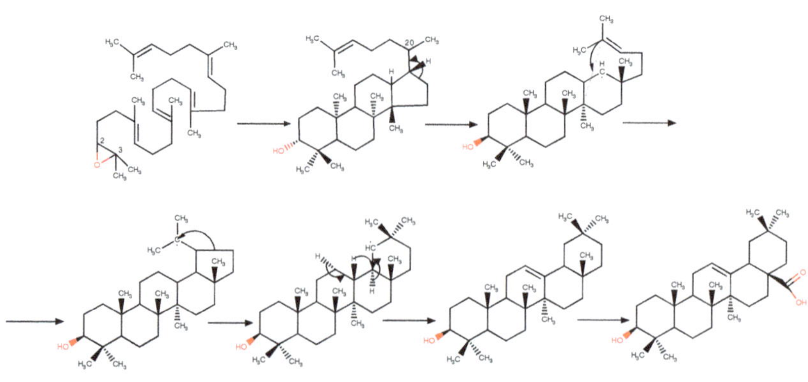

Figure 10. Biosynthetic pathway of calenduloside.

4. Materials and Methods

4.1. Plant Material, Solvents, Chemicals and Standards

Viola yedoensis Makino was collected from Bozhou, Anhui province, China, in December 2019. The plant was identified by Dr. P. Santhan, and a voucher specimen of the entire plant has also been deposited at the Herbarium facility at L'Oreal (Advanced Research, Bangalore), India, with the voucher specimen number Loreal R & I/17-06-2020. They were stored under frozen conditions at −20 °C and protected from light. To determine the moisture content and proportions of organic and inorganic fractions, a thermogravimetric investigation was carried out using a muffle furnace (Nabertherm GmbH, Lilienthal, Germany). The experimental protocol involved dehydration at 100 °C for 12 h, followed by calcination at 650 °C for 4 h. Aesculin (CAS No: 531-75-9) (Sigma-Aldrich $\geq$ 98%) and schaftoside (CAS No: 51938-32-0) (Shangai Tauto Biotech $\geq$ 98%, Shangai, China) of pharmaceutical primary standards were procured from Merck and Sigma-Aldrich, respectively. All solvents and reagents utilized in this study were purchased from Sigma-Aldrich (St. Louis, MO, USA). HPLC-MS-grade acetonitrile and methanol were purchased from Merck (Lowe, NJ, USA). Laboratory-grade ethanol was obtained from Merck Millipore (Darmstadt, Germany). Milli-Q Integral 15 system (Merck Millipore, Burlington, MA, USA) was used to prepare the HPLC-grade water. Standard solutions were prepared at a concentration of 2.5 mg/mL in DMSO.

4.2. Flash Chromatography

Flash chromatography of the extract was performed using the Sepacore Flash system X50, Buchi with a C18 column (250 mm × 2.5 mm). A total of 100 mg of the sample was solubilized in MeOH:Water (50:50, 5 mL) and filtered by 0.45 µm (Millipore Millex-HN). A gradient elution with ethanol:water mobile phase at the flow rate of 15 mL/min led to the preparation of a specific *Viola yedoensis* extract composition (Figure S7).

4.3. LC/HRMS/MS Instrumentation and Conditions

Acquisitions were realized on an Orbitrap Fusion high-resolution mass spectrometer (HRMS) equipped with an electrospray source (ESI), an Ultimate 3000 ultra-high-performance liquid chromatography (UPLC) system, and a photodiode-array detector (PDA) from Thermo Fisher Scientific. Separation was achieved on an ACQUITY BEH Shield RP18 (Waters, 100 × 2.1 mm, 1.8 µm) at 35 °C. Mobile phases used were water and acetonitrile (ACN), both acidified with 0.1% of formic acid. The gradient used was as follows: 0–1 min: 1% ACN, 1–15 min: 1 to 100% ACN, 15–20 min: 100% ACN, 20.1–25 min: 1% ACN, at a flow rate of 0.6 mL/min (direct hyphenation with HRMS). The autosampler was set at 10 °C. The injection volume was 2 µL. The wavelength range of the PDA detector was 210–700 nm. Acquisitions were performed on scan range *m/z* 100–1100 operated separately in positive (+3.8 kV) and negative (−3.4 kV) ion modes. ESI source parameters were set as follows: vaporizer temperature: 400 °C, capillary transfer tube temperature: 325 °C, sheath gas pressure: 30 arbitrary units, auxiliary gas: 10 arbitrary units, and sweep gas: 2 arbitrary units. The RF Lens was set at 60%, and AGC Target was in standard mode, with an injection time of 100 ms. Analyses were performed using full scan mode, and full scan MS—data-dependent MS2 mode (ddMS2). Collision experiments based on non-resonant excitation mode higher-energy collisional dissociation (HCD at NCE = 20%) and sequential MSn experiments based on resonant excitation mode (collision-induced dissociation, CID, at NCE = 25%) experiments were acquired. The mass spectrometer was calibrated prior to analysis with Pierce^{TM} FlexMix^{TM} calibration solution (Thermo Fisher Scientific, reference A39239). The mass resolution power of the analyzer was 120,000 m/Δm, full width at half maximum (FWHM) at *m/z* 200, for singly charged ions. All the data were manually inspected using Thermo Fisher Scientific software: Freestyle module of Xcalibur (version 4.4), Compound Discoverer (version 3.3) in combination with product ion spectral database mzCloud (2024 update) and an in-house mzVault database.

To estimate the difference in gas-phase acidity of OH groups between a dehydrated hexose and a phenolic aglycone, model molecules' values were used. The $\Delta H^{\circ}_{(acid)(M)}$ of model molecules M are considered as (i) tetrahydro-2-pyranol, 1540 kJ/mol or cyclohexanone, 1533 kJ/mol, used for a dehydrated hexose and (ii) phenol, 1462 kJ/mol or meta diphenol as resorcinol, 1450 kJ/mol used for an aglycone. Values obtained from the NIST WebBook.

4.4. Enzyme-Linked Immunosorbent (ELISA) Assays

The supernatant of extracted solution was injected into LC/HRMS system. Standard solutions were prepared at a concentration of 2.5 mg/mL in DMSO. The enzyme-linked immunosorbent (ELISA) assay Dulbecco's modified Eagle's medium (DMEM, optimized 1× DMEM with 4.5 g/L glucose, L-glutamine, sodium pyruvate, and HaCaT cell line) was purchased from AddexBio Technologies Inc., Krishgen. Human CD 44 ELISA kit was purchased from Abcam (Cambridge, UK), aquaporin-3 from Novus Biologicals, and keratin-10 and filaggrin from Abbexa (Cambridge, UK) and Cussabio (Houston, TX, USA). The cell lysis buffer (10×) was from Cell Signal Technology (Boston, MA, USA) and 6 well plates from Corning (New York, NY, USA). Envision 2105 Multimode Plate Reader from Perkin Elmer (Waltham, MA, USA) was used. TEWL and DSC analysis: PIE System, Falcom toolbox, 16 and 14 mm Die set, Punching Press, TEWL Sample holders, Rotary Shaker, ESPEC Temperature and Humidity Chamber, Mettler Toledo Analytical Balance, Circular Adhesive Tape. DSC: Mettler Toledo DSC 1 STAR e System (IN) (Mettler-Toledo India Private Limited, Mumbai, India) or DSC TA instrument Q200 (FR), Standard 40 µL Aluminum Crucible with Pin (IN) or Tzero Aluminium Hermetic Pans (FR).

4.5. Extraction and Fractionation of the Extract

The whole plant of *V. yedoensis* Makino (100 g) was dried, powdered, and subjected to hydroalcoholic extraction (50:50, EtOH:water, 65 °C, 3 cycles) to yield 30 g of the extract. The extract (30 g) was subjected to HP-20 resin fractionation [20:80 (EtOH:water), 60:40 (EtOH:water) and 100% EtOH] to yield three fractions VY-1, VY-2, and VY-3 (Figure 10). The fraction VY-1 and VY-2 comprised of the coumarins, aesculin, and esculetin, respectively. The fractions VY-3 rich in flavone glycosides was further fractionated by flash chromatography.

The fraction VY-3 was further fractionated by Flash chromatograph system. A total of 100 mg of the sample was solubilized in MeOH:Water (50:50, 5 mL) and filtered by 0.45 µm (Millipore Millex-HN). A gradient elution (Table S1) with methanol:water mobile phase and C18 column at the flow rate of 15 mL/min led to the isolation of fractions *Viola yedoensis* extract (*V. yedoensis* extract) enriched with flavonoids. A fraction of *V. yedoensis* extract was subjected to HRMS analysis for the identification of the compounds. Compounds **1–15** were previously reported from the same plant. Compounds **16–29** are reported for the first time from the genus *Viola*.

4.6. Cell Viability Assay

The effect of the compounds on cell viability after 24 h treatment was tested. Cells (10,000 per well) were seeded in a 96-well opaque plate and incubated overnight at 5% CO_2, 37 °C. The cells were treated with *V. yedoensis* extract 0.03%, aesculin 1%, schaftoside 0.1%, and sodium hyaluronate 5 kDa (positive control) 0.1% and incubated for 24 h at 5% CO_2, 37 °C. Cell Titer Glo Reagent (Promega-G7571) was added to the plate, and incubated, and luminescence was measured in Envision plate reader.

4.7. ELISA Assay

HaCaT cell line was trypsinized and seeded at the density of 1×10^6 cells/well in a 6-well plate and kept for overnight incubation at 37 °C, 5% CO_2 conditions. The following day, cells were treated with the following test compounds: *V. yedoensis* extract at 2%, aesculin at 1%, and hyaluronic acid at 0.1%, which were weighed and solubilized in culture media. The plant extract (0.03%) was further diluted at 1:1 ratio in DMEM to achieve test concentration. The test compounds were sterile-filtered with 0.2 μm syringe filter. The media were aspirated, and 1 mL/well of the prepared filtered dilutions was added to the 6-well plate as per the plate map and kept for overnight incubation at 37 °C, 5% CO_2 conditions. Following 24 h treatment with the test compounds, cells were harvested. The supernatant (spent media) was collected and centrifuged at 10,000 rpm for 10 min. The pellet containing the debris was discarded, and the supernatant was stored at −80 °C until further use. For the lysate, the adherent cells were trypsinized, the pellet was washed twice with PBS, and 50 μL of 1× cell lysis buffer was added to the cell pellet. The cells were incubated on ice for 30 min with intermittent mixing and stored at −80 °C for complete lysis. On the day of the experiment (ELISA), the frozen lysates were thawed on ice, centrifuged at 13,000 rpm at 4 °C for 15 min. The supernatant was collected and used for the ELISA.

The ELISA kits use the sandwich-ELISA principle. The micro-ELISA plates provided in the kits have been pre-coated with an antibody specific to target protein. Standards or samples are added to the micro-ELISA plate wells and combined with the specific antibody. Then, a biotinylated detection antibody specific for target protein and avidin-horseradish peroxidase (HRP) conjugate are added successively to each micro-plate well and incubated. Free components are washed away. The substrate solution is added to each well. Only those wells that contain target protein, biotinylated detection antibody, and avidin-HRP conjugate will appear blue in color. The enzyme–substrate reaction is terminated by the addition of stop solution, and the color turns yellow. The optical density (OD) is measured spectrophotometrically at a wavelength of 450 nm ± 2 nm. The OD value is proportional to the concentration of the target protein. The concentration of the target protein in the samples is calculated by comparing the OD of the samples to the standard curve. Data analysis: The average of duplicate readings for each standard and samples was taken. The average values were subtracted with the average zero standard optical density.

4.8. TEWL Measurements

The TEWL measurements were recorded in vitro on isolated stratum corneum using a motorized assembly. The stratum corneum samples were placed in the temperature and humidity control chamber for equilibrium at 30 °C and 30% RH. After overnight equilibrium, TEWL measurements "T1" were recorded. Measurement of flow of water crossing a sample of SC situated above a water tank was made in a glove box kept at 30 °C and 30% RH by means of a hygrometric probe coupled to a robotic sample holder and a data acquisition system Derma lab. The hygrometric probe (Figure S8) contains two humidity sensors located at different heights in the cylindrical probe chamber so that the humidity gradient can be measured and used to calculate the flux of water vapor. The measurements were performed before and after treatment for each sample across two different batches of stratum corneum with 4 replicates on each batch. Assessment of intercellular stratum corneum lipid perturbation by DSC: After TEWL measurements, the same samples were used for the DSC analysis after equilibrating them for an additional 4 h at 75% RH, 25 °C. The measurements were performed using aluminum hermetic-sealed pans and heating from 10 °C to 110 °C at a rate of 5 °C/min.

5. Conclusions

Our investigation aimed to screen a new composition of *Viola yedoensis* extract for its ability to restore epidermal barrier function through the expression of key epidermal protein markers, alongside enhanced skin hydration properties. In a primary screen for hydration benefits, the *V. yedoensis* extract was found to increase the protein expression levels of CD44 and aquaporin-3 (AQP3) in HaCaT cells by sandwich-ELISA assay. Owing to this new biological efficacy, we applied an integrated LC-HRMS/MS-based metabolomics approach under resonant (CID 25%) and non-resonant collisional (HCD 20%) excitation conditions, together with state-of-the-art computational tools, to provide acute insights regarding the characterization of bioactive components from the aerial parts of *V. yedoensis*. This methodology involved data analysis pipelines of untargeted metabolomics analysis that enabled the characterization and identification of 29 secondary metabolites, 14 of which are reported here for the first time in *V. yedoensis*, namely, esculetin, aesculin, apigenin, kaempferol C-glycosides, megastigmane glycosides, roseoside, platanionoside B, and eriojaposide B isomer, together with the rare calenduloside F and esculetin diglucoside, which are reported for the first time from the genus *Viola*.

Additionally, the study demonstrated for the first time that *V. yedoensis* extract, enriched with flavone glycosides, can be cosmetically employed for restoring the epidermal barrier function through the expression of key epidermal protein markers, CD44 and AQP3, with enhanced skin hydration properties. Furthermore, two major secondary metabolites, aesculin and schaftoside, were discovered to be the main components of *V. yedoensis* to upregulate the expression of the epidermal differentiation marker, K-10, which is chiefly responsible for maintaining the structural and mechanical integrity of the stratum corneum. The apigenin-C-glycoside, schaftoside, significantly increased the protein expression level of CD44, while the coumarin glycoside, aesculin, increased the AQP3 levels and upregulated the expression of keratin-10 protein, comparable to that of the positive reference, sodium hyaluronate. It is interesting to note that schaftoside significantly upregulated the expression of filaggrin, thereby suggesting that schaftoside could be a useful natural bioactive component in the future treatment of some of the most common dermatological diseases, such as atopic dermatitis and psoriasis. A comprehensive structure–activity study of the flavone C-glycosides and coumarin glycosides has yet to be reported, and it would be relevant in the future to examine more analogues from this class in this context. Most importantly, our results provide encouragement that the study of known glycosides from the *Viola* genus could lead to the discovery of additional novel bioactive components with enhanced skin barrier function and hydration benefits.

Supplementary Materials: The supporting information can be downloaded at: https://www.mdpi.com/article/10.3390/ijms252312770/s1.

Author Contributions: Conceptualization, S.T.P., S.T., J.E. and J.-C.T.; analytical methodology, S.T. and S.D.S.; analysis, S.D.S., A.D., A.S. and R.M.; data curation, S.T., S.D.S., A.D., S.T.P. and J.-C.T.; writing—original draft preparation, S.T.P., S.T., S.D.S., A.D., S.B., J.E. and J.-C.T., writing—review and editing, S.T.P., S.T., S.D.S., A.D., S.B., J.E. and J.-C.T.; visualization, S.T., S.D.S., A.D., S.B., J.E. and J.-C.T.; supervision, S.T.P.; project administration, S.T.P.; funding acquisition, S.J., D.R. and, J.E. All authors have read and agreed to the published version of the manuscript.

Funding: Financial support was provided by L'Oréal Innovative Ingredients department (D2I) and L'Oréal International Analysis department.

Institutional Review Board Statement: Not applicable.

Informed Consent Statement: Not applicable.

Data Availability Statement: Data is contained within the article and Supplementary Materials.

Conflicts of Interest: S.T., S.D.S., A.D., S.B., R.M., A.S., J.E., S.J., D.R. and S.T.P. have received research funding from L'Oréal. J.C.T. has been involved as an expert consultant for L'Oréal. A.D. received funding from L'Oréal to present at SMAP 2024 in Lille, France. Authors declare that they have no conflicts of interest with any other company or organization.

Abbreviations

SC	Stratum corneum
HaCaT	Human keratinocytes
TEWL	Trans-epidermal water loss
DSC	Differential scanning calorimetry
PBS	Phosphate buffered saline

DMEM	Dulbecco's modified Eagle medium
FBS	Fetal bovine serum
P13K	Phosphoinositide 3-kinase
CD44	Cell surface adhesion receptor
FLG	Filaggrin
AQP3	Aquaporin-3
K-10	Keratin-10
INVL	Involucrin
ELISA	Enzyme-linked immune sorbent assay
PPAR-γ	Peroxisome proliferator activated receptor-gamma
LXR	Liver X receptor
SD	Standard deviation

References

1. Scheuplein, R.J.; Blank, I.H. Permeability of the skin. *Physiol. Rev.* **1971**, *51*, 702–747. [CrossRef] [PubMed]
2. Madison, K.C. Barrier function of the skin: "la raison d'etre" of the epidermis. *J. Investig. Dermatol.* **2003**, *121*, 231–241. [CrossRef] [PubMed]
3. Verdier-Sevrain, S.; Bonte, F. Skin hydration: A review on its molecular mechanisms. *J. Cosmet. Dermatol.* **2007**, *6*, 75–82. [CrossRef] [PubMed]
4. Rajkumar, J.; Chandan, N.; Lio, P.; Shi, V. The Skin Barrier and Moisturization: Function, Disruption, and Mechanisms of Repair. *Ski. Pharmacol. Physiol.* **2023**, *36*, 174–185. [CrossRef]
5. Ishitsuka, Y.; Roop, D.R. Loricrin: Past, Present, and Future. *Int. J. Mol. Sci.* **2020**, *21*, 2271. [CrossRef]
6. Kim, B.E.; Leung, D.Y.; Boguniewicz, M.; Howell, M.D. Loricrin and involucrin expression is down-regulated by Th2 cytokines through STAT-6. *Clin. Immunol.* **2008**, *126*, 332–337. [CrossRef]
7. McAleer, M.A.; Irvine, A.D. The multifunctional role of filaggrin in allergic skin disease. *J. Allergy Clin. Immunol.* **2013**, *131*, 280–291. [CrossRef]
8. Randall, R.W.; Marty, O. Visscher. Structure and function of the epidermal barrier. *Am. J. Infect. Control* **2006**, *34*, S98–S110.
9. Mukherjee, S.; Mitra, R.; Maitra, A.; Gupta, S.; Kumaran, S.; Chakrabortty, A.; Majumder, P.P. Sebum and Hydration Levels in Specific Regions of Human Face Significantly Predict the Nature and Diversity of Facial Skin Microbiome. *Sci. Rep.* **2016**, *6*, 36062–36073. [CrossRef]
10. Rawlings, A.V.; Harding, C.R. Moisturization and skin barrier function. *Dermatol. Ther.* **2004**, *17*, 43–48. [CrossRef]
11. Rawlings, A.V.; Scott, I.R.; Harding, C.R.; Browser, P.A. Stratum corneum moisturization at the molecular level. *J. Investig. Dermatol.* **1994**, *103*, 731–740. [CrossRef] [PubMed]
12. Guo, L.; Chen, H.; Li, Y.; Zhou, Q.; Sui, Y. An aquaporin 3-notch1 axis in keratinocyte differentiation and inflammation. *PLoS ONE* **2013**, *8*, e80179. [CrossRef] [PubMed]
13. Bourguignon, L. Matrix Hyaluronan-Activated CD44 Signaling Promotes Keratinocyte Activities and Improves Abnormal Epidermal Functions. *Am. J. Pathol.* **2014**, *184*, 1912–1919. [CrossRef] [PubMed]
14. Kaya, G.; Rodriguez, I.; Jorcano, J.L.; Vassalli, P.; Stamenkovic, I. Selective suppression of CD44 in keratinocytes of mice bearing an antisense CD44 transgene driven by a tissue-specific promoter disrupts hyaluronate metabolism in the skin and impairs keratinocyte proliferation. *Genes Dev.* **1997**, *11*, 996–1007. [CrossRef] [PubMed]
15. Kezic, S.; Kemperman, P.M.; Koster, E.S.; de Jongh, C.M.; Thio, H.B.; Campbell, L.E.; Irvine, A.D.; McLean, I.W.; Puppels, G.J.; Caspers, P.J. Loss-of-function mutations in the filaggrin gene lead to reduced level of natural moisturizing factor in the stratum corneum. *J. Investig. Dermatol.* **2008**, *128*, 2117–2119. [CrossRef]
16. Available online: https://www.prnewswire.com/news-releases/valued-to-be-30-1-billion-by-2026--natural-and-organic-cosmetics-slated-for-robust-growth-worldwide-301379348.html (accessed on 10 September 2024).
17. Michalak, M. Plant Extracts as Skin Care and Therapeutic Agents. *Int. J. Mol. Sci.* **2023**, *24*, 15444. [CrossRef]
18. e Silva, V.R.L.; Schulman, M.A.; Ferelli, C.; Gimenis, J.M.; Ruas, G.W.; Baby, A.R.; Velasco, M.V.R.; Taqueda, M.E.; Kaneko, T.M. Hydrating effects of moisturizer active compounds incorporated into hydrogels: In vivo assessment and comparison between devices. *J. Cosmet. Dermatol.* **2009**, *8*, 32–39. [CrossRef]
19. Ryan, T.; Cornell, M.; Paldus, B.A. Compositions and Methods for Treating Skin. U.S. Patent US10682306B1, 30 October 2019.
20. Zheng, H.-R.; Wang, B.; Zhao, Z.; Zhang, Q.; Liang, M.-Y.; Yao, Y.-Q.; Bian, K.; Zhang, W.-R. Effects of *Viola yedoensis* Makino anti-itching compound on degranulation and cytokine generation and cytokine generation in RBL-2H3 mast cells. *J. Ethnopharmacol.* **2016**, *189*, 132–138. [CrossRef]
21. Jeong, Y.H.; Oh, Y.-C.; Cho, W.-K.; Shin, H.; Lee, K.Y.; Ma, J.Y. Anti-inflammatory effects of *Viola yedoensis* and the application of cell extraction methods for investigating bioactive constituents in macrophages. *BMC Complement. Altern. Med.* **2016**, *16*, 180. [CrossRef]
22. Cao, J.; Du, D.; Qin, Y.; Yin, C.; Cheng, Z. Microwave-assisted extraction coupled high performance liquid chromatography-electrospray ion trap mass spectrometry for the determination of three flavone di-C-glycosides in *Viola yedoensis*. *Curr. Anal. Chem.* **2015**, *11*, 225–233. [CrossRef]

23. Hong, J.L.; Zhou, H.Y.; Zhu, J.; Li, L.; Shu, P.; Qin, Z.Y.; Wu, G.; Lin, B.B.; Wang, G.-K.; Wang, Q.; et al. Comparative analysis of major constituents in *Viola yedoensis* Makino and different species from the genus Viola by High-Performance liquid chromatography with chemometrics methods. *J. Med. Plant Res.* **2011**, *21*, 5230–5239.
24. Zhang, L.; Li, M.Y.; Wang, L.W.; Gao, J.; Ma, C.M. Isolation, Identification, Quantification and Inhibitory activity on HCV protease of coumarins from *Viola* yedoensis, *Can. Chem. Trans.* **2013**, *1*, 157–164.
25. Sumner, L.W.; Amberg, A.; Barrett, D.; Beale, M.H.; Beger, R.; Daykin, C.A.; Fan, T.W.; Fiehn, O.; Goodacre, R.; Griffin, J.L.; et al. Proposed minimum reporting standards for chemical analysis Chemical Analysis Working Group (CAWG) Metabolomics Standards Initiative (MSI). *Metabolomics* **2007**, *3*, 211–221. [CrossRef] [PubMed]
26. Kebarle, P. A brief overview of the present status of the mechanisms involved in the electronspary mass spectrometry. *J. Mass Spectrom.* **2000**, *35*, 804–817. [CrossRef]
27. Ichou, F.; Schwarzenberg, A.; Lesage, D.; Alves, S.; Junot, C.; Machuron-Mandardc, X.; Tabet, J.C. Comparison of the activation time effects and the internal energy distributions for the CID, PQD and HCD excitation modes. *J. Mass Spectrom.* **2014**, *49*, 498–508. [CrossRef]
28. Chu, H.; Wang, J.; Wang, Q.; Chen, J.; Li, J.; Li, H.; Zhang, L. Protective Effect of n-Butanol Extract from *Viola yedoensis* on Immunological Liver Injury. *Chem. Biodivers.* **2021**, *18*, 2001043. [CrossRef]
29. Zhou, H.-Y.; Qin, M.-J.; Hong, J.-L.; Ni, Y.-J.; Wu, G. Chemical Constituents of *Viola yedoensis*. *Chin. J. Nat. Med.* **2009**, *7*, 290–292. [CrossRef]
30. Zhang, Q.; Wang, Q.; Chen, S. A comprehensive review of phytochemistry, pharmacology and quality control of plants from the genus Viola. *J. Pharm. Pharmacol.* **2023**, *75*, 1–32. [CrossRef]
31. Xie, C.; Veitch, N.C.; Houghton, P.J.; Simmonds, M.S.J. Flavone C-Glycosides from *Viola yedoensis* Makino. *Chem. Pharm. Bull.* **2003**, *51*, 1204–1207. [CrossRef]
32. Zhou, H.Y.; Hong, J.L.; Shu, P.; Ni, Y.J.; Qin, M.J. A new dicoumarin and anticoagulant activity from *Viola yedoensis* Makino. *Fitoterapia* **2009**, *80*, 283–285. [CrossRef]
33. Oshima, N.; Narukawa, Y.; Takeda, T.; Kiuchi, F. Collagenase inhibitors from *Viola yedoensis*. *J. Nat. Med.* **2013**, *67*, 240–245. [CrossRef] [PubMed]
34. Peng, M.; Watanabe, S.; Chan, K.W.K.; He, Q.; Zhao, Y.; Zhang, Z.; Lai, X.; Luo, D.; Vasudevan, S.G.; Li, G. Luteolin restricts dengue virus replication through inhibition of the proprotein convertase furin. *Antivir. Res.* **2017**, *143*, 176–185. [CrossRef] [PubMed]
35. Cole, R.B.; Tabet, J.C.; Salles, C.; Jallageas, J.C.; Crouzet, J. Structural "Memory Effects" Influencing decompositions of glucose alkoxide anions produced from monoterpene glycoside isomers in Tandem Mass Spectrometry. *Rapid Commun. Mass Spectrom.* **1989**, *3*, 59–61. [CrossRef]
36. Darii, E.; Gimbert, Y.; Alves, S.; Damont, A.; Perret, A.; Woods, A.S.; Fenaille, F.; Tabet, J.-C. First Direct Evidence of Interpartner Hydride/Deuteride Exchanges for Stored Sodiated Arginine/Fructose-6-phosphate Complex Anions within Salt-Solvated Structures. *J. Am. Soc. Mass Spectrom.* **2021**, *32*, 1424–1440. [CrossRef] [PubMed]
37. Fournier, F.; Remaud, B.; Blasco, T.; Tabet, J.C. Ion-Dipole Complex Formation from Deprotonated Phenol Fatty Acid Esters Evidenced by Using Gas-Phase Labeling Combined with Tandem Mass Spectrometry. *J. Am. Soc. Mass Spectrom.* **1993**, *4*, 343–351. [CrossRef]
38. Afonso, C.; Cole, R.B.; Tabet, J.-C. *Interscience Dissociations of Even Electron Ions*; Cole, R.B., Ed.; Wiley: Hoboken, NJ, USA, 2010; pp. 631–684.
39. Zhao, D.-L.; Zou, L.-B.; Lin, S.; Shi, J.-G.; Zhu, H.-B. 6,7-di-O-glucopyranosyl-esculetin protects SH-SY5Y cells from dopamine-induced cytotoxicity. *Eur. J. Pharmacol.* **2008**, *580*, 329–338. [CrossRef]
40. Guo, X.L.; Yang, L.; Yu, S.L.; Zhang, K.; Wang, J.H.; Wang, H.Y. Analysis of secondary metabolites in Xinjiang Morus nigra leaves using different extraction methods with UPLC-Q/TOF-MS/MS technology. *Open Chem.* **2023**, *21*, 20220320. [CrossRef]
41. Mahmoud, A.A.; Al-Shihry, S.S.; Hegazy, M.-E.F. Biological Activity of a Phloroglucinol Glucoside Derivative from Conyza aegyptiaca. *Z. Naturforsch.* **2009**, *64*, 513–517. [CrossRef]
42. Besson, E.; Dellamonica, G.; Chopin, J.; Markham, K.R.; Kim, M.; Koh, H.-S.; Fukami, H. C-Glycosylflavones from *Oryza sativa*. *Phytochemistry* **1985**, *24*, 1061–1064. [CrossRef]
43. Moniava, I.I.; Kemertelidze, E.P. Quercetin glycosides from *Onobrychis sosnowskyi* and *O. kaehetica*. *Khimiya Prir. Soedin.* **1971**, *6*, 833–834.
44. Zheng, J.; Kallio, H.; Yang, B. Sea Buckthorn (*Hippophae rhamnoides* ssp. rhamnoides) Berries in Nordic Environment: Compositional Response to Latitude and Weather Conditions. *J. Agric. Food Chem.* **2016**, *64*, 5031–5044. [CrossRef]
45. Zhang, Q.Y.; Wang, X.-Y.; Ying, H.P.; Cheng, T.-M.; Zhao, Y.-Y. Studies on the Chemical Constituents of *Carduus crispus* L. *China J. Chin. Mater. Medica* **2001**, *26*, 839–841.
46. Özden, S.; Dürüst, N.; Toki, K.; Saito, N.; Honda, T. Acylated Kaempferol glycosides from the flowers of *Delphinium formosum*. *Phytochemistry* **1998**, *49*, 241–245. [CrossRef]
47. Hasan Elrady, A. Saad. C-Glycosylflavones from Prosopsis chilensis. *Mansoura J. Pharm. Sci.* **1994**, *10*, 254–264.
48. Tsuka, I. Contribution to Flower Colors of Flavonoids Including Anthocyanins: A Review. *Nat. Prod. Commun.* **2015**, *10*, 529–544.
49. Tamaki, A.; Otsuka, H.; Toshinori, I. Platanionosides A-C, Megastigmane Diglycosides from the Leaves of *Alangium platanifolium*. *J. Nat. Prod.* **1999**, *62*, 1074–1076. [CrossRef]

50. Schliemann, W.; Kolbe, B.; Schmidt, J.; Nimtz, M.; Wray, V. Accumulation of apocarotenoids in mycorrhizal roots of leek (*Allium porrum*). *Phytochemistry* **2008**, *69*, 1680–1688. [CrossRef]
51. Charles, W.; Evans, D. *Trease and Evans' Pharmacognosy*, 6th ed.; Elsevier Health Science: Philadelphia, PA, USA, 2009; pp. 304–322.
52. Ju, S.; Tan, Y.; Wang, Q.; Zhou, L.; Wang, K.; Wen, C.; Wang, M. Antioxidant and anti-inflammatory effects of esculin and esculetin (Review). *Exp. Ther. Med.* **2024**, *27*, 248. [CrossRef]
53. Povetyeva, T.N.; Afanasyeva, O.G.; Zibareva, L.N.; Nesterova, Y.V.; Suslov, N.I.; Kulpin, P.V.; Zyuz'kov, G.N.; Krylova, S.G.; Vsyakikh, O.V. Investigation of wound healing activity of the flavonoid schaftoside isolated from *Lychnis chalcedonica*. *Res. Results Pharmacol.* **2023**, *9*, 99–104. [CrossRef]
54. Chen, Y.; Zhang, D.; Hu, D.; Li, X.; Luo, J.; Duan, X.; Zhang, Y.; Wang, Y.F. Alkaloids and flavonoids exert protective effects against UVB-induced damage in a 3D skin model using human keratinocytes. *Results Chem.* **2022**, *4*, 100298–100310. [CrossRef]
55. Strudwick, X.L.; Lang, D.L.; Smith, L.E.; Cowin, A.J. Combination of Low Calcium with Y-27632 Rock Inhibitor increases the Proliferative Capacity, Expansion Potential and Lifespan of Primary Human Keratinocytes while retaining Their Capacity to differentiate into Stratified Epidermis in a 3D Skin Model. *PLoS ONE* **2015**, *10*, e0123651. [CrossRef]
56. Lee, S.H.; Jeong, S.K.; Ahn, S.K. An update of the defensive barrier function of skin. *Yonsei Med. J.* **2006**, *47*, 293–306. [CrossRef] [PubMed]
57. Harding, C.R.; Watkinson, A.; Rawlings, A.V.; Scott, I.R. Dry skin, moisturization and corneodesmolysis. *Int. J. Cosmet. Sci.* **2000**, *22*, 21–52. [CrossRef] [PubMed]
58. Wertz, P.W. Stratum corneum lipids and water. *Exog. Dermatol.* **2004**, *3*, 53–56. [CrossRef]
59. Manuskiatti, W.; Maibach, H.I. Hyaluronic acid and skin: Wound healing and aging. *Int. J. Dermatol.* **1996**, *35*, 539–544. [CrossRef]
60. Verkman, A.S.; Mitra, A.K. Structure and function of aquaporin water channels. *Am. J. Physiol. Renal Physiol.* **2000**, *278*, F13–F28. [CrossRef]
61. Yang, R.; Chowdhury, S.; Choudhary, V.; Chen, X. Keratinocyte aquaporin-3 expression induced by histone deacetylase inhibitors is mediated in part by peroxisome proliferator-activated receptors (PPARs). *Exp. Dermatol.* **2020**, *29*, 380–386. [CrossRef]
62. Schmuth, M.; Moosbrugger-Martinz, V.; Blunder, S.; Dubrac, S. Role of PPAR, LXR, and PXR in epidermal homeostasis and inflammation. *Biochim. Biophys. Acta.* **2014**, *1841*, 463–473. [CrossRef]
63. Tian, X.; Peng, Z.; Luo, S.; Zhang, S.; Li, B.; Zhou, C.; Fan, H. Aesculin protects against DSS-Induced colitis though activating PPARγ and inhibiting NF-κB pathway. *Eur. J. Pharmacol.* **2019**, *857*, 172453–172461. [CrossRef]
64. Fowler, J. Understanding the Role of Natural Moisturizing Factor in Skin Hydration. *Practical Dermatol.* 2012. Available online: https://assets.bmctoday.net/practicaldermatology/pdfs/pd0712_ftr_nmfreview.pdf (accessed on 10 September 2024).
65. Hou, M.; Sun, R.; Hupe, M.; Kim, P.L.; Park, K.; Crumrine, D.; Lin, T.K.; Santiago, J.L.; Mauro, T.M.; Elias, P.M.; et al. Topical apigenin improves epidermal permeability barrier homoeostasis in normal murine skin by divergent mechanisms. *Exp. Dermatol.* **2013**, *22*, 210–213. [CrossRef]
66. Tsuji, G.; Hashimoto-Hachiya, A.; Kiyomatsu-Oda, M.; Takemura, M.; Ohno, F.; Ito, T.; Morino-Koga, S.; Mitoma, C.; Nakahara, T.; Uchi, H.; et al. Aryl hydrocarbon receptor activation restores filaggrin expression via OVOL1 in atopic dermatitis. *Cell Death Dis.* **2017**, *8*, e2931. [CrossRef]
67. Müller, F.B.; Huber, M.; Kinaciyan, T.; Hausser, I.; Schaffrath, C.; Krieg, T.; Hohl, D.; Korge, B.P.; Arin, M.J. A human keratin 10 knockout causes recessive epidermolytic hyperkeratosis. *Hum. Mol. Genet.* **2006**, *15*, 1133–1141. [CrossRef] [PubMed]
68. Totsuka, A.; Omori-Miyake, M.; Kawashima, M.; Yagi, J.; Tsunemi, Y. Expression of keratin 1, keratin 10, desmoglein 1 and desmocollin 1 in the epidermis: Possible downregulation by interleukin-4 and interleukin-13 in atopic dermatitis. *Eur. J. Dermatol.* **2017**, *27*, 247–253. [CrossRef] [PubMed]
69. Lee, C.M.; Maibach, H.I. Bioengineering analysis of water hydration: An overview. *Exog. Dermatol.* **2002**, *1*, 269–275. [CrossRef]
70. Alexander, H.; Brown, S.; Danby, S.; Flohr, C. Research techniques made simple: Transepidermal water loss measurement as a research tool. *J. Investig. Dermatol.* **2018**, *138*, 2295–2300.e1. [CrossRef]
71. Elkeeb, R.; Hui, X.; Chan, H.; Tian, L.; Maibach, H.I. Correlation of transepidermal water loss with skin barrier properties in vitro: Comparison of three evaporimeters. *Ski. Res. Technol.* **2010**, *16*, 9–15. [CrossRef]
72. Parrish, A.R. The impact of aging on epithelial barriers. *Tissue Barriers* **2017**, *5*, e1343172. [CrossRef]
73. Günata, Z. Biosynthesis of C13-norisoprenoids in vitis vinifera: Evidence of carotenoid cleavage dioxygenase (CCD) and secondary transformation of norisoprenoid compound. In *Carotenoid Cleavage Products*; ACS Symposium Series; American Chemical Society: Washington, DC, USA, 2013.

International Journal of
Molecular Sciences

MDPI

Article

Licochalcone D from *Glycyrrhiza uralensis* Improves High-Glucose-Induced Insulin Resistance in Hepatocytes

Yu Geon Lee [1,†], Hee Min Lee [2,†], Jin-Taek Hwang [1] and Hyo-Kyoung Choi [1,*]

[1] Personalized Diet Research Group, Korea Food Research Institute (KFRI), Wanju 55365, Republic of Korea; ugun2@kfri.re.kr (Y.G.L.); jthwang@kfri.re.kr (J.-T.H.)
[2] Kimchi Industry Promotion Division, World Institute of Kimchi, Gwangju 61755, Republic of Korea; hmlee@wikim.re.kr
[*] Correspondence: chkyoung@kfri.re.kr; Tel.: +82-63-219-9421
[†] These authors contributed equally to this work.

Abstract: This study investigated the therapeutic potential of licochalcone D (LicoD), which is derived from *Glycyrrhiza uralensis*, for improving glucose metabolism in AML12 hepatocytes with high-glucose-induced insulin resistance (IR). Ultra-high-performance liquid chromatography–mass spectrometry revealed that the LicoD content of *G. uralensis* was 8.61 μg/100 mg in the ethanol extract (GUE) and 0.85 μg/100 mg in the hot water extract. GUE and LicoD enhanced glucose consumption and uptake, as well as *Glut2* mRNA expression, in high-glucose-induced IR AML12 cells. These effects were associated with the activation of the insulin receptor substrate/phosphatidylinositol-3 kinase signaling pathway, increased protein kinase B α phosphorylation, and suppression of gluconeogenesis-related genes, such as *Pepck* and *G6pase*. Furthermore, GUE and LicoD promoted glycogen synthesis by downregulating glycogen phosphorylase. Furthermore, LicoD and GUE mitigated the downregulated expression of mitochondrial oxidative phosphorylation proteins in IR hepatocytes by activating the PPARα/PGC1α pathway and increasing the mitochondrial DNA content. These findings demonstrate the potential of LicoD and GUE as therapeutic options for alleviating IR-induced metabolic disorders by improving glucose metabolism and mitochondrial function.

Keywords: licochalcone D; *Glycyrrhiza uralensis*; insulin resistance; glucose metabolism; hepatocytes

1. Introduction

Insulin resistance (IR) is a serious condition that causes metabolic abnormalities that disrupt the balance between energy consumption and expenditure [1]. IR impairs the ability of muscle, fat, and liver cells to absorb glucose from the bloodstream in response to insulin, resulting in elevated blood glucose levels (hyperglycemia) [2]. Impaired glucose metabolism can trigger an inflammatory response that leads to the development of type 2 diabetes (T2DM), nonalcoholic fatty liver disease (NAFLD), and obesity [3]. Consequently, significant research has focused on identifying pharmaceutical options that can improve IR to prevent or treat various metabolic diseases [4].

Glucose metabolism in hepatocytes requires the activation of insulin receptor substrates (IRSs) and the phosphatidylinositol-3 kinase a(PI3K) signaling pathway [5]. This pathway plays a significant role in regulating the expression of metabolic genes involved in both glucose production and storage [5]. Furthermore, because the IRS/PI3K signaling pathway ensures that hepatocytes respond effectively to insulin to maintain glucose homeostasis, its disruption is closely associated with IR-induced metabolic disorders [6]. Therefore, improving insulin-mediated glucose metabolism by targeting the IRS/PI3K signaling pathway is an effective strategy for alleviating metabolic diseases, including T2DM [7].

The regulation of net glycogen synthesis is a critical physiological function of insulin signaling in hepatocytes. IR-induced hyperglycemia not only inhibits hepatic glycogen synthesis but also activates glycogen phosphorylase, the enzyme responsible for metabolizing glycogen into glucose [8]. This process induces muscle mass loss. One important function of protein kinase B α (Akt), which is activated by the IRS/PI3K signaling pathway, is to inactivate glycogen phosphorylase, thereby inhibiting glycogen breakdown [9]. Furthermore, insulin signaling through the IRS/PI3K/Akt pathway activates the expression of genes related to glycogen synthesis, including the regulation of glycogen synthase kinase 3 (GSK3) [10]. Decreased transcriptional repression of GSK3 occurs in patients with IR, indicating that GSK3 remains highly active, inhibiting glycogen synthase and further disrupting glycogen synthesis [11]. Therefore, targeted regulation of the activities of glycogen phosphorylase and glycogen synthase effectively prevent and improve IR-induced metabolic disorders.

Mitochondria are essential organelles for glucose metabolism that generate cellular energy via oxidative phosphorylation (OXPHOS) [12]. The interplay between mitochondrial function and metabolic disorders has gained significant attention, highlighting the crucial role of mitochondrial dysfunction in the progression of metabolic diseases, including T2DM [13]. In particular, IR in hepatocytes, which results in abnormal glucose disposal, is closely associated with mitochondrial dysfunction [14]. Thus, impaired glucose consumption in the IR state contributes to diminished mitochondrial energy production, thereby promoting the development of metabolic disorders [15]. Therefore, strategies targeting mitochondrial functions that enhance oxidative capacity or protein quality could help prevent and treat IR-induced metabolic diseases [16].

Glycyrrhiza species (approximately 30 species), commonly known as licorice, are widely distributed in Eastern and Western countries [17]. Among them, *Glycyrrhiza uralensis* (*G. uralensis*) accounts for over 90% of licorice production used in traditional medicine due to its numerous health benefits, including anti-oxidative, anti-inflammatory, and anti-cancer properties [18,19]. *G. uralensis* is particularly renowned for its rich content of flavonoids, especially licochalcones, which contribute to many of its potent pharmacological effects [20]. Licochalcone D (LicoD, Figure 1A) is a significant chalcone that exerts antitumor effects by regulating multiple signaling pathways [21]. However, few studies have focused on quantifying LicoD in licorice and investigating its effects on IR-induced metabolic disorders.

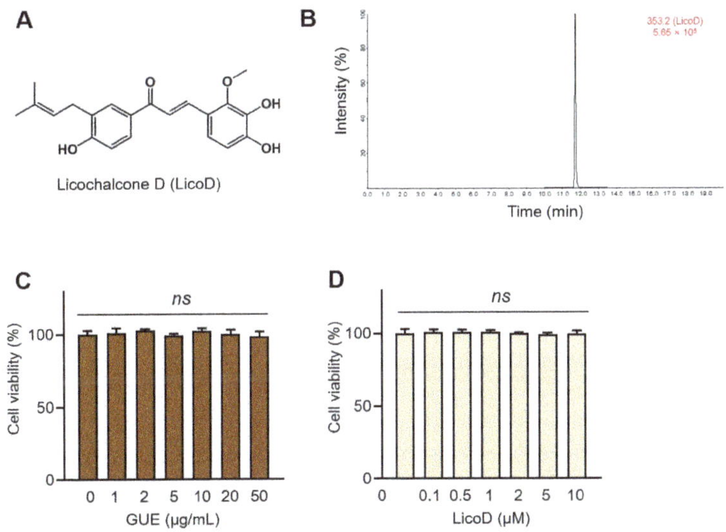

Figure 1. Quantitative analysis of licochalcone D (LicoD) from *Glycyrrhiza uralensis* ethanol extract (GUE) and their cytotoxicity to hepatocytes. (**A**) Chemical structure of LicoD. (**B**) Ultra-high-performance liquid chromatogram of LicoD and GUE; multiple reaction monitoring transition m/z 353.2 $[M–H]^-$ 150.0 $[M–H]^-$. (**C**) The cell viability under (**C**) GUE or (**D**) LicoD treatment was

measured using a water-soluble tetrazolium salt-1 kit. Data are expressed as mean ± standard deviation ($n \geq 3$). The different letters indicate significant differences ($p < 0.05$) as determined by one-way analysis of variance (ANOVA) followed by Tukey's post hoc test. ns, not significant.

This study quantified the LicoD content of *G. uralensis* ethanol extract (GUE) and hot water extract (GUH) and explored the effects of LicoD on glucose metabolism in high-glucose-induced IR hepatocytes.

2. Results

2.1. Determination of LicoD Contents in G. uralensis

We conducted UPLC–MS analysis to quantify the LicoD content in both GUE and GUH. LicoD had a retention time of 11.6 min on the total ion chromatogram (Figure 1B). The LicoD content was 8.61 ± 0.51 µg in GUE and 0.85 ± 0.06 µg in GUH, both expressed per 100 mg dry weight of *G. uralensis* roots (Table 1). Further experiments were performed using GUE to investigate whether *G. uralensis* and its active component, LicoD, could modulate glucose metabolism by alleviating IR-induced metabolic disorders. We also assessed the cytotoxicity of GUE and LicoD on hepatocytes using the water-soluble tetrazolium salt-1 (WST-1) assay. GUE (≤50 µg/mL) and LicoD (≤10 µM) did not reduce the viability of AML12 (Figure 1C,D).

Table 1. Contents of LicoD in *G. uralensis*.

Sample *	Contents ** (µg/100 mg dry wt.)
GUE	8.61 ± 0.51
GUH	0.85 ± 0.06

* GUE, *G. uralensis* ethanol extract; GUH, *G. uralensis* hot water extracts. ** Data are presented as the mean ± standard deviation (SD; $n \geq 3$).

2.2. Regulation of Glucose Metabolism by GUE or LicoD in IR Hepatocytes

Exposure of hepatocytes to 27 mM of glucose and insulin for 24 h significantly decreased glucose consumption and uptake, as well as *Glut2* mRNA expression, thereby successfully establishing the IR model of AML12 cells (Figure 2A–C), consistent with previous reports [22]. Treatment with GUE or LicoD dose-dependently increased glucose consumption and uptake in high-glucose-induced IR AML12 cells (Figure 2A,B). Furthermore, GUE (10 or 20 µg/mL) or LicoD (2 µM) mitigated the reduction in *Glut2* mRNA expression under IR conditions (Figure 2C). These results demonstrate that GUE or LicoD effectively alleviated abnormal glucose disposal in hepatocytes caused by high glucose.

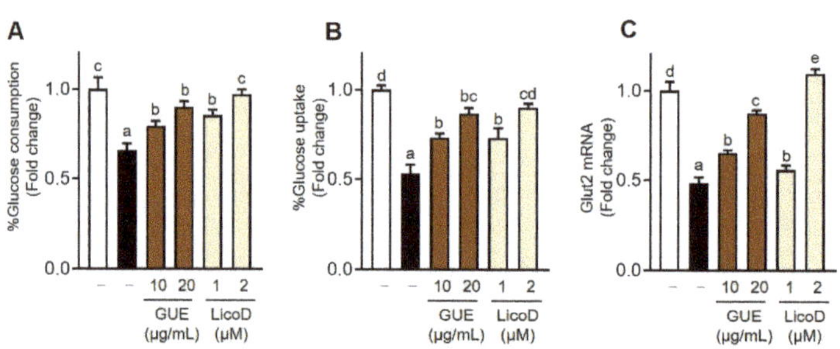

Figure 2. GUE and licochalcone D (LicoD) improved the glucose metabolism in insulin-resistant hepatocytes. AML12 cells were cultured in Dulbecco's Modified Eagle Medium/F12 containing 2% fetal bovine serum (FBS) without insulin for 24 h. Cells were then exposed to 27 mM glucose and insulin (1 nM) for 24 h with or without GUE (10 or 20 µg/mL) or LicoD (1 or 2 µM). (**A**) Glucose consumption, (**B**) glucose uptake, and (**C**) *Glut2* mRNA expression were determined.

Data are expressed as mean ± standard deviation ($n \geq 3$). The different letters indicate significant differences ($p < 0.05$) as determined by one-way ANOVA followed by Tukey's post hoc test. ns, not significant.

2.3. Enhancement of the IRS/PI3K Pathway by GUE or LicoD in IR Hepatocytes

To further elucidate the protective role of GUE and LicoD against high-glucose-induced abnormalities in glucose metabolism, we analyzed the expression levels of key insulin signaling proteins. Consistent with the patterns observed in glucose consumption and uptake, the phosphorylation levels of IRSs (Ser 307 or Ser 1011) and PI3K (Tyr458/Tyr199) significantly decreased in high-glucose-induced IR AML12 cells compared with normal cells (Figure 3). Interestingly, GUE or LicoD treatment dose-dependently increased phosphorylation in the IRS/PI3K pathway, indicating an association between the activation of IRS/PI3K signaling by GUE or LicoD and their therapeutic effect on the high-glucose-induced downregulation of glucose metabolism (Figure 3).

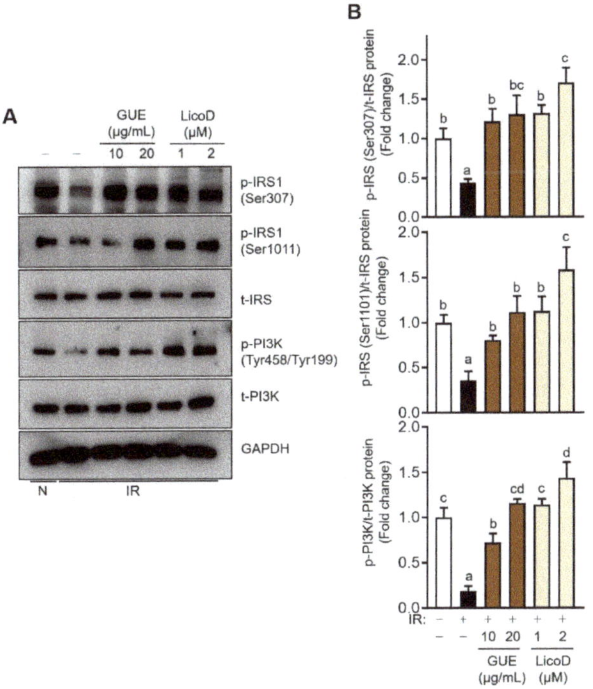

Figure 3. GUE and licochalcone D (LicoD) upregulated the IRS/PI3K signaling pathway in insulin-resistant hepatocytes. AML12 cells were cultured in Dulbecco's Modified Eagle Medium/F12 containing 2% FBS without insulin for 24 h. Cells were exposed to 27 mM glucose and insulin (1 nM) for 24 h with or without GUE (10 or 20 μg/mL) or LicoD (1 or 2 μM). (**A**) Expression levels of p-IRS1 (Ser307), p-IRS1 (Ser1011), t-IRS1, p-PI3K (Tyr458/Tyr199), t-PI3K, and GAPDH were analyzed using Western blot assay. (**B**) The quantitative expressions of p-IRS1 (Ser307)/t-IRS1, p-IRS1 (Ser1011)/t-IRS1, and p-PI3K (Tyr458/Tyr199)/t-PI3K are shown as bar graphs. Data are expressed as mean ± standard deviation ($n \geq 3$). The different letters indicate significant differences ($p < 0.05$) as determined by one-way ANOVA followed by Tukey's post hoc test. ns, not significant.

2.4. Inhibition of Akt-Mediated Gluconeogenesis by GUE or LicoD in IR Hepatocytes

Insulin signaling in hepatocytes regulates glycogen synthesis and gluconeogenesis through Akt activation, which is mediated by the IRS/PI3K signaling pathway. We examined the effect of GUE or LicoD on Akt phosphorylation and its downstream targets that are associated with gluconeogenesis (Figure 4). p-Akt (Ser437) levels decreased in high-glucose-induced IR AML12 cells. Consistent with the upstream IRS/PI3K pathway, p-Akt expression was significantly downregulated following GUE or LicoD treatment in high-glucose-induced IR AML12 cells (Figure 4A,B). Furthermore, p-FoxO1 (Ser256) levels decreased in high-glucose-induced IR AML12 cells, indicating the upregulation of gluconeogenesis, even under glucose-induced conditions (Figure 4A,B). However, GUE or LicoD treatment significantly increased p-FoxO1 levels and decreased *Pepck* and *G6pase* mRNA levels (Figure 4C,D). These results indicate that GUE or LicoD alleviated hepatic IR by suppressing the expression of genes associated with hepatic glucose production.

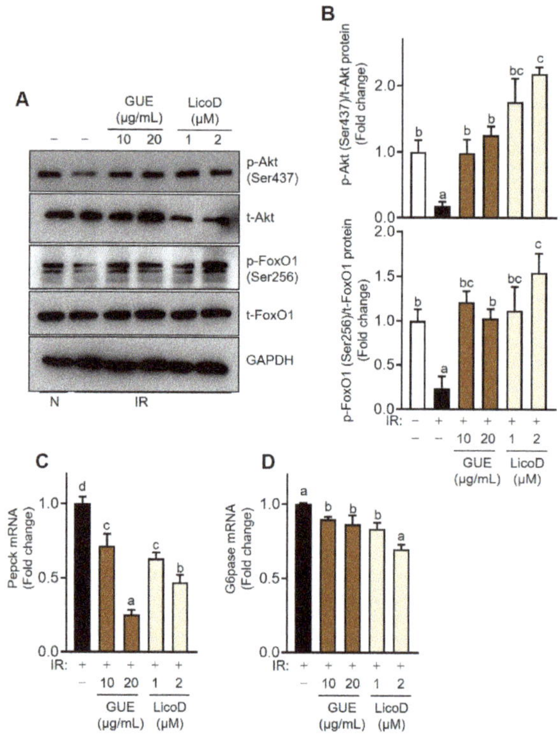

Figure 4. GUE and licochalcone D (LicoD) inhibited Akt-mediated gluconeogenesis in insulin-resistant hepatocytes. AML12 cells were cultured in Dulbecco's Modified Eagle Medium/F12 containing 2% FBS without insulin for 24 h. Cells were exposed to 27 mM glucose and insulin (1 nM) for 24 h with or without GUE (10 or 20 μg/mL) or LicoD (1 or 2 μM). (**A**) Expression levels of p-Akt (Ser437), t-Akt, p-FoxO1 (Ser256), t-FoxO1, and GAPDH were analyzed using Western blot assay. (**B**) The quantitative expressions of p-Akt (Ser437)/t-Akt and p-FoxO1 (Ser256)/t-FoxO1 are presented as bar graphs. mRNA expression of (**C**) *Pepck* and (**D**) *G6pase*. Data are expressed as mean ± standard deviation ($n \geq 3$). The different letters indicate significant differences ($p < 0.05$) as determined by one-way ANOVA followed by Tukey's post hoc test. ns, not significant.

2.5. Enhancement of Glycogen Synthesis by GUE or LicoD in IR Hepatocytes

To further explore the effects of GUE or LicoD on glycogen synthesis, we examined the levels of p-Gsk3β (Ser9) and p-GS (Ser641), key molecules involved in the regulation of glycogen synthase via insulin signaling (Figure 5). High glucose treatment inhibited glycogen synthesis, as evidenced by the upregulation of p-GS (Ser641) expression in treated cells (Figure 5A,B). Notably, GUE or LicoD treatment downregulated the high-glucose-induced increase in p-GS (Ser641) levels and dose-dependently increased p-Gsk3β (Ser9) expression (Figure 5A,B). Furthermore, qRT-PCR analysis revealed that *Gys2* and *Gsk3β* mRNA expression was significantly upregulated by GUE or LicoD treatment in high-glucose-induced IR AML12 cells (Figure 5C,D). These findings show that GUE and LicoD enhanced glycogen synthesis by activating enzymes involved in glycogen synthesis in IR hepatocytes.

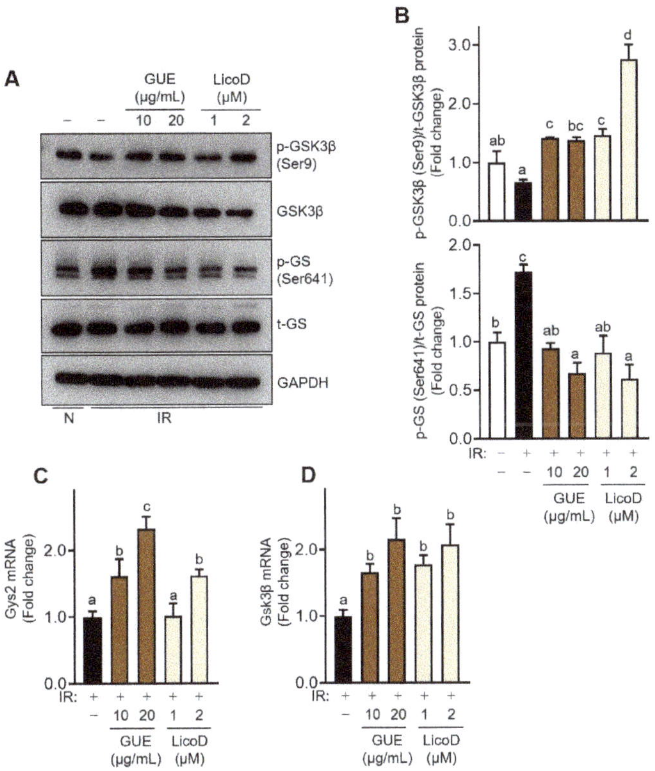

Figure 5. GUE and licochalcone D (LicoD) enhanced glycogen synthesis in insulin-resistant hepatocytes. AML12 cells were cultured in Dulbecco's Modified Eagle Medium/F12 containing 2% FBS without insulin for 24 h. Cells were exposed to 27 mM glucose and insulin (1 nM) for 24 h with or without GUE (10 or 20 μg/mL) or LicoD (1 or 2 μM). (**A**) Expression levels of p-GSK3β (Ser9), t-GSK3β, p-GS (Ser641), t-GS, and GAPDH were analyzed using Western blot assay. (**B**) The quantitative expressions of p-GSK3β (Ser9)/t-GSK3β (Ser641) and p-GS (Ser256)/t-GS are presented as bar graphs. mRNA expression of (**C**) *Gys2* and (**D**) *Gsk3β*. Data are expressed as mean ± standard deviation ($n \geq 3$). The different letters indicate significant differences ($p < 0.05$) as determined by one-way ANOVA followed by Tukey's post hoc test. ns, not significant.

2.6. Improvement of the PPARα/PGC1α Pathway by GUE or LicoD in IR Hepatocytes

Finally, we investigated the effects of GUE or LicoD on mitochondrial dysfunction in high-glucose-induced IR hepatocytes. The PPARα/PGC1α signaling pathway is crucial in mitochondrial energy metabolism and is closely linked to mitochondrial activation. Consistent with previous reports [23,24], PPARα and PGC1α expression significantly decreased in IR hepatocytes (Figure 6A,B). However, GUE or LicoD treatment dose-dependently mitigated this suppression of PPARα/PGC1α protein expression, as confirmed by mRNA analysis (Figure 6C,D). These data underscore the importance of PPARα/PGC1α expression, under induction by GUE or LicoD, in restoring mitochondrial function in IR hepatocytes. Furthermore, the mitochondrial DNA content encoding Cyb (mitochondrial complex III) and Co2 (mitochondrial complex IV) decreased in high-glucose-induced IR AML12 cells (Figure S1). In contrast, GUE or LicoD treatment promoted the increase in mitochondrial DNA mass in these cells (Figure S1). Furthermore, mitochondrial OXPHOS proteins (complexes II, III, and IV) were significantly downregulated in high-glucose-induced IR AML12 cells, which was reversed by treatment with GUE or LicoD (Figure 7 and Figure S2). Thus, our in vitro data show that GUE and LicoD effectively improved glucose metabolism in IR hepatocytes by activating PPARα/PGC1α signaling, thereby promoting mitochondrial biogenesis and protein quality control.

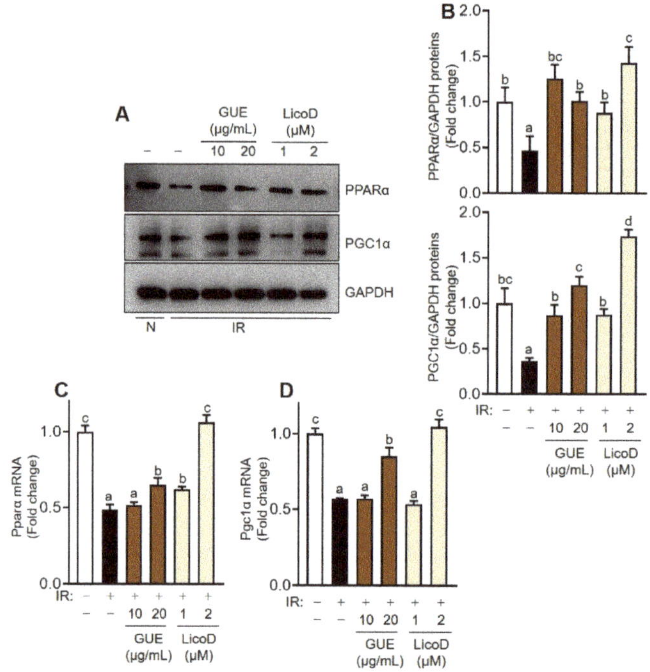

Figure 6. GUE and licochalcone D (LicoD) upregulated the PPARα/PGC1α pathway in insulin-resistant hepatocytes. AML12 cells were cultured in Dulbecco's Modified Eagle Medium/F12 containing 2% FBS without insulin for 24 h. Cells were exposed to 27 mM glucose and insulin (1 nM) for 24 h with or without GUE (10 or 20 μg/mL) or LicoD (1 or 2 μM). (**A**) Expression levels of PPARα, PGC1α, and β-actin were analyzed using Western blot assay. (**B**) The quantitative expressions of PPARα/β-actin and PGC1α/β-actin are presented as bar graphs. mRNA expression of (**C**) *Pparα* and (**D**) *Pgc1α*. Data are expressed as mean ± standard deviation ($n \geq 3$). The different letters indicate significant differences ($p < 0.05$) as determined by one-way ANOVA followed by Tukey's post hoc test. ns, not significant.

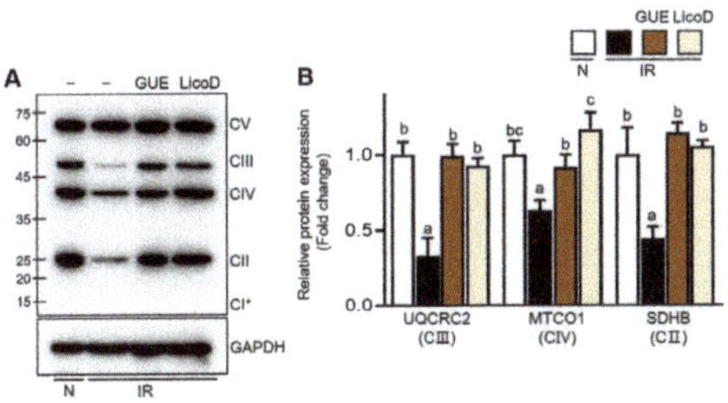

Figure 7. GUE and licochalcone D (LicoD) increased expression of oxidative phosphorylation (OXPHOS) complex proteins in insulin-resistant hepatocytes. AML12 cells were cultured in Dulbecco's Modified Eagle Medium/F12 containing 2% FBS without insulin for 24 h. Cells were exposed to 27 mM glucose and insulin (1 nM) for 24 h with or without GUE (20 μg/mL) or LicoD (2 μM). (**A**) Subunits of the oxidative phosphorylation (OXPHOS) complex (C) were analyzed using Western blot assay. (**B**) The quantitative expressions of UQCRC2/GAPDH, MTCO1/GAPDH, and SDHB/GAPDH are presented as bar graphs. Data are expressed as mean ± standard deviation ($n \geq 3$). The different letters indicate significant differences ($p < 0.05$) as determined by one-way ANOVA followed by Tukey's post hoc test. ns, not significant. CV, ATP synthase subunit alpha (ATP5A); CIII, ubiquinol-cytochrome C reductase core protein 2 (UQCRC2); CIV, mitochondrially encoded cytochrome C oxidase I (MTCO1); CII, succinate dehydrogenase subunit B (SDHB); CI, NADH dehydrogenase beta subcomplex subunit 8 (NDUFB8). The CI subunit (with a theoretical molecular weight of ~20 kD) was not detected.

3. Discussion

Few studies have investigated the effects of LicoD, a chalcone compound derived from licorice, on metabolic diseases, such as obesity and NAFLD, compared with other licochalcone derivatives, such as LicoA and LicoB. Previous studies have shown that LicoA inhibits fat accumulation and enhances FA transport in hepatocytes while promoting FA oxidation in differentiated adipocytes and lowering blood glucose levels in a high-fat diet (HFD)-induced obesity mouse model [25]. LicoB alleviates drug-induced hepatotoxicity in the mouse liver [26]. Furthermore, crude licorice extract, along with LicoA and LicoB, can enhance cell proliferation in aged C2C12 cells, indicating potential antiaging properties [27]. Although *G. uralensis* has been recognized for its various benefits against metabolic disorders, such as T2DM and obesity, the mechanisms by which its components improve IR-induced metabolic abnormalities have remained unclear [28,29]. In the present study, we investigated whether GUE and LicoD could ameliorate dysregulated glucose metabolism in IR hepatocytes and found that both enhanced glucose availability in high-glucose-induced IR AML12 cells by activating the IRS/PI3K/Akt signaling pathway, thereby promoting glycogen synthesis. Our results demonstrate the potential of these two products as natural therapeutic agents for alleviating IR-mediated metabolic diseases.

Following food intake, pancreatic β-cells detect elevated blood glucose levels and secrete insulin to promote anabolic processes in glucose-consuming organs, including the liver, adipose tissue, and skeletal muscle [30]. Anabolic activation increases glucose uptake from the bloodstream and promotes glucose utilization and storage [31]. Furthermore, insulin inhibits hepatic glucose production by suppressing the expression of gluconeogenic genes, such as *Pepck* and *G6pase*, thereby preventing hyperglycemia [32]. The intracellular actions of insulin begin with the activation of insulin receptor tyrosine kinase, which triggers the phosphorylation of IRS, which subsequently activates downstream targets, including the PI3K/Akt pathway, to maintain glucose homeostasis [33]. Recently, supplementation with natural compounds has been reported to prevent and improve IR-induced hyperglycemia. For example, ginsenoside Rg5, a major ginseng saponin, has been shown to improve hepatic glucose metabolism by activating the IRS/PI3K/Akt insulin signaling pathway, thereby resulting in increased glucose uptake and glycogen synthesis [34]. Cyanidin-3-*O*-glucoside was found to reverse IR and enhance glucose metabolism in IR hepatocytes and a diabetic mouse model by upregulating IRS phosphorylation [35]. In the present study, we found that the IRS-mediated PI3K pathway remained impaired even during glucose–insulin stimulation in IR hepatocytes. However, treatment with GUE or LicoD reactivated the downregulated phosphorylation of IRS/PI3K and suppressed the expression of gluconeogenesis-related genes in IR hepatocytes by modulating their transcriptional regulator, FoxO1. Furthermore, PI3K phosphorylates Akt and inhibits GSK3, thereby promoting glycogen synthase activity and glycogen synthesis. These findings indicate that GUE and LicoD can reduce blood glucose levels by promoting glucose storage and inhibiting glucose production in the hepatic IR state, thereby preventing hyperglycemia-induced metabolic diseases.

PPARα is a nuclear receptor that is critical in regulating lipid and glucose homeostasis [36]. Notably, PPARα contributes to the progression of IR-induced metabolic diseases, including T2DM [37]. PPARα activation enhances glucose uptake and utilization, particularly in the context of IR [38]. The significant role of PPARα in glucose metabolism has been the focus of pharmaceutical research on developing PPARα agonists as potential therapeutic agents for T2DM [39]. In hepatocytes, PPARα is essential for mitochondrial function, as it promotes mitochondrial biogenesis through its coactivator, PGC-1α [40]. Since mitochondria are central to cellular energy production and their dysfunction is influential to the development of metabolic disorders, including T2DM, impaired mitochondrial protein integrity can thus exacerbate IR and disrupt glucose homeostasis [41]. Therefore, activating the PPARα/PGC1α pathway can potentially mitigate IR and manage IR-induced metabolic diseases by enhancing mitochondrial quality [42]. We found that PPARα/PGC1α expression was significantly reduced in IR hepatocytes compared with normal ones. However, treatment with GUE or LicoD restored the expression of proteins and mRNAs in the PPARα/PGC1α pathway. Furthermore, the mitochondrial DNA content significantly increased in IR hepatocytes treated with GUE and LicoD. Unsurprisingly, since mitochondrial DNA encodes proteins essential for mitochondrial function, mitochondrial OXPHOS proteins were upregulated by GUE and LicoD treatment. These results indicate that GUE and LicoD enhance mitochondrial oxidative function under IR conditions by improving mitochondrial integrity, thereby helping mitigate IR-induced metabolic disorders.

The root of *G. uralensis*, one of the oldest medicinal plants, has been historically used to alleviate heat, cough, and pain [43]. Its pharmacological effects and molecular mechanisms of action have been extensively studied. For example, the $MeCl_2$ fraction of *G. uralensis* suppressed lipid accumulation and induced adipocyte browning in 3T3-L1 adipocytes, as demonstrated in an HFD-induced obesity mouse model [25]. *G. uralensis* root also exhibited significant hepatoprotective effects against alcohol-induced liver injury [44]. Furthermore, *G. uralensis* extract showed no hepatoxicity and reduced HFD-induced weight gain and blood glucose and insulin levels in mice, likely due to the suppression of de novo lipogenesis-related genes, including sterol regulatory element-binding protein 1 (*Srebp1*), FA synthase (*Fas*), and acetyl-CoA carboxylase (*Acc*) [45]. A recent study on a T2DM mouse model showed that *G. uralensis* root effectively suppressed high-glucose-induced TGF-β1 production by regulating Smad/Stat3-mediated mechanisms, potentially attenuating fibrosis-associated diabetic nephropathy [46]. In the present study, 20 μg/mL of GUE and 2 μM of LicoD exhibited comparable effects in improving glucose metabolism in IR hepatocytes. Our quantitative LC-MS data indicated that *G. uralensis* extract contains 8.61 μg/100 mg of dry weight LicoD. Therefore, the efficacy of *G. uralensis* extract in alleviating metabolic diseases may be partially attributed to LicoD. Further studies are necessary to evaluate the therapeutic potential of LicoD in treating metabolic diseases, such as NAFLD, T2DM, and obesity. Furthermore, the current findings should be validated using both in vivo and in vitro models.

4. Materials and Methods

4.1. Materials and Reagents

LicoD (purity: >99.8%) was purchased from MedChem Express (#HY−N4187; Monmouth Junction, NJ, USA). Mouse liver AML12 cell lines were obtained from American Type Culture Collection (Manassas, VA, USA). Glucose detection kits were obtained from Abcam (#ab136955; Cambridge, MA, USA). 2-(N-(7-nitrobenz-2-oxa-1,3-diazol-4-yl)amino)-2-deoxyglucose (2-NBDG, #N13195), dimethyl sulfoxide (#276855), D-(+)-glucose solution (100 g/L, #G8644), insulin (#I6634), and dexamethasone (#D1756) were purchased from Sigma-Aldrich (St. Louis, MO, USA). The primary antibodies used were p-IRS-1 (Ser307, #2381; Cell Signaling, Danvers, MA, USA); p-IRS-1 (Ser 1101, Cell Signaling); p-PI3K (Tyr458/Tyr199, #17366; Cell Signaling); PI3K (#4255; Cell Signaling); p-FoxO1 (Ser473, #4255, Cell Signaling); FoxO1 (#2880, Cell Signaling); p-Akt1 (Ser473, #05-736; Sigma-Aldrich); Akt1 (#sc-5298; Santa Cruz Biotechnology, Paso Robles, CA, USA); p-GSK-3β (Ser9, #5558; Cell Signaling); GSK-3β (#9315; Cell Signaling), p-GS, Ser641, #3891; Cell Signaling); GS (#sc-81173, Santa Cruz Biotechnology); peroxisome proliferator-activated receptor α (PPARα, #sc-9000; Santa Cruz Biotechnology); β-actin (#sc-47778; Santa Cruz Biotechnology); total OXPHOS cocktail (#ab110413; Abcam); GAPDH (#2118; Cell Signaling); and α/β-tubulin (#2148; Cell Signaling). Solvents used for LC-MS analysis were purchased from Fisher Scientific Korea Ltd. (Seoul, Republic of Korea).

4.2. Extraction and Identification of LicoD from G. uralensis

G. uralensis was cultivated in Yeongwol-gun (Gangwon province, Republic of Korea) and harvested in November 2017. *G. uralensis* extract was prepared as described previously [47]. Briefly, *G. uralensis* root was washed and dried. To prepare GUE, the dried sample was extracted twice using 70% EtOH (100 g/L) using a reflux cooling extraction method at 50 °C for 3 h. The extract was filtered, concentrated under reduced pressure, and freeze-dried. To prepare GUH, the dried sample was crushed and extracted twice by reflux cooling extraction at 50 °C for 3 h with distilled water 20 times the sample weight. Afterward, the extract was filtered and freeze-dried. The extract was stored at −20 °C until further use.

The freeze-dried samples were extracted with methanol for 3 h at 40 °C by sonication and filtered through a 0.22 μm pore filter syringe for UPLC–MS analysis (Waters, Milford, MA, USA). LicoD was isolated under the following conditions: column (C18, 1.7 μm, 2.1 × 100 mm, Waters); column temperature, 40 °C; and flow rate, 0.3 mL/min. The sample was eluted using a gradient system of H_2O (0.1% formic acid, solvent A) to acetonitrile (0.1% formic acid, solvent B), starting with 2% B and holding for 1.5 min; increasing to 15% B over 1.5 min and holding for 1 min; increasing to 35% B over 4 min and holding for 1 min; increasing to 90% B over 6 min and holding for 1.5 min; and decreasing to 2% B over 2 min and holding for 2 min. The mass spectrometer was set to multiple reaction monitoring mode to monitor the transition of LicoD: m/z 353.2 $[M - H]^-$ → 150.0 (quantifier) and 338.2 (qualifier). The optimal MS settings for quantification of LicoD were as follows: capillary voltage, 3.0 kV; cone voltage, 30 V; source temperature, 150 °C; desolvation temperature, 350 °C; desolvation gas flow rate, 650 L/h; cone gas flow, 150 L/h; and nebulizing gas flow, 7.0 L/min. Calibration curves ($y = 104.56x - 134.85$; $R^2 = 0.990$) were constructed using a standard of LicoD (1–50 μg/L). The LicoD content was quantitatively analyzed in triplicate. Data analysis was performed using MassLynx V4.1 software (Waters).

4.3. Cell Culture and Treatments

AML12 cells were cultured in Dulbecco's Modified Eagle Medium/F-12 (DMEM/F12, Gibco, Carlsbad, CA, USA) supplemented with 10% fetal bovine serum (FBS, Welgene, Gyeongsan, North Gyeongsang province, Republic of Korea), 1% antibiotic/antimycotic solution (Welgene) and 1% insulin–transferrin–selenium solution (10 μg/mL insulin, 5.5 μg/mL transferrin, and 5 ng/mL selenite; Invitrogen, Carlsbad, CA, USA) at 37 °C and 5% CO_2. The cells were sub-cultured before reaching 90% confluence. AML12 cells

(3×10^5 cells) were seeded into a 6-well plate and incubated for 24 h. To induce IR, the culture medium was replaced with DMEM/F12 containing 2% FBS without insulin for starvation. After 24 h of incubation, the cells were exposed to 27 mM glucose and insulin (1 nM) for 24 h with or without GUE or LicoD treatment at specified concentrations.

4.4. Cell Viability Assay

Cell viability was determined using the WST-1 assay (Roche Applied Science, Mannheim, Germany). After seeding cells into 96-well plates at a density of 2×10^4 cells per well, the cells were incubated with different concentrations of GUE (0–50 μg/mL) or LicoD (0–10 μM). After 24 h of incubation, 10% WST-1 solution was added to the culture medium. The cells were then incubated for an additional 3 h at 37 °C under a 5% CO_2 atmosphere. The resulting formazan products were measured at 450 nm using a microplate spectrophotometer (Molecular Devices, Sunnyvale, CA, USA).

4.5. Glucose Consumption Assay

We used a colorimetric glucose assay kit (#ab136955, Abcam) to determine cellular glucose consumption. Cells were seeded into a 6-well plate and cultured for 24 h. The culture medium was replaced with FBS-free medium for 30 min. 2-Deoxyglucose (2-DG) was added to the cells, and the mixture was incubated for 20 min at 37 °C to metabolize 2-DG into 2-DG-6-phosphate (2-DG6P). After washing with phosphate-buffered saline (PBS), cells were lysed with extraction buffer. Following a heating and cooling treatment, the cells were neutralized, and the cell supernatant was obtained using centrifugation at 10,000× g for 5 min. To generate NADPH from the accumulated 2-DG6P, a reaction mixture was added, and absorbance was analyzed at 412 nm at regular intervals for 10 min.

4.6. Glucose Uptake Assay

Cells were seeded into a 6-well plate and then exposed to FBS-free conditions for 1 h. Subsequently, the cells were treated with insulin (1 μM) and then with 0.1 mM 2-NBDG and incubated for 40 min. The cells were washed with PBS and collected through trypsinization. The fluorescence intensity of the cell pellet in FBS-free medium was measured using a microplate fluorescent reader (Ex/Em = 490/550 nm; Molecular Devices).

4.7. Quantitative Real-Time Polymerase Chain Reaction

Total RNA was extracted from cells using an RNeasy Mini Kit (Qiagen, Hilden, Germany) and used to synthesize cDNA using a cDNA reverse transcription kit (Toyobo Co., Ltd., Osaka, Japan) following the manufacturer's instructions. Quantitative real-time polymerase chain reaction (qRT-PCR) was performed using a SYBR Green Master Kit (#4913914001; Roche). The reaction was performed on a CFX Connect™ Real-Time PCR Detection System (Bio-Rad, Hercules, CA, USA). The primer sequences for target genes are listed in Table S1. The relative quantification of mRNA was calculated using the $2^{-\Delta\Delta Ct}$ method and normalized to housekeeping genes.

4.8. Western Blot Assay

Total proteins were extracted from cells using a radioimmunoprecipitation assay buffer (#R0278; Sigma-Aldrich) containing protease and phosphatase inhibitors (Roche). The protein concentration was determined using a bicinchoninic acid kit (#BCA1; Sigma-Aldrich). Equal amounts of proteins were separated by 8% sodium dodecyl sulfate–polyacrylamide gel electrophoresis and transferred to polyvinylidene fluoride membranes (Bio-Rad). After blocking the membranes with 5% nonfat dry milk in Tris-buffered saline containing Tween-20 (TBST) for 1 h, the membranes were incubated overnight with specific primary antibodies at 4 °C. The membranes were then washed with TBST for 1 h and incubated with the secondary antibody for 1 h at room temperature. Proteins were detected with an enhanced chemiluminescence reagent (Thermo Fisher Scientific, Sunnyvale, CA, USA). Band intensities were quantified using ImageJ (version 1.54d).

4.9. Mitochondrial DNA Quantification

DNA was isolated from the cells using a DNA purification kit (Bioneer Corp., Daejeon, Korea). The ratio of mitochondrial DNA (mtDNA) to genomic DNA was determined by qPCR. The RT-qPCR primers used in this study are listed in Table S1. All primer pairs were run in individual reactions. The final mtDNA/nucDNA ratio for each sample was calculated by averaging the ratios obtained from each primer pair. Expression data were analyzed using the $2^{-\Delta\Delta CT}$ method.

4.10. Statistical Analysis

All statistical analyses were performed using Prism 9 (GraphPad Software, La Jolla, CA, USA). Data were expressed as mean $\pm$ standard deviation. Statistical differences were analyzed using one-way analysis of variance (ANOVA), followed by Tukey's post hoc test for multiple comparisons or by unpaired two-tailed Student's t-test. Statistical significance was considered at p-value < 0.05.

5. Conclusions

In conclusion, this study demonstrated the therapeutic potential of LicoD, a chalcone compound derived from *G. uralensis* root, in alleviating various metabolic disorders. LicoD and GUE enhanced glucose metabolism by activating the IRS/PI3K/Akt signaling pathway, improving mitochondrial function, and promoting glycogen synthesis in IR hepatocytes. These results indicate that LicoD from *G. uralensis* is a potential natural therapeutic agent for managing IR-induced metabolic diseases. Further research, including in vitro and in vivo studies, is required to fully validate the therapeutic efficacy of LicoD in metabolic disorders.

Supplementary Materials: The following supporting information can be downloaded at https://www.mdpi.com/article/10.3390/ijms251810066/s1: Figure S1: *Glycyrrhiza uralensis* ethanol extract (GUE) and licochalcone D (LicoD) increased mitochondrial DNA (mtDNA) content in insulin-resistant hepatocytes; Figure S2: Effects of *GUE* and licochalcone D LicoD) on oxidative phosphorylation (OXPHOS) complex proteins in insulin-resistant hepatocytes.

Author Contributions: Conceptualization, J.-T.H., H.-K.C. and Y.G.L.; writing—original draft. H.M.L. and Y.G.L.; methodology, H.M.L., J.-T.H. and Y.G.L.; data curation, H.M.L. and Y.G.L. and J.-T.H.; formal analysis, H.M.L. and Y.G.L.; funding acquisition, H.-K.C.; supervision, H.-K.C.; writing—review and editing, Y.G.L. and H.-K.C. All authors have read and agreed to the published version of the manuscript.

Funding: This work was supported by the Main Research Program (E0210400 and E0210601) of the Korea Food Research Institute and funded by the Ministry of Science, ICT, and Future Planning. The authors thank members of the Personalized Diet Research Group for their support, collegiality, and critical review of the manuscript.

Institutional Review Board Statement: Not applicable.

Informed Consent Statement: Not applicable.

Data Availability Statement: The data presented in this study are available upon request from the corresponding author.

Conflicts of Interest: Author 'Hee Min Lee' was employed by the company 'World Institute of Kimchi'. The remaining authors declare that the research was conducted in the absence of any commercial or financial relationships that could be construed as a potential conflict of interest.

References

1. Lee, S.H.; Park, S.Y.; Choi, C.S. Insulin Resistance: From Mechanisms to Therapeutic Strategies. *Diabetes Metab. J.* **2022**, *46*, 15–37. [CrossRef] [PubMed]
2. Li, M.; Chi, X.; Wang, Y.; Setrerrahmane, S.; Xie, W.; Xu, H. Trends in insulin resistance: Insights into mechanisms and therapeutic strategy. *Signal Transduct. Target. Ther.* **2022**, *7*, 216. [CrossRef] [PubMed]

3. Frankowski, R.; Kobierecki, M.; Wittczak, A.; Różycka-Kosmalska, M.; Pietras, T.; Sipowicz, K.; Kosmalski, M. Type 2 Diabetes Mellitus, Non-Alcoholic Fatty Liver Disease, and Metabolic Repercussions: The Vicious Cycle and Its Interplay with Inflammation. *Int. J. Mol. Sci.* **2023**, *24*, 9677. [CrossRef] [PubMed]
4. Tabatabaei-Malazy, O.; Larijani, B.; Abdollahi, M. Targeting metabolic disorders by natural products. *J. Diabetes Metab. Disord.* **2015**, *14*, 57. [CrossRef] [PubMed]
5. Niswender, K.D.; Morrison, C.D.; Clegg, D.J.; Olson, R.; Baskin, D.G.; Myers, M.G., Jr.; Seeley, R.J.; Schwartz, M.W. Insulin Activation of Phosphatidylinositol 3-Kinase in the Hypothalamic Arcuate Nucleus: A Key Mediator of Insulin-Induced Anorexia. *Diabetes* **2003**, *52*, 227–231. [CrossRef]
6. Han, H.S.; Kang, G.; Kim, J.S.; Choi, B.H.; Koo, S.H. Regulation of glucose metabolism from a liver-centric perspective. *Exp. Mol. Med.* **2016**, *48*, e218. [CrossRef]
7. Feng, Y.; Ren, Y.; Zhang, X.; Yang, S.; Jiao, Q.; Li, Q.; Jiang, W. Metabolites of traditional Chinese medicine targeting PI3K/AKT signaling pathway for hypoglycemic effect in type 2 diabetes. *Front. Pharmacol.* **2024**, *15*, 1373711. [CrossRef]
8. Petersen, M.C.; Vatner, D.F.; Shulman, G.I. Regulation of hepatic glucose metabolism in health and disease. *Nat. Rev. Endocrinol.* **2017**, *13*, 572–587. [CrossRef]
9. Huang, X.; Liu, G.; Guo, J.; Su, Z. The PI3K/AKT pathway in obesity and type 2 diabetes. *Int. J. Biol. Sci.* **2018**, *14*, 1483. [CrossRef]
10. Teli, D.M.; Gajjar, A.K. Glycogen synthase kinase-3: A potential target for diabetes. *Bioorganic Med. Chem.* **2023**, *92*, 117406. [CrossRef]
11. Wang, L.; Li, J.; Di, L.J. Glycogen synthesis and beyond, a comprehensive review of GSK3 as a key regulator of metabolic pathways and a therapeutic target for treating metabolic diseases. *Med. Res. Rev.* **2022**, *42*, 946–982. [CrossRef] [PubMed]
12. Suomalainen, A.; Nunnari, J. Mitochondria at the crossroads of health and disease. *Cell* **2024**, *187*, 2601–2627. [CrossRef] [PubMed]
13. Rocha, M.; Apostolova, N.; Diaz-Rua, R.; Muntane, J.; Victor, V.M. Mitochondria and T2D: Role of autophagy, ER stress, and inflammasome. *Trends Endocrinol. Metab.* **2020**, *31*, 725–741. [CrossRef] [PubMed]
14. Shannon, C.E.; Ragavan, M.; Palavicini, J.P.; Fourcaudot, M.; Bakewell, T.M.; Valdez, I.A.; Ayala, I.; Jin, E.S.; Madesh, M.; Han, X.; et al. Insulin resistance is mechanistically linked to hepatic mitochondrial remodeling in non-alcoholic fatty liver disease. *Mol. Metab.* **2021**, *45*, 101154. [CrossRef] [PubMed]
15. Kim, J.-a.; Wei, Y.; Sowers, J.R. Role of mitochondrial dysfunction in insulin resistance. *Circ. Res.* **2008**, *102*, 401–414. [CrossRef] [PubMed]
16. James, A.M.; Collins, Y.; Logan, A.; Murphy, M.P. Mitochondrial oxidative stress and the metabolic syndrome. *Trends Endocrinol. Metab.* **2012**, *23*, 429–434. [CrossRef]
17. Zhang, Q.; Ye, M. Chemical analysis of the Chinese herbal medicine Gan-Cao (licorice). *J. Chromatogr. A* **2009**, *1216*, 1954–1969. [CrossRef]
18. So-Young, L.; So-Lim, P.; Sung-Hun, Y.; Young-Do, N.; Seong-Il, L. Quality Characteristics of Low-Salt Gochujang Added with Glycyrrhiza uralensis and Brassica juncea. *Prev. Nutr. Food Sci.* **2011**, *16*, 348–356.
19. Liao, W.C.; Lin, Y.-H.; Chang, T.-M.; Huang, W.-Y. Identification of two licorice species, Glycyrrhiza uralensis and Glycyrrhiza glabra, based on separation and identification of their bioactive components. *Food Chem.* **2012**, *132*, 2188–2193. [CrossRef]
20. Fu, Y.; Chen, J.; Li, Y.-J.; Zheng, Y.-F.; Li, P. Antioxidant and anti-inflammatory activities of six flavonoids separated from licorice. *Food Chem.* **2013**, *141*, 1063–1071. [CrossRef]
21. Zhang, Y.; Wang, L.; Dong, C.; Zhuang, Y.; Hao, G.; Wang, F. Licochalcone D exhibits cytotoxicity in breast cancer cells and enhances tumor necrosis factor-related apoptosis-inducing ligand-induced apoptosis through upregulation of death receptor 5. *J. Biochem. Mol. Toxicol.* **2024**, *38*, e23757. [CrossRef] [PubMed]
22. Yudhani, R.D.; Sari, Y.; Nugrahaningsih, D.A.A.; Sholikhah, E.N.; Rochmanti, M.; Purba, A.K.R.; Khotimah, H.; Nugrahenny, D.; Mustofa, M. In Vitro Insulin Resistance Model: A Recent Update. *J. Obes.* **2023**, *2023*, 1964732. [CrossRef] [PubMed]
23. Guerre-Millo, M.; Rouault, C.; Poulain, P.; André, J.; Poitout, V.; Peters, J.M.; Gonzalez, F.J.; Fruchart, J.-C.; Reach, G.; Staels, B. PPAR-α–null mice are protected from high-fat diet–induced insulin resistance. *Diabetes* **2001**, *50*, 2809–2814. [CrossRef] [PubMed]
24. Qian, L.; Zhu, Y.; Deng, C.; Liang, Z.; Chen, J.; Chen, Y.; Wang, X.; Liu, Y.; Tian, Y.; Yang, Y. Peroxisome proliferator-activated receptor gamma coactivator-1 (PGC-1) family in physiological and pathophysiological process and diseases. *Signal Transduct. Target. Ther.* **2024**, *9*, 50. [PubMed]
25. Lee, H.E.; Yang, G.; Han, S.-H.; Lee, J.-H.; An, T.-J.; Jang, J.-K.; Lee, J.Y. Anti-obesity potential of Glycyrrhiza uralensis and licochalcone A through induction of adipocyte browning. *Biochem. Biophys. Res. Commun.* **2018**, *503*, 2117–2123. [CrossRef]
26. Teng, H.; Chen, M.; Zou, A.; Jiang, H.; Han, J.; Sun, L.; Feng, C.; Liu, J. Hepatoprotective effects of licochalcone B on carbon tetrachloride-induced liver toxicity in mice. *Iran. J. Basic. Med. Sci.* **2016**, *19*, 910.
27. Ahmad, K.; Lee, E.J.; Ali, S.; Han, K.S.; Hur, S.J.; Lim, J.H.; Choi, I. Licochalcone A and B enhance muscle proliferation and differentiation by regulating Myostatin. *Phytomedicine* **2024**, *125*, 155350. [CrossRef]
28. Mae, T.; Kishida, H.; Nishiyama, T.; Tsukagawa, M.; Konishi, E.; Kuroda, M.; Mimaki, Y.; Sashida, Y.; Takahashi, K.; Kawada, T.; et al. A licorice ethanolic extract with peroxisome proliferator-activated receptor-gamma ligand-binding activity affects diabetes in KK-Ay mice, abdominal obesity in diet-induced obese C57BL mice and hypertension in spontaneously hypertensive rats. *J. Nutr.* **2003**, *133*, 3369–3377. [CrossRef]

29. Zhang, Y.; Xu, Y.; Zhang, L.; Chen, Y.; Wu, T.; Liu, R.; Sui, W.; Zhu, Q.; Zhang, M. Licorice extract ameliorates hyperglycemia through reshaping gut microbiota structure and inhibiting TLR4/NF-κB signaling pathway in type 2 diabetic mice. *Food Res. Int.* **2022**, *153*, 110945. [CrossRef]
30. Rahman, M.S.; Hossain, K.S.; Das, S.; Kundu, S.; Adegoke, E.O.; Rahman, M.A.; Hannan, M.A.; Uddin, M.J.; Pang, M.G. Role of Insulin in Health and Disease: An Update. *Int. J. Mol. Sci.* **2021**, *22*, 6403. [CrossRef]
31. Małkowska, P. Positive Effects of Physical Activity on Insulin Signaling. *Curr. Issues Mol. Biol.* **2024**, *46*, 5467–5487. [CrossRef] [PubMed]
32. Onyango, A.N. Excessive gluconeogenesis causes the hepatic insulin resistance paradox and its sequelae. *Heliyon* **2022**, *8*, e12294. [CrossRef] [PubMed]
33. Boucher, J.; Kleinridders, A.; Kahn, C.R. Insulin receptor signaling in normal and insulin-resistant states. *Cold Spring Harb. Perspect. Biol.* **2014**, *6*, a009191. [CrossRef] [PubMed]
34. Zhu, Y.; Yang, H.; Deng, J.; Fan, D. Ginsenoside Rg5 Improves Insulin Resistance and Mitochondrial Biogenesis of Liver via Regulation of the Sirt1/PGC-1α Signaling Pathway in db/db Mice. *J. Agric. Food Chem.* **2021**, *69*, 8428–8439. [CrossRef]
35. Ye, X.; Chen, W.; Huang, X.F.; Yan, F.J.; Deng, S.G.; Zheng, X.D.; Shan, P.F. Anti-diabetic effect of anthocyanin cyanidin-3-O-glucoside: Data from insulin resistant hepatocyte and diabetic mouse. *Nutr. Diabetes* **2024**, *14*, 7. [CrossRef]
36. Lin, Y.; Wang, Y.; Li, P.F. PPARα: An emerging target of metabolic syndrome, neurodegenerative and cardiovascular diseases. *Front. Endocrinol.* **2022**, *13*, 1074911. [CrossRef]
37. Corrales, P.; Vidal-Puig, A.; Medina-Gómez, G. PPARs and Metabolic Disorders Associated with Challenged Adipose Tissue Plasticity. *Int. J. Mol. Sci.* **2018**, *19*, 2124. [CrossRef]
38. Lalloyer, F.; Vandewalle, B.; Percevault, F.; Torpier, G.; Kerr-Conte, J.; Oosterveer, M.; Paumelle, R.J.; Fruchart, J.-C.; Kuipers, F.; Pattou, F. Peroxisome proliferator–activated receptor α improves pancreatic adaptation to insulin resistance in obese mice and reduces lipotoxicity in human islets. *Diabetes* **2006**, *55*, 1605–1613. [CrossRef]
39. Rodríguez-Vilarrupla, A.; Laviña, B.; García-Calderó, H.; Russo, L.; Rosado, E.; Roglans, N.; Bosch, J.; García-Pagán, J.C. PPARα activation improves endothelial dysfunction and reduces fibrosis and portal pressure in cirrhotic rats. *J. Hepatol.* **2012**, *56*, 1033–1039. [CrossRef]
40. Brunmeir, R.; Xu, F. Functional Regulation of PPARs through Post-Translational Modifications. *Int. J. Mol. Sci.* **2018**, *19*, 1738. [CrossRef]
41. Pinti, M.V.; Fink, G.K.; Hathaway, Q.A.; Durr, A.J.; Kunovac, A.; Hollander, J.M. Mitochondrial dysfunction in type 2 diabetes mellitus: An organ-based analysis. *Am. J. Physiol. Endocrinol. Metab.* **2019**, *316*, E268–E285. [CrossRef] [PubMed]
42. Duncan, J.G. Peroxisome proliferator activated receptor-alpha (PPARα) and PPAR gamma coactivator-1alpha (PGC-1α) regulation of cardiac metabolism in diabetes. *Pediatr. Cardiol.* **2011**, *32*, 323–328. [CrossRef] [PubMed]
43. Jong-Hyun, H.; Kyong-Haeng, L.; Kwang-Won, Y. Characterization of the Immunologically Active Components of Glycyrrhiza uralensis Prepared as Herbal Kimchi. *Prev. Nutr. Food Sci.* **2003**, *8*, 29–35.
44. Jung, J.C.; Lee, Y.H.; Kim, S.H.; Kim, K.J.; Kim, K.M.; Oh, S.; Jung, Y.S. Hepatoprotective effect of licorice, the root of Glycyrrhiza uralensis Fischer, in alcohol-induced fatty liver disease. *BMC Complement. Altern. Med.* **2016**, *16*, 19. [CrossRef] [PubMed]
45. Hong, M.K.; Han, Y.; Park, H.J.; Shin, M.R.; Roh, S.S.; Kwon, E.Y. The Synergistic Action of Metformin and Glycyrrhiza uralensis Fischer Extract Alleviates Metabolic Disorders in Mice with Diet-Induced Obesity. *Int. J. Mol. Sci.* **2023**, *24*, 936. [CrossRef] [PubMed]
46. Lin, H.C.H.; Paul, C.R.; Kuo, C.H.; Chang, Y.H.; Chen, W.S.T.; Ho, T.J.; Day, C.H.; Velmurugan, B.K.; Tsai, Y.; Huang, C.Y. Glycyrrhiza uralensis root extract ameliorates high glucose-induced renal proximal tubular fibrosis by attenuating tubular epithelial-myofibroblast transdifferentiation by targeting TGF-β1/Smad/Stat3 pathway. *J. Food Biochem.* **2022**, *46*, e14041. [CrossRef]
47. Park, S.K.; Lee, J.; Park, S.H.; Lee, Y.G. Anti-obesity effects of Glycyrrhiza uralensis ethanol extract on the inhibition of 3T3-L1 adipocyte differentiation in high-fat diet-induced C57BL/6J mice. *Food Sci. Preserv.* **2023**, *30*, 716–728.

International Journal of *Molecular Sciences*

MDPI

Article

Computational Investigation of the Therapeutic Potential of *Detarium senegalense* in the Management of Erectile Dysfunction

Jerius Nkwuda Ejeje [1,2], Emmanuel Ayodeji Agbebi [3,4], Makhosazana Siduduzile Mathenjwa-Goqo [5], Obinna Aru Oje [2], Precious Eseose Agboinghale [1,6], Ikechukwu Theophilus Ebe [7], Tajudeen Olabisi Obafemi [1], Ezekiel Adewole [8], Omaka N. Omaka [9], Sunday Amos Onikanni [1], Basiru Olaitan Ajiboye [10], Olaposi Idowu Omotuyi [3,11] and Babatunji Emmanuel Oyinloye [1,3,5,*]

1 Phytomedicine, Biochemical Toxicology and Biotechnology Research Laboratories, Department of Biochemistry, College of Sciences, Afe Babalola University, PMB 5454, Ado-Ekiti 360001, Nigeria; ejeje.jerius@funai.edu.ng (J.N.E.); preciousagboinghale@gmail.com (P.E.A.); oobafemi@abuad.edu.ng (T.O.O.); onikannisa@abuad.edu.ng (S.A.O.)

2 Department of Biochemistry, Faculty of Biological Sciences, Alex Ekwueme Federal University Ndufu Alike Ikwo, P.O. Box 1010, Abakaliki 480101, Nigeria; obinna.oje@funai.edu.ng

3 Institute of Drug Research and Development, S.E. Bogoro Center, Afe Babalola University, PMB 5454, Ado-Ekiti 360001, Nigeria; agbebiea@abuad.edu.ng (E.A.A.); olaposi.omotuyi@abuad.edu.ng (O.I.O.)

4 Department of Pharmacognosy and Natural Products, College of Pharmacy, Afe Babalola University, Ado-Ekiti 360001, Nigeria

5 Biotechnology and Structural Biology (BSB) Group, Department of Biochemistry and Microbiology, University of Zululand, KwaDlangezwa 3886, South Africa; mathenjwam@unizulu.ac.za

6 Department of Chemistry and Biomolecular Sciences, Faculty of Science, University of Ottawa, Ottawa, ON K1N 6N5, Canada

7 Department of Medical Biochemistry, Faculty of Basic Medical Sciences, David Umahi Federal University of Health Sciences, Uburu 491105, Nigeria; ikechukwuebe1@gmail.com

8 Department of Chemical Sciences, College of Sciences, Afe Babalola University, PMB 5454, Ado-Ekiti 360001, Nigeria; adewolee@abuad.edu.ng

9 Department of Chemistry, Faculty of Physical Sciences, Alex Ekwueme Federal University Ndufu Alike Ikwo, P.O. Box 1010, Abakaliki 480101, Nigeria; omaka.omaka@funai.edu.ng

10 Phytomedicine and Molecular Toxicology Research Laboratory, Department of Biochemistry, Federal University Oye-Ekiti, Oye-Ekiti 360101, Nigeria; basiru.ajiboye@fuoye.edu.ng

11 Department of Pharmacology and Toxicology, College of Pharmacy, Afe Babalola University, PMB 5454, Ado-Ekiti 360001, Nigeria

* Correspondence: babatunjioe@abuad.edu.ng

Abstract: Erectile dysfunction (ED) is a multifactorial social problem affecting men worldwide. While phosphodiesterase type 5 inhibitors (PDE5) like sildenafil are commonly used, they often present side effects, underscoring the need for alternative therapies. Therefore, this study investigated the potential of phytochemicals from *Detarium senegalense* in the management of ED. A library of phytochemicals from *Detarium senegalense* was generated, prepared, and interacted with six key enzymes implicated in ED, including PDE5, using the Schrödinger Maestro suite. The results identified catechin, epicatechin, and gallic acid as the leading compounds with significant binding affinities for the targeted enzymes. Catechin and epicatechin (−9.877 and −11.408 kcal/mol, respectively) exhibited comparable binding affinities to sildenafil (−11.926 kcal/mol) on PDE5. The MD simulation results also revealed superior stability and ability to maintain interaction with key amino acids at the active site of PDE5 over the entire simulation period for these compounds. These compounds also demonstrated favorable ADMET profiles over sildenafil, including high gastrointestinal absorption and no violation of Lipinski's rule, indicating good bioavailability and drug likeness. These findings suggest that flavonoids from *Detarium senegalense*, especially catechin and epicatechin, have potential in the management of ED by interacting with multiple targets involved in its pathogenesis.

Keywords: erectile dysfunction; *Detarium senegalense*; phosphodiesterase 5; sildenafil; catechin; epicatechin

1. Introduction

Erectile dysfunction (ED) is a multifactorial condition that is characterized by a persistent inability to achieve or sustain penile erection sufficiently for satisfactory sexual intercourse despite the presence of proper erotic stimuli [1–3]. It was previously/synonymously referred to as impotence [1,4]. Though this condition is usually common in men aged ≥40 years, with the prevalence increasing with increasing age and co-morbidities (e.g., diabetes and other cardiometabolic disorders), it is a social problem that can affect men of all ages, cultures, and ethnic groups, with a significant effect on the quality of life of the person and their partners [1]. Its prevalence has been estimated to be between 14% and 48% [1]. Specifically, the 2021 National Survey of Sexual Wellbeing estimated the prevalence of erectile dysfunction in the United States as 24.2%, with the prevalence increasing with age. However, this value could have been underestimated because of male reservedness and privacy [1,5,6]. It has been known to negatively impact the quality of life and can lead to loss of self-esteem, anxiety, and depression [7,8]. The etiology of ED is multifactorial, with aging, psychological, and neurological disorders and other co-morbidities/organic causes like cardiovascular and metabolic disorders the key contributing factors. Benign prostate hyperplasia (BPH) and some medications (antipsychotic, antidepressant, etc.) can also cause ED. The relaxation of the intracavernosal smooth muscle is the key critical process in penile erection [2,9]. It involves a series of process involving nitric oxide (NO), which stimulates cyclic guanosine monophosphate (cGMP) production, which in turn activates protein kinase G, closing the calcium channel. This causes the intracavernosal smooth muscle to relax and increases the arterial blood flow, resulting in a rigid erection. The degradation of the cGMP by penile phosphodiesterase (PDE) reverses this process by contracting the corporal smooth muscle. Therefore, any problem along these pathway/process can lead to erectile dysfunction [2,9–11]. Current treatments in the market target key players along this pathway to help in the management of ED. An example is sildenafil (Viagra®), a very commonly used drug. It is a phosphodiesterase inhibitor that prevents cGMP degradation by inhibiting penile PDE, thus leading to sustained erection. However, these drugs are bedeviled by serious side effects, including headache, dizziness, flushing, and cardiovascular complications [1,12,13]. Therefore, there is a need for new therapy with fewer side effects and comparable efficacy. The use of medicinal plants with aphrodisiac properties was the mainstay for treatment of ED in previous centuries [1,14]. Medicinal plants have made key contributions to the drug development process [15], and several medicinal plants have shown promising potential in the management of ED, including ginseng, *Lepidium meyenii*, *Pinus pinaster*, *Tribulus terrestris*, and saffron, among others [16]. The use of medicinal plants in the management of ED is gaining attention in underdeveloped countries, mainly because of the affordability and reported minimal side effects with the use of medicinal plants in complementary therapies [17–19]. Studies have shown that some phenolic compounds (particularly flavonoids) possess PDE5-inhibitory activity [20]. Moreover, *Garcinia kola* [21] and *Ocimum gratissimum* [22] have been studied and found to inhibit key markers of ED. Aside from PDE5 inhibitors, other viable therapeutic options should be considered for the management of ED. From various reports, adenosine deaminase, arginase, O-GlcNAc transferase (OGT), phosphodiesterase 5, protein kinase C (PKC), and steroidogenic cytochrome P450 17A1 have all been implicated in the etiology of erectile dysfunction [23,24]. The depletion of adenosine, L-arginine, OGT, testosterone; dihydrotestosterone (DHT), second messenger molecules (cAMP and cGMP), and nitric oxide (NO) occurs with the activation of these enzymes. Therefore, the inhibition of these enzymes will play a key role in the management of erectile dysfunction [23,25]. This study, therefore, aims to explore the potential of *Detarium senegalense* in the management of erectile dysfunction via its inhibitory potential on these enzymes.

Detarium senegalense J.F.Gmel. (Figure 1) is a medicinal plant native to the west African region and is commonly known as the African star apple or tallow tree [26,27]. It is a member of the family Fabaceae, and various parts of the plant, including the bark, leaves, and fruits, have been used traditionally for culinary and medicinal purposes. Recent scientific investigations have unveiled the rich phytochemical composition of *Detarium senegalense*, which includes flavonoids, tannins, saponins, alkaloids, and phenolic compounds [26]. These bioactive constituents are endowed with diverse pharmacological properties, such as antidiabetic, antidiarrheal, antioxidant, anti-inflammatory, antimicrobial, and vasodilatory effects [26,28,29]. Since the pathophysiology of erectile dysfunction involves an intricate interplay among vascular, neural, hormonal, and psychological factors, several phytochemicals present in this plant exhibit promising pharmacological activities that may address the underlying mechanisms contributing to ED. Flavonoids, such as quercetin and kaempferol, possess antioxidant properties and can enhance endothelial function by promoting nitric oxide (NO) production and inhibiting oxidative stress-induced endothelial dysfunction [23,30]. Additionally, saponins found in the plant may modulate hormonal pathways implicated in erectile function, including the regulation of testosterone levels. The vasodilatory effects of the phytochemicals present in the plant are particularly relevant to erectile function, as adequate penile blood flow is indispensable for achieving and maintaining erection. Flavonoids exert vasodilatory effects through various mechanisms, including the upregulation of endothelial nitric oxide synthase (eNOS) expression and the inhibition of vasoconstrictor pathways. Furthermore, saponins may enhance cavernosal smooth muscle relaxation by modulating intracellular calcium levels and potassium channel activity [31]. This makes the plant a promising one and led to our investigation of its potential in the management of ED via its interaction with key enzymes involved in erectile dysfunction pathogenesis.

Figure 1. *Detarium senegalense* J.F.Gmel. in its natural habitat.

2. Results

Molecular docking was performed to generate the glide scores, which is predictive of the binding affinity of the ligands with each receptor and consequently their activity. The Glide scores of the ligands at each receptor's active site are presented as a heatmap in Figure 2 below.

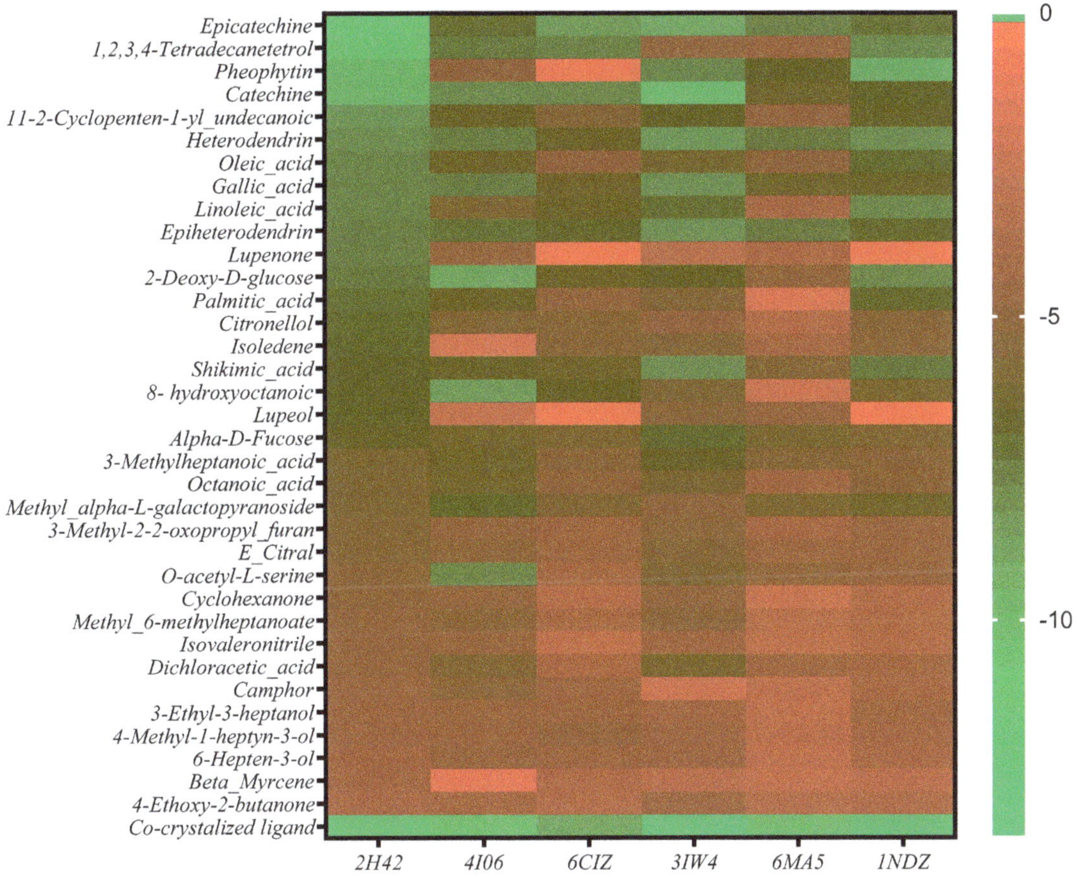

Figure 2. Heatmap showing the Glide scores of phytochemicals from *Detarium senegalense* against human PDE5, arginase 2, steroidogenic cytochrome P450 17A1, protein kinase C alpha, OGT, and adenosine deaminase.

From the results above, it was observed that catechin and its isomer, epicatechin; gallic acid, and pheophytin had good binding affinity with the receptors, and these were therefore selected as our hit compounds. The chemical structures of these hit compounds and sildenafil are provided in Figure 3. Across most of the receptors, catechin and epicatechin had better binding affinity than the other hit compounds and a comparable binding affinity to the co-crystallized ligands. For example, on the PDE5 enzyme, catechin and epicatechin had binding affinity of −9.877 and −11.408 kcal/mol compared with the −11.926 kcal/mol value of the co-crystallized ligand, sildenafil, a standard drug in the market for the treatment of erectile dysfunction. These compounds also exhibited similar binding affinity to the co-crystallized ligand at the active site of the steroidogenic cytochrome P450 17A1 enzyme with binding affinity of −7.921 and −8.547 kcal/mol, respectively, compared with the −9.029 kcal/mol value of the co-crystallized ligand. These values are presented in Table 1 below.

Table 1. Showing the Glide scores (kcal/mol), MM/GBSA (kcal/mol), and interacting residues of each target with their co-crystallized ligands and best hit compounds.

Target	Compound	Glide Score	MM/GBSA	Key Interaction
2H42	CCL	−11.926	−85.1	H-bond-GLN817 Pi-stacking-TYR612, PHE820
	Catechin	−9.877	−36.42	H-bond-GLN775, GLN817 Pi-stacking-TYR612, HIS613
	Epicatechin	−11.408	−41.31	H-bond-ASN661, THR723, GLN817 Pi-stacking-HIS613, PHE820
	Pheophytin	−9.97	−52.17	H-bond-ASN662, THR723 Pi-stacking-PHE820
3IW4	CCL	−13.574	−75.51	H-bond-THR401, GLU418, VAL420, ASP467 Pi-cation-ASP424, ASP467
	Catechin	−10.458	−48.93	H-bond-GLU387, VAL420
	Epicatechin	−9.072	−43.69	H-bond-GLU418, ASP467, ASN468
	Pheophytin	−7.834	−41.5	H-bond-GLU418, GLY423 Pi-stacking-PHE350
6MA5	CCL	−10.236	−47.56	H-bond-GLN839, ALA896, LYS898 Pi-stacking-HIE558, HIE901 Salt bridge-LYS842
	Catechin	−6.276	−33.24	H-bond-ASN557, PHE837, ALA896
	Epicatechin	−7.586	−32.16	H-bond-HIE562, ALA896, LYS898 Pi-stacking-HIE901
	Pheophytin	−6.573	−52.16	H-bond-HIE558, PHE837, GLN839, HIE920, THR922
1NDZ	CCL	−12.244	−77.9	H-bond-HIS17, ASP19, GLU217, ASH296 Pi-stacking-HIS17, PHE61, PHE65
	Catechin	−6.667	−29.13	H-bond-GLY184, GLU217, ASH296
	Epicatechin	−7.019	−31.54	H-bond-LEU56, GLU217
	Pheophytin	−9.682	−71.91	H-bond-HIS17, ASP19, SER103, ASP185, GLU217
4IO6	CCL	−10.414	−7.95	H-bond-ASP147, SER156, ASN158, HIS160, ASP202, GLU205
	Catechin	−8.048	−21.31	H-bond-HIS160, GLY161, ASP202, GLU205, ASP253 Pi-stacking-HIE145, HIS160
	Epicatechin	−6.935	−13.95	H-bond-THR154, ASN158, GLY161, ASP200, GLU296 Pi-stacking-HIE145, HIS160
	Pheophytin	−4.408	−22.5	H-bond-SER155, ASP200, VAL201, ASP202, GLU205
6CIZ	CCL	−9.029	−35.21	H-bond-ASN202
	Catechin	−7.921	−29.08	H-bond-ASP298, VAL482
	Epicatechin	−8.547	−18.77	H-bond-ASP298, VAL482

The interactions of these ligands with key amino acids at the active site of each receptor was also recorded and are presented in Table 1 above and Figure 4 below. Figure 4a depicts the docking validation, where the co-crystallized ligands were extracted and redocked to the active site of the proteins. The docked pose is superimposed with that of the co-crystallized state to confirm that the co-crystallized ligands could reenact their poses, and the root mean square deviations (RMSDs) were recorded. This was reported for only 2H42 and 3IW4 (as shown in Figure 4a), because the work thenceforth focused on these two receptors, mainly 2H42. Figure 4b shows the 2D ligand interactions of the co-crystallized ligand (CCL), catechin, epicatechin, and pheophytin with human PDE5, protein kinase C alpha, OGT, and adenosine deaminase, while Figure 4c shows the 2D ligand interactions of the co-crystallized ligand (CCL), catechin, epicatechin, and pheophytin with arginase 2 and steroidogenic cytochrome P450 17A1. It was observed that these hit compounds exhibited similar interaction to the amino acids at the active sites, just like the co-crystallized ligands upon binding to the receptors. For instance, on the PDE5 enzyme, catechin and epicatechin exhibited hydrogen bonds with GLN817 and pi-stacking with TYR612 and PHE820, respectively, just like the co-crystallized ligand sildenafil. In addition to these, catechin exhibited an additional H-bond with GLN775 and epicatechin with ASN661 and THR723. Also, on the O-GlcNAc transferase enzyme, epicatechin displayed similar interactions with key amino acids to the co-crystallized ligand. For example, it exhibited an H-bond with both ALA896 and LYS898, just like the co-crystallized ligand, and displayed pi-stacking with HIE901. They also exhibited similar hydrophobic interaction, as shown in Figure 4. These compounds did not display similar interactions to the co-crystallized ligand at 6CIZ; therefore, we focused on the promising targets thenceforth.

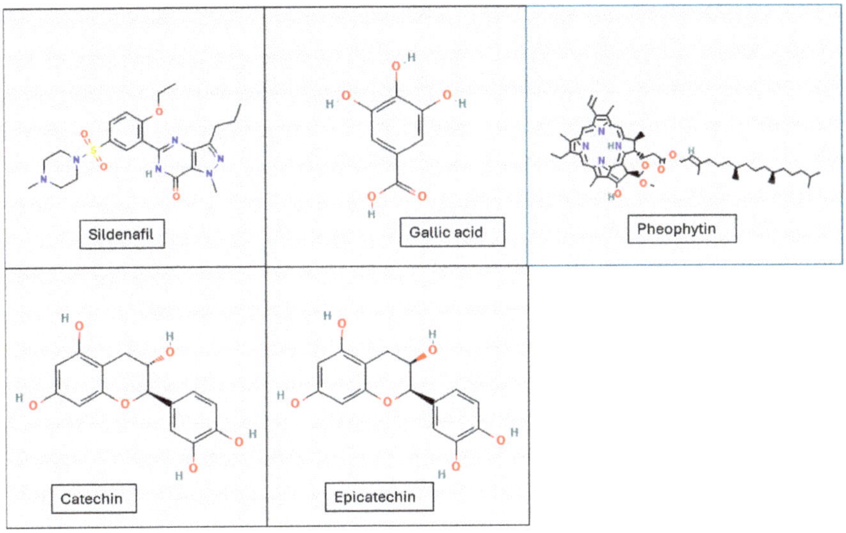

Figure 3. Structures of sildenafil, gallic acid, pheophytin, catechin, and epicatechin (source: PubChem online database https://pubchem.ncbi.nlm.nih.gov/).

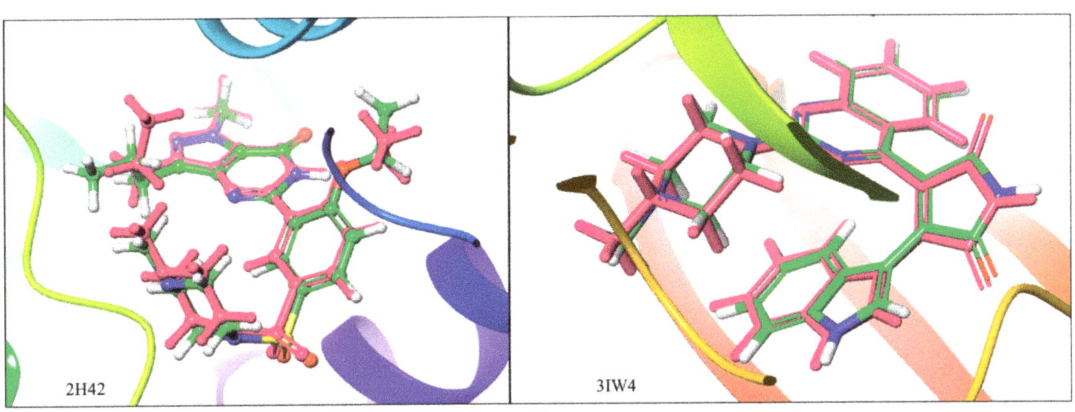

(**a**)

Figure 4. *Cont.*

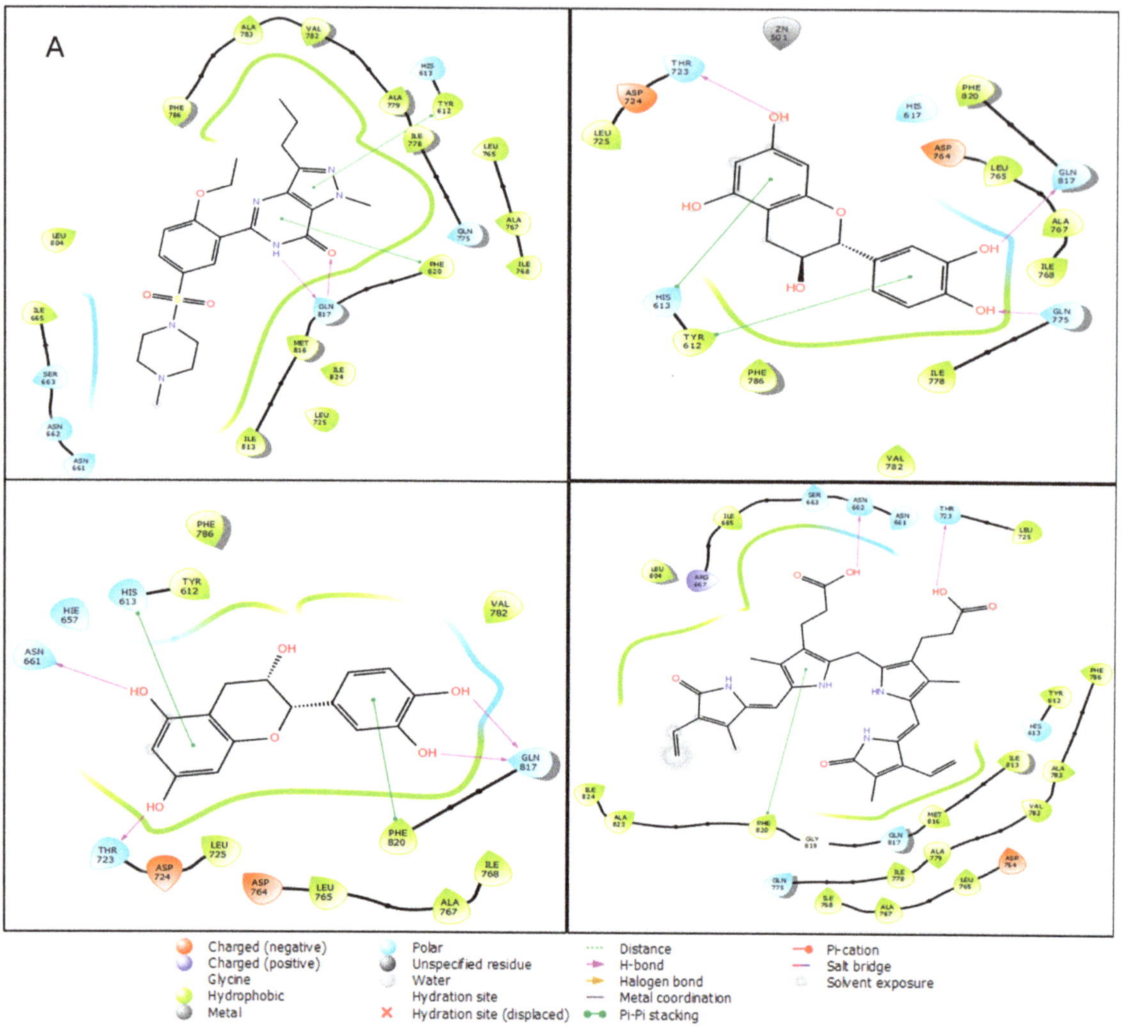

Figure 4. *Cont.*

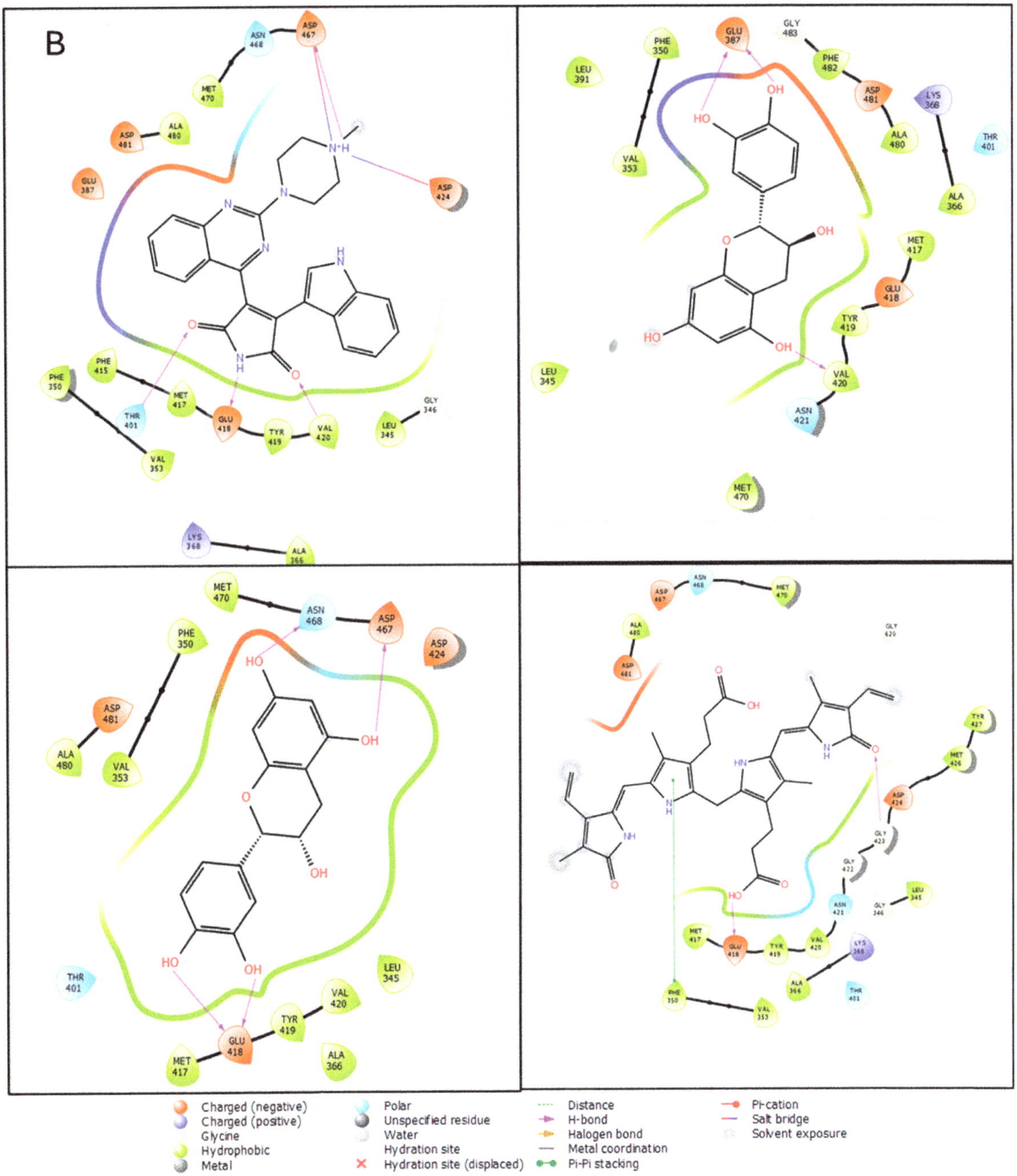

Figure 4. *Cont.*

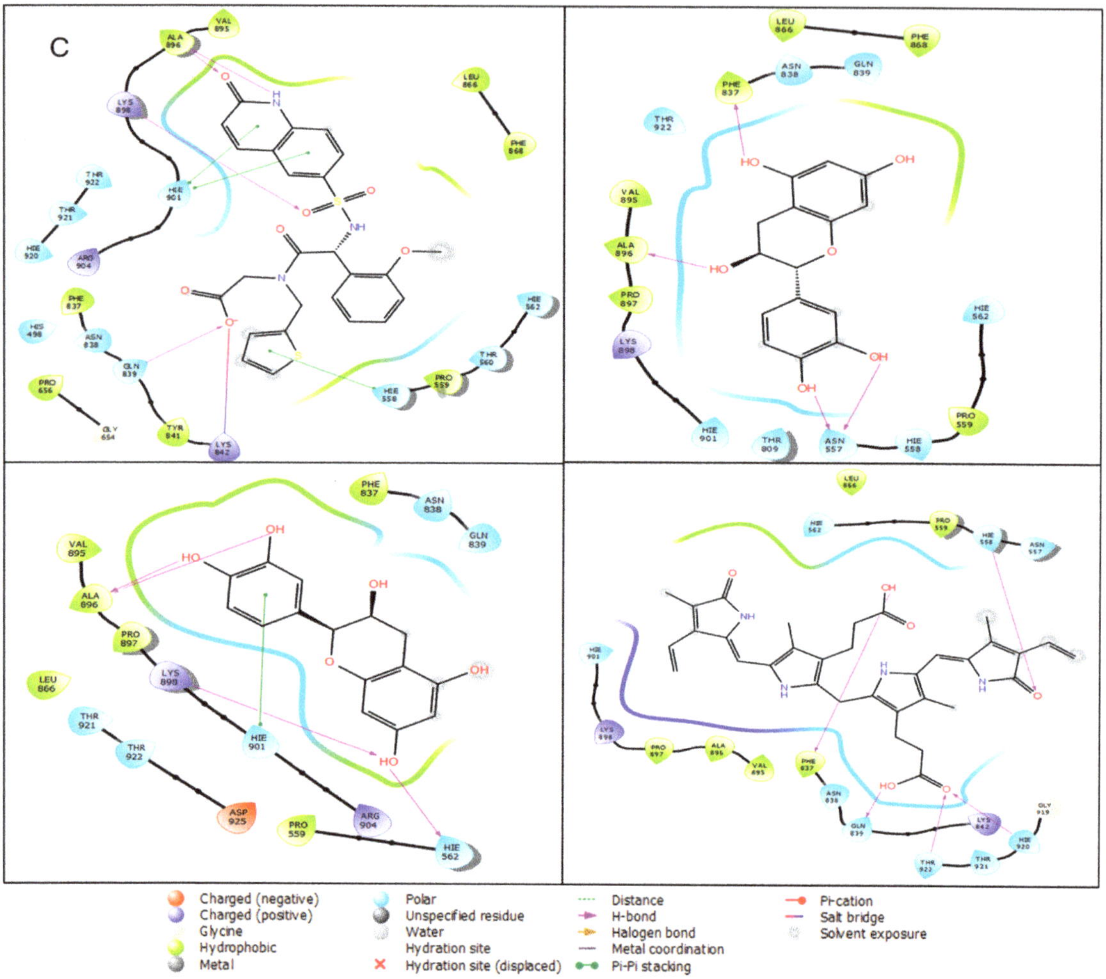

Figure 4. *Cont.*

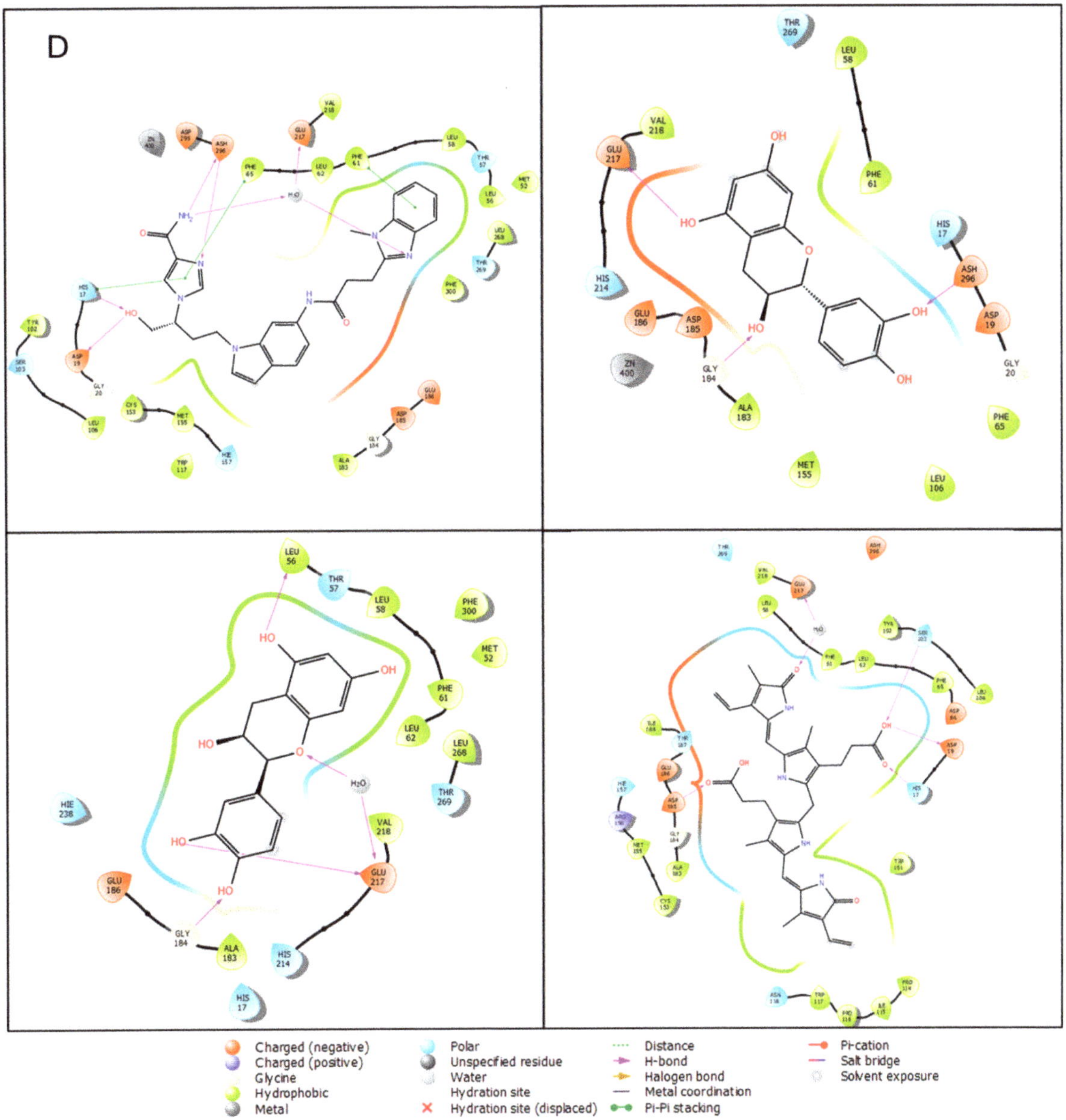

(b)

Figure 4. *Cont.*

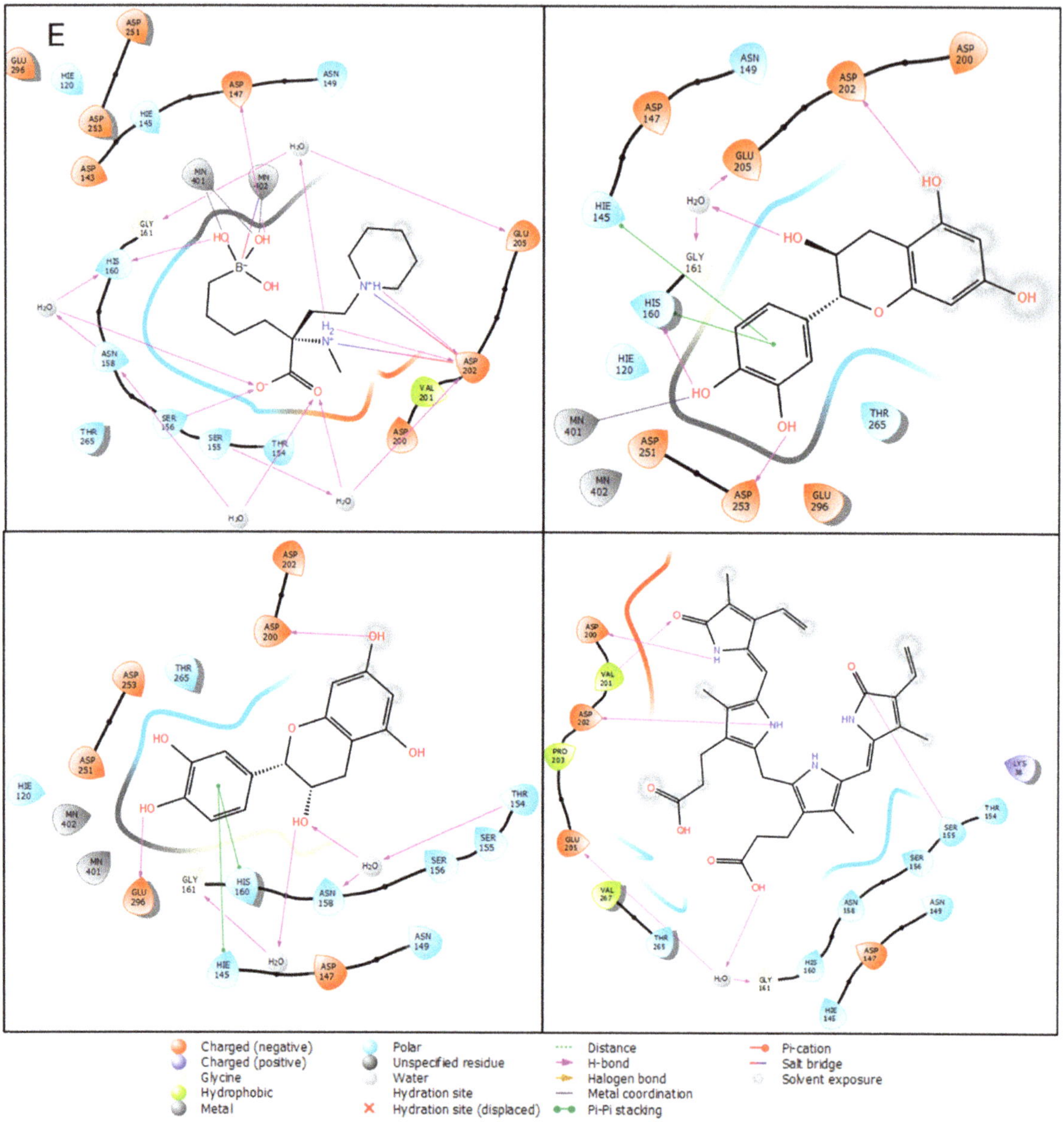

Figure 4. *Cont.*

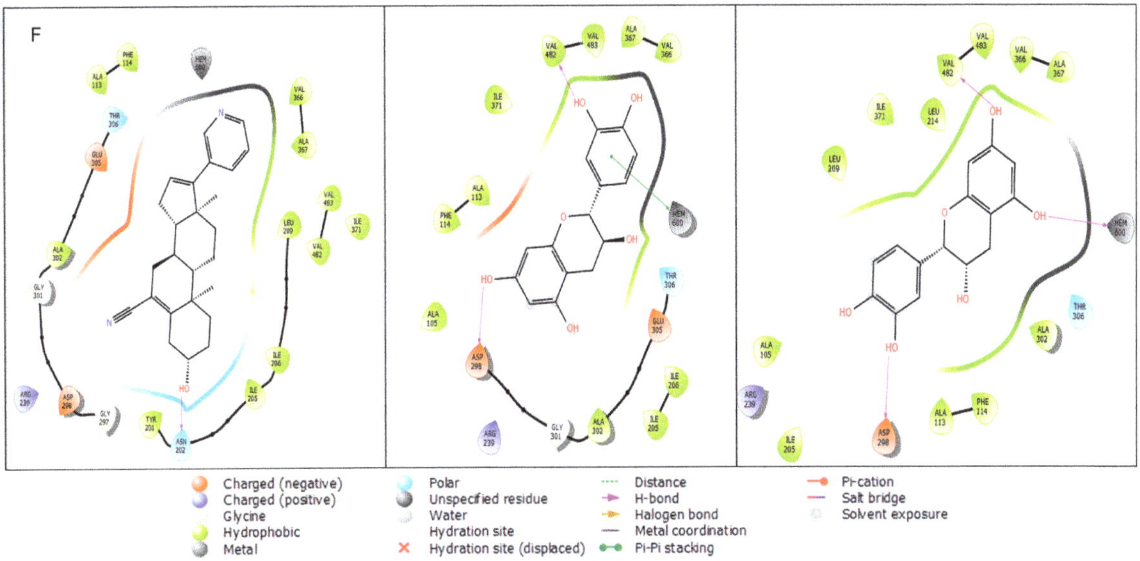

(c)

Figure 4. (**a**) Superimposed structures of the co-crystalized ligands in their co-crystallized (green) and re-docked poses (magenta) at the active sites of the receptors. (RMSD = 0.65 and 0.26 A for 2H42 and 3IW4, respectively). (**b**) 2D ligand interactions of the co-crystallized ligand (CCL), catechin, epicatechin, and pheophytin with (**A**) human PDE5, (**B**) protein kinase C alpha, (**C**) OGT, and (**D**) adenosine deaminase. (**c**) 2D ligand interactions of the co-crystallized ligand (CCL), catechin, epicatechin, and pheophytin with (**E**) arginase 2 and (**F**) steroidogenic cytochrome P450 17A1.

The molecular mechanics general Born surface (MM/GBSA) offers an effective means of calculating ligand–protein binding free energy. It is a measure of the amount of free energy involved in a particular set of interactions. For this study, MM/GBSA calculations were conducted using the Prime module of the Schrödinger suite. Our hit compounds showed comparable binding energy to the co-crystallized ligands. For instance, pheophytin showed similar binding energy to the adenosine deaminase enzyme as the co-crystallized ligand, with values of −71.91 and −77.9 Kcal/mol, respectively. Catechin exhibited superior binding energy to the arginase 2 enzyme compared with the co-crystallized ligand (−21.31 vs. −7.95 Kcal/mol). However, the co-crystallized ligands had better binding energies with PDE5 and protein kinase C alpha, as shown in Table 1 and Figure 5.

The two compounds (epicatechin and catechin) with the best interactions with the human PDE5 receptor were used for the molecular dynamic study (Figure 6). The root mean square deviation (RMSD), root mean square fluctuation (RMSF), radius of gyration (rGyr), molecular surface area (MolSA), solvent-accessibility surface area (SASA), and polar surface area (PSA), among other parameters, were determined to predict the stability and interraction of these compounds at the receptor's active site. The results are presented as means ± SEM in Armstrong units (Å), as shown in Table 2. While the RMSD, RMSF, and rGyr give information about protein–ligand stability, SASA gives information about the surface area of the molecule that is accessible by water, and PSA gives information about the solvent-accessible surface area in a molecule that is contributed by nitrogen and oxygen atoms only. Catechin was found to have a lower RMSF, RMSD, and radius of gyration (0.74, 1.517, and 3.58 vs. 0.83, 1.73, and 4.15 respectively) in comparison with sildenafil, while Epicatechin had a comparable RMSD and SASA (1.729 and 56.81 vs. 1.73 and 54.20) when compared with sildenafil, as shown in Table 2.

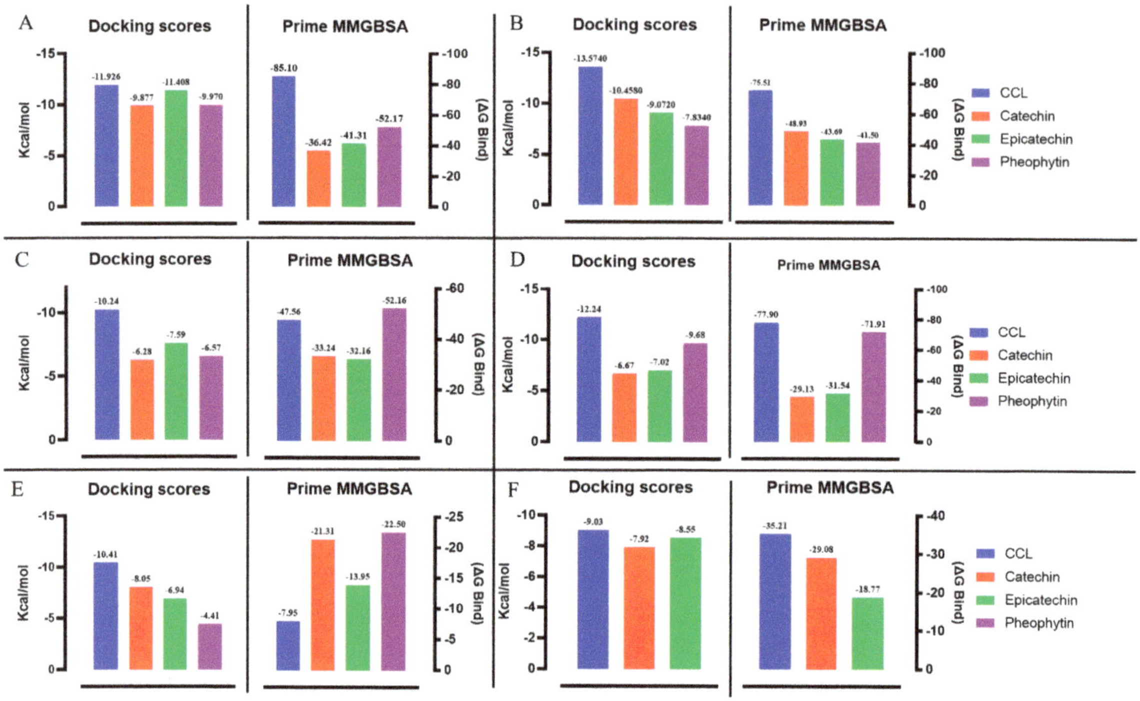

Figure 5. Docking and Prime MM/GBSA scores of the co-crystallized ligand (CCL), catechin, epicatechin, and pheophytin with (**A**) human PDE5, (**B**) protein kinase C alpha, (**C**) OGT, (**D**) adenosine deaminase, (**E**) arginase 2, and (**F**) steroidogenic cytochrome P450 17A1.

Table 2. Interactive properties of MDs of the PDE5 receptors and protein–ligand interactions.

Receptor	Ligand	P_RMSF	RMSD	rGyr	MolSA	SASA	PSA
2H42	Sildenafil	0.83 ± 0.035	1.73 ± 0.008	4.15 ± 0.001	410.0 ± 0.17	54.20 ± 0.34	140.8 ± 0.12
	Epicatechin	0.91 ± 0.027	1.729 ± 0.007	3.77 ± 0.001	255.7 ± 0.04	56.81 ± 0.33	245.8 ± 0.08
	Catechin	0.74 ± 0.023	1.517 ± 0.005	3.58 ± 0.002	253.5 ± 0.05	26.88 ± 0.27	259.8 ± 0.11

Note: Values are presented as means ± standard error of mean (SEM) measured in Armstrong units (Å). RMSD: complex root mean square deviation; P_RMSF: protein root mean square fluctuation; rGyr: radius of gyration; MolSA: molecular surface area; SASA: solvent-accessibility surface area; PSA: polar surface area.

The MM/GBSA calculations based on the poses from the molecular dynamics for the phosphodiesterase 5 complexes were performed to provide a more comprehensive assessment of the binding energies. The results showed that the co-crystallized ligand, sildenafil, had higher binding energy (MM/GBSA) than catechin and epicatechin (Table 3). This result agrees with that observed with the docked poses.

Table 3. Calculated MM/GBSA values of the MD trajectories of the PDE5 protein–ligand complex poses.

Receptor	Ligand	MM/GBSA
2H42	Sildenafil	−76.29 ± 0.38
	Epicatechin	−38.84 ± 0.69
	Catechin	−45.72 ± 0.46

Note: Values are presented as mean ± standard error of mean (SEM) measured in kcal/mol.

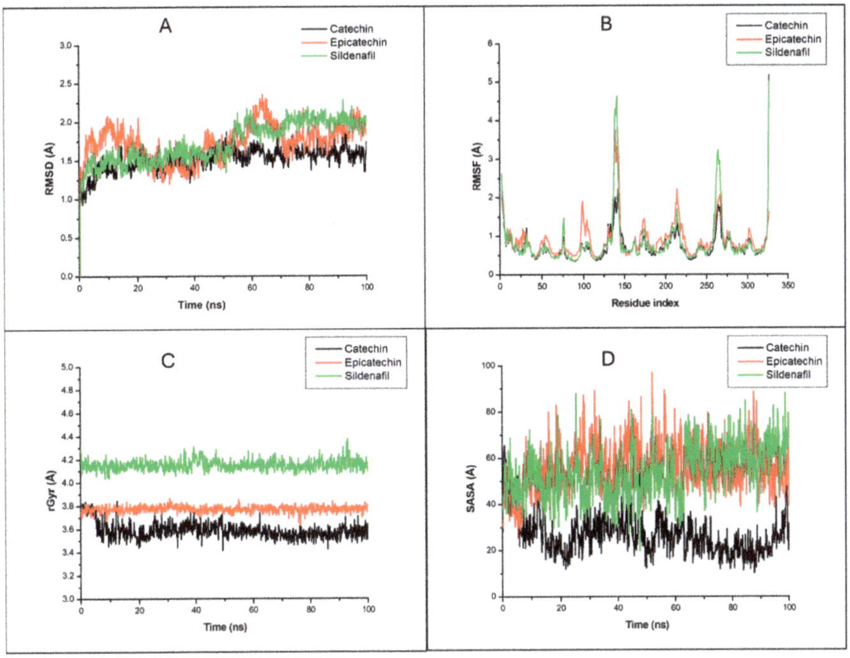

Figure 6. MD simulation results of 2H42 complexed to catechin, epicatechin, and sildenafil. (**A**) RMSD, (**B**) RMSF, (**C**) rGyr, and (**D**) SASA graphical plot. All simulations were carried out using Schrödinger's Maestro suite (Desmond).

The absorption, distribution, metabolism, and excretion profile of a compound helps in understanding the physicochemical, pharmacokinetic, and toxicological properties of the compound. For this purpose, the SwissADME web server was used. Catechin and epicatechin (shown in Figure 6) had the same profile save for their consensus log P value, with a slight difference (0.83 vs. 0.85). Aside from that, all other values were the same. It was observed that these compounds had favorable profiles as they did not violate Lipinski's rule and had a favorable molecular weight, which yielded them favorable synthetic accessibility scores. They also possessed a favorable/acceptable topological polar surface area (TPSA) and consensus log P (except pheophytin), making them promising drug candidates (drug likeness), as shown in Table 4. Regarding their pharmacokinetic properties, bioavailability, and cytochrome P450-metabolizing enzyme-inhibitory potentials, the compounds were highly water-soluble and possessed a high gastrointestinal absorption profile. They did not permeate the blood–brain barrier, had a favorable bioavailability profile, and did not inhibit the CYP1A2, CYP2C19, CYP2C9, CYP2D6, or CYP3A4 enzymes (except gallic acid, which is an inhibitor of CYP3A4). Both catechin and epicatechin are P-glycoprotein substrates, while gallic acid is not, as shown in Table 5.

Table 4. In-silico drug-likeness prediction of the catechin and gallic acid.

Compound	MW	#H-Bond Acceptors	#H-Bond Donors	TPSA	Consensus Log P	#Lipinski Violation	Synthetic Accessibility
Catechin	290.27	6	5	110.38	0.83	0	3.5
Gallic acid	170.12	5	4	97.99	0.21	0	1.22
Pheophytin	871.2	8	2	121.94	9.91	2	10

Table 5. The pharmacokinetic properties, bioavailability, and cytochrome P450-metabolizing enzyme-inhibitory potentials of the compounds.

	GI Absorption	BBB Permeant	P-Glycoprotein Substrate	CYP1A2 Inhibitor	CYP2C19 Inhibitor	CYP2C9 Inhibitor	CYP2D6 Inhibitor	CYP3A4 Inhibitor	Bioavailability Score
Catechin	High	No	Yes	No	No	No	No	No	0.55
Gallic acid	High	No	No	No	No	No	No	Yes	0.56
Pheophytin	Low	No	Yes	No	No	No	No	No	0.17

Synthetic accessibility and bioavailability scores ranged from 1 (very easy to synthesize) to 10 (very difficult to synthesize) and 0 (not bioavailable) to 1 (100% bioavailable). For synthetic accessibility scores, according to the scale, the lower the value, the easier it is to be synthesized; therefore, a lower value is desirable. For bioavailability scores, usually, any compound with a bioavailability score of ≥0.55 is considered ideal and absorbed very well by the body. Therefore, these compounds (except pheophytin) had favorable/acceptable pharmacokinetic and drug-likeness profiles. Overall, pheophytin did not exhibit a favorable or desirable ADME profile and therefore was not included in further studies.

3. Discussion

Erectile dysfunction (ED) is increasingly a social problem that affects not only the elderly but men of all ages and significantly affects their quality of life and that of their partners [32]. Though phosphodiesterase 5 inhibitors (PDE5is) are available in the market for the management of ED, there is a need for further development of safer and more effective therapies for this purpose owing to their side effects, including headache and cardiovascular problems [19,33]. There is also a need to explore other plausible targets that may contribute to the initiation and sustenance of penile erection [34,35]. Natural sources (natural products) have contributed to drug development and offer a safe source of lead compounds for drug design, discovery, and development [15,36]. Medicinal plants have been known to contribute to the treatment of various diseases, including erectile dysfunction. Several medicinal plant extracts have shown promising potential in the management of ED, including ginseng, *Garcinia kola*, *Lepidium meyenii*, *Pinus pinaster*, *Tribulus terrestris*, and saffron, among others [16,19], thus making this path of investigation a plausible one. In silico studies, which involve computer-based simulations and molecular modeling, have become powerful tools for predicting the binding affinity of compounds to specific proteins, and are encouraged at the onset of the research process to identify lead compounds from a large library of compounds/database. They are a fast, cost-effective, and integrative means of predicting pharmacological outcomes and improving the drug discovery process.

To understand the interactions between the phytochemicals from *Detarium senegalense* and key enzymes involved in the pathogenesis of ED, molecular docking simulations were conducted, with sildenafil citrate used as the standard. The results showed that catechin and epicatechin had favorable binding energies with most of the selected targets, especially PDE5, protein kinase C alpha, and steroidogenic cytochrome P450 17A1, indicating potential interactions with these enzymes (Table 1). Regarding the PDE5 enzyme, catechin and epicatechin exhibited hydrogen bonds with GLN817 and pi-stacking with TYR612, just like the standard drug, sildenafil citrate. In addition to this, catechin exhibited an additional H-bond with GLN775 and epicatechin with ASN661 and THR723. These interactions (coupled with binding energy, as shown in Table 1) suggests its inhibitory potential on the PDE enzyme, which is responsible for the degradation of cGMP. This finding is in tandem with that reported by [19,21] that flavonoids from *Anonna senegalensis* and *Garcinia kola* inhibit the PDE5 enzyme.

Overactivity of protein kinase C has been shown to contribute to the pathophysiology of ED in diabetes [37]. The inhibition of this enzyme is known to relieve NO/cGMP pathway impairment in penile vascular tissues, thus alleviating ED symptoms [37,38]. Catechin shows strong binding affinity with PKC, with a binding score of −10.46 kcal/mol compared to the −13.57 kcal/mol of the co-crystallized ligand (Table 1 and Figure 5). This shows that catechin has the potential to bind to this target strongly and inhibit its activity/overactivity in ED, thereby aiding sustained erection for satisfactory intercourse via modulation of the NO/cGMP pathway.

In addition, gallic acid, a phenolic compound isolated from this plant [39], also exhibits good binding afinity with PDE5 and PKC, as shown in Figure 1. This further supports the report by [40] that phenolics from natural sources exhibit potential to inhibit key ED enzymes. Also, [41] reported the potential of gallic acid in the management of ED.

The interaction of amino acid residues of the PDE5, PKC, and steroidogenic cytochrome P450 17A1 enzymes with their respective standards and catechin were similar, making these worthy of further investigation. Therefore, molecular dynamic (MD) simulation analysis was performed to evaluate structural and functional relationships with the protein–ligand complex. The MD simulation emulates the biological system and gives information about the stability of the complex, conformational changes, and the individual residue flexibility/fluctuation during simulation [42,43]. The root mean square deviation (RMSD), root mean square fluctuation (RMSF), radius of gyration (rGyr), pressure swing adsorption (PSA), and solvent-accessibility surface area (SASA) were analyzed, as shown in Table 2 (means ± SEM) and Figure 6.

The root mean square deviation (RMSD) measures the degree of structural/conformational variation in a ligand–protein complex over time [43]. A constant and low ligand–RMSD value shows that the ligand maintains a similar pose to its docking/starting pose during the simulation, while a fluctuating RMSD value indicates that there is frequent alteration in the ligand pose in the binding pocket [44]. From our results (Table 2, Figure 6), it was observed that our compounds (epicatechin and catechin) possessed low RMSD values (1.729 ± 0.007 and 1.517 ± 0.005, respectively) throughout the 100 ns simulation, indicating their stability at the receptor site. They also possessed a similar RMSD to the standard drug, sildenafil (1.73 ± 0.008). This result supports our findings from the molecular docking study, where the ligands showed very high binding affinity.

RMSF plots indicate which part of the protein constantly moves throughout a 100 ns simulation and reveals the key residues involved in the strongest interactions with a ligand [43,44]. The compounds had similar RMSF to the standard drug, sildenafil. Also, both compounds maintained a low RMSF value at their most important residues (Figure 6). This indicates that there were minimal movements in the binding region, indicating stable ligand binding.

The rGyr gives information about changes in protein compactness and stability, with lower average rGyr values indicating a protein's compactness and stability and vice versa [42,43,45]. Our compounds of investigation (epicatechin and catechin) possessed a relatively lower rGyr (3.77 ± 0.001 and 3.58 ± 0.002, respectively) compared to the standard drug, sildenafil (4.15 ± 0.001). The plot showed stability throughout the simulation period for epicatechin-2H42, indicating it did not cause any distortion in the structure of the protein. This supports our molecular docking and RMSD findings that epicatechin and catechin may be effective inhibitors of the human phosphodiesterase enzyme.

While the binding affinity of a compound is important for its activity, it is critical to consider the pharmacokinetic profile and drug likeness of the compounds [46,47]. The Lipinski rule of five (RO5) has been postulated in the drug discovery process to help determine the potential of a molecule to become a drug [48]. As postulated, it is important that a molecule does not violate more than one of these RO5 [49]. The standard drug (sildenafil citrate) violated two of these RO5 [19], while catechin did not violate any of the RO5 (Table 4), thus suggesting that it may well be a promising candidate with better activity and fewer side effects than sildenafil. Its high bioavailability, water solubility, and gastrointestinal absorption profile compared to that of sildenafil position it as a viable lead compound. It did not permeate the blood–brain barrier (Table 5); therefore, it is likely to be devoid of the headache side effects observed with sildenafil. Catechin, epicatechin, and gallic acid also had synthetic accessibility scores of 3.5, 3.5, and 1.22, respectively, making them very easy to synthesize, and their molecular weight of 290.27, 290.27, and 170.12 g/mol, respectively, gives room for possible structural modification/optimization to improve their efficacy. However, pheophytin, though having good binding affinity, possessed a poor ADMET profile and was therefore excluded from further studies. It can therefore be said that the epicatechin- and catechin-rich fraction of *Detarium senegalense* could lead to improved sexual performance with fewer side effects expected compared to sildenafil. Ref. [50] reported on the antidiabetic activity of different fractions of *Detarium senegalense* stem bark extracts. Diabetes mellitus has been known to sometimes lead to erectile dysfunction [51,52]. Therefore, it can be said that this plant has the potential to ameliorate diabetes-associated erectile dysfunction. Regarding the toxicity/safety of this plant, an acute toxicity study revealed no significant signs in toxicity in rats up to 5000 mg/kg [53]. The oral LD50 of catechin and epicatechin has been reported to be >10,000 mg/kg in rodent studies [54,55], while that of gallic acid is >5000 mg/kg [56,57]. This suggests a favorable safety profile of this plant and these compounds.

4. Materials and Methods

4.1. Virtual Screening and Docking Platform

Literature reviews were conducted to retrieve the phytochemicals that have previously been characterized from *Detarium senegalense* [26,29,50]. These compounds were collected from the PubChem online database (https://pubchem.ncbi.nlm.nih.gov/, accessed on 11 September 2024) and docked to the active sites of the selected targets to predict compounds with the best inhibitory potential to block these targets that have been implicated in erectile dysfunction. Schrödinger Maestro 11.5 was used for the docking study using the standard molecular docking principles, while the SwissADME tool was used for the pharmacokinetic and physicochemical properties, and toxicity prediction.

4.2. Phytochemical Library Generation and Preparation

A library of phytochemicals that have previously been characterized from *Detarium senegalense* and reported in the literature was created [26,29,50]. The two-dimensional (2D) structures of these phytochemicals (in SDF format) were retrieved from the PubChem online database (https://pubchem.ncbi.nlm.nih.gov/). The 2D structures were transformed into 3D structures using the ligprep tool in Schrödinger by adding hydrogen atoms, ionizing at pH (7.2 $\pm$ 0.2), and removing salt using Ep2i/UNEP/-Zk. The OPLS3e force field was utilized for ionization and tautomeric state formation, as previously described [58].

4.3. Receptor Retrieval and Preparation

The three-dimensional (3D) X-ray crystal structure of the selected receptors, human phosphodiesterase 5 (PDB ID: 2H42) [59], arginase 2 (PDB ID: 4I06) [60], steroidogenic cytochrome P450 17A1 (PDB ID: 6CIZ) [61], protein kinase C alpha (PDB ID: 3IW4) [62], O-GlcNAc transferase (PDB ID: 6MA5) [63], and adenosine deaminase (PDB ID: 1NDZ) [64], were retrieved from the Protein Data Bank (https://www.rcsb.org), with their corresponding bound ligands. The PyMOL molecular graphics system (version 2.5, Schrödinger, LLC., New York, NY, USA) was used for visualization of the proteins. The protein preparation wizard tool in Maestro was used to prepare the protein before performing the molecular docking, as previously described [58]. Briefly, bond order assignment was conducted, hydrogens added, zero-order metal bonds made, disulfide bonds created, water molecules removed, and het states generated using Epik at pH 7.0 $\pm$ 0.2 during protein production. Protein refinement was performed by optimizing the H-bond assignment, and then the OPLS3e force field was used to minimize the protein.

4.4. Receptor Grid Generation

The Receptor Grid Generation tool was used to create the prepared protein grid on the binding site (Glide Grid). The receptor grid depicts the area where the ligand and protein interact. The coordinate of the co-crystallized ligand was used to specify and generate the receptor grid/active site for docking. By selecting the co-crystallized ligand at the active site of the receptor, the binding location was automatically mapped (by a cubic grid box), covering all of the amino acid residues at the active site. The default Glide Grid setting was used, and the produced grid's three-dimensional coordinates X, Y, and Z, respectively, for each of the proteins were 2H42: (12.42, −3.89, 2.05) Å, 3IW4: (5.47, 29.95, 52.13) Å, 6CIZ: (29.4, 142.65, 40.05) Å, 6MA5: (−0.23, −45.62, 15.45) Å, 1NDZ: (48.6, 52.65, 19.22) Å, 4I06: (34.04, 85.71, 72.04) Å respectively.

4.5. Molecular Docking

Docking was achieved on Maestro 11.5 with the Glide tool using extra-precision (XP) docking techniques. The default setting of the Glide tool was used, with the ligand sampling set to be flexible, no constraints set, and post-docking minimization enabled. The co-crystallized ligands were extracted and re-docked into the active site to validate the molecular docking study.

4.6. Molecular Mechanics/Generalized Born Surface Area (MM/GBSA)

The potential binding free energy of the receptor–ligand docked complexes was calculated using Prime MM/GBSA in the Schrödinger suite. For the free binding energy calculation of the docked complexes, the solvent model and force field were set to VSGB and OPLS3, respectively, while other options were left at the default settings, as previously reported [58].

4.7. Molecular Dynamic (MD) Simulation and Trajectory Analysis

The molecular dynamic (MD) simulation for the native phosphodiesterase 5 enzyme (2H42) and the two complexes (2H42-epicatechin and 2H42-catechin) was executed using the Desmond module of the Schrödinger suite. The system setup, MD production, and trajectory analysis were performed as previously reported [65]. All simulations were carried out using the OPLS2005 force field. The protein–ligand complexes were bound in an orthorhombic box, with the box size calculation method set as buffer, all three distances set at 10 Å, then the volume of the box minimized. The TIP3P water model was used as the solvent model. Sodium and chloride ions were added to neutralize the overall charge of the system, and the salt concentration was set to 0.15 M to mimic physiological conditions. The standard protocols within the Maestro environment were employed to initially prepare and minimize the system. System relaxation was undertaken in an NPT ensemble at 300 K and 1 atm using a Nosé–Hoover thermostat and a Martyna–Tobias–Klein barostat, respectively. The MD simulation was performed for 100 nanoseconds (ns), and the trajectory sampling was set at an interval of 100 ps with 1000 frame numbers, allowing for extensive sampling of the conformational space. During the MD simulation, the long-range electrostatic interactions were calculated using the particle mesh Ewald (PME) method. The outputs of the simulation were visuallized and analyzed by MS-MD trajectory analysis and a simulation interaction diagram. The data were plotted using Origin version 6.0.

4.8. Absorption, Distribution, Metabolism, Excretion, and Toxicological (ADMET) Prediction

The SwissADME online server (http://www.swissadme.ch/index.php#, accessed on 25 May 2024) was used to estimate the physicochemical, pharmacokinetic, and toxicological properties of the lead compounds, which predict their ADMET profile in the human body [66].

5. Conclusions

Based on the analysis of the molecular docking results, this study revealed that some phytochemicals from *Detarium senegalense* possess a favorable binding affinity with human PDE5, protein kinase C alpha, and steroidogenic cytochrome P450 17A1. The docking analysis showed that catechin, epicatechin, and gallic acid possess a high affinity for these receptors. Catechin stands out from all these phytochemicals because of its superior binding energies, favorable pharmacokinetic profile, and good binding conformation/interaction at the receptor site. Our results therefore suggest that catechin can act as a multi-target inhibitor of key enzymes involved in the pathogenesis of erectile dysfunction. Therefore, this study suggests that the flavonoid-rich fraction of *Detarium senegalense* could lead to improved sexual performance and sustained erection for satisfactory sexual intercourse. Also, catechin could serve as a potential lead compound for development of a drug for the management of ED, owing to its satisfactory pharmacokinetic profile, non-violation of the Lipinski rule of five, good binding affinity, and good interaction at the receptor sites of the enzymes. It is important to note that while in silico studies provide valuable insights, experimental validation is necessary to confirm the inhibitory activity of these compounds. Also, there has been no report on the quantitative analysis of these compounds (to determine the amount of catechin and epicatechin) in *Detarium senegalense*. Therefore, there is a need to quantify these compounds that are found in the plant to support their role in the management of ED. Moreover, further research is needed to optimize their pharmacological properties and evaluate their effectiveness in animal models and clinical trials. If successful, these compounds could open new avenues for the development of cost-effective and safe drugs for the management of erectile dysfunction.

Overall, this computational investigation supports the potential of *Detarium senegalense* phytochemicals as promising candidates for developing new, safer, and more effective therapies for erectile dysfunction, offering an alternative to current treatments with fewer side effects.

Author Contributions: Conceptualization, B.E.O., B.O.A., T.O.O. and O.I.O.; methodology, J.N.E. and E.A.A.; validation, E.A. and M.S.M.-G.; investigation, J.N.E., E.A.A., B.E.O., S.A.O., O.A.O., O.N.O. and P.E.A.; data curation, J.N.E., M.S.M.-G., O.N.O. and E.A.A.; writing—original draft, J.N.E. and E.A.A.; writing—review and editing, E.A. and I.T.E.; supervision, B.E.O., B.O.A., T.O.O. and O.I.O.; project administration, B.E.O. All authors have read and agreed to the published version of the manuscript.

Funding: This research received no external funding.

Institutional Review Board Statement: Not applicable.

Informed Consent Statement: Not applicable.

Data Availability Statement: The data presented in this study are available upon request from the corresponding author.

Acknowledgments: The authors acknowledge the Institute of Drug Research and Development, S.E. Bogoro Research Center, Afe Babalola University, for making their biocomputational facility available for this study. They also greatly appreciate the continued support of the University of Zululand Research Committee.

Conflicts of Interest: The authors declare no conflicts of interest.

References

1. Mazzilli, F. Erectile Dysfunction: Causes, Diagnosis and Treatment: An Update. *J. Clin. Med.* **2022**, *11*, 6429. [CrossRef] [PubMed]
2. Leslie, S.W.; Sooriyamoorthy, T. *Erectile Dysfunction*; StatPearls Publishing LLC: St. Petersburg, FL, USA, 2024.
3. Muneer, A.; Kalsi, J.; Nazareth, I.; Arya, M. Erectile Dysfunction. *BMJ* **2014**, *348*, g129. [CrossRef] [PubMed]
4. Lewis, R.W.; Fugl-Meyer, K.S.; Corona, G.; Hayes, R.D.; Laumann, E.O.; Moreira, E.D.; Rellini, A.H.; Segraves, T. Definitions/Epidemiology/Risk Factors for Sexual Dysfunction. *J. Sex. Med.* **2010**, *7*, 1598–1607. [CrossRef] [PubMed]
5. Shamloul, R.; Ghanem, H. Erectile Dysfunction. *Lancet* **2013**, *381*, 153–165. [CrossRef]

6. Mark, K.P.; Arenella, K.; Girard, A.; Herbenick, D.; Fu, J.; Coleman, E. Erectile Dysfunction Prevalence in the United States: Report from the 2021 National Survey of Sexual Wellbeing. *J. Sex. Med.* **2024**, *21*, 296–303. [CrossRef]
7. MacDonagh, R.; Ewings, P.; Porter, T. The effect of erectile dysfunction on quality of life: Psychometric testing of a new quality of life measure for patients with erectile dysfunction. *J. Urol.* **2002**, *167*, 212–217. [CrossRef]
8. Agaba, P.; Ocheke, A.; Akanbi, M.; Gimba, Z.; Ukeagbu, J.; Mallum, B.; Agaba, E. Sexual Functioning and Health-Related Quality of Life in Men. *Niger. Med. J.* **2017**, *58*, 96. [CrossRef]
9. Dean, R.C.; Lue, T.F. Physiology of Penile Erection and Pathophysiology of Erectile Dysfunction. *Urol. Clin. N. Am.* **2005**, *32*, 379–395. [CrossRef]
10. Ignarro, L.J.; Bush, P.A.; Buga, G.M.; Wood, K.S.; Fukuto, J.M.; Rajfer, J. Nitric Oxide and Cyclic GMP Formation upon Electrical Field Stimulation Cause Relaxation of Corpus Cavernosum Smooth Muscle. *Biochem. Biophys. Res. Commun.* **1990**, *170*, 843–850. [CrossRef]
11. Giuliano, F. Neurophysiology of Erection and Ejaculation. *J. Sex. Med.* **2011**, *8*, 310–315. [CrossRef]
12. Pofi, R.; Giannetta, E.; Feola, T.; Galea, N.; Barbagallo, F.; Campolo, F.; Badagliacca, R.; Barbano, B.; Ciolina, F.; Defeudis, G.; et al. Sex-Specific Effects of Daily Tadalafil on Diabetic Heart Kinetics in RECOGITO, a Randomized, Double-Blind, Placebo-Controlled Trial. *Sci. Transl. Med.* **2022**, *14*, eabl8503. [CrossRef] [PubMed]
13. Salonia, A.; Bettocchi, C.; Boeri, L.; Capogrosso, P.; Carvalho, J.; Cilesiz, N.C.; Cocci, A.; Corona, G.; Dimitropoulos, K.; Gül, M.; et al. European Association of Urology Guidelines on Sexual and Reproductive Health—2021 Update: Male Sexual Dysfunction. *Eur. Urol.* **2021**, *80*, 333–357. [CrossRef] [PubMed]
14. Saxena, A.; Prakash, P.; Sissodia, N.; Sharma, P.; Porwal, M. Erectile Dysfunction: A Review and Herbs Used for Its Treatment. *Int. J. Green Pharm.* **2012**, *6*, 109. [CrossRef]
15. Agbebi, E.A.; Omotuyi, O.I.; Oyinloye, B.E.; Okeke, U.B.; Apanisile, I.; Okor, B.; Adefabijo, D. Ethnomedicine, Phytochemistry, and Pharmacological Activities of Uvaria Chamae P. Beauv.: A Comprehensive Review. *Naunyn. Schmiedebergs. Arch. Pharmacol.* **2024**, *397*, 5421–5436. [CrossRef] [PubMed]
16. Lee, H.W.; Lee, M.S.; Kim, T.-H.; Alraek, T.; Zaslawski, C.; Kim, J.W.; Moon, D.G. Ginseng for Erectile Dysfunction. *Cochrane Database Syst. Rev.* **2021**, *2021*, CD012654. [CrossRef]
17. Abdillahi, H.S.; Van Staden, J. South African Plants and Male Reproductive Healthcare: Conception and Contraception. *J. Ethnopharmacol.* **2012**, *143*, 475–480. [CrossRef]
18. Chauhan, N.S.; Sharma, V.; Dixit, V.K.; Thakur, M. A Review on Plants Used for Improvement of Sexual Performance and Virility. *Biomed Res. Int.* **2014**, *2014*, 868062. [CrossRef]
19. Folawiyo, M.A.; Omotuyi, I.O.; Ajao, F.O.; Besong, E.; Adelusi, T.I.; Ajayi, A.F. Catechin from Anonna Senegalensis Is a Potential Inhibitor of Erectile Dysfunction: Implication for Its Use in Male Sexual Enhancement. *Appl. Biochem. Biotechnol.* **2023**, *195*, 4936–4964. [CrossRef]
20. Ojo, O.A.; Ojo, A.B.; Maimako, R.F.; Rotimi, D.; Iyobhebhe, M.; Alejolowo, O.O.; Nwonuma, C.O.; Elebiyo, T.C. Exploring the Potentials of Some Compounds from Garcinia Kola Seeds towards Identification of Novel PDE-5 Inhibitors in Erectile Dysfunction Therapy. *Andrologia* **2021**, *53*, e14092. [CrossRef]
21. Ojo, A.B.; Ojo, O.A.; Okesola, M.A.; Ajiboye, B.O.; Oyinloye, B.E. Garcinia Kola Extracts Improve Biochemical Markers Associated with Erectile Function: Possible Applications in Clinical Treatment? *Acta Fac. Medicae Naissensis* **2019**, *36*, 15–26. [CrossRef]
22. Ojo, O.A.; Ojo, A.B.; Oyinloye, B.E.; Ajiboye, B.O.; Anifowose, O.O.; Akawa, A.; Olaiya, O.E.; Olasehinde, O.R.; Kappo, A.P. Ocimum Gratissimum Linn. Leaves Reduce the Key Enzymes Activities Relevant to Erectile Dysfunction in Isolated Penile and Testicular Tissues of Rats. *BMC Complement. Altern. Med.* **2019**, *19*, 71. [CrossRef] [PubMed]
23. Adefegha, S.A.; Oboh, G.; Fakunle, B.; Oyeleye, S.I.; Olasehinde, T.A. Quercetin, Rutin, and Their Combinations Modulate Penile Phosphodiesterase-5′, Arginase, Acetylcholinesterase, and Angiotensin-I-Converting Enzyme Activities: A Comparative Study. *Comp. Clin. Path.* **2018**, *27*, 773–780. [CrossRef]
24. Burris-Hiday, S.D.; Scott, E.E. Steroidogenic Cytochrome P450 17A1 Structure and Function. *Mol. Cell. Endocrinol.* **2021**, *528*, 111261. [CrossRef] [PubMed]
25. Goswami, S.K.; Inamdar, M.N.; Jamwal, R.; Dethe, S. Effect of Cinnamomum Cassia Methanol Extract and Sildenafil on Arginase and Sexual Function of Young Male Wistar Rats. *J. Sex. Med.* **2014**, *11*, 1475–1483. [CrossRef] [PubMed]
26. Dassou, G.H.; Favi, G.A.; Salako, K.V.; Ouachinou, J.M.-A.S.; Trekpo, P.; Akouete, P.; Agounde, G.; Djidohokpin, D.; Dansi, M.; Kouyaté, A.M.; et al. An Updated Review of the African Multipurpose Tree Species Detarium Senegalense J.F.Gmel. (Fabaceae). *South Afr. J. Bot.* **2023**, *157*, 525–539. [CrossRef]
27. Janick, J.; Paull, R.E. (Eds.) *The Encyclopedia of Fruit & Nuts*; CABI: Wallingford, UK, 2008; ISBN 9780851996387.
28. Sanni, F.S.; Onyeyili, P.A.; Hamza, H.G.; Sanni, S.; Enefe, N.G. Effects of Detarium Senegalense JF Gmelin Aqueous Stem Bark Extract on Castor Oil Induced Diarrhoea in Albino Rats. *Sokoto J. Vet. Sci.* **2018**, *16*, 41. [CrossRef]
29. Sowemimo, A.A.; Pendota, C.; Okoh, B.; Omotosho, T.; Idika, N.; Adekunle, A.A.; Afolayan, A.J. Chemical Composition, Antimicrobial Activity, Proximate Analysis and Mineral Content of the Seed of Detarium Senegalense JF Gmelin. *African J. Biotechnol.* **2011**, *10*, 9875–9879. [CrossRef]
30. Adeyi, A.O.; Ajisebiola, B.S.; Sanni, A.A.; Oladele, J.O.; Mustapha, A.-R.K.; Oyedara, O.O.; Fagbenro, O.S. Kaempferol Mitigates Reproductive Dysfunctions Induced by Naja Nigricollis Venom through Antioxidant System and Anti-Inflammatory Response in Male Rats. *Sci. Rep.* **2024**, *14*, 3933. [CrossRef]

31. Kunjiappan, S.; Pandian, S.R.K.; Panneerselvam, T.; Pavadai, P.; Kabilan, S.J.; Sankaranarayanan, M. Exploring the Role of Plant Secondary Metabolites for Aphrodisiacs. In *Plant Specialized Metabolites: Phytochemistry, Ecology and Biotechnology*; Springer Nature: Cham, Switzerland, 2023; pp. 1–19.
32. Elterman, D.S.; Bhattacharyya, S.K.; Mafilios, M.; Woodward, E.; Nitschelm, K.; Burnett, A.L. The Quality of Life and Economic Burden of Erectile Dysfunction. *Res. Reports Urol.* **2021**, *13*, 79–86. [CrossRef]
33. ElHady, A.K.; El-Gamil, D.S.; Abdel-Halim, M.; Abadi, A.H. Advancements in Phosphodiesterase 5 Inhibitors: Unveiling Present and Future Perspectives. *Pharmaceuticals* **2023**, *16*, 1266. [CrossRef]
34. Andersson, K.-E. Mechanisms of Penile Erection and Basis for Pharmacological Treatment of Erectile Dysfunction. *Pharmacol. Rev.* **2011**, *63*, 811–859. [CrossRef] [PubMed]
35. Akorede, B.A.; Hassan, S.A.; Akhigbe, R.E. Penile Erection and Cardiovascular Function: Effects and Pathophysiology. *Aging Male* **2024**, *27*, 2336627. [CrossRef] [PubMed]
36. Chaachouay, N.; Zidane, L. Plant-Derived Natural Products: A Source for Drug Discovery and Development. *Drugs Drug Candidates* **2024**, *3*, 184–207. [CrossRef]
37. El Assar, M.; La Fuente, J.M.; Sosa, P.; Fernández, A.; Pepe-Cardoso, A.J.; Martínez-Salamanca, J.I.; Rodríguez-Mañas, L.; Angulo, J. PKC Inhibition Improves Human Penile Vascular Function and the NO/CGMP Pathway in Diabetic Erectile Dysfunction: The Role of NADPH Oxidase. *Int. J. Mol. Sci.* **2024**, *25*, 3111. [CrossRef]
38. Angulo, J.; González-Corrochano, R.; Cuevas, P.; Fernández, A.; La Fuente, J.M.; Rolo, F.; Allona, A.; Sáenz de Tejada, I. Diabetes Exacerbates the Functional Deficiency of NO/CGMP Pathway Associated with Erectile Dysfunction in Human Corpus Cavernosum and Penile Arteries. *J. Sex. Med.* **2010**, *7*, 758–768. [CrossRef]
39. Nafissatou, D.N.; Sylvie, M.; Yves, P.; Ric, B.F.E.D.E.; Christian, M.; Marc, L.; Claudie, D.M.; Manuel, D. Comparison of Phenolic and Volatile Profiles of Edible and Toxic Forms of Detarium Senegalense J. F. GMEL. *African J. Biotechnol.* **2016**, *15*, 622–632. [CrossRef]
40. Oboh, G.; Ademiluyi, A.O.; Ademosun, A.O.; Olasehinde, T.A.; Oyeleye, S.I.; Boligon, A.A.; Athayde, M.L. Phenolic Extract from Moringa Oleifera Leaves Inhibits Key Enzymes Linked to Erectile Dysfunction and Oxidative Stress in Rats' Penile Tissues. *Biochem. Res. Int.* **2015**, *2015*, 175950. [CrossRef]
41. Behdarvand-Margha, Z.; Ahangarpour, A.; Shahraki, M.; Komeili, G.; Khorsandi, L. The Effects of Gallic Acid and Metformin on Male Reproductive Dysfunction in Diabetic Mice Induced by Methylglyoxal: An Experimental Study. *Int. J. Reprod. Biomed.* **2021**, *19*, 715–724. [CrossRef]
42. Oyinloye, B.E.; Shamaki, D.E.; Agbebi, E.A.; Onikanni, S.A.; Ubah, C.S.; Aruleba, R.T.; Dao, T.N.P.; Owolabi, O.V.; Idowu, O.T.; Mathenjwa-Goqo, M.S.; et al. In Silico Comparison of Bioactive Compounds Characterized from Azadirachta Indica with an FDA-Approved Drug Against Schistosomal Agents: New Insight into Schistosomiasis Treatment. *Molecules* **2024**, *29*, 1909. [CrossRef]
43. Saibu, O.A.; Singh, G.; Omoboyowa, D.A.; Oyejoke, A.K.; Olugbodi, S.A.; Bamisaye, A.; Adeniji, C.B.; Ajayi, T.M.; Akinpelu, Y.I.; Ogunwole, C.A.; et al. Discovery of Putative Natural Compounds Inhibitor of the Germinant Spore Receptor CspC in Clostridioides Difficile Infection: Gaining Insights via In Silico and Bioinformatics Approach. *Informatics Med. Unlocked* **2023**, *42*, 101339. [CrossRef]
44. Macalalad, M.A.B.; Gonzales, A.A. In Silico Screening and Identification of Antidiabetic Inhibitors Sourced from Phytochemicals of Philippine Plants against Four Protein Targets of Diabetes (PTP1B, DPP-4, SGLT-2, and FBPase). *Molecules* **2023**, *28*, 5301. [CrossRef] [PubMed]
45. Borjian Boroujeni, M.; Shahbazi Dastjerdeh, M.; Shokrgozar, M.; Rahimi, H.; Omidinia, E. Computational Driven Molecular Dynamics Simulation of Keratinocyte Growth Factor Behavior at Different PH Conditions. *Informatics Med. Unlocked* **2021**, *23*, 100514. [CrossRef]
46. Saliu, T.P.; Umar, H.I.; Ogunsile, O.J.; Okpara, M.O.; Yanaka, N.; Elekofehinti, O.O. Molecular Docking and Pharmacokinetic Studies of Phytocompounds from Nigerian Medicinal Plants as Promising Inhibitory Agents against SARS-CoV-2 Methyltransferase (Nsp16). *J. Genet. Eng. Biotechnol.* **2021**, *19*, 172. [CrossRef] [PubMed]
47. Omotuyi, O.; Oyinloye, B.; Agboola, S.; Agbebi, A.E.; Afolabi, E.O.; Femi-Oyewo, M. Bridelia Ferruginea Phytocompounds Interact with SARS-CoV-2 Drug Targets: Experimental Validation of Corilagin Contribution. *Sci. African* **2023**, *22*, e01920. [CrossRef]
48. Bickerton, G.R.; Paolini, G.V.; Besnard, J.; Muresan, S.; Hopkins, A.L. Quantifying the Chemical Beauty of Drugs. *Nat. Chem.* **2012**, *4*, 90–98. [CrossRef]
49. Benet, L.Z.; Hosey, C.M.; Ursu, O.; Oprea, T.I. BDDCS, the Rule of 5 and Drugability. *Adv. Drug Deliv. Rev.* **2016**, *101*, 89–98. [CrossRef]
50. Ekene, E.N.; Odigie, M.O.; Edward, U.F.; Enebeli, S.K.; Akuodor, G.C.; Chilaka, K.C. Comparative Antihyperglycemic Potentials of Different Fractions of Detarium Senegalense Stem Bark Extract on Streptozotocin-Induced Diabetic Wistar Rats. *GSC Biol. Pharm. Sci.* **2023**, *23*, 249–259. [CrossRef]
51. Defeudis, G.; Mazzilli, R.; Tenuta, M.; Rossini, G.; Zamponi, V.; Olana, S.; Faggiano, A.; Pozzilli, P.; Isidori, A.M.; Gianfrilli, D. Erectile Dysfunction and Diabetes: A Melting Pot of Circumstances and Treatments. *Diabetes. Metab. Res. Rev.* **2022**, *38*, e3494. [CrossRef]

52. Parmar, R.; Verma, S.; Neelkamal; Pathak, V.; Bhadoria, A. Prevalence of Erectile Dysfunction in Type 2 Diabetes Mellitus (T2DM) and Its Predictors among Diabetic Men. *J. Fam. Med. Prim. Care* **2022**, *11*, 3875. [CrossRef]
53. Nwachukwu, E.O.; Akuodor, G.C.; Ilo, C.E.; Anele, D.O.; Ofor, C.C.; Ramalan, M.A.; Ezeokpo, B.C. The Anti-Seizure and Anti-Nociceptive Potential of Hexane Fraction from the Leaves of Detarium Senegalense. *Magna Sci. Adv. Res. Rev.* **2022**, *6*, 005–013. [CrossRef]
54. Santa Cruz Biotechnology, Inc. Material Safety Data Sheet. (+)-Catechin (hydrate) Sc-205673. 2010. Available online: https://datasheets.scbt.com/sc-204673.pdf (accessed on 8 November 2024).
55. Santa Cruz Biotechnology, Inc. Material Safety Data Sheet. (-)Epicatechin Sc-205672. 2010. Available online: https://datasheets.scbt.com/sc-205672.pdf (accessed on 8 November 2024).
56. Variya, B.C.; Bakrania, A.K.; Madan, P.; Patel, S.S. Acute and 28-Days Repeated Dose Sub-Acute Toxicity Study of Gallic Acid in Albino Mice. *Regul. Toxicol. Pharmacol.* **2019**, *101*, 71–78. [CrossRef] [PubMed]
57. Ray, S.D.; Krmic, M.; Hussain, A.; Marvilli, C.; Fabian, R.; Niha, A.; Danai, M.; Smith, Z.; Jalshgari, A.; Malik, N.; et al. Toxicity of Natural Products. In *Encyclopedia of Toxicology*; Elsevier: Amsterdam, The Netherlands, 2024; pp. 257–282.
58. Oyinloye, B.E.; Agbebi, E.A.; Agboola, O.E.; Ubah, C.S.; Owolabi, O.V.; Aruleba, R.T.; Onikanni, S.A.; Ejeje, J.N.; Ajiboye, B.O.; Omotuyi, O.I. Skin Anti-Aging Potentials of Phytochemicals from Peperomia Pellucida against Selected Metalloproteinase Targets: An In Silico Approach. *Cosmetics* **2023**, *10*, 151. [CrossRef]
59. Wang, H.; Liu, Y.; Huai, Q.; Cai, J.; Zoraghi, R.; Francis, S.H.; Corbin, J.D.; Robinson, H.; Xin, Z.; Lin, G.; et al. Multiple Conformations of Phosphodiesterase-5. *J. Biol. Chem.* **2006**, *281*, 21469–21479. [CrossRef] [PubMed]
60. Van Zandt, M.C.; Whitehouse, D.L.; Golebiowski, A.; Ji, M.K.; Zhang, M.; Beckett, R.P.; Jagdmann, G.E.; Ryder, T.R.; Sheeler, R.; Andreoli, M.; et al. Discovery of (R)-2-Amino-6-Borono-2-(2-(Piperidin-1-Yl)Ethyl)Hexanoic Acid and Congeners as Highly Potent Inhibitors of Human Arginases I and II for Treatment of Myocardial Reperfusion Injury. *J. Med. Chem.* **2013**, *56*, 2568–2580. [CrossRef]
61. Fehl, C.; Vogt, C.D.; Yadav, R.; Li, K.; Scott, E.E.; Aubé, J. Structure-Based Design of Inhibitors with Improved Selectivity for Steroidogenic Cytochrome P450 17A1 over Cytochrome P450 21A2. *J. Med. Chem.* **2018**, *61*, 4946–4960. [CrossRef]
62. Wagner, J.; von Matt, P.; Sedrani, R.; Albert, R.; Cooke, N.; Ehrhardt, C.; Geiser, M.; Rummel, G.; Stark, W.; Strauss, A.; et al. Discovery of 3-(1 H -Indol-3-Yl)-4-[2-(4-Methylpiperazin-1-Yl)Quinazolin-4-Yl]Pyrrole-2,5-Dione (AEB071), a Potent and Selective Inhibitor of Protein Kinase C Isotypes. *J. Med. Chem.* **2009**, *52*, 6193–6196. [CrossRef]
63. Martin, S.E.S.; Tan, Z.-W.; Itkonen, H.M.; Duveau, D.Y.; Paulo, J.A.; Janetzko, J.; Boutz, P.L.; Törk, L.; Moss, F.A.; Thomas, C.J.; et al. Structure-Based Evolution of Low Nanomolar O-GlcNAc Transferase Inhibitors. *J. Am. Chem. Soc.* **2018**, *140*, 13542–13545. [CrossRef]
64. Terasaka, T.; Kinoshita, T.; Kuno, M.; Nakanishi, I. A Highly Potent Non-Nucleoside Adenosine Deaminase Inhibitor: Efficient Drug Discovery by Intentional Lead Hybridization. *J. Am. Chem. Soc.* **2004**, *126*, 34–35. [CrossRef]
65. Onikanni, S.A.; Lawal, B.; Fadaka, A.O.; Bakare, O.; Adewole, E.; Taher, M.; Khotib, J.; Susanti, D.; Oyinloye, B.E.; Ajiboye, B.O.; et al. Computational and Preclinical Prediction of the Antimicrobial Properties of an Agent Isolated from Monodora Myristica: A Novel DNA Gyrase Inhibitor. *Molecules* **2023**, *28*, 1593. [CrossRef]
66. Daina, A.; Michielin, O.; Zoete, V. SwissADME: A Free Web Tool to Evaluate Pharmacokinetics, Drug-Likeness and Medicinal Chemistry Friendliness of Small Molecules. *Sci. Rep.* **2017**, *7*, 42717. [CrossRef]

International Journal of
Molecular Sciences

MDPI

Article

Computational Screening of T-Muurolol for an Alternative Antibacterial Solution against *Staphylococcus aureus* Infections: An In Silico Approach for Phytochemical-Based Drug Discovery

Soham Bhattacharya [1,*], Pijush Kanti Khanra [2], Adrish Dutta [3], Neha Gupta [3], Zahra Aliakbar Tehrani [4], Lucie Severová [5], Karel Šrédl [5], Marek Dvořák [6] and Eloy Fernández-Cusimamani [3,*]

1 Department of Agroecology and Crop Production, Faculty of Agrobiology, Food and Natural Resources, Czech University of Life Sciences Prague, Kamýcká 129, Suchdol, 16500 Prague, Czech Republic
2 Department of Biosciences and Bioengineering, Indian Institute of Technology Guwahati, Guwahati 781039, India; k.pijush@iitg.ac.in
3 Department of Crop Sciences and Agroforestry, Faculty of Tropical AgriSciences, Czech University of Life Sciences Prague, Kamýcká 129, Suchdol, 16500 Prague, Czech Republic; duttaa@ftz.czu.cz (A.D.); guptan@ftz.czu.cz (N.G.)
4 Institute of Biotechnology of the Czech Academy of Sciences, BIOCEV, Prumyslova 595, 25250 Vestec, Czech Republic; mah.tehrani7@gmail.com
5 Department of Economic Theories, Faculty of Economics and Management, Czech University of Life Sciences Prague, Kamýcká 129, 16500 Prague, Czech Republic; severova@pef.czu.cz (L.S.); sredl@pef.czu.cz (K.Š.)
6 Department of Trade and Finance, Faculty of Economics and Management, Czech University of Life Sciences Prague, Kamýcká 129, 16500 Prague, Czech Republic; dvorakmarek@pef.czu.cz
* Correspondence: bhattacharya@af.czu.cz (S.B.); eloy@ftz.czu.cz (E.F.-C.)

Abstract: *Staphylococcus aureus* infections present a significant threat to the global healthcare system. The increasing resistance to existing antibiotics and their limited efficacy underscores the urgent need to identify new antibacterial agents with low toxicity to effectively combat various *S. aureus* infections. Hence, in this study, we have screened T-muurolol for possible interactions with several *S. aureus*-specific bacterial proteins to establish its potential as an alternative antibacterial agent. Based on its binding affinity and interactions with amino acids, T-muurolol was identified as a potential inhibitor of *S. aureus* lipase, dihydrofolate reductase, penicillin-binding protein 2a, D-Ala:D-Ala ligase, and ribosome protection proteins tetracycline resistance determinant (RPP TetM), which indicates its potentiality against *S. aureus* and its multi-drug-resistant strains. Also, T-muurolol exhibited good antioxidant and anti-inflammatory activity by showing strong binding interactions with flavin adenine dinucleotide (FAD)-dependent nicotinamide adenine dinucleotide phosphate (NAD(P)H) oxidase, and cyclooxygenase-2. Consequently, molecular dynamics (MD) simulation and recalculating binding free energies elucidated its binding interaction stability with targeted proteins. Furthermore, quantum chemical structure analysis based on density functional theory (DFT) depicted a higher energy gap between the highest occupied molecular orbital and lowest unoccupied molecular orbital ($E_{HOMO-LUMO}$) with a lower chemical potential index, and moderate electrophilicity suggests its chemical hardness and stability and less polarizability and reactivity. Additionally, pharmacological parameters based on ADMET, Lipinski's rules, and bioactivity score validated it as a promising drug candidate with high activity toward ion channel modulators, nuclear receptor ligands, and enzyme inhibitors. In conclusion, the current findings suggest T-muurolol as a promising alternative antibacterial agent that might be a potential phytochemical-based drug against *S. aureus*. This study also suggests further clinical research before human application.

Keywords: density functional theory; phytochemicals; pharmacokinetics; molecular docking; MD simulation

1. Introduction

The population growth and invasion dynamics of *S. aureus* are undoubtedly opportunistic, as they subvert the human immune surveillance machinery by evoking the metastatic colony dissemination from superficial to life-threatening levels [1]. The source of bacterial dissemination mainly evolved from the mucosal contacts, as the mucosal layer mainly favours harbouring of bacterial strains. Therefore, the mucosal surfaces of the nose, throat, vaginal wall, open wound, and gastrointestinal tract impart favourable colonization carriages of *S. aureus* bacteraemia (SAB) [2]. Colonization is also frequent in human juveniles and patients who have already been vulnerable to Human immunodeficiency virus (HIV) and diabetes [3]. Despite the backdrop of prevalence and prognostic outcomes of SAB in the last decade, it has been witnessed the emergence of multi-drug-resistant (MDR) strains to empiric therapies, antibiotics, and large-scale treatment options [4]. Although, the present array of antibiotics, including methicillin, vancomycin, daptomycin, and linezolid, multi-drug-resistant *S. aureus* has emerged as the most formidable bacterial strain, causing life-threatening infections and illnesses in humans [5]. The estimated numbers of invasive disease cases and morbidity–mortality in the United States due to methicillin-resistant *S. aureus* (MRSA) pathogenesis were 94,360 and 18,650 in 2005, respectively [2]. Similarly, MRSA infection is associated with annual healthcare costs of about $3 billion a year and is set to increase over time [6]. Additionally, SAB-associated infections are also reported in the context of immunotherapies and oncological treatments [7]. The rampant and escalating severity of diseases, coupled with the failure of rational therapeutic approaches, due to the evolution of MDR *S. aureus* strains, might lead one to speculate that humans could become highly vulnerable and unmatched in facing the uncontrolled infectious threat posed by *S. aureus* [3]. Therefore, considering the systemic toxicity and limited efficacy of current antibacterial medications, identifying the underlying mechanisms of this epidemic and developing alternative antibacterial drugs should be the prime goal of the current research. These new drugs should aim to achieve high success rates against *S. aureus* and prevent obsolescence, thereby mitigating the uncertainty of public health and improving patient care for SAB-associated infections.

Plant-derived phytochemicals can be adopted as sustainable alternative antibacterial agents. Over the last three decades, phytochemicals have undergone extensive experimental validation, establishing themselves as emerging antibacterial agents with significant potential [8]. In the treatment of penicillin G-resistant strains of *S. aureus*, many phytochemicals have successfully passed experimental tests against various MDR strains [9]. Phytochemicals-based antibacterials not only possess antibacterial activity but also act as an anti-inflammatory and antioxidant agent that can dramatically improve the targeting of the infection site while minimizing systemic exposure and associated toxicity [10]. Therefore, different classes of phytochemicals, such as phenols, alkaloids, terpenoids, flavonoids, carotenoids, organosulfur compounds, and coumarins, etc., are actively available in the market as food and drug administration (FDA)-approved drugs and used in herbal medicinal products [11]. Phytochemicals have been shown to impede major bacterial MDR factors, including bacterial virulence factors, ion pumps, bacterial cell wall toxins, cellular microstructure, replication machinery, and membrane permeability [12]. Therefore, this study focuses on the in silico repurposing of a new generation of phytochemicals against *S. aureus*, aiming to explore alternative antibacterial therapeutic approaches.

T-muurolol is a cadinene sesquiterpenoid with versatile characteristics, as several studies have demonstrated its diverse biological activities, including antimicrobial, antioxidant, anti-inflammatory, and antitermitic effects [13–15]. Although T-muurolol is a potential candidate with various reported antimicrobial activities and could serve as an alternative solution to combat *S. aureus* infections, the phytochemical-based drug-reprofiling approach is still in its early stages. Currently, there are no such research articles detailing the re-profiling of T-muurolol as an alternative treatment for suppressing SAB-associated infections, and its molecular interaction with *S. aureus*-associated proteins remains largely unexplored. Therefore, this study investigates the potential of T-muurolol as an anti-staphylococcal agent through molecular docking, pharmacokinetics, molecular dynamics, and chemical structure analysis.

2. Results

2.1. Molecular Docking and Interaction Analysis

The molecular interactions between T-muurolol and tested bacteria-specific proteins for *S. aureus*, along with various proteins related to antioxidant activity and inflammation in humans, are summarized in Tables 1 and 2. T-muurolol exhibited a good binding affinity with all the proteins used in the study between −7.5 kcal/mol and −4.3 kcal/mol. T-muurolol demonstrated a strong binding affinity for *S. aureus* lipase (6KSI), showing a binding energy of −7.1 kcal/mol, which is notably higher than its affinity for other tested pathogenic proteins. A hydrogen bond of bond length 3.78 Å with A chains of amino acid residue Asp236, along with non-weak interactions with A chains of Leu250, Lys249, and Tyr240, were found in this interaction (Figure 1C,D).

Table 1. Binding free-energy values (kcal/mol) of T-muurolol as ligand along with ascorbic acid, ibuprofen, and 2-oxazolidinone as references for good antioxidant, anti-inflammatory, and antibacterial agents, respectively.

Contents	PDB Entry	Ligands			
		T-Muurolol	Ascorbic Acid	Ibuprofen	2-Oxazolidinone
Anti-inflammatory Proteins	4YYZ	−5.7	-	−6.2	-
	1N8Q	−6.5	-	−7.7	-
	2PE4	−5.8	-	−6.2	-
	3V92	−6.7	-	−6.3	-
	4CX7	−5.7	-	−5.7	-
	1CX2	−6.3	-	−6.4	-
Antioxidant Proteins	1HCK	−6.2	−4.8	-	-
	2CDU	−6.4	−5.4	-	-
	2F8A	−4.9	−4.4	-	-
	3HFF	−4.3	−4.1	-	-
SA Pathogenesis Proteins	2O8L	−6.2	-	-	−3.4
	4AE5	−5.7	-		−3
	6KSI	−7.1	-	-	−3.9
Universal Bacterial Proteins for their Survival	1JZQ	−7.2	-	-	−3.6
	1KZN	−6.1	-	-	−3.7
	2VEG	−5.3	-	-	−3.9
	2ZDQ	−5.8	-	-	−3.9
	3RAE	−6	-	-	−4
	3SRW	−7.5	-	-	−3.7
	3UDI	−5.8	-	-	−3.5
	3TTZ	−5.9	-	-	−3.3
MRSA	1GHI	−6	-	-	−3.3
	1MWU	−6.1	-	-	−3.6
VRSA	3N8D	−5.9	-	-	−3.9
TetRSA	2FJ1	−5.7	-	-	−3.5
	3J25	−6.6	-	-	−3.3
	7Y58	−5.8	-	-	−3.7

Table 2. Top hit interactions of T-muurolol with proteins and their hydrogen bonds and several other interactions.

PDB Entry	No. of H-Bonds	H-Bonds and Interacting Residues	No. of Other Interactions	Other Interactions and Numbers	Other Interaction and Interacting Residues
1CX2	0		8	Alkyl and Pi-alkyl (8)	Ile112 (2), Tyr115 (2), Val89 (3), Val116 (1)
1MWU	1	Glu294 (1)	6	Alkyl and Pi-alkyl (5), Pi-sigma (1)	Ala276 (1), Lys289 (3), Tyr272 (2)
2CDU	0		11	Alkyl and Pi-alkyl (10), Pi-sigma (1)	Ile160 (2), Leu299 (2), Pro298 (1), Tyr159 (3), Tyr188 (3)
3J25	0		10	Alkyl and Pi-alkyl (10)	Ala554 (2), Cys603 (2), Ile544 (2), Leu593 (4)
3N8D	0		2	Alkyl (2)	Val19 (2)
3SRW	1	Phe93 (1)	11	Alkyl and Pi-alkyl (10), Pi-sigma (1)	Ala8 (2), Leu6 (2), Leu21 (3), Phe93 (2), Val32 (2)
6KSI	1	Asp236 (1)	6	Alkyl and Pi-alkyl (6)	Leu250 (1), Lys249 (3), Tyr240 (2)

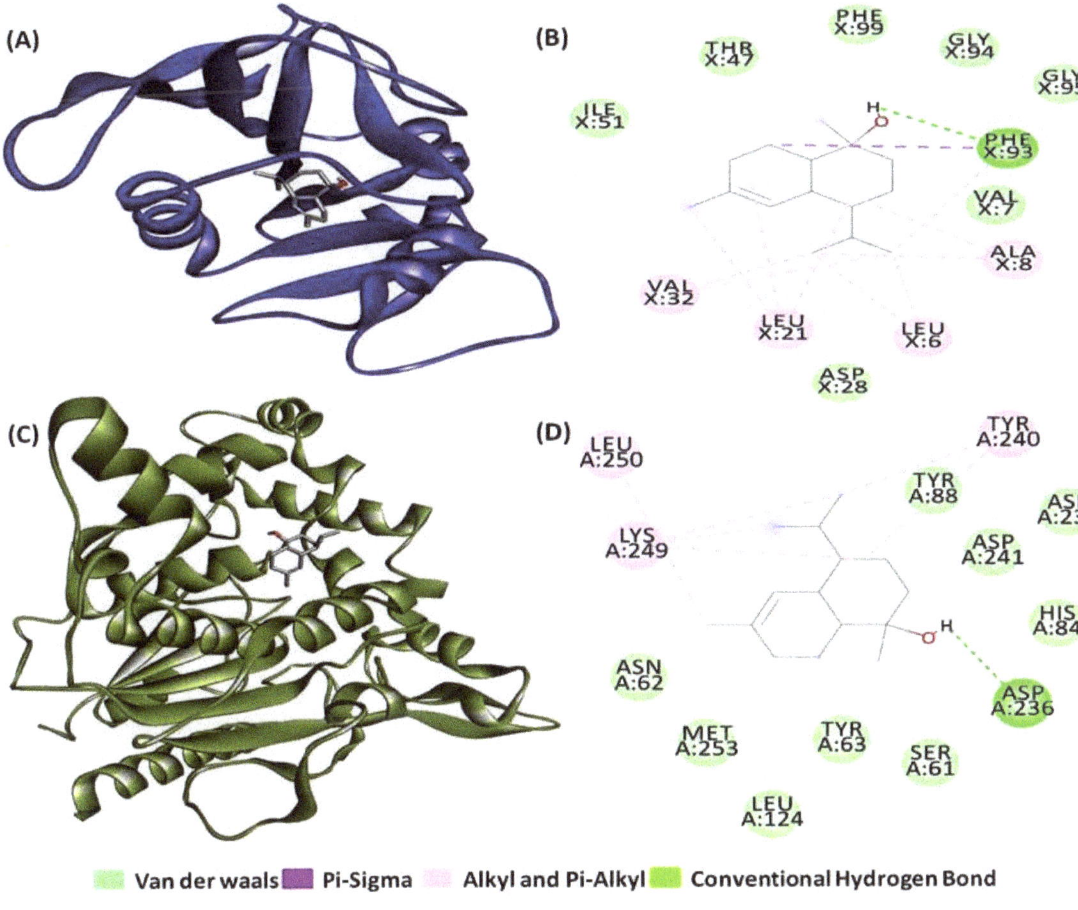

Figure 1. Three-dimensional and two-dimensional interactions of T-muurolol as ligand with (**A**) and (**B**) dihydrofolate reductase (PDB ID:3SRW), and (**C**,**D**) *S. aureus* lipase (PDB ID: 6KSI) as proteins.

T-muurolol also demonstrated significant binding interactions (with a binding energy of −7.5 kcal/mol) with dihydrofolate reductase (3SRW), a key enzyme responsible for bacterial DNA synthesis. Notably, T-muurolol formed a hydrogen bond of 5.03 Å with amino acid Phe93, along with one group of electrostatic and two groups of hydrophobic interactions (Pi-Sigma, Alkyl, and Pi-Alkyl) with the amino acid residues of Ala8, Leu6, Leu21, Phe93, and Val32, which are mainly responsible for strong interaction (Figure 1A,B). Similarly, MDR *S. aureus* penicillin-binding protein 2a (1MWU) is a key drug target responsible for broad-spectrum beta-lactam resistance. T-muurolol exhibited a binding affinity of −6.1 kcal/mol with 1MWU, forming a 4.42 Å hydrogen bond with Glu294, along with electrostatic and hydrophobic interactions involving Ala276, Lys289, and Tyr272, contributing to strong binding interactions (Figure 2A,B). The robust interactions between PBPs and T-muurolol can effectively inhibit MRSA by disrupting peptidoglycan cross-linking and impeding bacterial cell wall synthesis. Likewise, D-Ala: D-Ala ligase (3N8D), a potential target for vancomycin-resistant *Staphylococcus aureus* (VRSA) due to its role in ligase impairment, is affected by T-muurolol, evidenced by significant binding energy with 3N8D (−5.9 kcal/mol). Furthermore, T-muurolol demonstrates substantial binding affinities with RPP TetM (3J25) with a binding energy of −6.6 kcal/mol. These proteins serve as key targets in the context of tetracycline-resistant *Staphylococcus aureus* (TetRSA), emphasizing the potential of T-muurolol as a therapeutic agent against resistant bacterial strains.

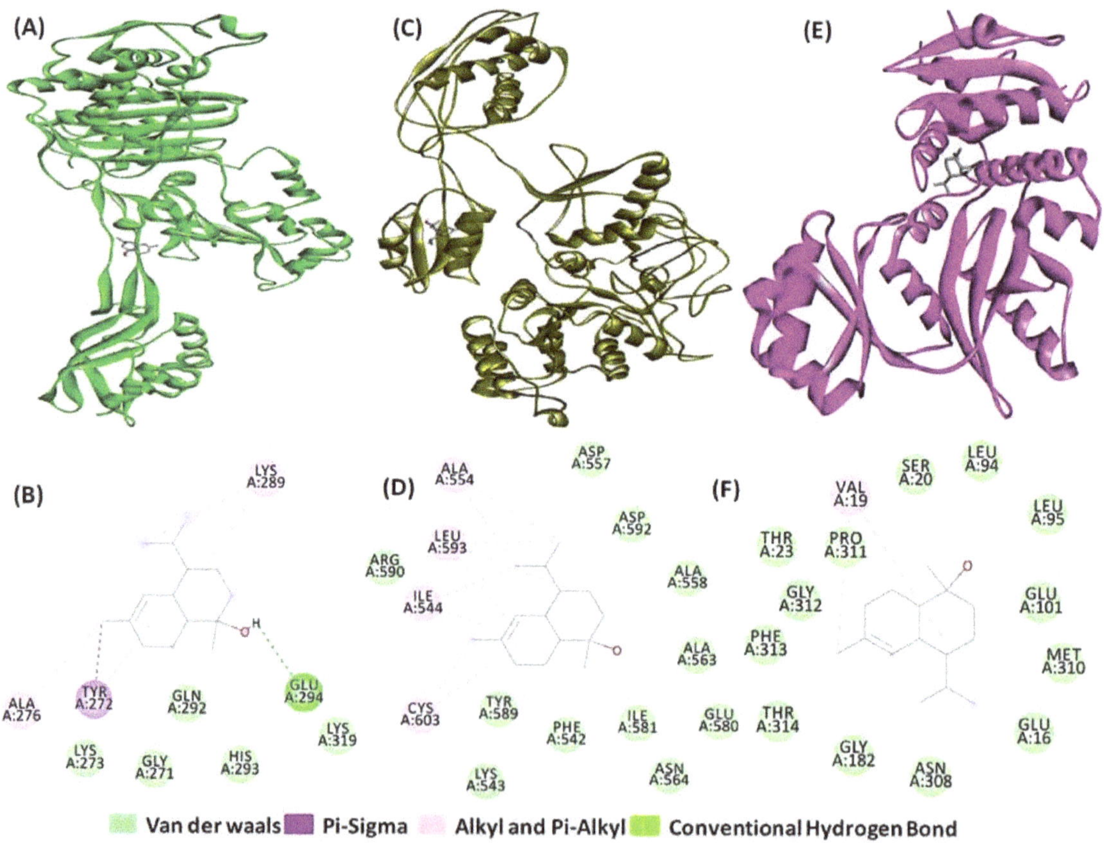

Figure 2. Three-dimensional and two-dimensional interactions of T-muurolol as ligand with (**A**) and (**B**) penicillin-binding protein 2a (PDB ID: 1MWU), (**C**,**D**) RPP TetM in complex with the 70S ribosome (PDB ID: 3J25), and (**E**,**F**) D-Ala:D-Ala ligase (PDB ID: 3N8D) as proteins.

Moreover, T-muurolol displayed favourable binding interactions with FAD-dependent NAD(P)H oxidase(2CDU) (−6.4 kcal/mol) among all the tested antioxidant proteins. Additionally, it demonstrated strong binding affinity with cyclooxygenase-2 (1CX2) (−6.3 kcal/mol), which was higher compared to the reference anti-inflammatory drug ibuprofen (Figure 3) (Table 1).

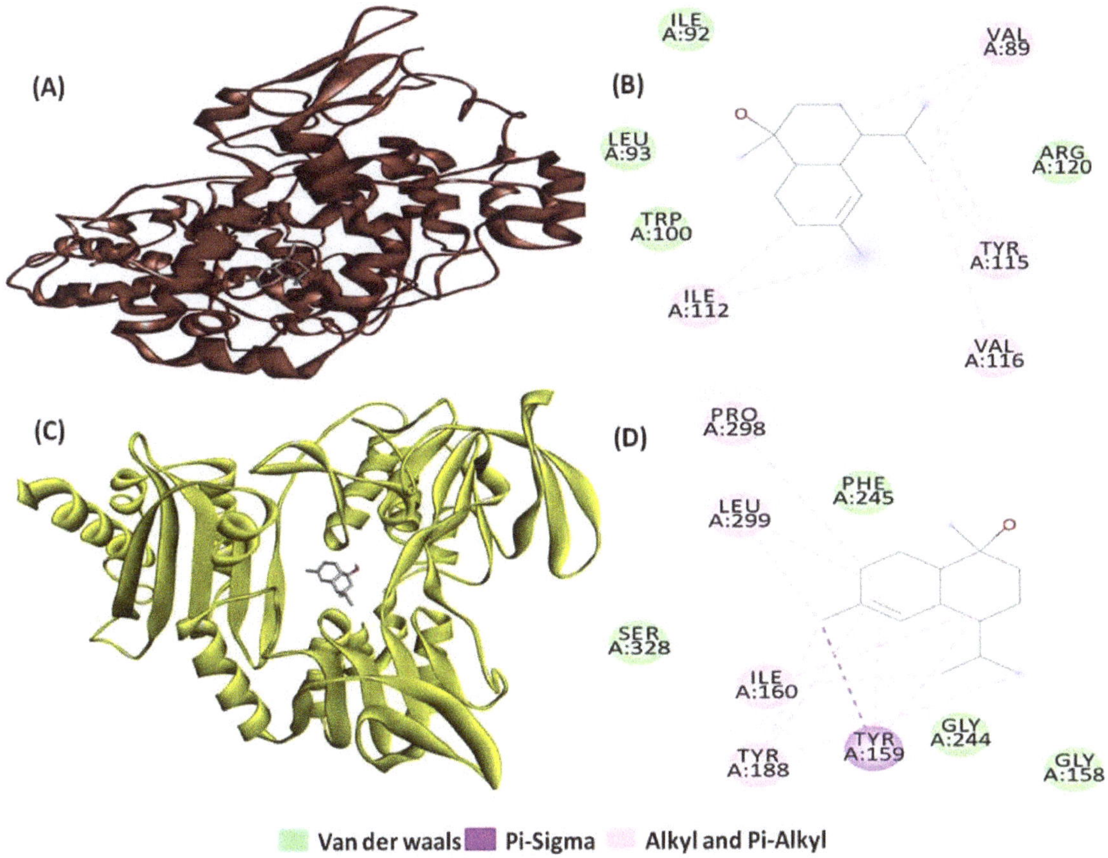

Figure 3. Three-dimensional and two-dimensional interactions of T-muurolol as ligand with (**A**,**B**) cyclooxygenase-2 (PDB ID: 1CX2), and (**C**,**D**) FAD-dependent NAD(P)H oxidase (PDB ID: 2CDU) as proteins.

2.2. ADMET Pharmacokinetic Analysis

2.2.1. Absorption

The water solubility of tested compounds, cell permeability utilizing the colon carcinoma (Caco-2) cell line, human intestinal absorption, skin permeability, and whether the molecule is a P-glycoprotein substrate or inhibitor are the primary parameters for evaluating drug absorption criteria. The current results suggested that T-muurolol had a lower solubility than water, whereas the reference antibiotic drug 2-oxazolidinone seemed to be highly soluble (Table 3). Caco-2 cell permeability influences the ultimate bioavailability; according to Chandra et al. [16], a drug with a value of > 0.90 is deemed highly permeable. The obtained results exhibited the high permeability of T-muurolol in the Caco-2 cell line with a value of 1.479. Similarly, all the tested compounds showed a higher gastrointestinal (GI) absorption percentage, which was more than 90% (Table 3). A study by Saha et al. [17] explained that the human intestine is primarily an active location for drug absorption, with more than 30% being considered rapidly absorbed. Our results also indicated that T-muurolol, along with reference drugs, is neither a P-glycoprotein substrate nor an inhibitor (Table 3).

Table 3. Predicted ADMET pharmacokinetic properties.

Property	Model Name	Compounds		
		T-Muurolol	Ibuprofen	2-Oxazolidinone
Absorption	Water solubility (log mol/L)	−4.073	−3.696	0.243
	Caco2 permeability (log Papp in 10^{-6} cm/s)	1.479	1.729	1.582
	GI absorption (%)	94.296	94.064	98.901
	P-glycoprotein substrate	No	No	No
	P-glycoprotein I inhibitor	No	No	No
	P-glycoprotein II inhibitor	No	No	No
Distribution	VDss (human) (log L/kg)	0.42	−0.803	−0.092
	Fraction unbound (human)	0.28	0.239	0.813
Metabolism	CYP2D6 substrate	No	No	No
	CYP3A4 substrate	No	No	No
	CYP1A2 inhibitor	No	No	No
	CYP2C19 inhibitor	No	No	No
	CYP2C9 inhibitor	No	No	No
	CYP2D6 inhibitor	No	No	No
	CYP3A4 inhibitor	No	No	No
Excretion	Total clearance (log mL/min/kg)	1.085	0.263	0.496
	Renal OCT2 substrate	No	No	No
Toxicity	AMES toxicity	No	No	Yes
	Max. tolerated dose (human) (log mg/kg/day)	0.343	1.015	1.249
	hERG I inhibitor	No	No	No
	hERG II inhibitor	No	No	No
	Oral Rat Acute Toxicity (LD50) (mol/kg)	1.918	2.303	2.084
	Hepatotoxicity	No	Yes	No
	Skin Sensitisation	Yes	Yes	No
	Minnow toxicity (log mM)	0.743	0.619	2.872

2.2.2. Distribution

Here, drug distribution is generally analyzed based on the steady-state volume of distribution (VDss), and fraction of unbound value to human plasma. The steady-state volume of distribution (VDss) is a key pharmacokinetic parameter for determining drug dosages, indicating the theoretical volume needed to achieve the same blood plasma concentration as the administered dose. Higher VDss values suggest greater distribution into tissues rather than plasma. For antibiotics and antivirals, extensive tissue distribution is often desired. VDss is considered low if log(VDss) < -0.15 and high if >0.45 [17]. T-muurolol showed a moderate VDss value of 0.42 log L/kg, which suggests its great distribution over tissue (Table 3). A drug's effectiveness is influenced by its binding to blood proteins; more binding facilitates membrane crossing [18]. The fraction unbound to human plasma should be between 0.02 and 1.0 [18]. T-muurolol showed good fraction unbound values with 0.28 (Table 3).

2.2.3. Metabolism

The cytochrome P450 (CYP) enzyme group includes isozymes that metabolize drugs, fatty acids, steroids, bile acids, and carcinogens. Drug metabolism depends on whether a substance is a CYP substrate or inhibitor. A study found that T-muurolol and the reference drugs tested are neither substrates nor inhibitors of CYP enzymes (Table 3), indicating they will be metabolized effectively and not hindered by the body's biological processes.

2.2.4. Excretion

Organic cation transporter 2 (OCT2) is a renal uptake transporter crucial for drug deposition and clearance from the kidneys [16]. Excretion is assessed by total clearance and whether the compound is a substrate of OCT2 or not. The predicted results showed that T-muurolol is not a substrate of OCT2 (Table 3), suggesting it may be eliminated through a different route. Also, T-muurolol showed a moderate excretion clearance efficacy with a log(CLtot) value of 1.085 mL/min/kg.

2.2.5. Toxicity

The toxicity of the drug compounds was assessed using AMES results, human maximum tolerated dose, oral rat-acute toxicity, hepatotoxicity, skin sensitization, minnow toxicity, and human ether-a-go-go gene (hERG) inhibition. The AMES result indicated that T-muurolol is non-mutagenic and non-carcinogenic (Table 3). The maximum recommended tolerance dose (MRTD) estimates human toxic doses and is considered low if below log 0.477 (mg/kg/day) [17]. So, T-muurolol showed less toxicity, with an MRTD value of log 0.343 mg/kg/day, whereas the reference drugs seemed to be highly toxic. The results also showed that none of the tested compounds are non-hERG inhibitors. However, T-muurolol was predicted to be skin-sensitive, although no hepatotoxic prediction was established. All of the tested compounds showed higher oral rat-acute toxicity (LD_{50}) (Table 3) which indicated less lethality compared to one with a lower LD_{50} value [19]. If a molecule's log LC50 (concentration causing 50% fathead minnow mortality) is below 0.5 mM (log $LC_{50} < -0.3$), it is deemed highly acutely toxic. Current findings suggest that T-muurolol and reference drugs are less toxic, with significantly higher scores than the mentioned LC_{50} threshold (Table 3).

2.3. Drug-Likeness and Bio-Activity Analysis

The drug-likeness property of the compounds is based on Lipinski's rule of five, which encompasses crucial molecular properties that impact a drug's pharmacokinetics in the human body. Current results showed that none of the tested compounds violated Lipinski's parameters, such as molecular weight, number of hydrogen bond donors and acceptors, and octanol-water partition coefficient (Table 4). The topological polar surface area (TPSA) calculation indicates a drug's bioavailability and its hydrogen bonding potential. All tested compounds have a TPSA range of 20.23–38.33 Å (Table 4), significantly lower than the typical range of 160 Å [20]. Our results also indicated an ideal bioavailability score of 0.55 for T-muurolol which will be well-absorbed by the human body [21].

Table 4. Drug likeness properties.

Compounds	Lipinski's Rule of Five	Molecular Weight (g/mol)	H-Bond Acceptors	H-Bond Donors	LogP	TPSA (Å2)	Bioavailability Score
T-muurolol	Yes; zero violation	222.37	1	1	3.42	20.23	0.55
Ibuprofen	Yes; zero violation	206.28	2	1	3.01	37.3	0.85
2-Oxazolidinone	Yes; zero violation	87.08	2	1	−0.04	38.33	0.55

Bioactivity scores for the drug compound can be calculated based on parameters including kinase inhibition, protease inhibition, enzyme activity inhibition, binding to G protein-coupled receptors (GPCRs) and nuclear receptors, and ion channel modulation. The bioactivity score indicated that T-muurolol is highly active toward ion channel modulators, nuclear receptor ligands, and enzyme inhibitors with moderate activity toward GPCR ligands, but at the same time, it is inactive toward kinase inhibitors and protease inhibitors (Table 5).

Table 5. Prediction of bioactivity score of compounds.

Compounds	* GPCRs	ICM	KI	NRL	PI	EI
T-muurolol	−0.09	0.05	−0.87	0.39	−0.63	0.4
Ibuprofen	−0.17	−0.01	−0.72	0.05	−0.21	0.12
2-Oxazolidinone	−3.47	−3.26	−3.67	−3.78	−3.23	−3.4

* GPCRs—G protein-coupled receptors, ICM—ion channel modulator, KI—kinase inhibitor, NRL—nuclear receptor ligand, PI—protease inhibitor, EI—enzyme inhibitor. A molecule with a bioactivity score above 0.00 is likely to exhibit significant biological activities; scores ranging from −0.50 to 0.00 are considered moderately active, while those below −0.50 are presumed inactive.

2.4. MD Simulation Analysis

Structural integrity, stability, compactness protein folding of apo-protein (ligand-free protein), and holo-proteins (proteins with ligands) had been critically analyzed based on some variable characteristics, namely RMSD (root mean square deviation), RMSF (root mean square fluctuation), R_g (radius of gyration), hydrogen bonds (HB) formation, SASA (solvent accessible surface area), and binding free energy decomposition (MM-PBSA), after 100 ns of MD run. The superimposition of T-muurolol on the targeted proteins was shown in Figure 4 to observe the time-dependent protein–drug interaction.

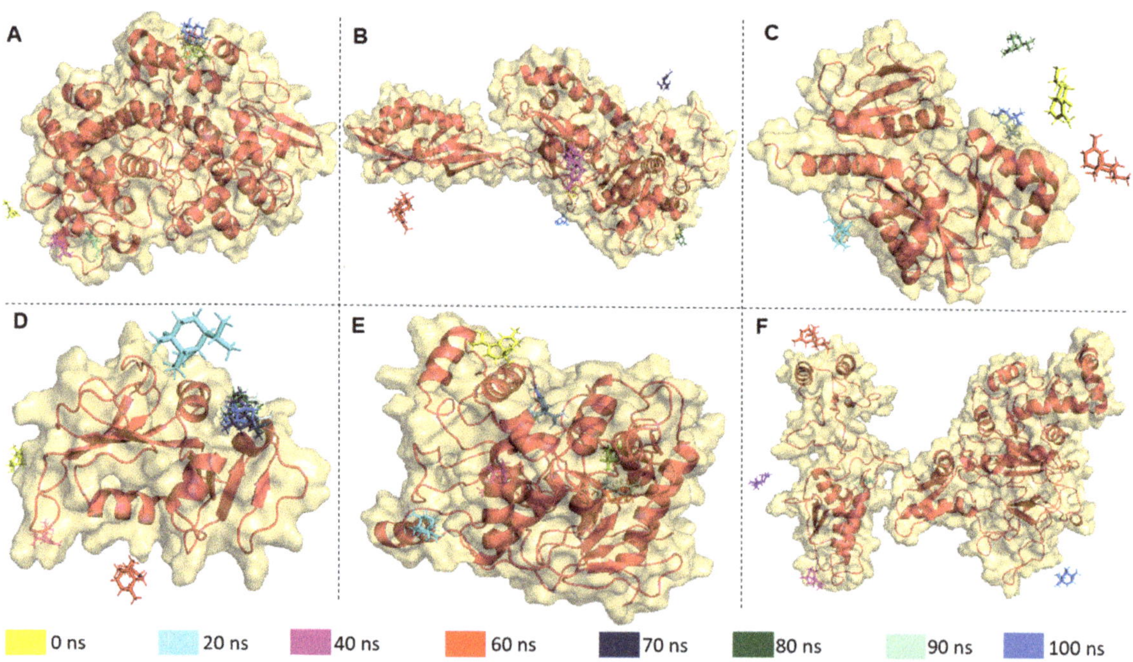

Figure 4. Superimposition of the selected drug T-muurolol on the targeted proteins (**A**) 1CX2 (**B**) 1MWU (**C**) 3N8D (**D**) 3SRW (**E**) 6KSI (**F**) 3J25 in the time-dependent manner of 10/20 ns intervals to observe the concomitant protein-drug interactions.

2.4.1. Monitoring Conformational Stability of the Backbone

The root mean square deviation (RMSD) of the six apo-protein and holo-protein system was illustrated in Figure 5, where the apo-protein and holo-protein system of 1CX2, 1MWU, 3N8D, 3SRW, 6KSI, and 3J25 range in between ~(0.15–0.45) nm (Figure 5A) ~(0.18–0.64) nm (Figure 5B), ~(0.10–0.50) nm (Figure 5C), ~(0.05–0.27) nm (Figure 5D), ~(0.10–0.30) nm (Figure 5E), and ~(0.50–2.0) nm (Figure 6F), with an average apo-proteins' RMSD of ~0.27 nm (1CX2), ~0.34 nm (1MWU), ~0.27 nm (3N8D), ~0.15 nm (3SRW), ~0.18 nm (6KSI), and ~0.93 nm (3J25). In MD simulations, RMSD indicates the conformational stability of the C_α backbone and protein–ligand interactions. As shown in Figure 5A, holo-protein systems 1CX2-ibuprofen and 1CX2-T-muurolol deviated after 10 ns. The RMSD for 1CX2-ibuprofen was more erratic, suggesting greater backbone stability for 1CX2-T-muurolol throughout the MD trajectory. Similarly, in Figure 5E,F, aberrant fluctuations were observed for 6KSI and 3J25 holo-proteins between 0 and 60 ns and 10–100 ns, respectively, compared to their apo-proteins. This indicates that significant ligand–protein interactions can cause instability and deviations, leading to decreased conformational backbone stability. Conversely, Figure 5B,D show that the RMSD trajectories of 1MWU and 3SRW holo-proteins aligned closely with their apo-proteins, indicating that these proteins maintained the backbone stability upon ligand interaction. In Figure 5C, the comparative RMSD analysis reveals that the dip in RMS deviation of 3N8D-T-muurolol beyond that of 3N8D after 40 ns suggests π–π stacking between T-muurolol and 3N8D [22].

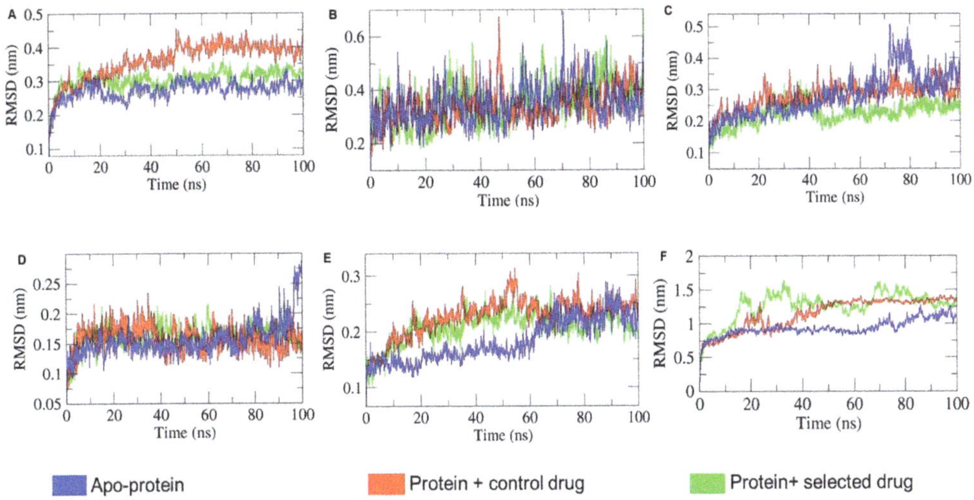

Figure 5. Comparative analysis of root mean square deviation (RMSD) of the consecutive apo-proteins and holo-proteins combinations (proteins with control drug and selected drug) mentioned in different subfigures: (**A**) 1CX2, (**B**) 1MWU, (**C**) 3N8D, (**D**) 3SRW, (**E**) 6KSI, and (**F**) 3J25 combinations.

2.4.2. Residual Flexibility Analysis

The thermodynamic stability of protein–ligand complexes is determined by residual fluctuations (RMSF) of proteins in the presence of ligands. Comparing the RMSF of apo-proteins and holo-proteins characterizes the conformational stability and flexibility of amino acid $C\alpha$ atoms. Larger RMSF peaks in holo-protein residues indicate greater flexibility, while smaller peaks indicate rigidity, reflecting maximal and minimal interactions with their ligands and surrounding polar molecules, respectively [22]. Figure 6 gives us the rationale insights about the comparative analysis of apo-protein and holo-protein specific average RMSF values which ranges as 0.17 ± 0.02 nm for 1CX2, 0.26 ± 0.001 nm for 1MWU, 0.21 ± 0.04 nm for 3N8D, 0.13 ± 0.01 nm for 3SRW, 0.13 ± 0.005 nm for 6KSI, and 0.32 ± 0.24 nm for 3J25. From the average RMSF values of combined apo-protein and holo-protein systems, 3J25 holo-protein showed the largest increase in fluctuation, while 6KSI holo-protein showed the smallest change compared to their apo-proteins. This indicates that 3J25 maintained residual flexibility, whereas 6KSI retained rigidity in the presence of specific ligands. Among the six holo-proteins, 6KSI demonstrated the best interactions with ligands and polar molecules, while 3J25 showed the least interaction, as supported by binding free energy calculations in Table 6 for T-muurolol and protein interactions. Figure 6 shows that higher RMSF values for specific holo-protein residues 80–100 for 1CX2-ibuprofen,

390–410 for 1MWU-T-muurolol, 75–78 for 3N8D-2-Oxazolidinone, 175–240 for 6KSI-2-Oxazolidinone, 400–550 for 3J25-2-Oxazolidinone, and 550–620 for 3J25-T-muurolol indicate that proteins associated with T-muurolol maintained greater residual rigidity compared to their control drug complexes. This suggests that T-muurolol exhibited more stable interactions with all proteins over the 100 ns MD timespan compared to the control drugs.

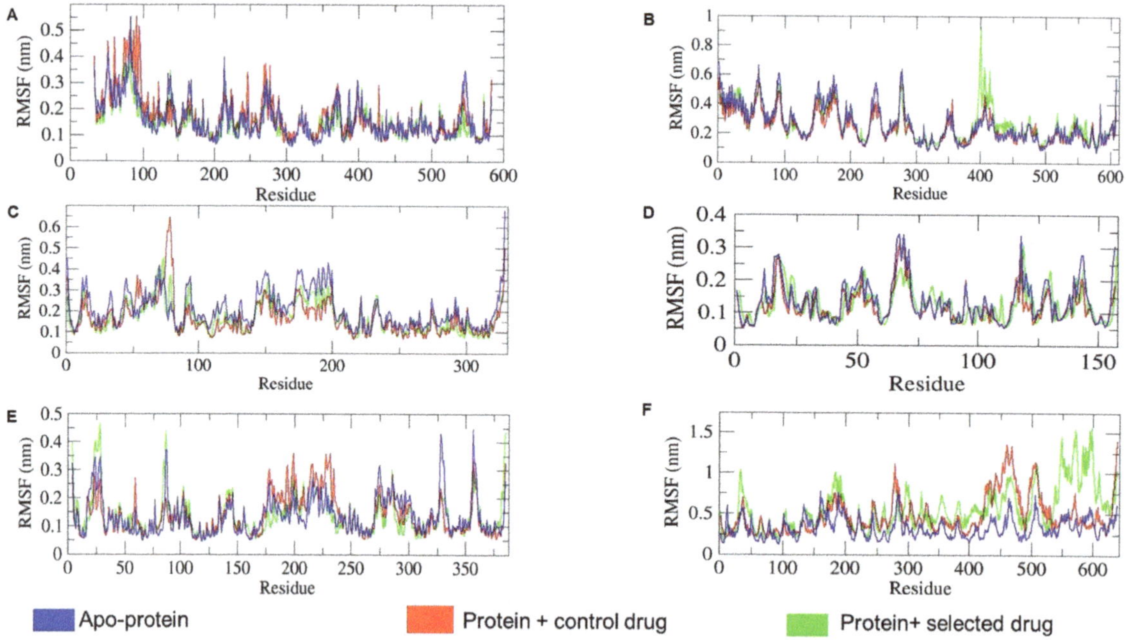

Figure 6. Comparative analysis of root mean square fluctuation (RMSF) values of the consecutive apo-proteins and holo-proteins combinations (proteins with control drug and selected drug) mentioned in different subfigures: (**A**) 1CX2, (**B**) 1MWU, (**C**) 3N8D, (**D**) 3SRW, (**E**) 6KSI, and (**F**) 3J25 combinations.

Table 6. Binding free energy (KJ/mol) calculation of the following proteins based on MD interaction with T-muurolol.

Macromolecules	Van der Waals Energy (KJ/mol)	Electrostatic Energy (KJ/mol)	Polar Solvation Energy (KJ/mol)	SASA Energy (KJ/mol)	Binding Free Energy (KJ/mol)
1CX2	−53.91	−2.68	23.93	−6.82	−39.48
1MWU	−0.05	0.46	−55.05	−0.93	−55.58
3N8D	−12.92	0.58	−6.58	−2.46	−21.39
3SRW	−29.99	−1.68	24.85	−4.21	−11.03
6KSI	−71.82	−4.36	36.27	−10.72	−50.64
3J25	−5.164	−0.14	2.14	−1.84	−5.01

2.4.3. Structural Compactness Analysis

Structural compactness is analyzed by R_g (radius of gyration) values. A steady and less fluctuating R_g value indicates a more compact protein structure. R_g, the distance between the centre of mass of all protein atoms, is crucial for assessing stable protein folding [23]. As shown in Figure 7, owing to the volumetric size of proteins, the R_g values for the apo-protein and holo-protein systems varied as follows: 2.44–2.54 nm for 1CX2, 3.55–3.80 nm for 1MWU, 2.10–2.40 nm for 3N8D, 1.54–1.65 nm for 3SRW, 2.05–2.18 nm for 6KSI, and 2.80–3.70 nm for 3J25. The comparative R_g trajectory of apo-protein and holo-protein systems in Figure 7A, B, D shows that 1CX2, 1MWU, and 3SRW systems consistently maintained steady R_g values with moderate fluctuations. The holo-proteins closely overlapped with their corresponding apo-proteins, indicating that holo-proteins retained compactness similar to apo-proteins despite ligand interactions. Conversely, Figure 7C,E show that 3N8D-T-muurolol and 6KSI-T-muurolol exhibited greater compactness and integrity compared to their apo-protein and protein–control drug counterparts. The R_g trajectories for 3N8D-T-muurolol and 6KSI-T-muurolol were steadier and declined gradually after 40 and 60 ns, respectively, over the 100 ns MD run. Additionally, in Figure 7F, the 3J25 apo-protein showed steady behaviour with moderate R_g fluctuations compared to its holo-proteins. This indicates that 3J25 holo-proteins progressively lost compactness and stability in protein folding after successive ligand interactions.

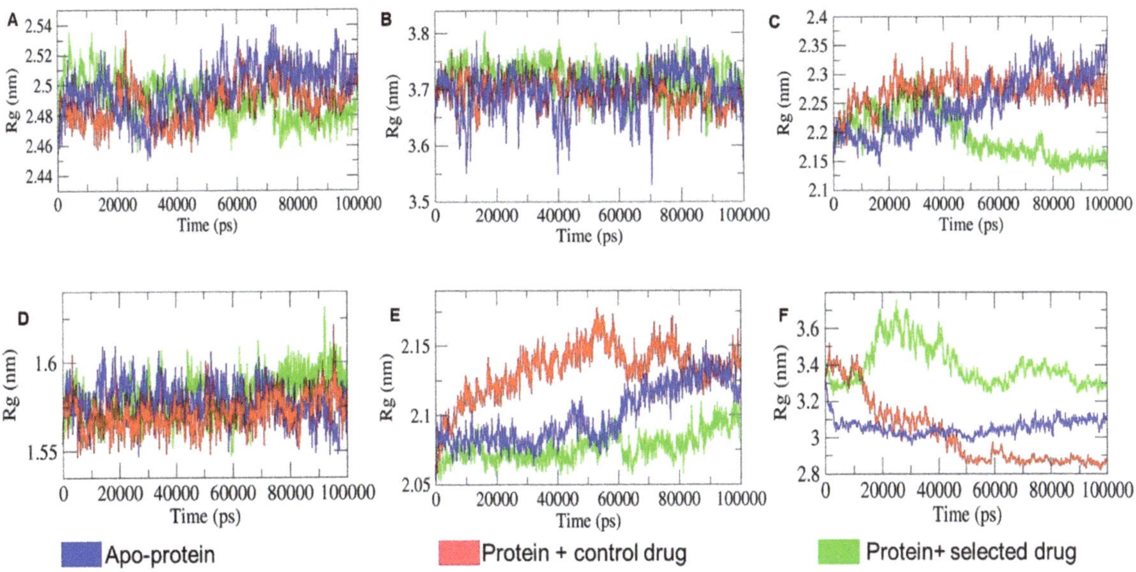

Figure 7. Comparative analysis of radius of gyration (R_g) values of the respective combinations of apo-proteins and holo-proteins and the depicted combinations in different subfigures: (**A**) 1CX2, (**B**) 1MWU, (**C**) 3N8D, (**D**) 3SRW, (**E**) 6KSI, and (**F**) 3J25 combinations.

2.4.4. SASA Landscape Analysis

SASA measures the protein surface area accessible to solvent molecules. With TIP3P water, hydrophilic apo-proteins showed increased SASA, while hydrophobic ones showed decreased SASA due to unfolding and folding. Ligand binding can shield surface residues, altering hydrophilicity and hydrophobicity, thus impacting the SASA and relative size of holo-proteins compared to apo-proteins [24]. As illustrated in Figure 8, SASA varied among apo-protein and holo-protein systems according to protein size: ~(240–280) nm^2 for 1CX2, ~(308–337) nm^2 for 1MWU, ~(160–185) nm^2 for 3N8D, ~(90–100) nm^2 for 3SRW, ~(165–195) nm^2 for 6KSI, and ~(335–420) nm^2 for 3J25. In the 1CX2, 3N8D, and 6KSI systems, SASA of T-muurolol-bound holo-proteins significantly decreased during 50–100 ns, 40–100 ns, and 10–80 ns MD intervals, respectively, compared to their apo-proteins and protein–control drug complexes. This suggests substantial T-muurolol binding to surface residues during these intervals, leading to increased hydrophobicity, size shrinkage, and protein folding for 1CX2-T-muurolol, 3N8D-T-muurolol, and 6KSI-T-muurolol. Conversely, in Figure 8F, 3J25-T-muurolol showed increased SASA, indicating higher hydrophilicity than its apo-protein and protein–control drug complex during 10–85 ns. This suggests that ligand interactions led to greater hydrophilicity, protein unfolding, and relative size expansion for 3J25-T-muurolol compared to 3J25 apo-protein and 3J25-2-Oxazolidinone.

Furthermore, substantial overlapping SASA trajectories and moderate alterations in time-dependent SASA values between apo-protein and holo-protein systems of 1MWU and 3SRW suggest comparatively less significant ligand–protein interactions in these systems. Except for 1MWU holo-proteins, the differing SASA profiles of T-muurolol-protein interaction complexes indicate that hydrophobic and Van Der Waals interactions predominantly contributed to binding affinity in T-muurolol holo-proteins, as supported by the MM-PBSA andscape of T-muurolol–protein interactions in Table 6.

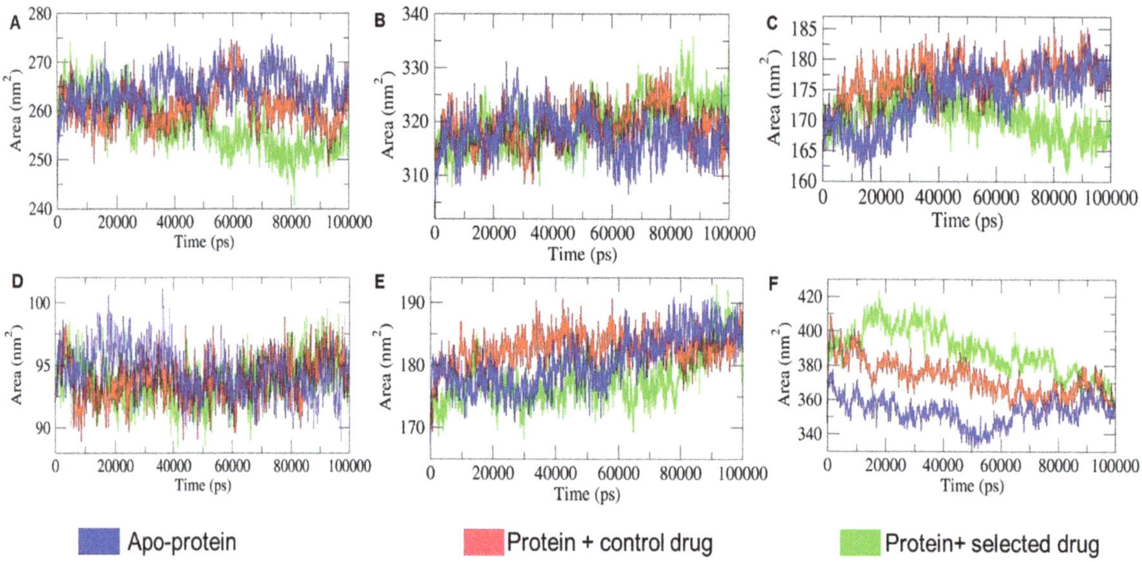

Figure 8. Comparative analysis of SASA (solvent accessible surface area) profiles of respective combinations of apo-proteins and holo-proteins and the depicted combinations in different subfigures: (**A**) 1CX2, (**B**) 1MWU, (**C**) 3N8D, (**D**) 3SRW, (**E**) 6KSI, and (**F**) 3J25 combinations.

2.4.5. H Bonds Formation Analysis

GROningen MOlecular Simulation (GROMACS) analysis predicts ligand–protein complex stability by assessing hydrogen bond formation. The occupancy of H bonds in the interaction complexes of protein–ligands aids in heightening electrostatic energy flow [23]. In Figure 9, 6KSI-T-muurolol consistently formed one intermolecular hydrogen bond throughout the MD interval ~(26–93) ns, indicating strong bond stability. Conversely, 1MWU-T-muurolol displayed intermittent bond formation at ~85 ns, suggesting less stable interactions. In this case, the H bond is not sufficient to serve as significant evidence of a stable binding interaction between T-muurolol and 1MWU. Additionally, 1CX2-T-muurolol showed regular and irregular bond formation patterns at specific intervals, while 3N8D-T-muurolol exhibited intermittent bond formation across multiple MD time spans. 3SRW-T-muurolol maintained regular bond formation, whereas 3J25-T-muurolol formed bonds infrequently at various MD intervals.

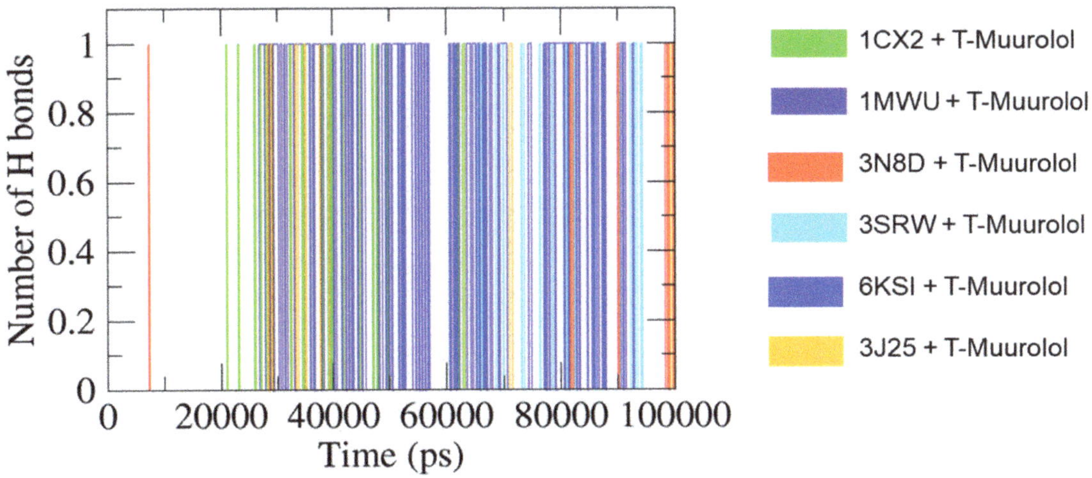

Figure 9. Illustration of the hydrogen bonds' intensity up to MD time span of 100,000 ps (picoseconds) formed between T-muurolol and respective proteins depicted in the figure itself.

2.5. Evaluation of MM-PBSA Binding Free Energy Decomposition

Table 6 presents the binding free energy decomposition for protein–ligand interactions involving T-muurolol with six macromolecules (1CX2, 1MWU, 3N8D, 3SRW, 6KSI, and 3J25). The MM-PBSA binding free energies (ΔG_{BA}) are approximately ~(−39.48) KJ/mol (1CX2), ~(−55.58) KJ/mol (1MWU), ~(−21.39) KJ/mol (3N8D), ~(−11.03) KJ/mol (3SRW), ~(−50.64) KJ/mol (6KSI), and ~(−5.01) KJ/mol (3J25). Van der Waals interactions contributed the most to ΔG_{BA}, indicating that short-range dipole–dipole, permanent, and transient hydrophobic interactions are crucial in the protein–ligand binding stability. In addition to van der Waals energy, electrostatic and polar solvation energies reflect hydrogen bond formation, desolvation, and solvation effects among charged dipoles and polar and non-polar molecules in the MD environment. The protein–ligand complexes are surrounded by the polar solvent TIP3P (water) molecules and ions (K^+ and Cl^-). Thus, the conformational stability and folding of the proteins are heavily influenced by polar solvation and electrostatic energies. Polar solvation impacts the interactions by shielding the proteins' polar hydrophilic residues, decreasing the hydrophobic binding affinity of ligands. This indicates that an increase in polar solvation energy negatively correlates with ligand–protein interactions, as it reduces the effectiveness of other binding energies (SASA, van der Waals, hydrophobic, and electrostatic). Consequently, higher polar solvation energy negatively impacts the overall stability of strong protein–ligand interactions [25]. Strong protein–ligand interactions influenced by polar solvation release positive energy, while SASA and electrostatic energies favour negative energy flow, indicating exothermic binding. Polar solvents like TIP3P enhance SASA interactions for hydrophilic surfaces, promoting protein folding and increasing hydrophobic interactions. Lower SASA in protein–ligand complexes correlates with better binding affinity, strengthening hydrophobic interactions and stabilizing the complex [24,26]. As shown in Table 6, the 1MWU-T-muurolol complex had the highest ΔG_{BA} due to increased polar solvation energy and decreased electrostatic, van der Waals, and hydrophobic energies. This indicates significant polar solvation with TIP3P molecules interacting with surface hydrophilic residues, reducing SASA and other binding energies. Conversely, 6KSI maintained the strongest interaction with a ΔG_{BA} of ~(−50.64) KJ/mol through maximized hydrophobic and electrostatic interactions and minimized polar solvation effects over a 100 ns MD interval. This suggests that higher polar solvation energy and ΔG_{BA} are not definitive indicators of strong protein–ligand interactions [25]. Thus, it is suggested that the combined contributions of van der Waals, electrostatic, SASA, and hydrophobic interactions in enhancing ΔG_{BA} support robust ligand–protein interactions.

2.6. Quantum Chemical Computational Analysis

2.6.1. Frontier Molecular Orbital (FMO) Analysis

Molecular orbital (MO) analysis, especially the examination of frontier molecular orbitals (FMOs) such as the highest occupied molecular orbital (HOMO) and the lowest unoccupied molecular orbital (LUMO), is essential for understanding a molecule's electronic structure and reactivity. FMOs offer valuable insights into molecular interactions and help predict the chemical reactivity and stability of compounds. Figure 10A displays the HOMO and LUMO molecular orbitals for the optimized geometry of T-muurolol. The HOMO and LUMO energies for T-muurolol are −6.37 eV and 0.52 eV, respectively, with an energy gap (ΔE_{Gap}) of 6.9 eV. In the FMO plots, the green and red lobes represent the orbital phases, with one colour indicating a positive phase and the other a negative phase. Hyperconjugation stabilizes molecular orbitals through overlap between an occupied orbital and a neighbouring electron-deficient orbital. As shown in Figure 10A, the HOMO is localized over the double bond region, typical of high electron density in alkenes, with some electron density around the hydroxyl group, indicating potential nucleophilic attack sites. The LUMO, on the other hand, shows electron density around the carbon atoms adjacent to the double bond and the hydroxyl group. The LUMO lobes, like those of the HOMO, also display phases (green and red) but are located in different regions of the molecule. The significant electron density around the double bond in both orbitals highlights this region as crucial for the molecule's reactivity.

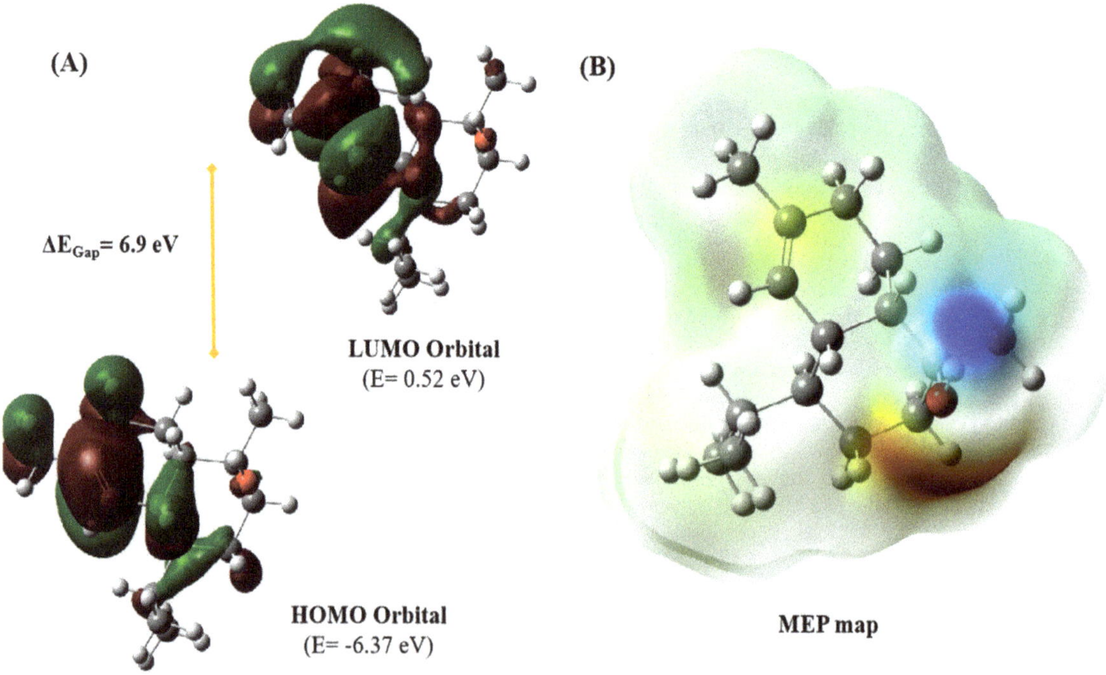

Figure 10. (**A**) HOMO and LUMO molecular orbitals of the T-muurolol molecule, along with their energy gap reported in electron volts (eV) using an iso-surface value of 0.02 electron. (**B**) Molecular electrostatic potential map of T-muurolol molecule (red atom: O; blue atom: N; grey atom: C; white atom: H). These plots are derived from optimizing the T-muurolol molecule using the DFT/B3LYP/6 −311G(d,p) basis set level. Molecular electrostatic potential analysis.

A Molecular electrostatic potential (MEP) map visually represents the charge distribution across a molecule, highlighting positive and negative potential areas. This is useful for predicting molecular interactions, particularly electrostatic interactions. The MEP map of T-muurolol (Figure 10B) emphasizes the hydroxyl group as a key functional site due to its distinct polar regions. The map shows regions of positive potential (blue, 4.528×10^{2} −e) that are likely to attract nucleophiles and areas of negative potential (red, -4.528×10^{-2} e) that are likely to attract electrophiles, with the rest of the molecule being relatively neutral (green). The pronounced blue and red regions around the hydroxyl group indicate significant polarity, making this group in T-muurolol a reactive site for hydrogen bonding and polar interactions. The neutral green regions, covering most of the hydrocarbon skeleton, suggest that these parts of the molecule are less reactive but contribute to its overall shape and steric properties (Figure 10B).

2.6.2. Global Chemical Reactivity Parameters

In computational chemistry, Koopmans' theorem [27] serves as a powerful tool for estimating key electronic properties of molecules, enabling the derivation of essential global reactivity parameters. By applying this theorem, we can derive several crucial parameters that provide quantitative insights into the behaviour of the T-muurolol molecule, including electronegativity ($\chi = -1/2\ (E_{HOMO} + E_{LUMO})$), chemical potential ($\mu = -\chi$), global hardness index ($\eta = 1/2\ (E_{LUMO} - E_{HOMO})$), global softness index ($\sigma = 1/\eta$), and global electrophilicity index ($\omega = \mu^2/2\eta$). All parameters listed above (Table 7) provide a quantitative framework for predicting and rationalizing the chemical behaviour of T-muurolol molecules, aiding in the design and synthesis of new compounds with desired reactivity profiles.

Table 7. Molecular orbital energy and Koopman's parameters for T-muurolol molecule obtained at DFT/B3LYP/6-311G(d,p) basis set level.

List of Parameters	Values Calculated at the DFT Level
Dipole moment (in debye or D)	1.67
E_{HOMO} (in eV)	−6.37
E_{LUMO} (in eV)	0.52
electronegativity index (χ, in eV)	2.92
chemical potential index (μ, in eV)	−2.92
global hardness index (η, in eV)	3.45
softness index (σ, in eV^{-1})	0.29
global electrophilicity index (ω, in eV)	1.24

3. Discussion

T-muurolol is a bioactive sesquiterpene predominantly found as a major compound in the essential oils of several plants, including *Alpinia zerumbet*, *Pulicaria somalensis*, *Calocedrus formosana*, *Calendula officinalis*, and *Chamaecyparis obtusa* with reported antifungal, antibacterial, and antioxidant activities [13–15,28,29]. Sesquiterpenes, due to their lipophilicity, integrate into bacterial lipid membranes and passively diffuse through the bilayer, disrupting membrane integrity. This increases permeability and destabilizes membrane proteins, compromising bacterial function. In Gram-positive bacteria, such as *S. aureus*, the porous peptidoglycan layer allows sesquiterpenes like T-muurolol to diffuse through and penetrate the cytoplasmic membrane, where they inhibit specific bacterial targets [30]. With a documented history of bioactivity, T-muurolol could be an effective agent in inhibiting *S. aureus*-associated infections. For that, it was screened against various drug targets of *S. aureus* to evaluate its efficacy as an alternative antibacterial agent.

T-muurolol exhibited a strong binding interaction with *S. aureus* lipase (SAL) which is a triacylglycerol esterase, a key virulence factor that degrades immune-responsive lipids, inhibits innate immune cell activation, and disrupts host immune recognition [31]. So, SAL is a major drug target for suppressing *S. aureus* pathogenesis. Strong interactions with SAL indicate its potency as an inhibitor. However, few SAL inhibitors have been introduced until now, suggesting that T-muurolol could be an effective agent for targeting SAL in further drug development. Similarly, dihydrofolate reductase, a key protein in the thymidine synthesis pathway essential for DNA synthesis, is crucial for bacterial survival [32]. The strong binding interaction of T-muurolol with this enzyme suggests its efficacy in inhibiting bacterial replication.

MDR *S. aureus* strains are major contributors to *S. aureus*-associated infections, posing significant management challenges and life-threatening risks to humans. T-muurolol has demonstrated strong molecular interactions with various drug targets specific to MDR *S. aureus* strains. For MRSA, T-muurolol targets penicillin-binding protein 2a, thereby inhibiting cell wall biosynthesis. It also exhibits a high binding affinity for D-Ala:D-Ala ligase, a critical drug target in VRSA. Additionally, T-muurolol interacts strongly with the RPP TetM protein in complex with the 70S ribosome, which is implicated in tetracycline resistance in TetRSA by mutating the interaction point of domain IV with the tetracycline binding site [33]. These interactions suggest that T-muurolol can inhibit TetRSA susceptibility by effectively targeting this protein. Furthermore, T-muurolol exhibits effective binding interactions with FAD-dependent NAD(P)H oxidase, which is involved in the ROS mechanism in the human body [34], and cyclooxygenase-2, which catalyzes the biosynthesis of pro-inflammatory prostanoids. These interactions establish its antioxidant and anti-inflammatory activities. These results consequently indicate the potential of T-muurolol as an effective agent for mitigating *S. aureus*-associated inflammation and cellular oxidative stress. Moreover, MD simulation analysis elucidated the stability of T-muurolol interactions with selected proteins. RMSD analysis revealed enhanced backbone stability across all T-muurolol interactions, compared to protein–control drug interactions. Similarly, RMSF analysis indicated stable interaction of T-muurolol with all proteins over the 100 ns MD timespan, surpassing control drugs. Except for the penicillin-binding protein 2a complex, differing SASA profiles of T-muurolol–protein interaction complexes suggested that hydrophobic and van der Waals interactions predominantly contributed to T-muurolol's binding affinity. R_g values of proteins and protein–ligand complexes highlighted increased compactness and structural integrity in the presence of T-muurolol relative to control drugs. Hydrogen bonds remained consistent with docking scores, exhibiting minimal alterations in protein–ligand interaction dynamics. Our simulation study, consistent with the docking results, reveals that the 1MWU and 3SRW complexes exhibit less stable interactions compared to other T-muurolol–protein complexes, with a strong binding affinity driven primarily by hydrophobic and electrostatic interactions, rather than hydrogen bonding. Furthermore, the recalculated binding energies of selected drugs using the MM-PBSA method after a 100 ns MD simulation demonstrated T-muurolol's superior binding energy toward targeted proteins.

Additionally, quantum chemical structure analysis revealed a ΔE_{Gap} of 6.9 eV for T-muurolol, indicating greater stability and lower polarizability, resulting in higher chemical hardness [35]. A higher hardness value suggests lower reactivity [20], indicating the less reactive nature of T-muurolol. However, molecules with smaller HOMO–LUMO energy gaps are more interactive and inclined to participate in chemical events involving bond breaking or formation [36]. T-muurolol, with higher HOMO–LUMO values, signifies less anticipation in bond formations like covalent, hydrogen, and ionic bonds, correlating with molecular interaction studies showing fewer bond formations during protein–ligand interactions. A lower global electrophilicity index suggests a strong nucleophilic molecule, while a higher value indicates a strong electrophilic characteristic [35,37]. T-muurolol exhibits a moderate electrophilic nature with a value of 1.24 eV, indicating moderate facilitation of bond formation with biomolecules. Chemical potential reflects the electron escaping tendency from the molecule, with a lower value implying greater stability, indicating the greater structural stability of T-muurolol. Additionally, MEP analysis highlights the hydroxyl group in T-muurolol as the functional active site for hydrogen bonding and polar interactions.

Based on ADMET pharmacokinetics analysis, T-muurolol exhibits lower water solubility but possesses good absorption and distribution properties within the human body. Toxicity parameter analysis indicates that T-muurolol can be a promising drug candidate for inhibiting *S. aureus* bacterial-specific proteins, with very low toxicity observed. These findings suggest T-muurolol as an antibacterial agent against *S. aureus*, which has the benefit of reducing infection-related inflammation and oxidative stress. Several studies have demonstrated the antibacterial potential of T-muurolol, as a key bioactive compound in essential oils. For instance, Su et al. [38] identified T-muurolol as a major bioactive compound in *Diospyros discolor* essential oil, exhibited strong antibacterial activity against pathogens, including *S. aureus*, with MICs ranging from 31.25 to 500 μg/mL. In *Fraxinus angustifolia* essential oil, T-muurolol was found as an abundant compound and effective against both Gram-positive and Gram-negative bacteria, with MICs between 20 and 300 μg/mL [39]. Additionally, *Hirtellina lobelia* essential oil, rich in T-muurolol, showed activity against MDR bacteria and fungi, with MICs ranging from 4 to 512 μg/mL [40]. These findings underscore T-muurolol as a significant contributor to the antibacterial efficacy of these essential oils, highlighting its potential as a potent antibacterial agent.

Phytochemical-based drug discovery using molecular docking and ADMET analysis provides a cost-effective method for identifying novel compounds and addressing multidrug resistance. However, in silico predictions are limited by structural data and may not fully capture accurate, effective drug concentration, pharmacokinetics, or toxicity. Future research should incorporate deep learning and advanced computational models to improve accuracy. Additionally, further in vitro and in vivo studies will be performed to validate T-muurolol's efficacy against *S. aureus* and its MDR strains and compare them with in silico findings. This study lays the groundwork for subsequent investigations, and the follow-up studies can provide insight into diverse prevention strategies for *S. aureus* infections.

4. Materials and Methods

4.1. Target Protein and Ligand Preparation

For molecular drug targets of *S. aureus*, several pathogeneses causing bacterial-specific proteins were identified, such as V8 protease (PDB ID: 2O8L), signal Transduction Protein Trap (PDB ID: 4AE5), and *S. aureus* lipase (PDB ID: 6KSI). Along with them, several MDR *S. aureus*-specific proteins are also targeted, such as beta-lactamase (PDB ID: 1GHI) and penicillin-binding protein 2a (PDB ID: 1MWU) for MRSA, Tet repressor protein (PDB ID: 2FJ1), RPP TetM in complex with 70S ribosome (PDB ID: 3J25), QacA antiporter protein (PDB ID: 7Y58) for TetRSA, and D-Ala: D-Ala ligase (PDB ID: 3N8D) for VRSA, respectively [41]. Additionally, several universal bacterial proteins were identified, as described by Jianu et al. [32].

Further, to assess the bioactivity of the phytochemicals, we targeted several proteins involved in antioxidant and anti-inflammatory activities. For anti-inflammatory activities, cyclooxygenase-2 (PDB ID: 1CX2), lipoxygenase with protocatechuic acid (PDB ID: 1N8Q), hyaluronidase (PDB ID: 2PE4), lipoxygenase (PDB ID: 3V92), inducible nitric oxide synthase (PDB ID: 4CX7), and 11β-hydroxysteroid dehydrogenase 1 (PDB ID: 4YYZ) proteins were targeted, whereas, for antioxidant activities, human cyclin-dependent kinase 2 complex (PDB ID: 1HCK), FAD-dependent NAD(P)H oxidase (PDB ID: 2CDU), glutathione peroxidase (PDB ID: 2F8A), and superoxide dismutase (PDB ID: 3HFF) were targeted.

The crystal structures of all proteins were retrieved from the RCSB PDB (accessed on March 12, 2024) and were prepared for docking studies by removing unwanted chains and residues using UCSF Chimera 1.16.

For ligand preparation, the 3D SDF structures of T-muurolol, ascorbic acid, ibuprofen, and 2-oxazolidinone were downloaded from PubChem (accessed on March 12, 2024). In this study, 2-oxazolidinone was selected as the reference compound due to its role as the fundamental structural backbone of the oxazolidinone class of antibiotics, which is primarily responsible for the antibacterial efficacy of all derivatives within this group and effective against all MDR *S. aureus*. These compounds were imported into Avogadro 1.2.0

and subjected to energy minimization using the MMFF94 force field with the steepest descent algorithm. The structures were then converted to PDBQT format for docking.

4.2. Molecular Docking and Binding Analysis

AutoDock Tools 1.5.7 (Scripps Research Institute, La Jolla, CA, USA) was used to predict protein–ligand interactions. Water molecules were removed, and polar hydrogens and Kollman charges were added to the protein structures. The protein and ligand files were converted to PDBQT format. Active sites were selected based on blind-docking and enclosed within a 3D affinity grid centred on the active site residues. Docking was performed via command prompt, as described by Gupta et al. [42]. Binding energies were recorded, and initial visualizations were performed with BIOVIA Discovery Studio Visualizer (BIOVIA, San Diego, CA, USA). The follow-up detailed analysis of amino acid and ligand interactions was also performed using the same. Compounds with the best binding affinities were selected for further molecular dynamics simulation.

Furthermore, for cross-checking the binding affinities of the ligands with proteins, docking was performed in the CB-Dock2 online server (https://cadd.labshare.cn/cb-dock2/index.php, accessed on 12 March 2024).

4.3. Pharmacokinetics Prediction and Bio-Activity Analysis

Pharmacokinetics parameters related to absorption, distribution, metabolism, excretion, and toxicity (ADMET) play a substantial role in the detection of novel drug candidates. To predict candidate molecules using in silico methods, the pkCSM (https://biosig.lab.uq.edu.au/pkcsm/prediction, accessed on 12 March 2024) online tool was used. Parameters such as AMES toxicity, maximum tolerated dose (human), hERG I and hERG II inhibitory effects, oral rat acute toxicities, hepatotoxicity, skin sensitization, and fathead minnow toxicity were explored. In addition to these, molecular weight, hydrogen bond acceptor, hydrogen bond donor, number of rotatable bonds, topological polar surface area, octanol/water partition coefficient, and number of violations of Lipinski's rule of five were also surveyed using SwissADME (http://www.swissadme.ch/, accessed on 18 March 2024). The molecular properties and bioactivity score of the ligands were obtained from the molinspiration web server (https://www.molinspiration.com/cgi/properties, accessed on 18 March 2024).

4.4. Molecular Dynamics and Simulation

The six best protein–ligand complexes from the molecular docking study, selected based on the lowest binding energy and optimal docked pose, were chosen for MD simulation. The macromolecular structures (PDB ID_1CX2, 1MWU, 6KSI, and 3N8D) consisted of multiple subunits that were segregated; their additional homodomains had been removed via Discovery Studio Visualizer v20.1.0.19195 (https://www.3ds.com/products/biovia/discovery-studio, accessed on 18 April 2024), followed by the removal of adjacent heteroatoms via similar fashion. Then, the modified structures were further modified to become compatible with in silico virtual screening and molecular dynamic simulation (MDS). A comparison map of the dynamic characteristics of targeted proteins and their protein–ligand complexes using GROMACS 2020.1 according to the described methods by Kandasamy et al. [40]. In this study, the PDB structures of the proteins were transformed to gmx (gromacs) format using the CHARMM36 force field (http://www.charmm-gui.org/, accessed on 18 April 2024). The parameter files for the docked complexes were created using the guidelines outlined in the GROMACS course. The topology and parameter files for protein and ligand were generated on the CHARMM-GUI server, and selected protein–ligand complex files were prepared using the CHARMM-GUI Multicomponent Assembler (MCA). The cube enclosing the system was sized based on 'Calculated Volume Fraction', 'Minimum Recommended Size Length', and 'Maximum Volume Fraction', with a 20 Å buffer to prevent periodic image overlap during MD simulation. Following the final system size determination, the solvent composition was calculated, and the system was solvated with TIP3P water and 150 mM K^+ and Cl^- ions per the Solution Builder protocol. GROMACS files for proteins and protein–ligand complexes were extracted from CHRMM-GUI files for 100 ns (nanoseconds) with 2 femtoseconds (fs) steps.

The analysis of root means square deviation (RMSD), root means square fluctuations (RMSF) of backbone atoms, solvent accessible surface area (SASA), radius of gyration (R_g), and hydrogen bond (HB) formation were analyzed after the successful completion of MD simulation of specific protein–ligand interaction complexes with a time interval of 20 ns throughout the whole 100 ns of MD trajectory. All the graphs (RMSD, RMSF, SASA, R_g, and HB) were plotted by using Qtgrace (https://qtgrace.sourceforge.io/, accessed on 18 April 2024).

4.5. Binding Free Energy Analysis

Apart from in silico docking-based binding affinity analysis between proteins and specific ligands, the strength of protein–ligand interactions was further assessed using MD simulation-specific binding free energy analysis. This was accomplished by calculating the binding free energies with the MM-PBSA (Molecular Mechanics Poisson–Boltzmann Surface Area) method. Using this method, the GROMACS function "g_mmpbsa" captured MD trajectory files of protein–ligand interaction complexes throughout 100 ns at 5 ns intervals by using the following equation:

$$\Delta G_{BA} = \Delta E_{MM} + \Delta G_{PBSA} - T\Delta S_{MM}$$

where ΔG_{BA}, ΔE_{MM}, ΔG_{PBSA}, and $T\Delta S_{MM}$ are denoted as average free energy, average molecular mechanics energy, solvation energy, and solute configuration entropy, respectively [43].

4.6. Computational Chemical Structure Analysis

The chemical structure of the T-muurolol molecule was optimized using the DFT/B3LYP method with the 6-311G(d,p) basis set in the Gaussian-09 programme (Gaussian 09, Revision A.02) [44]. This optimized geometry was then used to generate the frontier molecular orbitals (FMOs) and molecular electrostatic potential (MEP) plots for the T-muurolol molecule. Additionally, the chemical potential of the structure was examined using DFT.

5. Conclusions

The present study explored the potential of T-muurolol as an alternative antibacterial agent against *S. aureus*. Several in silico approaches, such as molecular docking, MD simulation, quantum chemical structure analysis, and pharmacokinetic prediction, have been utilized to screen the efficacy of T-muurolol as a drug. As a result, T-muurolol was identified as a potential antibacterial agent that can inhibit several *S. aureus-associated* bacterial proteins responsible for pathogenesis, bacterial survival, and multi-drug resistance. Along with it, T-muurolol is also identified as an antioxidant and anti-inflammatory agent that signifies its role in mitigating bacterial infection-associated inflammation and cellular oxidative stress in the human body. Further, quantum chemical structure analysis revealed its structural stability and less reactive nature. The results from the ADMET and bioactivity analysis strongly indicate that T-muurolol has the potential to act as a drug and demonstrate significant biological activities. Moreover, this study suggests that T-muurolol has the potential to counteract *S. aureus* infections as an alternative antibacterial solution. However, further detailed in vitro and in vivo experimental validation is required to confirm its activity.

Author Contributions: Conceptualization, S.B.; methodology, S.B., P.K.K., A.D. and N.G.; formal analysis, S.B., P.K.K., A.D., N.G., Z.A.T., L.S., K.Š., M.D. and E.F.-C.; investigation, S.B., P.K.K., A.D., N.G. and Z.A.T.; data curation, S.B., P.K.K., A.D., N.G. and Z.A.T.; writing—original draft preparation, S.B.; writing—review and editing, S.B., P.K.K., A.D., N.G., Z.A.T. and E.F.-C.; visualization, S.B., P.K.K., A.D., N.G. and Z.A.T.; supervision, E.F.-C.; project administration, L.S., K.Š. and M.D.;

funding acquisition, L.S., K.Š. and M.D. All authors have read and agreed to the published version of the manuscript.

Funding: The research was funded by the Faculty of Economics and Management, Czech University of Life Sciences in Prague, grant number 2022B0004.

Institutional Review Board Statement: Not applicable.

Informed Consent Statement: Not applicable.

Data Availability Statement: Data are contained within the article.

Acknowledgments: The authors acknowledge the project "e-Infrastruktura CZ" (e-INFRA ID:90254) provided within the program Projects of Large Research, Development and Innovations Infrastructures in the Czech Republic for supplying computational resources.

Conflicts of Interest: The authors declare no conflicts of interest.

References

1. Pollitt, E.J.G.; Szkuta, P.T.; Burns, N.; Foster, S.J. *Staphylococcus aureus* Infection Dynamics. *PLoS Pathog.* **2018**, *14*, e1007112. [CrossRef]
2. Tong, S.Y.C.; Davis, J.S.; Eichenberger, E.; Holland, T.L.; Fowler, V.G. *Staphylococcus aureus* Infections: Epidemiology, Pathophysiology, Clinical Manifestations, and Management. *Clin. Microbiol. Rev.* **2015**, *28*, 603–661. [CrossRef]
3. Liu, G.Y. Molecular Pathogenesis of *Staphylococcus aureus* Infection. *Pediatr. Res.* **2009**, *65*, 71–77. [CrossRef]
4. Gatadi, S.; Madhavi, Y.V.; Chopra, S.; Nanduri, S. Promising Antibacterial Agents against Multidrug Resistant *Staphylococcus aureus*. *Bioorg. Chem.* **2019**, *92*, 103252. [CrossRef]
5. Drew, R.H. Emerging Options for Treatment of Invasive, Multidrug-Resistant *Staphylococcus aureus* Infections. *Pharmacotherapy* **2007**, *27*, 227–249. [CrossRef]
6. Shoaib, M.; Aqib, A.I.; Muzammil, I.; Majeed, N.; Bhutta, Z.A.; Kulyar, M.F.-A.; Fatima, M.; Zaheer, C.-N.F.; Muneer, A.; Murtaza, M.; et al. MRSA Compendium of Epidemiology, Transmission, Pathophysiology, Treatment, and Prevention within One Health Framework. *Front. Microbiol.* **2023**, *13*, 1067284. [CrossRef]
7. Cosio, T.; Coniglione, F.; Flaminio, V.; Gaziano, R.; Coletta, D.; Petruccelli, R.; Dika, E.; Bianchi, L.; Campione, E. Pyodermitis during nivolumab treatment for non-small cell lung cancer: A case report and review of the literature. *Int. J. Mol. Sci.* **2023**, *24*, 4580. [CrossRef]
8. Borges, A.; Saavedra, M.J.; Simoes, M. Insights on Antimicrobial Resistance, Biofilms and the Use of Phytochemicals as New Antimicrobial Agents. *Curr. Med. Chem.* **2015**, *22*, 2590–2614. [CrossRef]
9. Dhama, K.; Tiwari, R.; Chakraborty, S.; Saminathan, M.; Kumar, A.; Karthik, K.; Wani, M.Y.; Amarpal, A.; Singh, S.; Rahal, A. Evidence Based Antibacterial Potentials of Medicinal Plants and Herbs Countering Bacterial Pathogens Especially in the Era of Emerging Drug Resistance: An Integrated Update. *Int. J. Pharmacol.* **2014**, *10*, 1–43. [CrossRef]
10. Solanki, R.; Jodha, B.; Prabina, K.E.; Aggarwal, N.; Patel, S. Recent Advances in Phytochemical Based Nano-Drug Delivery Systems to Combat Breast Cancer: A Review. *J. Drug Deliv. Sci. Technol.* **2022**, *77*, 103832. [CrossRef]
11. Egbuna, C.; Kumar, S.; Ifemeje, J.C.; Ezzat, S.M.; Kaliyaperumal, S. *Phytochemicals as Lead Compounds for New Drug Discovery*; Elsevier: Amsterdam, The Netherlands, 2019; ISBN 978-0-12-817891-1.
12. Suganya, T.; Packiavathy, I.A.S.V.; Aseervatham, G.S.B.; Carmona, A.; Rashmi, V.; Mariappan, S.; Devi, N.R.; Ananth, D.A. Tackling Multiple-Drug-Resistant Bacteria with Conventional and Complex Phytochemicals. *Front. Cell. Infect. Microbiol.* **2022**, *12*, 883839. [CrossRef] [PubMed]
13. Elzaawely, A.A.; Xuan, T.D.; Koyama, H.; Tawata, S. Antioxidant Activity and Contents of Essential Oil and Phenolic Compounds in Flowers and Seeds of *Alpinia zerumbet* (Pers.) B.L. Burtt. & R.M. Sm. *Food Chem.* **2007**, *104*, 1648–1653. [CrossRef]
14. Cheng, S.-S.; Wu, C.-L.; Chang, H.-T.; Kao, Y.-T.; Chang, S.-T. Antitermitic and Antifungal Activities of Essential Oil of *Calocedrus formosana* Leaf and Its Composition. *J. Chem. Ecol.* **2004**, *30*, 1957–1967. [CrossRef] [PubMed]
15. Sahingil, D. GC/MS-Olfactometric Characterization of the Volatile Compounds, Determination Antimicrobial and Antioxidant Activity of Essential Oil from Flowers of Calendula (*Calendula officinalis* L.). *J. Essent. Oil Bear. Plants* **2019**, *22*, 1571–1580. [CrossRef]
16. Chandra, A.; Chaudhary, M.; Qamar, I.; Singh, N.; Nain, V. In Silico Identification and Validation of Natural Antiviral Compounds as Potential Inhibitors of SARS-CoV-2 Methyltransferase. *J. Biomol. Struct. Dyn.* **2022**, *40*, 6534–6544. [CrossRef] [PubMed]
17. Saha, S.; Nandi, R.; Vishwakarma, P.; Prakash, A.; Kumar, D. Discovering Potential RNA Dependent RNA Polymerase Inhibitors as Prospective Drugs against COVID-19: An In Silico Approach. *Front. Pharmacol.* **2021**, *12*, 634047. [CrossRef]
18. Bhowmik, D.; Nandi, R.; Jagadeesan, R.; Kumar, N.; Prakash, A.; Kumar, D. Identification of Potential Inhibitors against SARS-CoV-2 by Targeting Proteins Responsible for Envelope Formation and Virion Assembly Using Docking Based Virtual Screening, and Pharmacokinetics Approaches. *Infect. Genet. Evol.* **2020**, *84*, 104451. [CrossRef]

19. Sinha, M.; Jagadeesan, R.; Kumar, N.; Saha, S.; Kothandan, G.; Kumar, D. In-Silico Studies on Myo Inositol-1-Phosphate Synthase of *Leishmania donovani* in Search of Anti-Leishmaniasis. *J. Biomol. Struct. Dyn.* **2022**, *40*, 3371–3384. [CrossRef]
20. Abdelrheem, D.A.; Rahman, A.A.; Elsayed, K.N.M.; Abd El-Mageed, H.R.; Mohamed, H.S.; Ahmed, S.A. Isolation, Characterization, In Vitro Anticancer Activity, Dft Calculations, Molecular Docking, Bioactivity Score, Drug-Likeness and Admet Studies of Eight Phytoconstituents from Brown Alga *Sargassum platycarpum*. *J. Biomol. Struct. Dyn.* **2021**, *1225*, 129245. [CrossRef]
21. Martin, Y.C. A Bioavailability Score. *J. Med. Chem.* **2005**, *48*, 3164–3170. [CrossRef]
22. Chaudhary, N.; Aparoy, P. Deciphering the Mechanism behind the Varied Binding Activities of COXIBs through Molecular Dynamic Simulations, MM-PBSA Binding Energy Calculations and per-Residue Energy Decomposition Studies. *J. Biomol. Struct. Dyn.* **2017**, *35*, 868–882. [CrossRef] [PubMed]
23. Choudhary, S.; Kesavan, A.K.; Juneja, V.; Thakur, S. Molecular Modeling, Simulation and Docking of Rv1250 Protein from Mycobacterium Tuberculosis. *Front. Bioinform.* **2023**, *3*, 1125479. [CrossRef]
24. Arooj, M.; Shehadi, I.; Nassab, C.N.; Mohamed, A.A. Physicochemical Stability Study of Protein–Benzoic Acid Complexes Using Molecular Dynamics Simulations. *Amino Acids* **2020**, *52*, 1353–1362. [CrossRef] [PubMed]
25. Naïm, M.; Bhat, S.; Rankin, K.N.; Dennis, S.; Chowdhury, S.F.; Siddiqi, I.; Drabik, P.; Sulea, T.; Bayly, C.I.; Jakalian, A.; et al. Solvated Interaction Energy (SIE) for Scoring Protein−Ligand Binding Affinities. 1. Exploring the Parameter Space. *J. Chem. Inf. Model.* **2007**, *47*, 122–133. [CrossRef] [PubMed]
26. Verma, A.; Kumar, A.; Debnath, M. Molecular Docking and Simulation Studies to Give Insight of Surfactin Amyloid Interaction for Destabilizing Alzheimer's Aβ42 Protofibrils. *Med. Chem. Res.* **2016**, *25*, 1616–1622. [CrossRef]
27. Koopmans, T. Über Die Zuordnung von Wellenfunktionen Und Eigenwerten Zu Den Einzelnen Elektronen Eines Atoms. *Physica* **1934**, *1*, 104–113. [CrossRef]
28. Assaeed, A.; Elshamy, A.; El Gendy, A.E.-N.; Dar, B.; Al-Rowaily, S.; Abd-ElGawad, A. Sesquiterpenes-Rich Essential Oil from Above Ground Parts of *Pulicaria somalensis* Exhibited Antioxidant Activity and Allelopathic Effect on Weeds. *Agronomy* **2020**, *10*, 399. [CrossRef]
29. Górski, K.M.; Kowalczyk, T.; Picot, L.; Rijo, P.; Ghorbanpour, M.; Sitarek, P. The Precious Potential of the Sacred Tree *Chamaecyparis obtusa* (Siebold & Zucc.) Endl. as a Source of Secondary Metabolites with Broad Biological Applications. *Int. J. Mol. Sci.* **2024**, *25*, 2723. [CrossRef]
30. Radulovic, N.S.; Blagojevic, P.D.; Stojanovic-Radic, Z.Z.; Stojanovic, N.M. Antimicrobial plant metabolites: Structural diversity and mechanism of action. *Curr. Med. Chem.* **2013**, *20*, 932–952.
31. Hu, C.; Xiong, N.; Zhang, Y.; Rayner, S.; Chen, S. Functional Characterization of Lipase in the Pathogenesis of *Staphylococcus aureus*. *Biochem. Biophys. Res. Commun.* **2012**, *419*, 617–620. [CrossRef]
32. Jianu, C.; Stoin, D.; Cocan, I.; David, I.; Pop, G.; Lukinich-Gruia, A.T.; Mioc, M.; Mioc, A.; Șoica, C.; Muntean, D.; et al. In silico and in vitro evaluation of the antimicrobial and antioxidant potential of *Mentha × smithiana* R. Graham essential oil from Western Romania. *Foods* **2021**, *10*, 815. [CrossRef] [PubMed]
33. Dönhöfer, A.; Franckenberg, S.; Wickles, S.; Berninghausen, O.; Beckmann, R.; Wilson, D.N. Structural Basis for TetM-Mediated Tetracycline Resistance. *Proc. Natl. Acad. Sci. USA* **2012**, *109*, 16900–16905. [CrossRef]
34. Cipriano, A.; Viviano, M.; Feoli, A.; Milite, C.; Sarno, G.; Castellano, S.; Sbardella, G. NADPH Oxidases: From Molecular Mechanisms to Current Inhibitors. *J. Med. Chem.* **2023**, *66*, 11632–11655. [CrossRef] [PubMed]
35. Pearson, R.G. Chemical Hardness and Density Functional Theory. *J. Chem. Sci.* **2005**, *117*, 369–377. [CrossRef]
36. Klingmüller, U.; Schilling, M.; Depner, S.; D'Alessandro, L. Biological Foundations of Signal Transduction, Systems Biology and Aberrations in Disease. In *Computational Systems Biology: From Molecular Mechanisms to Disease*, 2nd ed.; Academic Press: Amsterdam, The Netherlands, 2014; pp. 45–64. ISBN 978-0-12-405926-9.
37. Haroon, M.; Akhtar, T.; Shaikh, Q.; Mehmood, H.; Khalid, M.; Asghar, M.A.; Alshehri, S.M.; Ojha, S.C. Facile Synthesis and DFT Analysis of Novel Thiazole-Based Hydrazones: An Experimental and Theoretical Perspective. *ACS Omega* **2023**, *8*, 27488–27499. [CrossRef] [PubMed]
38. Su, Y.C.; Hsu, K.P.; Wang, E.I.C.; Ho, C.L. Composition, in vitro cytotoxic, and antimicrobial activities of the flower essential oil of *Diospyros discolor* from Taiwan. *Nat. Prod. Commun.* **2015**, *10*, 1934578X1501000744. [CrossRef]
39. Zouaghi, N.; Bensiradj, N.H.; Cavaleiro, C.; Nadjemi, B.; Trari, M. Phytochemical study and antibacterial effects of *Fraxinus angustifolia* Vahl (Algeria): Experimental and computational investigations. *Waste Biomass Valor.* **2021**, *12*, 3605–3616. [CrossRef]
40. Khoury, M.; El Beyrouthy, M.; Ouaini, N.; Eparvier, V.; Stien, D. *Hirtellina lobelii* DC. essential oil, its constituents, its combination with antimicrobial drugs and its mode of action. *Fitoterapia* **2019**, *133*, 130–136. [CrossRef]
41. Bhattacharya, S.; Dutta, A.; Khanra, P.K.; Gupta, N.; Dutta, R.; Tzvetkov, N.T.; Milella, L.; Ponticelli, M. In silico exploration of 4 (α-l-rhamnosyloxy)-benzyl isothiocyanate: A promising phytochemical-based drug discovery approach for combating multi-drug resistant *Staphylococcus aureus*. *Comput. Biol. Med.* **2024**, *179*, 108907. [CrossRef]
42. Gupta, N.; Bhattacharya, S.; Urbanová, K.; Dutta, A.; Hazra, A.K.; Fernández-Cusimamani, E.; Leuner, O. Systematic Analysis of Antimicrobial Activity, Phytochemistry, and in Silico Molecular Interaction of Selected Essential Oils and Their Formulations from Different Indian Spices against Foodborne Bacteria. *Heliyon* **2023**, *9*, e22480. [CrossRef]

43. Kandasamy, T.; Sen, P.; Ghosh, S.S. Multi-Targeted Drug Repurposing Approach for Breast Cancer via Integrated Functional Network Analysis. *Mol. Inform.* **2022**, *41*, 2100300. [CrossRef] [PubMed]
44. Gaussian 09 Citation. Available online: https://gaussian.com/g09citation/ (accessed on 28 May 2024).

International Journal of
Molecular Sciences

MDPI

Review

Stilbenes from Vine Extracts: Therapeutic Potential and Mechanisms

Luís P. Brás [1,2], Ângelo Luís [1], Gregory Chatel [2,*], Sílvia Socorro [1] and Ana Paula Duarte [1,*]

1 RISE-Health, Department of Medical Sciences, Faculty of Health Sciences, University of Beira Interior, Av. Infante D. Henrique, 6200-506 Covilhã, Portugal; luis.bras@ubi.pt (L.P.B.); angelo.luis@ubi.pt (Â.L.); ssocorro@fcsaude.ubi.pt (S.S.)
2 EDYTEM, CNRS, University Savoie Mont Blanc, F-73000 Chambéry, France
* Correspondence: gregory.chatel@univ-smb.fr (G.C.); apduarte@ubi.pt (A.P.D.)

Abstract

The wine industry represents a significant economic sector; however, it generates large volumes of waste that can be valorized due to the presence of bioactive compounds, particularly stilbenes. These naturally occurring stilbenes exhibit remarkable potential in the prevention and treatment of various diseases, including cardioprotection, neuroprotection, antidiabetic properties, anti-inflammatory activity, and cancer prevention and therapy. This review discusses biosynthesis, structures, extraction methods, and mechanisms of action of stilbenes, with a particular emphasis on cancer prevention and treatment. Evidence from in vitro, in vivo, and clinical studies demonstrate that stilbenes modulate multiple molecular pathways by promoting apoptosis, inhibiting cell proliferation, and regulating inflammation, oxidative stress, and metabolism. However, the clinical application of stilbenes is limited by their low bioavailability. To overcome this, pharmaceutical formulations have been developed to enhance their stability and bioavailability, reduce side effects, and improve target interactions. These advances are expected to increase the therapeutic efficacy of stilbenes. Furthermore, information on the health benefits of less common stilbenes remains limited, highlighting the need for further research on these compounds.

Keywords: waste viticulture; extraction; stilbenes; biological activities

1. Introduction

Viticulture is one of the most economically and culturally significant agricultural sectors worldwide. Wine is among the most widely produced and consumed beverages, with 94% of global production concentrated in just 29 countries [1]. Global consumption has reached approximately 221 million hectolitres, according to the International Organisation of Vine and Wine (OIV) [2]. Alongside this economic value, the wine industry generates large volumes of waste and byproducts estimated at 20–30% of the initial grape mass during winemaking [3,4].

These residues arise at two key stages: during harvesting (solid waste) and during processing (liquid waste) [5]. Solid waste includes grape pomace (45%), grape stalks (7.5%), grape pips (6%), grape marc (30%), and yeast (41.5%) [6–8]. Yeast lees alone account for ~5% of grape weight and contain ethanol, tartaric acid, phenolic compounds, and microbial biomass [9]. The fermentation and pressing steps contribute to an additional 20–25% of total waste [10].

In addition to winery waste, vineyard maintenance also contributes significantly to biomass generation. Annual pruning, essential for balanced plant development, produces an estimated 6–18 million tons of biomass each year, including shoots, branches, and leaves [11,12]. This accumulation results from vineyard renewal aimed at ensuring balanced growth. However, this material is often underutilized despite its potential for bio-based valorization. The biomass derived from vine waste and grape byproducts contains bioactive molecules important for human health [13].

While viticulture brings substantial economic benefit to many countries [14], the growth in grape production has led to increased vineyard surface area and value, along with a rise in the accumulation of both organic and inorganic waste [15,16]. This poses challenges to both economic and ecological management.

From an economic standpoint, inefficiencies persist, i.e., energy from renewable sources is poorly recycled, large quantities of waste are generated, and passive architectural strategies to optimize space and energy efficiency are underused [15–17]. Ecologically, the accumulation of waste increases the sector's environmental footprint, which can be quantified using life cycle assessment (LCA) methodologies [18,19].

Despite these constraints, grapevines and their byproducts are rich in bioactive phytochemicals of potential value to the food, pharmaceutical, and cosmetics industries. These include dietary fiber, phenolics, proteins, and lipids [9,20–26], as well as stilbenes, which show promise as natural additives, antimicrobial and antioxidant agents, and even fillers in sustainable food packaging materials [15,27–29]. Leaves, in particular, contain organic acids, lipids, and polyphenols, and are already exploited in cosmetics [15,30]. Additionally, grapevine residues can be valorized as soil fertilizers or for bioenergy production [31].

The profile of polyphenols present in grape-based materials is subject to considerable fluctuation, reflecting not only genetic diversity and environmental variation, but also the interplay of cultivation and processing strategies. Each step from grape ripening to yeast selection shapes the ultimate spectrum of these compounds, with maceration time standing out for its impact on concentration [16,19]. Despite being a rich reservoir of bioactives like flavonols and anthocyanins, the grape skin's direct contribution to wine composition is intentionally moderated; winemakers routinely remove these skins early to balance sensory qualities, inadvertently discarding a source of potential nutritional and therapeutic interest. This reality raises the question of how conventional practices might be optimized to both reserve product appeal and maximize the exploitation of health-promoting polyphenols, a challenge still largely unaddressed in mainstream oenology [28].

Grape seeds also contain high concentrations of antioxidants, including fiber, proteins, carbohydrates, lipids, minerals, phytosterols, phenolic compounds, vitamin E, and melatonin [15,32]. These polyphenols display activities relevant to stress response and chronic diseases. Notably, they intervene in cancer-related mechanisms such as tumor cell proliferation, metastasis, and drug resistance [33,34].

Among polyphenols, stilbenes are particularly interesting. As phytoalexins, they protect grapevines from pathogens and are mainly located in woody tissues like pruning canes and stalks [35]. The best known stilbene, trans-resveratrol (5-[(1E)-2-(4-hydroxyphenyl)ethenyl]-1,3-benzenediol; CAS: 501-36-0), has been extensively studied for its antioxidant, cardioprotective, and anticancer properties [36]. Stilbenes are ideal candidates for high-value valorization in the field of cancer prevention and anticancer therapy since they are found in pruning waste [15].

This review explores the therapeutic potential of vine-derived stilbenes particularly those found in pruning residues as anticancer agents. It covers their biosynthesis, extraction, chemical structures, biological activities, and mechanisms of action, while also addressing current limitations and future perspectives for their medical application.

2. Stilbenes from Grapevine: Biosynthesis, Structures, and Extraction

2.1. Polyphenols in Grapevine: General Classification and Biological Roles

Polyphenols, fundamentally, are the product of plant metabolism resulting in molecules that incorporate multiple phenolic motifs [37]. The chemical identity of a phenol, a compound consisting of an aromatic ring bearing at least one hydroxyl group, serves as the foundational structure for this diverse family. However, reducing these compounds to a structural formula overlooks the evolutionary rationale behind their biosynthesis, as well as their often-underestimated functional relevance in plant defense and, potentially, in human health [38].

The sheer heterogeneity of polyphenols, manifesting across grapevine tissues from root to leaf, defies easy classification. While chemists typically subdivide them by structure, highlighting families such as stilbenes or flavonoids, this approach captures only part of a dynamic and context-dependent reality [39,40]. Grape-derived plant material alone is home to a remarkable spectrum of compounds, containing not only well-characterized subclasses but also a long list of less-understood entities. This taxonomic richness, while scientifically intriguing, complicates efforts to standardize bioactive ingredient extraction for practical use [35].

Flavonoids are a class of polyphenols and are defined by a common structure composed of two aromatic rings joined by an oxygenated heterocyclic ring. Based on this core, they are subdivided into nine subclasses: chalcones, aurones, flavones, isoflavones, flavanols, anthocyanins, isoflavonoids, and bioflavonoids. Additionally, polyphenols can form associations with various carbohydrates and organic acids, contributing to their solubility, bioactivity, and interaction with other biomolecules [41].

Natural polyphenols function as secondary metabolites, enabling plants to respond to environmental stressors. These include ultraviolet (UV) radiation, pathogen attacks, nutrient-deficient soils, temperature extremes, drought conditions and herbivory [39,42,43]. As noted by Sies, polyphenol-rich plants and spices, especially those containing flavonoids, have been used for millennia in traditional Eastern medicine, although they are less common in Western therapeutic practices [44].

The extensive scientific literature now demonstrates that dietary polyphenols may confer health benefits across multiple biological systems. Their activities have been linked to protective effects against cancer, neurodegenerative diseases, cardiovascular dysfunctions, metabolic syndrome, diabetes, aging, and chronic inflammation [44–63].

2.2. Focus on Stilbenes: Structure, Diversity, Biosynthesis, and Natural Functions

Stilbenes are a class of polyphenols formed through the condensation of three C2 carbon residues with an activated hydroxycinnamic acid, a mechanism similar to that observed in flavonoid biosynthesis. Structurally, they consist of two aromatic rings linked by an ethylene bridge, forming the core diphenylethylene structure (C6-C2-C6) [64,65]. These compounds exist in both trans and cis isomeric forms, with the trans configuration being more stable and commonly found in nature [64–66].

According to the study by Goufo et al., stilbenes can be categorized based on polymerization degree, including monomers, dimers, trimers, tetramers, pentamers, and hexamers [35]. The most widely studied stilbenes such as resveratrol, pterostilbene, and piceatannol exhibit a range of biological properties including antioxidant, anti-inflammatory, and anticancer activities [67–69]. Pterostilbene additionally shows analgesic properties, while resveratrol has demonstrated protective effects against atherosclerosis [67,68].

In grapevine, stilbene production is notably induced by mechanical damage, such as pruning of fresh grapevine canes, which stimulates the biosynthesis of compounds like resveratrol and piceatannol [70]. In a study by Guerrero et al., ε-viniferin was identified as the predominant stilbene (26–52%) in one-year-old grapevine canes, followed by other compounds such as trans-resveratrol, piceatannol, pinosylvin, rhapontigenin, pterostilbene, and isorhapontigenin, whose levels varied over time (Figure 1) [71]. These molecules vary structurally, and this variation determines their chemical behavior and bioactivity. For instance, hydroxylated and glycosylated stilbenes have improved water solubility, enhancing incorporation into hydrophilic systems. Methoxylated stilbenes, by contrast, are more lipophilic and bioavailable. Glycosylated forms tend to be less stable and more prone to isomerization than their aglycone counterparts. Ortho-hydroxylated stilbenes, due to the ability to form stable semiquinone radicals, display greater antioxidant, anti-inflammatory, and anticancer effects compared to meta-hydroxylated isomers [72].

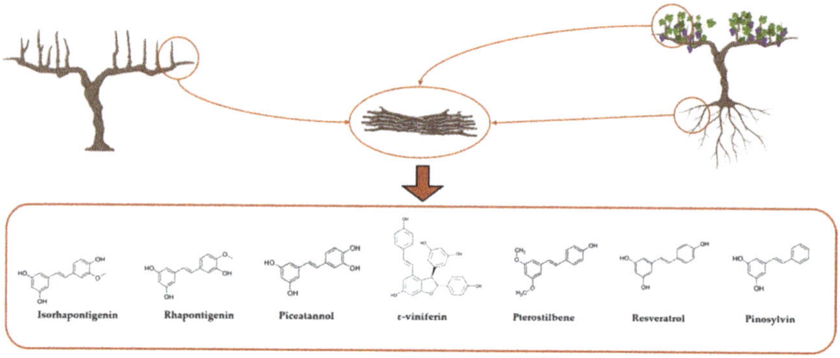

Figure 1. Schematic of the grapevine: stilbenes are primarily located in the wood, roots, branches, and stems, where they help prevent wood degradation and play defensive roles. These molecules exhibit structural variability, which determines their chemical behavior and bioactivity.

Stilbenes occupy a dual role within the plant kingdom, acting both as an ever-present line of defense and as rapid responders to environmental threat. Their involvement in shielding plants from infectious agents or harsh climatic conditions illustrates the evolutionary value of chemical versatility [71,73–75]. Not uniquely confined to grapevines, these compounds are distributed across a wide botanical landscape, from staple crops to regional medicinal herbs. The notable presence of pterostilbene in berries and residues underlines the untapped reservoir of such molecules in both dietary and agricultural contexts. This cross-species distribution implicates stilbenes as a convergent solution to common biological challenges but also complicates efforts to establish source-specific therapeutic claims [71,76].

The biosynthesis of stilbenes is catalyzed by stilbene synthase (STS), an enzyme that evolved from chalcone synthases (CHS) through convergent evolution. This enzyme family displays tissue-specific and development-specific expression patterns. For example, STS gene expression is lower in young grapevine leaves but increases significantly in mature leaves and grape skins during ripening especially in varieties such as Cabernet Sauvignon and Norton peaking at harvest time (Figure 2) [73,77].

Grapevine byproducts like pomace and winemaking residues are rich in polyphenols, including stilbenes [78]. In viticultural waste, total stilbene concentrations range from 2400 to 5800 mg/kg dry weight (DW) [71]. Cebrián et al. observed the accumulation of phenolic compounds in grapevine buds after 1, 3, and 6 months of post-pruning storage in two Vitis vinifera varieties [79]. In grapevine canes, trans-resveratrol levels range from 441 to 7532 mg/kg DW, while trans-ε-viniferin concentrations vary from 1218 to 5341 mg/kg DW, depending on cultivar, vintage, and storage conditions.

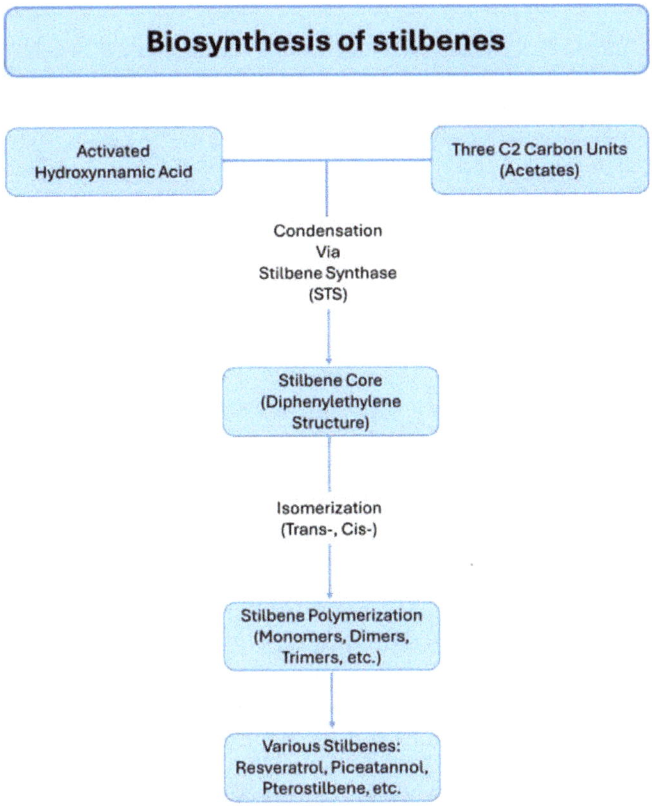

Figure 2. Flow diagram of stilbenes' biosynthesis.

Dormancy in grapevine, including endodormancy and ecodormancy, which occur during cold climates, significantly influences the biochemical composition of canes [80]. Hybrid cultivars adapted to northern latitudes, such as those grown in Estonia, have been found to accumulate high levels of dietary stilbenes like resveratrol and viniferin [81,82].

Grapevine canes, rich in bioactive stilbenes, demonstrate notable antifungal and antioxidant properties, making them valuable for nutraceutical applications [71]. Bud extracts contain up to 29% stilbenes and their incorporation in winemaking presents an eco-friendly strategy to reduce sulfur dioxide (SO_2) use, particularly in white wines [83]. Toasted grapevine canes also enhance the antioxidant properties of wine by contributing high levels of prodelphinidins and stilbenes [84].

Furthermore, grapevine leaves and cane extracts are a source of bioactive antioxidant molecules and other valuable metabolites for pharmaceutical, nutraceutical, and food applications [81,85]. The abundance of raw materials from viticultural waste provides a compelling opportunity for industrial valorization, given the commercial and pharmacological potential of stilbenes [86,87].

2.3. Extraction Methods of Polyphenols and Stilbenes

The pursuit of efficient polyphenol extraction continues to evolve, reflecting the chemical diversity and sensitivity of these compounds. While early protocols relied on harsh conditions and extended processing times, contemporary approaches are trending toward gentler and rapid techniques, such as microwave-assisted extraction (MAE), ultrasound-assisted extraction (UAE), and ultrahigh-pressure extraction, which preserve and selectively recover targets from complex matrices. A critical, though often underestimated, factor in yield and purity is the nature of the solvent system that balances polarity to suit the solubility profile of the molecules of interest. Complexities such as acid–base stability and matrix interference persist, challenging researchers to continually fine-tune protocols for industrial and pharmacological scalability [88].

Conventional methods for extracting stilbenes particularly trans-resveratrol and trans-ε-viniferin from grape canes primarily rely on conventional solid–liquid extraction (CSLE). This involves maceration in a solvent, with or without stirring, under atmospheric pressure and temperatures ranging from ambient to reflux conditions. Extraction efficiency is influenced by multiple factors, including the solvent type, solvent-to-solid ratio, temperature, extraction time, pH, and light exposure. Hydroalcoholic mixtures such as ethanol/water are commonly used due to their efficacy and non-toxicity [89]. Angel et al. showed that ethanol/water mixtures outperform other solvents in yielding stilbenes [90,91]. An increase in the solvent-to-solid ratio generally improves yield, with optimal ratios typically between 5 and 50 mL/g. Higher temperatures also enhance extraction, with optimal results achieved between 20 °C and 80 °C without degrading the compounds. Prolonged extraction times may increase concentration but can also cause compound degradation. Additionally, alkaline conditions and UV light exposure negatively affect stilbene stability.

Emerging techniques such as UAE, MAE, and pressurized solvent extraction (PSE) offer several advantages over conventional methods. UAE employs acoustic cavitation generated by ultrasonic waves to enhance mass transfer, disrupt plant tissues, and increase surface area, thus improving yields and reducing extraction time and energy use [92]. Despite its benefits, UAE requires careful optimization of parameters like frequency, power, and temperature. MAE, on the other hand, uses microwave energy to heat solvents, promoting efficient cell disruption and reducing extraction times. This method enables uniform heating, rapid energy transfer, and lower solvent consumption. Compared to conventional heating, MAE reduces energy waste and enhances extraction of bioactive molecules in an environmentally friendly manner [92–96]. Optimized MAE conditions have been shown to increase stilbene recovery, though comparisons across studies are often difficult due to variability in extraction parameters [33,89]. PSE employs pressures between 10 and 15 MPa to maintain solvents in a liquid state above their boiling points, improving solubility and mass transfer. For stilbenes, ethanol/water (25:75, v/v) at a 30 mL/g ratio, 105 °C, 1 mL/minute flow rate, and 5.2 MPa pressure has demonstrated higher diffusivity and shorter extraction times compared to conventional methods. One study reduced extraction time from 8 h (Soxhlet) to 10 min using PSE, with slightly increased yields [89].

A key trend in extraction science is the shift towards processes that minimize time and energy consumption without compromising yield. As demonstrated by recent comparative studies, techniques leveraging microwave or pressure are especially effective, and may even obviate the need for added solvents when the inherent moisture of the sample is sufficient. Despite universal gains in speed and ecological footprint, not all advanced workflows are equally user-friendly; fine-tuning parameters, particularly for ultrasound-based approaches, remain a nuanced, trial-and-error exercise. This reality tempers the initial enthusiasm for high-tech solutions with the need for context-sensitive optimization in practical settings [36].

Extraction yields are affected by grapevine variety and environmental factors such as UV exposure, soil type, and vineyard practices like leaf removal or fertilization. Even within the same variety, yields may vary due to biological and climatic influences. Post-pruning parameters such as storage duration, cane provenance, and physiological stage also play a significant role in extraction optimization [89].

Trans-resveratrol and trans-ε-viniferin are the dominant stilbenes in grape canes, but other oligomeric stilbenes such as hopeaphenol, ampelopsins, vitisins, pterostilbene, piceatannol, and polydatin are also present and bioactive [97–99]. Nonetheless, stilbenes account for only 3.1% of the dry mass of grape canes, meaning over 95% of the biomass remains underutilized after extraction [89].

3. Biological Activities of Stilbenes and Health Implications

3.1. Bioavailability and Metabolism of Stilbenes

Despite their reputation for limited absorption, some stilbenes and their derivatives have been documented to attain concentrations in vivo sufficient for biological relevance. The extent to which this translates into tangible physiological effects, however, is still fiercely debated, and often contingent on the nuances of both compound structure and formulation [100]. However, the most researched stilbene, resveratrol, has a short half-life because to its quick metabolism (14 min), limited oral bioavailability (20–30%), and low water solubility (<0.05 mg/mL) [101,102]. Oral absorption in humans is around 75%, but extensive intestinal and hepatic first-pass metabolism (mainly glucuronidation and sulfation) reduces systemic availability [103]. Stilbenes' complexation with cyclodextrins improves the phenolic compounds' solubility but not their bioavailability within target tissues [104]. However, other formulation strategies such as incorporation into bile acid complexes, liposomes, nanoparticles, or co-administration with absorption enhancers like piperine can significantly increase plasma concentrations and tissue delivery [105–110].

Pterostilbene exhibits superior oral bioavailability (~80%), attributed to its increased lipophilicity (due to two methoxy groups) and higher metabolic stability (one free hydroxyl), resulting in slower glucuronidation and a longer half-life compared to resveratrol (1.73 vs. 1.48 h) [111–116]. Gnetol possesses lower bioavailability (~6.6%) but a longer half-life (4.2 h) and sustains serum levels of its glucuronide metabolite, likely supporting pharmacological activity [117,118]. In contrast, ε-viniferin and other stilbenes (piceatannol, pinostilbene) display very low oral bioavailability (<1%, or 0.8% for ε-viniferin), attributed to poor absorption and/or extensive metabolism [119].

Stilbenes are mainly metabolized via phase II reactions (glucuronidation, sulfation) at phenolic groups [120–127]. In human plasma, phase II conjugates predominate (resveratrol-3-sulfate being the most abundant), whereas in animal models, glucuronide forms are often higher. Some metabolites, especially conjugated forms, may serve as reservoirs, being deconjugated to release active stilbene, and enterohepatic recirculation has been observed. Dihydro-resveratrol, formed by intestinal hydrogenation, retains relevant biological activity, particularly in the colon [128,129].

Tissue distribution studies show that pterostilbene and its sulfate accumulate in tissues, particularly in the central nervous system, unlike resveratrol, whose plasma and tissue levels decrease quickly. Piceatannol and pinostilbene are efficiently glucuronidated and show high hepatic concentrations but low systemic exposure and rapid clearance [130–134]. Piceatannol can also serve as a metabolic product of resveratrol via CYP1B1, suggesting a prodrug relationship [135].

Nanotechnological strategies including PEGylated liposomes and lipid nanocapsules can significantly enhance brain delivery and systemic concentrations of stilbenes, addressing issues of chemical instability and photosensitivity [78,105] and novel soluble galenic formulations also markedly increase resveratrol's bioavailability and biological effects. In clinical settings, stilbenes are generally well tolerated even at high doses; adverse effects are mild and mostly gastrointestinal, and doses above 1 g/day are avoided due to tolerability limits [136–140].

Metabolite profiles are complex, including a wide array of glucuronide and sulfate forms, as well as hydrogenated and methylated derivatives [120–127,133]. For some stilbenes, tissue- or adipose-localized conjugates may be reconverted in situ, maintaining local activity. Excretion occurs mainly via feces rather than urine, as seen for ε-viniferin. Detection of certain stilbenes (e.g., ε-viniferin) in rat brain tissue indicates the ability to cross the blood–brain barrier, suggesting potential for central nervous system effects [141,142].

In summary, stilbenes demonstrate varied pharmacokinetic profiles, mainly limited by low bioavailability and rapid metabolism. Pterostilbene stands out for its higher oral bioavailability, tissue accumulation, and longer half-life, whereas other stilbenes may benefit from novel delivery systems to enhance clinical efficacy and expand therapeutic potential.

3.2. Antioxidant and Anti-Inflammatory Properties

The preventive effects of stilbenes on various diseases are largely attributed to their antioxidant activities, including anti-cyclooxygenase action and modulation of lipid and lipoprotein metabolism (Figure 3) [77]. Stilbenes counter oxidative stress by inhibiting reactive oxygen and nitrogen species (ROS and RNS), a property influenced by hydroxyl group position and number. Ortho-dihydroxylated compounds, like piceatannol, are more potent due to enhanced stabilization of semiquinone radicals, particularly at the R4's position of resveratrol. Their antioxidant activity is also modulated by pH and protonation states. While direct radical scavenging is limited in vivo, stilbenes regulate antioxidant enzymes such as catalase, superoxide dimustase (SOD), nicotinamide adenine dinucleotide phosphate (NADPH) oxidase, glutathione peroxidase (GPx), glutathione S-transferase (GST), and NQO. For instance, resveratrol upregulates catalase, GPx1, SOD1, and SOD3, while suppressing NOX2 and NOX4 in ApoE-KO mice [82].

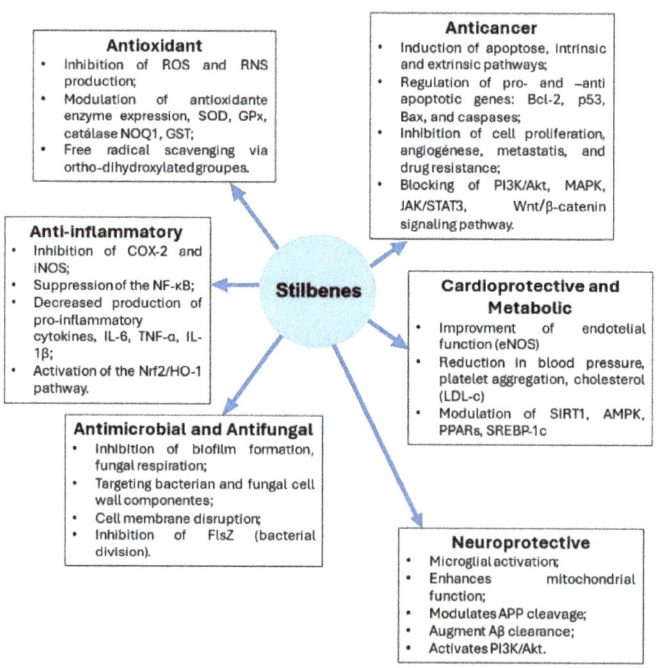

Figure 3. Diagram of mechanisms of biological activities of stilbenes.

Stilbenes also exert strong anti-inflammatory effects. They inhibit cyclooxygenase (COX) enzymes and nitric oxide synthases, reduce cytokine production, and suppress nuclear factor κB (NF-κB) signaling. Stilbenes such as piceatannol, pterostilbene, pinosylvin, desoxyrhapontigenin, and rhapontigenin inhibit inducible nitric oxide synthase (iNOS), reduce prostaglandin synthesis, and COX-2. Oxyresveratrol and piceatannol also inhibit nitrite production and iNOS, and resveratrol, oxyresveratrol, and piceatannol block NF-κB activation. Pterostilbene inhibits nuclear translocation of NF-κB and reduces pro-inflammatory cytokine production. Piceatannol is more effective than resveratrol in inhibiting COX-2, NF-κB activation, and cytokine production, while pinosylvin inhibits COX-2 with nearly twice the potency of resveratrol. Rhapontigenin is less active than desoxyrhapontigenin due to a missing hydroxyl at R3′ [82].

Piceatannol may contribute to anti-inflammatory responses by activating pathways involving heme oxygenase-1 (HO-1). In human endothelial cells, it upregulates HO-1 expression by activating the nuclear factor erythroid 2-related factor 2 (Nrf2). Inhibition of HO-1 negates piceatannol's effects, restoring the expression of tumor necrosis factor-α (TNF-α), interleukins-6 (IL-6), and NF-κB and p65 phosphorylation [143]. Piceatannol has also shown the ability to increase HO-1 expression in macrophages, endothelial cells, mammary epithelial cells, rat liver, and neuronal cells [144–149]. However, it is worth noting that HO-1's role in metabolic disorders is not well understood and, in some cases, has shown pro-inflammatory effects [150]. Thus, it is possible that piceatannol's anti-inflammatory effects are mediated through the regulation of the upstream activator of HO-1, Nrf2 [147].

Several studies have demonstrated piceatannol's ability to inhibit inflammatory pathways. Free fatty acids (FFAs) promote inflammation, with palmitic acid having pro-inflammatory properties. Piceatannol prevents the inhibitory effects of palmitic acid on insulin receptor substrate 1 (IRS-1) phosphorylation, glucose uptake, eNOS phosphorylation, and nitric oxide production in human endothelial cells [143]. Furthermore, piceatannol reduces lipopolysaccharide (LPS)-induced protein expression of IL-6 and TNF-α, attenuates NF-κB and signal transducer and activator of transcription 3 (STAT3) signaling in macrophages, and prevents IL-6 secretion and STAT3 and STAT5 signaling in human lymphocytes [144,151–153].

In addition, resveratrol enhances tetrahydrobiopterin (BH4) synthesis in cell and animal models, showing reduced oxidation [154]. BH4 is reduced by oxidative stress, negatively affecting various physiological functions [155]. Resveratrol can inhibit cyclic adenosine monophosphate (cAMP) phosphodiesterase, thereby increasing cellular levels of cAMP in cell and animal models [156]. Phosphodiesterases (PDEs) are enzymes that degrade cAMP and cyclic guanosine monophosphate (cGMP), which are second messengers involved in regulating numerous genes and cellular functions [157].

Stilbenes, particularly resveratrol, can activate cAMP signaling pathways. For instance, resveratrol boosts antioxidant defenses in aging cells, where cAMP induces Nrf2 expression. This pathway is also activated by alpha-melanocyte-stimulating hormone, a hormonal activator of melanocortin receptors coupled to the G-protein–cAMP cascade [158]. Protein kinase A and cAMP response element-binding protein (CREB) mediate Nrf2 activation through cAMP response elements (CREs) biding in their promoters, resulting in Nrf2 and Nrf2-related gene (e.g., glutathione S-transferase pi 1 and NOS) transactivation [159–161].

One of the more intriguing avenues in stilbene pharmacology concerns the interplay between resveratrol and energy-sensing molecular machinery. By interfering with phosphodiesterases, resveratrol alters intracellular calcium and thereby activates a cascade that converges on Adenosine monophosphate-activated protein kinase (AMPK), a critical metabolic checkpoint [162]. These events ripple outward, promoting shifts in both NAD^+ pools and sirtuin activity, which could, in theory, recalibrate key metabolic regulators implicated in chronic disease. Yet, a challenge persists, i.e., drawing causal links from these signatures in model systems back to consistent outcomes in whole organisms remains elusive [163–165].

3.3. Antimicrobial and Antifungal Activities

Stilbenes have demonstrated antibacterial, antiviral, and antifungal activity (Figure 3) [105,166–168]. Resveratrol and its analogues act as inhibitors of fungal tyrosinase [106,107]. In addition, stilbenes inhibit fungal cellular respiration and the lipid peroxidation of fungal membranes [136]. The antifungal activity of resveratrol inhibits a broad spectrum of fungi, including *Pyricularia oryzae, Plasmopara viticola, Cladosporium cucumerinum, Sphaeropsis sapinea, Phytophthora capsici, Phytophthora colocasiae, Botrytis cinerea, Candida albicans*, and *Colletotrichum gloeosporioides* [137–139,169]. Pinosylvin shows the highest efficacy against Gram-negative bacteria, followed by resveratrol, piceatannol, oxyresveratrol, and pterostilbene. Against Gram-positive bacteria, pinosylvin and pterostilbene are more effective than resveratrol. Pterostilbene's hydrophobicity and methylation improve membrane penetration and antifungal potency up to five times higher than resveratrol in inhibiting conidial germination and in vitro mycelial growth [72].

The first mechanism for antimicrobial activities involves their ability to inhibit conventional targets, such as the cell's membrane and wall, cell division, DNA, mitochondria, the calmodulin–calcineurin pathway, and the phosphoenolpyruvate (PEP)-dependent phosphotransferase system. The second mechanism relates to their ability to act as antibiofilm and antivirulence agents. Thus, stilbenes inhibit biofilm formation and assist in the eradication of already formed biofilms. Furthermore, another mechanism of stilbenes is their ability to reverse drug resistance through inhibition of alternative targets, target-modifying enzymes, and antibiotic-modifying enzymes [170]. Stilbenes have an antimicrobial mechanism that involves disrupting bacterial membranes, resulting in the leakage of cellular contents and leading to cell death. Pterostilbene, resveratrol, and toremifene (an anticancer agent approved by the Food and Drug Administration (FDA) and synthesized in 1981) have demonstrated antimicrobial activity through cell membrane disruption [171–173]. The membrane-associated protein phosphatidylglycerophosphate synthase (PgsA) catalyzes the substitution of cytidine monophosphate with glycerol phosphate to produce phosphatidylglycerol phosphate (PG-P). Subsequently, PG-P is dephosphorylated by PgpP to yield phosphatidylglycerol (PG), which is one of the main components of the cell membrane [174].

Another mechanism involves the action on the bacterial cell wall, which consists of a complex peptidoglycan polymer net composed of N-acetylmuramic acid and N-acetylglucosamine, with an attached pentapeptide. The biosynthesis of bacterial peptidoglycan is catalyzed by sequential Muramyl ligases in the intracellular steps. Stilbenes have the ability to inhibit MurD, an enzyme that catalyzes the reaction from UDP- N-acetylmuramic-Ala to UDP- N-acetylglucosamine-dipeptide [175].

Wall teichoic acids are anionic glycopolymers anchored in the cell walls of Gram-positive bacteria that play critical roles in bacterial physiology. Wall teichoic acids' biosynthesis initiates with a reaction is catalyzed by the membrane-associated glycosyltransferase TagA [111]. Stilbenes have shown promising activity against Gram-positive bacteria by targeting cell wall teichoic acids [111].

The fungal cell wall is composed of complex components, mainly polysaccharides (chitins, glucans, and mannans), lipids, and proteins. Stilbenes inhibit the activity of chitin synthases [176]. NADPH–cytochrome P450 reductase chemokine receptor

(CCR1) is involved in the yeast cell wall. Stilbenes have the ability to inhibit the activity of NADPH–cytochrome P450 reductase CCR1, leading to defects in yeast cell wall integrity [177].

Stilbenes have shown highly promising results in reducing the mycelial growth of phytopathogenic fungi such as *Pythium aphanidermatum*, *Rhizoctonia solani Kuhn*, *Exserohilum turcicum*, and *Fusarium oxysporum*, as well as the yeast *Saccharomyces cerevisiae*, in both agar medium and broth microdilution assays, using chlorothalonil as a positive control [178]. A series of resveratrol oligomers were purified from the canes of Vitis vinifera. These compounds were tested against *Candida albicans* [179,180]. Pterostilbene was shown to be a more potent inhibitor compared to 5-fluorocytosine, a reference antifungal agent used against *Candida albicans*. Moreover, pterostilbene proved to be a promising inhibitor against 12 non-albicans *Candida* (NAC) species, including emerging pathogens such as *Candida guilliermondii* and *Candida famata*. These findings support the idea that naturally occurring stilbene-type phytoalexins, produced by plants and certain bacterial species, serve as defense mechanisms against pathogens and represent a rich source for the discovery of novel fungicides [181].

Stilbenes can inhibit DNA synthesis. For example, trans-dihydroresveratrol dimer has been shown to block the ATP-binding site of DNA gyrase, inhibiting, therefore, the enzyme's activity and consequent DNA synthesis [182]. Another mechanism is causing DNA damage. For instance, oxyresveratrol binds directly to DNA, inducing the molecules' cleavage and subsequent mitochondria-mediated apoptosis in *Candida albicans* [183]. Resveratrol's prooxidant activity induces DNA damage through an increase in ROS, malondialdehyde accumulation, and glutathione depletion in *Salmonella typhimurium* [184].

Beyond their central role in cellular energetics, mitochondria stand at the crossroads of survival and programmed cell death. Loss of membrane integrity precipitates a sequence that leads inevitably to apoptosis [185]. Evidence implicates resveratrol as a potent modulator of this process, which is able to destabilize mitochondrial function and tip the scales towards cell death, as seen in both pathogen and tumor models. This mechanistic insight, while compelling, raises further questions about tissue selectivity and the balance between therapeutic efficacy and toxicity [186].

The cell membrane and wall, DNA, and mitochondria are common targets in antimicrobial therapies and, therefore, stilbenes' effects on the mentioned cellular structures are of interest when designing new treatment strategies for pathologies caused by microorganisms. For example, resveratrol and piceatannol perform a reversible inhibition in ATPase activity and consequent ATP synthesis in *Escherichia coli* [187]. Moreover, resveratrol's antibacterial activity extends to the downregulation of FtsZ expression and inhibition of Z-ring formation, a dynamic structure essential to cell division in prokaryotic organisms since it recruits division-related proteins and directs septal peptidoglycan synthesis [188,189].

Conventional antibiotics are becoming less effective overtime due to an increase in the number of multidrug-resistant microorganisms. Virulence and its associated factors determine pathogenicity and allow pathogens to acquire desired characteristics, such as immune evasion or modulation, colonization, and tissue damage. Toxins are also implicated in pathogenesis and include superantigens, surface proteins, hemolysins, and leukocidins. The pathological mechanism of α-hemolysin (Hla) secreted by *Staphylococcus aureus*, for example, relies on the direct binding to erythrocytes' cell membrane and subsequent pore formation and lysis. Thus, resveratrol and trans-stilbene have the ability to inhibit hemolysis caused by *Staphylococcus aureus* through the repression of the *hla* gene expression, thereby reducing the microorganism's virulence [140]. Furthermore, the two-component system SaeRS exerts essential functions in virulence factors' production [108]. Resveratrol can target the mentioned system, resulting in the downregulation of *saeRS* and consequent reduction in α-hemolysin secretion [109].

It is estimated that between 40% and 80% of the existing bacteria and archaea reside in biofilms, a macrocolony of microorganisms attached to a surface that constitutes a contributing factor in the development of chronic infections [110,112,113]. Biofilms offer microorganisms an advantage under various environmental challenges [114].

Biofilm formation occurs in five progressive stages, which include initial reversible attachment, irreversible attachment, first layer formation, mushroom- or tower-shaped structure formation, and dispersion and reattachment [112]. Stilbenes are capable of disturbing the hallmarks in biofilm formation, arresting their maturation, and allowing for the elimination of the mentioned structure. For example, resveratrol inhibits the production of flagellin and, consequently, reduces the motility of *Proteus mirabilis*, thereby preventing swarmer cell invasion into human uroepithelial cells [115]. Resveratrol also impairs the motility and adhesion of *Salmonella typhimurium* to HeLa cells through downregulation of flagellar genes [190]. Furthermore, resveratrol suppresses *fimA* and *xadA* expression, resulting in low levels of afimbrial adhesin and impaired adhesion capacity of *Xylella fastidiosa* to host cell surfaces [191]. The downregulation of fimbriae-associated gene expression by resveratrol in *Porphyromonas gingivalis* blocks biofilm production and increases in a directly proportional manner to the stilbene's concentration [192]. The other stilbene that is able to impair biofilm production and disrupt mature biofilms is pterostilbene, which suppresses filamentation-related gene expression through the Ras/cAMP pathway [180].

Biofilms release planktonic cells or small clusters of pathogens through dispersion in a continuous manner [193]. Stilbenes interfere with biofilm propagation through targeting the enzymes involved, such as micrococcal nuclease, whose production is reduced following resveratrol administration, resulting in more efficient *Staphylococcus aureus* biofilm clearance [194].

3.4. Anticancer Effects: Preclinical and Clinical Evidence

In a recent epidemiological study, it was estimated that approximately 1,280,000 deaths in the European Union and 618 120 deaths in the United States were caused by cancer in 2025 [120,195]. Cancer is the second leading cause of death since the beginning of the 21st century, preceded by cardiovascular pathologies and followed by diabetes and chronic respiratory diseases [196]. The increase in patient care, therapy efficiency, and awareness of the impact of healthier lifestyles in cardiovascular diseases has caused a drop in its mortality rate, falling behind that of cancer [121,122,197].

Increasing attention has been given to chemoprevention as an alternative approach to cancer control. There is growing evidence that oxidative stress induced by ROS is linked to carcinogenesis at multiple stages [123]. ROS are abundant free radicals in cells and are associated with various degrees of tissue damage. Oxidative stress arises from an imbalance between ROS production and neutralization or elimination by the organisms' naturally occurring antioxidant machinery. Consequently, oxidative stress can cause multiple issues, such as damage to proteins, lipids, and DNA, triggering or modulating the initiation, promotion, and progression of cancer [124].

Resveratrol has demonstrated potential anticancer activity, first reported by Jang et al., and has sparked significant interest in its chemopreventive and chemotherapeutic properties (Figure 3) [125]. Resveratrol has been tested in both in vitro and in vivo carcinogenesis assays for various cancer types, including leukemia, prostate, breast, lung, colon, ovary, liver, oral cavity, thyroid, and non-melanoma and melanoma skin cancer (Table 1) [126,198–206]. The chemopreventive properties of resveratrol rely on the compound's antioxidant activity. All stages of carcinogenesis are affected by the stilbene's anticancer activity, with the inhibition of COX-2 representing the major hallmark [125].

To date, three different isoforms of COX have been described, i.e., COX-1, which is expressed in normal tissues and participates in tissue homeostasis, COX-2, which results from overexpression in cases of inflammation or the development of neoplasms, and COX-3, a variant of COX-1 [207]. Many studies show that COX-2 plays a significant role in tumor progression, as its levels are elevated in premalignant and malignant tissues and are associated with a reduced survival rate in cancer patients, making it an unfavorable prognostic factor [116,208–210]. Clinical trials suggest that COX-2 inhibitors may be a solution for preventing the development adenomas and, potentially, carcinomas in the colon [207]. However, the clinical efficacy of these inhibitors is questionable due to increased cardiovascular risks [130].

Thus, resveratrol, which has shown positive effects in delaying cancer progression due to its significant potential in inhibiting COX-2, does not demonstrate toxicity effects [127]. Prostaglandins, produced through COX activity, have also been linked to cancer development and progression [116,211].

The increased production of prostaglandins influences the metabolism of carcinogens, tumor cell proliferation, and metastatic potential [131,212]. Thus, the inhibition of prostaglandin synthesis has been proposed as a strategy to prevent tumor development [131,132,211]. Numerous studies confirm that resveratrol has the capacity to inhibit COX-2, thereby reducing the synthesis of prostaglandin E2 (PGE2). Cianciulli et al. reported that resveratrol negatively regulates COX-2 and PGE2 in human intestinal Caco-2 cells treated with LPS, likely related to the inhibition of the NF-κB transcription factor [133]. NF-κB is associated with inflammatory, immune, and oncogenic responses [134].

Resveratrol activates AMPK, effectively preventing tumorigenesis [133,204,213,214]. Studies on redox status and antioxidant mechanisms' function indicate that resveratrol acts as a potent chemoprotector in in vivo cancer models. In rats treated with the hepatotoxic carcinogen azoxymethane (AOM), which induces oxidative imbalance, resveratrol partially reversed effects such as lipid peroxidation, glutathione (GSH) depletion, and increased NO levels in the liver [215]. Additionally, resveratrol induces the expression of SOD and catalase through a mechanism involving the tumor suppressor gene phosphatase and TENsin homolog (PTEN) and protein kinase B (PKB) signaling pathways. PTEN-mediated inhibition of the phosphoinositide 3-kinase (PI3K)/PKB pathway leads to increased activity of SOD, GSH peroxidase, and catalase [216].

Studies showed that resveratrol induces apoptosis in several cancer types. This pro-apoptotic stimulation is related to changes in the cell cycle, caspase activation, and downregulation of XIAP, Survivin, Bcl-2, and Bcl-xL, and upregulation of Bax, Bak, Bim, Noxa, PUMA, p21, and TRAIL-R2/DR5 [143–145,217–220]. These effects are also correlated with p53 activation [143,144,218,220]. For instance, resveratrol and piceatannol increase cytoplasmic calcium concentrations in human breast cancer cells (MDA-MB-231), activating p53 and triggering pro-apoptotic gene transitions [145]. In prostate cancer cells with p53 mutations (Du145), resveratrol induces p53 phosphorylation, restoring wild-type-like DNA binding and promoting pro-apoptotic events [146–148].

Pterostilbene, a natural analog of resveratrol with greater bioavailability [101,146], shares significant similarity with resveratrol but exhibits stronger in vitro antioxidant activity and demonstrates substantial clinical potential across various diseases [149,150,221]. Pterostilbene shows chemopreventive properties against cancer in different in vitro and in vivo assays (Table 1). Studies have demonstrated that pterostilbene, which can act as an active apoptotic agent, also inhibits growth, adhesion, and metastasis development [118,152–154]. These effects have been observed in several types of cancer, including breast, lung, stomach, prostate, pancreatic, melanoma, and/or colon cancers, highlighting its role as a chemopreventive agent [152,155–159,217,221,222].

Rimando et al. analyzed the antioxidant activity of pterostilbene, finding that it inhibits preneoplastic lesion formation caused by carcinogenic agents in a rat mammary organ culture model [149]. Moreover, Chiou et al. demonstrated that pterostilbene is more potent than resveratrol in preventing AOM-induced colon tumorigenesis by activating the Nrf2-mediated antioxidant pathway [160].

In a similar experimental model, pterostilbene has shown to regulate the expression of inflammation-related genes such as COX-2 and iNOS [161,223]. Likewise, in immortalized human keratinocytes (HaCaT), pterostilbene enhances Nrf2 nuclear translocation and the expression of Nrf2-dependent oxidative stress-associated molecules, reinforcing the central role of Nrf2 in the chemoprevention promoted by pterostilbene [127]. In HT-29 colon cancer cells, pterostilbene inhibited cytokine induction through the p38 transcription factor pathway, as well as other anti-inflammatory pathways such as ERK, JAK-STAT, NF-κB, c-Jun NH2-terminal kinase, and PI3K [161]. This inhibition is attributed to the reduced expression of COX-2 and iNOS, suggesting that the MAPK cascade activation by p38 is an essential pathway for the anti-inflammatory action of pterostilbene [161].

Furthermore, pterostilbene is equally potent to resveratrol in inhibiting NF-κB, COX-2, iNOS, and activator protein 1 (AP-1) as demonstrated in a skin carcinogenesis model induced by tissue-type plasminogen activator (TPA) in mice [224]. Pterostilbene induces PTEN expression in prostate cancer, leading to decreased levels of miR-106b, miR-17, and miR-20a. The effect of pterostilbene on restoring PTEN mRNA and protein levels to normal values was greater than that of resveratrol, suggesting that pterostilbene exhibits greater in vivo activity due to the methoxy groups occupying the place of hydroxyl groups [225].

In a study involving a mice model of UVB-induced skin carcinogenesis, pterostilbene proved superior to resveratrol in skin damage of acute and chronic nature [127]. This study demonstrated that pterostilbene has anticarcinogenic effects, maintaining skin antioxidant defenses (e.g., GSH peroxidase, catalase, and superoxide dismutase activities and GSH levels) and reducing oxidative damage to DNA, proteins, and lipids induced by UVB radiation [127].

Several studies validate pterostilbene as an efficient anticancer therapeutical agent acting on multiple signaling pathways. In an rat model of AOM-induced colon carcinogenesis, a pterostilbene-enriched diet resulted in reduced aberrant crypt foci, decreased transcriptional activation of COX-2 and iNOS, inhibition of glycogen synthase kinase-3β (GSK-3β) phosphorylation, suppression of the Wnt/β-catenin signaling pathway, and downregulation of cyclin D1, vascular endothelial growth factor (VEGF), and matrix metalloproteinases (MMPs), and activation of Ras, PI3K/PKB, and epidermal growth factor receptor (EGFR) pathways [160]. Additionally, pterostilbene reduces pro-inflammatory cytokines such as IL-1β, IL-4, and TNF-α, as well as nuclear phospho-p65 levels [226].

McComarck et al. demonstrated that pterostilbene inhibits breast cancer cell proliferation stimulated by leptin through reduced JAK/STAT3 signaling [227]. In pancreatic cancer cells treated with pterostilbene, there was an upregulation of pro-apoptotic genes, changes in phosphorylated STAT3 levels, increased antioxidant activity by manganese superoxide dismutase (MnSOD), and enhanced expression of Smac/DIABLO and cytochrome c [152]. Additionally, Liu et al. proposed that pterostilbene inhibits JAK2/STAT3 signaling, reducing the expression of STAT3 target genes, including anti-apoptotic proteins Mcl-1 and Bcl-xL, while increasing proteins related to mitochondrial apoptosis (Bak, Bax, cleaved caspase-3, and cytosolic cytochrome c) and cyclin-dependent kinase inhibitors (e.g., p21 and p27) in osteosarcoma cells [228].

Pterostilbene acts as a chemopreventive agent, with effects extending beyond its anti-inflammatory and antioxidant properties or apoptosis-promoting effects. Studies have proposed that pterostilbene can induce cell death through autophagy [156,158,159,229,230]. Recently, it was observed that pterostilbene induces tumor autophagy through a mechanism dependent on lysosomal membrane permeabilization mediated by Hsp70 [217].

In recent years, the traditional model of cancer progression has been revised to emphasize the importance of tumoral heterogeneity in the development of chemo/radiotherapy resistance and post-treatment relapse. Cancer stem cells (CSCs) are attractive targets due to their self-renewal capacity, ability to generate heterogeneous tumor cell lineages, and high tumorigenicity [231–233]. Studies suggest that resveratrol and pterostilbene promote Argonaute-2 expression and activity, a central RNAi component, inhibiting cancer stem cell-like characteristics in breast cancer by increasing tumor-suppressor miRNAs such as miR-200c, miR-141, and miR-16 [234]. Pterostilbene suppresses CSC generation and metastatic potential in various experimental models, modulating epithelial–mesenchymal transition pathways and preventing the enrichment of CD133(+) CSCs in irradiated hepatoma [235,236].

Piceatannol, a hydroxylated analog of resveratrol, exhibits health benefits like resveratrol, though fewer studies are available (Table 1) [237–239]. Li et al. found that the anticancer properties of piceatannol may arise from its prooxidant effects due to the presence of copper (Cu)(II), which induces hydroxyl radical formation via Fenton and Haber—Weiss reactions, leading to DNA damage [240,241]. Piceatannol inhibits hydrogen peroxide (H_2O_2)-induced NF-κB activation, ceramide, LPS, phorbol 12-myristate 13-acetate, and okadaic acid [242]. Additionally, it inhibits p65 phosphorylation and its nuclear translocation, and TNF-induced Iκ-Bα phosphorylation. Moreover, the hydroxyl groups at the 3′ and 4′ positions appear to play a crucial role in stilbene's anticancer activity, as the same effects were observed for treatment with resveratrol treatment [242].

Piceatannol could reduce iNOS expression, thereby decreasing NO production. Furthermore, it inhibits COX-2 expression in RAW 264.7 cells stimulated with LPS and BV2 microglia cells [243,244]. Additionally, piceatannol increases protein levels of HO-1 in MCF10A, a human cell line of the mammary epithelium. The underlying mechanism involves the release of Nrf2 from Kelch-like ECH-associated protein 1 and its translocation to the nucleus, with consequent binding to the antioxidant response element, promoting HO-1 expression [245].

The effects of ε-viniferin on cancer prevention and treatment have been studied in over 20 cell models, including neural, breast, skin, liver, lung, bladder, stomach, colon, and hematological cancers. A published study demonstrated that, in approximately half of the total cases, the determined IC_{50} values for ε-viniferin were below 60 μM, highlighting the compound's great potential in reducing the growth of cancer cells. Furthermore, comparative approaches on the anticancer activity of resveratrol and ε-viniferin have indicated that the latter is more active in one-third of the total cases [141].

In certain studies, the mechanisms underlying the cytotoxic and antiproliferative activities of ε-viniferin have been explored, with most being linked to apoptosis. This form of programmed cell death is essential for the development of the organism and tissue homeostasis. Apoptosis can be initiated by two main mechanisms: the intrinsic pathway, mediated by mitochondria, and the extrinsic pathway, mediated by death receptors. Most pro-apoptotic stimuli for the intrinsic pathway are associated with the permeabilization of the outer mitochondrial membrane. This process is regulated by a balance between pro- and anti-apoptotic members of the Bcl-2 family, leading to the release of cytochrome c into the cytosol. This apoptotic factor activates caspases and proteases responsible for the apoptotic phenotype [246].

ε-viniferin can induce apoptotic cell death in various tumor cell lines, including leukemia, myeloma, glioma, malignant melanoma, and hepatocellular carcinoma [247–251]. It induces apoptosis by disrupting normal mitochondrial transmembrane potential and activating pro-apoptotic proteases. Caspase-3 activation has been highlighted in leukemia and hepatocellular carcinoma cells [248,252]. Barjot et al. demonstrated that caspase-8 activation by ε-viniferin mediates a significant portion of this compound's cytotoxic effect in U266 myeloma cell lines. In C6 glioma cells, activation of caspases-8, -9, and -3 has been observed [247,253].

The antiproliferative effect results in cell cycle arrest. ε-viniferin's ability to halt the cell cycle has been investigated. Barjot et al. analyzed that treatment with ε-viniferin led to an accumulation of U266 myeloma cells in the G2-M phases and a reduction in cells in the G0/G1 phases [247]. Furthermore, ε-viniferin arrests the cell cycle of melanoma cells in the S phase and reduces the percentage of cells in the G1 phase, increasing regulatory proteins that control cell cycle progression through the S phase, such as cyclin E1 [249].

In addition, ε-viniferin, like other resveratrol oligomers, has been studied for its in vitro anticancer activities (Table 1). The cytotoxic activities vary depending on the tumor cell line and the structure of the resveratrol oligomer studied. Ito et al. investigated 18 oligomers ranging from dimers to octamers and tested them on five human tumor cell lines [254]. Seven of these compounds exhibited cytotoxic effects, primarily oligomers with more than four basic resveratrol units. However, other studies have suggested that the number of resveratrol units does not influence cytotoxic activity [255–257].

Although larger molecules are more active, their efficacy may be limited due to low bioavailability in vivo studies. To date, no in vivo trials have been conducted to compare ε-viniferin with resveratrol in studying its potential beneficial effects on the promotion, progression, or treatment of cancer. Several studies on ε-viniferin have not isolated the compound but instead used it in plant extracts or compound mixtures. These studies have shown that such mixtures or combinations are sometimes more active than the isolated molecules due to the synergistic effect between compounds [141].

For example, Vineatrol® has demonstrated more efficient antiproliferative effects than isolated resveratrol or ε-viniferin in leukemia and hepatoma cells [248,258]. Additionally, ε-viniferin has been combined with anticancer drugs such as vincristine or cisplatin. Özdemir et al. demonstrated that ε-viniferin can enhance cell sensitivity to vincristine treatment in HepG2 cells [250]. Furthermore, the same author reported a strong apoptotic effect when combining cisplatin with ε-viniferin. Thus, ε-viniferin could potentially be used as a combined treatment with multiple anticancer drugs to reduce drug resistance and lower the required doses, thereby limiting side effects in living organisms [141].

Nivelle et al. achieved very positive results, as they observed reduced toxicity in normal fibroblasts compared to melanoma cancer cell lines [249]. Moreover, in non-transformed hepatocytes (HH4), higher concentrations of the stilbene dimer were required to induce toxicity compared to hepatocellular carcinoma cell lines such as HepG2 and Hep3B [259].

In conclusion, most studies have shown that ε-viniferin can reduce cancer cell growth, inducing apoptosis, and modifying the cell cycle [141]. Antioxidants act as agents that slow cancer progression by scavenging free radicals, inhibiting or preventing oxidative damage, and reducing oxidative stress [260].

Table 1. Overview of in vitro studies on the antitumor activity of representative stilbenes.

Stilbenes	Experimental In Vitro Model	Methodology	Experimental Conditions	Significant Results	Ref
Resveratrol	D407 cells	H_2O_2-induced cytotoxicity or MTT assay	0, 25, 50, and 100 μM for 24 h	Resveratrol offered protection to D407 retinal pigment epithelial cells against H_2O_2-induced cytotoxicity, resulting in reduced cytotoxicity.	[261]
	N2a cells	Fluorescein diacetate assay	1.5, 3.125, 6.25, 12.5, 25, 50, and 100 μM for 48 h	Plasma membrane integrity was compromised in resveratrol-treated N2a cells, reducing cell viability.	[262]
	T24 cells	Cell Proliferation Kit II	50, 100, 150, 200, and 250 μM for 24 h	Resveratrol exposure led to a decrease in cell viability, which was greater for s higher concentration of the stilbene. The determined IC_{50} was 178.73 μM.	[263]
	MCF-7 cells	MTS assay	20, 40, 60, 80, and 100 μM for 48 h	Resveratrol was cytotoxic in MCF-7 cells, aligning with the compound's anticancer properties. Changes in membrane fluidity and, consequently, in cellular signaling pathways were detected.	[264]
	HRT cells	MTT assay	25, 50, 100, 200, and 300 μM for 72 h	Resveratrol inhibits HRT cell growth in a concentration-dependent manner.	[265]
	MCF-7 cells	Sulforhodamine B (SRB) assay	15 μg/mL for 24 h	Co-treatment with resveratrol and doxorubicin drastically lowered doxorubicin's IC_{50} from 0.417 μg/mL to 0.035 μg/mL, showing it to be a good adjuvant in antitumor therapy.	[266]
Trans-resveratrol	PC-3 cells	MTS assay	0, 3, 10, 30, and 100 μM for 72 h	Trans-resveratrol exhibited an inhibitory effect on cell viability in PC3 cell line at concentrations ranging from 10 to 100 μM.	[267]
	MCF-7 cells JURKAT E.6, and THP-1	MTT or XTT assay	10, 30, 50, 70, 90, and 100 μM for 24, 48, 72, and 96 h	As trans-resveratrol concentrations increased, cell viability showed a greater or lesser pronounced decrease, depending on the cell line. Therefore, the effect of trans-resveratrol depends not only on dose and treatment duration but also on the cell type considered.	[268]
	MCF-7, Du145, and PC-3 cells	MTT assay	1×10^{-15}–1×10^{-3} M for 48 h	Trans-resveratrol exhibited a cytotoxic effect on cancer cells at concentrations ranging from 1×10^{-7} to 1×10^{-4} M.	[269]
	MCF-7 cells	MTT or neutral red uptake (NRU) assay	1, 5, 10, 25, and 50 μM for 24 h	Solely an administration of trans-resveratrol did not lead to a decrease in cell viability in MCF-7 cell line. However, a pre-treatment with trans-resveratrol conferred protection against rotenone-induced toxicity.	[270]
	HepG2, Vero, and MCF-7 cells	MTT assay	0, 0.2, 0.4, 2, 4, 6.25, 12.5, 25, 50, and 100 μM for 24, 48, 72 h	Trans-resveratrol demonstrated cytotoxicity in all mentioned cell types at concentrations equal to or greater than 50 μM after 48 h.	[271]
	PC12 cells	MTT or NRU assay	0, 5, 10, and 25 μM for 24 h	Trans-resveratrol decreases the viability of P12 cells, as shown by both performed assays.	[272]

Table 1. *Cont.*

Stilbenes	Experimental In Vitro Model	Methodology	Experimental Conditions	Significant Results	Ref
Trans-resveratrol	HCT-116, HCT-116/p53(−/−), HepG2, and Hep3B cells	CellTiter-Blue® and SRB assay	0, 1, 10, and 100 μM for 72 h	Trans-resveratrol demonstrated a significant ability to decrease cell viability at a concentration of 100 μM after 72 h.	[273]
	HepG2 cells	WST-1 assay	0.5–100 g/mL for 24, 36, and 48 h	Trans-resveratrol decreased the viability of HepG2 cells, evidencing the compound's anticancer activity. The calculated IC_{50} was shown to decreased over time.	[273]
	MCF-10A, MCF-7, MDA-MB-231, and ZR-75-1 cells	CellTiter-Glo® Luminescent Cell Viability assay	1–350 μM for 48 h	Trans-resveratrol was proven to be an efficient inhibitor of the cancer cell lines MCF-7, ZR-75-1, and MDA-MB-231, with IC_{50} values of 68.3 ± 2.6, 82.2 ± 4.8, and 67.6 ± 4.1 μM, respectively. Moreover, it was three times more potent in the MCF-10A cell line, with an IC_{50} of 20.0 ± 2.9 μM.	[274]
	SW480 cells	MTT assay	30 μM for 48, 72, and 96 h	There was a time- and dose-dependent decline in cell survival.	[275]
	MCF-7 cells	Annexin V-FITC and propidium iodide assay	6.25–50 μg/mL for 24 h	The viability of MCF-7 cells was suppressed following trans-resveratrol treatment.	[276]
	ARPE-19 cells, transmitochondrial normal RPE cybrid, and transmitochondrial AMD RPE cybrid cells	MTT assay	1000 μM for 48 h	Viability assays demonstrated that trans-resveratrol has beneficial properties for cybrid cells, increasing their viability compared to untreated cells.	[277]
	HepG2 cells	MTT assay	2.5, 10, 30, 50, 70, 100, 140, and 200 μM for 24, and 48 h	Trans-resveratrol has inhibitory effects on the cell viability of HepG2 cells, having greater impact in higher concentrations and prolonged exposure times.	[278]
	LTC-14 cells	MTT assay	0, 0.1, 1, 10, 100, and 1000 μM for 24, and 48 h	LTC-14 cells experienced a decrease in cell viability to below 50% in the presence of trans-resveratrol at a concentration of 200 μM.	[279]
	NCTC clone 929 cells	NRU assay	15.63, 31.25, 62.5, 125, and 250 μM for 24 h	Trans-resveratrol caused significant cell injury and death with an associated IC_{50} of 50 μM.	[280]
	A549 cells	MTT assay	0, 5.5, 11, 21.9, 32.9, 43.8, 87.6, 131.4, and 175.2 μM for 24 h	Trans-resveratrol treatment led to cell viability inhibition in a dose-dependent manner. The IC_{50} was determined to be 85.5 μM.	[281]
Pterostilbene	MCF-7, MDA-MD-231, and ZR-75-1	MTT assay	10 μmol/L, 20 μmol/L, 30 μmol/L (pterostilbene) + 5 μmol/L (Tamoxifen) for 24, 48, and 72 h	Combined therapy with pterostilbene and tamoxifen reduced cell viability in all cell lines. A greater decrease in viability was observed for the 24 h treatment.	[282]

Table 1. *Cont.*

Stilbenes	Experimental In Vitro Model	Methodology	Experimental Conditions	Significant Results	Ref
Pterostilbene	MDA-MB-231 and T-47D cells	MTT assay	10–100 μM for 48 h	A decrease in cell viability and significant morphological changes were observed in both cell lines following the treatment with pterostilbene. The IC_{50} concentrations for MDA-MB-231 and T-47D cells were 45.7 ± 0.01 and 63.1 ± 0.11 μM, respectively.	[283]
	MCF-7, SK-BR-3, and MDA-MB-468 cells	American Type Culture Collection	0–100 μM for 72 h	Treatment with pterostilbene arrested cells growth in a dose-dependent manner for all three cell lines, exhibiting a greater impact in MDA-MB-468 cells. The calculated IC_{50} values were 87.6 ± 9.0 μM for MCF-7 cells, 64.4 ± 4.6 μM for SK-BR-3 cells, and 45.7 ± 5.2 μM for MDA-MB-468 cells.	[284]
	HeLa, CaSki, and SiHa cells	MTT assay	0–200 μM for 72 h	All cell lines' proliferation was inhibited by pterostilbene in a manner that varied in a directly proportional way to concentration. The IC_{50} for each cell line was calculated as follows: IC_{50} = 32.67 μM for HeLa, IC_{50} = 14.83 μM for CaSki, and IC_{50} = 34.17 μM for SiHa, indicating growth-inhibitory effects.	[285]
	TC-1 mouse cells after co-transformation with HPV16-E6, HPV16-E7, and c-Ha-Ras oncogenes	WST-1 assay	0–100 μM for 72 h	Pterostilbene exhibits significant cytotoxicity, evidenced by the formation of cytoplasmic blebs observed after 48 h. The number of apoptotic cells increased in a dose-dependent manner and the obtained IC_{50} of pterostilbene was 15.61 μM.	[286]
	CL187, C COLO 205, HCT-8, SW480, Lovo, and HCT-116 cells	Cell Counting Kit-8 (CCK-8) assay	1–100 μM for 24, 48, and 72 h	Pterostilbene inhibited cellular activity of all cell lines in a dose-dependent manner. After 72 h of treatment, the IC_{50} of pterostilbene for most of the cell lines used (except SW480 cells) was determined to be below 30 μM.	[287]
	HT-29 cells	SRB assay	5–100 μM for 48 h	A significant decrease in cell growth was only observed at concentrations equal to or greater than 10 μM. The extent of inhibitory effects was shown to be dependent on pterostilbene dosage.	[288]
	HT-29 cells	MTT assay	5 and 20 μM for 24 and 48 h	Treatment with pterostilbene at 20 μM inhibited the metabolic activity of HT-29 cells up to 80.2 ± 5.9%.	[289]
	HEC-1A and ECC-1 cells	MTS assay	0, 18.75, 37.5, 75, 150, and 300 μM for 48 h	Pterostilbene treatment significantly reduced cell viability in a dose-dependent manner, with associated IC_{50} values ranging between 72 and 78 μM for both cell lines.	[290]

Table 1. *Cont.*

Stilbenes	Experimental In Vitro Model	Methodology	Experimental Conditions	Significant Results	Ref
Pterostilbene	Kuramochi, Caov-3, OVCAR-4, OVCAR-8, and SKOV3 cells	MTT assay	0, 37.5, 75, 150, and 300 μM for 48 h	Cell viability was markedly reduced by pterostilbene in a dose-related way, with the IC_{50} for each cell line as follows: 161.2 μM for Kuramochi, 100.6 μM for Caov-3, 143.8 μM for OVCAR-4, 74.8 μM for OVCAR-8, and 95.2 μM for SKOV3 cells.	[291]
	LNCaP and PC3 cells	MTT assay	0, 20, 40, 60, 80, and 100 μM for 48 h	Pterostilbene reduced cell viability for both cell lines and in a dose-dependent manner. The IC_{50} values ranged between 70–80 μM for LNCaP cells, and 80–100 μM for PC3 cells.	[292]
	MIA PaCa-2 and PANC-1 cells	MTT assay	10–100 μM for 24, 48, and 72 h	Pterostilbene inhibited cell viability in a dose- and time-dependent manner in both cell lines. The IC_{50} concentration values varied depending on cell type and selected time points, with MIA PaCa-2 showing 72 μM at 24 h, 51 μM at 48 h, and 32 μM at 72 h, while PANC-1 showed 84 μM at 24 h, 33 μM at 48 h, and 29 μM at 72 h.	[293]
	A375, A549, HT29, and MCF-7 cells	Countess Automated Cell Counter and SRB Toxicology Assay	0–200 μM for 24, 48, and 72 h	Pterostilbene inhibited cell viability in a dose- and time-dependent manner in both cell lines. The IC_{50} values determined were cell type-dependent, being much lower for HT29 (IC_{50} = 60.3 mmol/L) and MCF7 (IC_{50} = 44.0 mmol/L) cells than for A375 (IC_{50} = 14.7 mmol/L) and A549 (IC_{50} = 28.6 mmol/L) cells.	[217]
	11–18, HCC827, HCC4006, H1975, and PC9 cells	MTT assay	0–150 μM for 72 h	Pterostilbene inhibited cell viability in all cell lines, with IC_{50} values ranging between 23.8 and 40.7 μM.	[294]
	HepG2 cells	MTT assay and CCK-8 assay	12.5–100 μM for 24 h	Cell viability and proliferation were reduced for all concentrations considered in a dose-dependent manner.	[295]
	HT29, MKN74, and CT26 cells	MTS assay	10, 50, and 100 μM for 48 h	Pterostilbene reduces cell viability in all three cell lines. The determined IC_{50} values were 21 μM for CT26, 63 μM for HT29, and 65 μM for MKN74 cells.	[296]
	CAR cells	MTT assay	5, 10, 25, 50, 75, and 100 μM for 24, 48, and 72 h	Pterostilbene induces cytotoxicity in a time- and dose-dependent manner. The IC_{50} values after 24, 48, and 72 h of incubation were 78.26 ± 4.33, 48.04 ± 3.68, and 20.65 ± 4.88 μM, respectively.	[297]
	MDA-MB-231 cells	MTT assay	1, 5, 20, 30, and 50 μg/mL for 48 h	Pterostilbene exhibits an inhibitory associated with an IC_{50} value of 79.5 ± 6.36 μg/mL.	[298]

Table 1. *Cont.*

Stilbenes	Experimental In Vitro Model	Methodology	Experimental Conditions	Significant Results	Ref
Pterostilbene	AsPC-1, BxPC-3, MIA PaCa-2, and PANC-1 cells	MTT assay	0, 50, 75, 100, 125, and 150 μM for 48 h	Increasing concentrations of pterostilbene reduced viability in all cell lines tested, pointing towards a dose-dependent sensitivity to the mentioned compound. The IC_{50} values ranged from 110 to 130 μM.	[299]
	MDA-MB-231 cells	MTT assay	2.5, 5, 10, 20, 40, and 80 μM for 24 h	A 24 h treatment with pterostilbene at 5 μM resulted in a 12% reduction in survival of MDA-MB-231 cells. The IC_{50}, IC_{80}, and IC_{85} doses against MDA-MB-231 cells were 30.4, 12.1, and 9.7 μM, respectively, confirming selective anticancer toxicity.	[300]
	GBC-SD, NOZ, and SGC-996 cells	CCK-8 assay	0–80 μmol/L for 48 h	Pterostilbene exhibits cytotoxic effects on all three cell lines. The estimated IC_{50} for GBC-SD cells was above 80 μmol/L, between 40–60 μmol/L for NOZ cells, and approximately 80 μmol/L for SGC-996 cells.	[301]
	C6, LN18, LN229, T98G, U87, and HUVECs cells	MTT assay	0, 20, 40, 80, and 100 μM for 24, 48, and 72 h	Pterostilbene inhibited cell viability on C6, LN18, LN229, T98G, and U87 cells. The IC_{50} values of pterostilbene treatment for 48 h were 30.10 μM for C6 cells, 22.30 μM for LN18 cells, 37.56 μM for LN229 cells, 32.93 μM for T98G cells, and 46.18 μM for U87 cells. Pterostilbene had a minimal impact on HUVEC cells compared to the previously mentioned cell lines.	[302]
Piceatannol	AGS, SK-MES-1, and J82 cells	MTT assay	0–100 μg/mL for 72 h	Besides enhancing gemcitabine's cytotoxic and apoptotic effects, piceatannol actively inhibited SK-MES-1 cell viability. The synergistic combination increased the expression of the Bcl-2 pro-apoptotic protein family. IC_{50} concentrations for AGS, SK-MES-1, and J82 cells were 10.8 ± 0.7, 7.64 ± 0.5, and 6.7 ± 0.3 μg/mL, respectively.	[303]
	T24 and HT1376 cells	XTT assay	0.5, 2.5, 5, and 10 μM for 48 h	Piceatannol showed a dose-dependent inhibitory effect on the proliferation of both T24 and HT1376 cell lines. The IC_{50} values were 3.9 and 4.6 μM, respectively.	[304]
	HL-60 cells	MTT assay	10–200 μM for 24, 48, and 72 h	Piceatannol significantly inhibited HL-60 cell growth in a time- and dose-dependent manner. A moderate inhibition of HL-60 cells viability was observed after a 72 h treatment with piceatannol at 10, 20, and 50 μM. The highest inhibition was observed after 24, 48, and 72 h treatment with 100–200 μM concentration range.	[305]

Table 1. *Cont.*

Stilbenes	Experimental In Vitro Model	Methodology	Experimental Conditions	Significant Results	Ref
Piceatannol	WM266-4 and A2058 cells	MTT assay	0, 1, 10, 20, 40, 100, and 200 μM for 36 h	Both cell lines exhibited decreased viability following piceatannol treatment. The calculated IC_{50} was 29.4 μM for WM266-4 cells, and 15.6 μM for A2058 cells.	[306]
	LNCaP, Du145, and PC3M cells	MTS assay	1, 5, 10, 25, 50, and 100 μM for 6 days.	All cell lines were susceptible to piceatannol treatment, exhibiting declining cellular activity. The IC_{50} values obtained were 31.7 μM for LNCaP, 23.2 μM for Du145, and 34.6 μM for PC3M cells.	[307]
	B16 cells	MTT assay	5–400 μM for 24 h	Piceatannol exhibited cytotoxicity effects, resulting in decreased cell viability. The obtained IC_{50} was 1.53 μM.	[308]
	U937 cells	MTT assay	0–100 μM for 24 h	Exposure to piceatannol inhibited cell viability, with an associated IC_{50} of 5 μM.	[309]
	NCI-H522 cells	WST-8 assay	10, 30, 50, 80, and 100 μM for 24, 48, and 72 h	Piceatannol treatment notably decreased NCI-H522 cell viability. The IC_{50} values at each timepoint were 53, 23, and 17 μM, respectively.	[310]
	Caco-2 and HCT-116 cells	Crystal violet assay	12.5, 25, 50, 100, and 200 μM for 24, 48, and 72 h	Piceatannol cytotoxic effects led to a decrease in cell viability fin both cell lines after a 72 h treatment. The obtained IC_{50} of piceatannol in Caco-2 and HCT-116 cells was 50 μM.	[311]
	L1210, K562, and HL-60 cells	Trypan blue dye exclusion	0–500 μM for 24 h	All cell lines were sensible to piceatannol's cytotoxic effects. The calculated IC_{50} values of piceatannol were 50 μmol/L, <10 μmol/L, and <20 μmol/L for K562, HL-60, and L1210 cells, respectively.	[312]
	RAW 264.7 cells	MTT assay	0–50 μg/mL for 48 h	Piceatannol exhibits inhibitory activity, with an associated IC_{50} value of 5.7 μg/mL.	[313]
	HSG, HL-60 HSC-2, and HSC-3, (tumor cell lines) HPC, HGF, and HPLF (normal cells line),	MTT assay (HGF, HPC, HPLF, HSC-2, HSC-3, AND HSG) Trypan blue dye exclusion (HL-60)	10–1000 μM for 24 h	Piceatannol exhibits greater inhibitory effects on cancer cells compared to normal cells. The IC_{50} values for cancer cell lines were 63 μM for HSC-2, 232 μM for HSC-3, 373 μM for HSG, and 11 μM for HL-60 cells. In contrast, the IC_{50} values for normal cells were 367 μM for HGF, 414 μM for HPC, and >1000 μM for HPLF cells.	[314]
	SW1990 and PANC-1 cells	CCK-8 assay	1, 10, 20, 40, 100, and 200 μM for 72 h	Piceatannol inhibited up to 50% cell proliferation for both cell lines. The IC_{50} value for SW1990 cells was 30.69 μM, while for PANC-1 cells it was 21.82 μM.	[315]
	MOLT-4 cells	NRU assay	0.05, 15, 25, 50, and 100 μM for 48 h	Piceatannol reduced cellular viability with a calculated IC_{50} of 45.5 μM.	[316]
	HeLa cells	MTT assay	0–250 μM for 48 h	Piceatannol decreased cell viability and the associated IC_{50} value was 375.20 μM.	[317]

Table 1. *Cont.*

Stilbenes	Experimental In Vitro Model	Methodology	Experimental Conditions	Significant Results	Ref
Piceatannol	Mouse embryonic stem cells (ESCs)	MTT assay	1–20 μM for 72 h	High concentrations of piceatannol exhibited cytotoxicity. The obtained IC_{50} value was 13.5 μM.	[318]
	C6 cells (proliferating and growth arrested)	Lowry method	1–100 μM for 72 h in proliferating cells and 24 h in growth-arrested cells	Piceatannol exhibits cytotoxic effects on both growth-arrested and proliferating cells. The IC_{50} concentration for growth-arrested cells was 20 ± 2 μM, while for proliferating cells it was 28 ± 4 μM.	[319]
	10ScNCr/23, A-431, RAW 264.7, and CCR-CEM cells	Trypan blue dye exclusion	0–50 μM for 24 h	Piceatannol exhibits inhibitory effects on all cell lines. The IC_{50} concentration in RAW 264.7 cells were 1.30 ± 0.12 μM.	[320]
	THP-1 cells	Light microscopy	10, 20, 30, 40, and 50 μM for 48 h	Significant cytotoxic effects with noticeable cell shrinkage were observed at concentrations above 30 μM.	[321]
(-)-ε-viniferin	HSC-2, HSC-3, HCF, HPC, HPLF, HSG, and HL-60 cells	MTT assay in adherent cells Trypan blue dye exclusion in non-adherent cells	0–1000 μM for 24 h	The four tumor cell lines (HSC-2, HSC-3, HSG, and HL-60) were more sensitive to (-)-ε-viniferin than the remaining normal cell lines. The IC_{50} values were 42 μM for HSC-2 cells, 84 μM for HSC-3 cells, 111 μM for HCF cells, 146 μM for HPC cells, 94 μM for HPLF cells, 110 μM for HSG cells, and 31 μM for HL-60 cells.	[314]
	P-388 cells	MTT assay	0–100 μM for 48 h	ε-viniferin moderately inhibited cell viability in comparison to hopeaphenol, which exhibited a greater effect. The IC_{50} measured at 18.1 ± 0.7 μM.	[255]
	HepG2 and Chang cells	MTT assay	1.56–200 μg/mL for 72 h	No cytotoxic effect was detected in either cell lines.	[322]
(+)-ε-viniferin	RAW 264.7 cells	MTT assay	1, 5, and 10 μM for 12 h	Cell viability was significantly reduced to 60% after exposure of 10 μM. IC_{50} was not determined.	[323]
trans-ε-viniferin	K562, L1210, and HCT116 cells	MTT assay	0–50 μM for 48 h	No cytotoxicity was detected. The IC_{50} was assumed to be above 50 μM.	[324]
	AGS, MRC-5, SK-MES-1, and J82 cells	MTT assay	0–100 μg/mL for 72 h	Cytotoxicity was observed for all cell lines tested. The IC_{50} values were 42.6 ± 1.7 μM for AGS cells, 49.9 ± 3 μM for MRC-5 cells, 78.8 ± 3.3 μM in SK-MES-1 cells, for 56.7 ± 1.2 μM in J82 cells.	[303]
	Mouse primary astrocytes and neurons co-culture	CellTitel 96 ® Aqueous assay	1, 5, 10, 20, 50, and 100 μM for 24 h	Cell viability was significantly reduced when cells were exposed to concentrations of 50 and 100 μM.	[325]

Table 1. *Cont.*

Stilbenes	Experimental In Vitro Model	Methodology	Experimental Conditions	Significant Results	Ref
trans-ε-viniferin	AGS, COLO 205, HepG2, HL-60, and HT-29 cells	MTT assay	0–100 μg/mL for 48 h	Dose-dependent cytotoxicity was reported, with a greater effect observed in HL-60 cells. The determined IC_{50} values were: 9.3 ± 0.3 μM in AGS cells, 85.5 ± 8.1 μM in COLO 205 cells, 7.7 ± 0.2 μM in HepG2 cells, 5.6 ± 1.4 μM in HL-60 cells, and 13.9 ± 0.1 μM in HT-29 cells.	[326]
	Hep3B, HepG2, and HH4 cells	Crystal violet assay	0–200 μM for 24, 48, and 72 h	It was more cytotoxic to Hep3B cells and reduced cell quantity in a dose- and time-dependent manner. Higher amounts were required to cause toxicity in HH4 cells. The IC_{50} values obtained were the following: - Hep3B cells: 108.1 ± 31.8 μM (24 h), 73.9 ± 17.3 μM (48 h), 63.1 ± 10.8 μM (72 h). - HepG2 cells: 140 ± 39.7 μM (24 h), 103.8 ± 19.2 μM (48 h), 94.8 ± 28.3 μM (72 h). - HH4 cells: >200 μM (24 h), 192.7 ± 21.1 μM (48 h), 177.9 ± 20.5 μM (72 h).	[326]
	HepG2 and Caco-2 cells	MTS assay, NRU, and protein content	0–100 μg/mL for 24 and 48 h	For every endpohint examined, both cell lines showed a time-dependent decline in cell viability. The IC_{50} values were: - HepG2: 28.28 ± 2.15 μg/mL 24 h and 17.85 ± 3.03 μg/mL for 48 h. - Caco-2 cells: 36.72 ± 3.01 μg/mL for 24 h and 20.63 ± 1.25 μg/mL 48 h.	[327]
trans-ε-viniferin and cis-ε-viniferin	HeLa, MCF-7, C6, HepG2, and HT-29 cells	MTT assay	0–100 μM for 70 h	Cis- and trans-ε-viniferin to all cell lines, although greater significance was registered for C6 and HeLa cells. The IC_{50} values for trans-ε-viniferin were: - 20.4 μM in HeLa cells, 44.8 μM in MCF-7 cells, 18.4 μM in C6 cells, 74.3 μM in HepG2 cells, and 88.4 μM in HT-29 cells. The IC_{50} values for cis-ε-viniferin were: - 21.5 μM in HeLa cells, and 47.2 μM in MCF-7 cells, 20.1 μM in C6 cells, 76.2 μM in HepG2 cells, and 90.2 μM in HT-29 cells.	[328]
ε-viniferin	WSU-CLL cells	Trypan blue dye exclusion	0–100 μM for 24, 48, and 72 h	A concentration- and time-dependent decrease in cell viability was observed, with resveratrol overperforming ε-viniferin. Inhibited cell proliferation was accompanied by a reduction in DNA synthesis. The IC_{50} value determined at 72 h was 60 μM.	[248]
	HL-60 cells	MTT assay	10–200 μM for 24 h	Cell viability decreased in a concentration-dependent manner. The IC_{50} was 33 μM.	[329]

Table 1. *Cont.*

Stilbenes	Experimental In Vitro Model	Methodology	Experimental Conditions	Significant Results	Ref
ε-viniferin	HepG2 cells	Trypan blue dye exclusion	30 μM for 24, 48, and 72 h. 1, 5, 10, 30, 60, and 100 μM for 48 h	At 60 μM, ε-viniferin completely blocks cell proliferation. After 48 h, the toxicity potential of ε-viniferin was lower than resveratrol. The IC_{50} for 48 h was 58.4 μM.	[258]
	SW480 cells	Trypan blue dye exclusion and MTT assay	30 μM for 24, 48, 72, and 96 h in trypan blue dye exclusion. 3, 30, 60, and 100 μM for 48 h in coulter counter	Cells exposed to ε-viniferin grew similarly to the control group, with a reduced growth rate and increasing percentage of cell inhibition. In the MTT assay, no significant inhibition of cell proliferation was recorded.	[275]
	VSMCs cells	MTS assay	10, 20, and 30 μM for 48 h	The potential for arresting cell proliferation rate of ε-viniferin at 20 μM was significantly higher than resveratrol's at 20 and 30 μM.	[330]
	SK-MEL-25 and HT-144 cells	MTT assay Trypan blue dye exclusion	25–200 μM for 24, 48, and 72 h	Both melanoma lines showed time- and dose-dependent reduction in survival. The IC_{50} for 48 h was 60 μM.	[249]
	C6 cells	WST-1 assay	95 and 130 μM 12, 24, and 48 h	Proliferation decreased at all doses and times tested in C6 cells.	[253]
	Caco-2 cells	MTT and NRU assays	1.56, 3.12, 6.25, 12.5, 25, 50, and 100 μM for 24 h	At and above 25 μM, cell viability in Caco-2 cells decreased. ε-viniferin was slightly more effective than resveratrol.	[331]
	Vascular endothelial cells (VECs)	H_2O_2-induced cytotoxicity	10, 20, and 30 μM for 24 h	ε-viniferin effectively protected cells from cytotoxic effects of H_2O_2. A 24 h pre-treatment with ε-viniferin reduced intracellular ROS.	[332]
	VECs	H_2O_2-induced cytotoxicity	5 and 10 μM for 24 h	At 10 μM, a pre-treatment with ε-viniferin conferred VEC with resistance against H_2O_2-induced oxidative stress.	[333]
	A2058, A549, HOS, U2OS, and MCF-10A cells	MTT assay	1–15 μM for 24, 48, and 72 h	ε-viniferin exhibited a time- and dose-dependent decrease in the viability of HOS, U2OS, and A549 cells, but not in A2058 cells.	[334]
ε-viniferin glucoside	PC12 cells	MTT assay	0–10 μM for 24 h	Cell viability was not significantly altered following the exposure to the stilbene.	[335]

3.5. Cardiovascular and Metabolic Benefits

Stilbenes act on various factors involved in the bioavailability of NO, such as arginase, dimethylargininase (DDAH), and asymmetric dimethylarginine (ADMA), thereby modulating NO production during oxidative stress [163,336–338]. Also, stilbenes influence metabolic pathways like adipogenesis, lipogenesis, lipolysis, thermogenesis, and fatty acid oxidation. They regulate PPARγ, sterol regulatory element-binding protein 1c (SREBP-1c), uncoupling (UCPs), sirtuin 1 (SIRT1), lipoprotein lipase (LPL), fatty acid synthase (FAS), and acetyl-CoA carboxylase (ACC). Piceatannol and pterostilbene modulate lipid metabolism via ACC, SREBP1, PPAR-γ, and PPAR-α [72]. Pterostilbene demonstrates greater in vivo efficiency than resveratrol due to its higher bioavailability and greater capacity to bind three amino acids of PPARα versus two for resveratrol [339]. Resveratrol has vasodilatory properties, enhancing endothelial NO production and increasing the expression of eNOS in rat arteries while decreasing the expression of NADPH oxidase [340,341]. Studies show that resveratrol activates membrane-associated structures, such as estrogen receptors, which trigger an intracellular signaling cascade targeting the AMPK pathway. This activation can phosphorylate eNOS at serine 1177 [342]. Additionally, resveratrol activates SIRT1, reducing acetylation and leading to eNOS activation. Resveratrol improves SIRT1-dependent NO bioavailability induced by insulin in HUVEC cells cultured under high glucose conditions for 48 h [343]. Another study shows that resveratrol suppresses SIRT1 inhibition, reducing eNOS acetylation in HUVEC cells exposed to H_2O_2-induced oxidative stress [344].

Furthermore, resveratrol increases NO production in blood platelets via the PI3K/protein kinase B (Akt) pathway, activates vasodilator-stimulated phosphoprotein, and inhibits p38MAPK. This leads to reduced platelet activation and ROS production [345]. Resveratrol also dose-dependently increases RNA synthesis of VEGF and eNOS and reduces endothelin secretion in HUVEC cells, lowering blood pressure and improving vasodilation [346]. In human coronary smooth muscle cells incubated with resveratrol, cGMP synthesis increased, activating cGMP-dependent protein kinase, which in turn activated potassium channels, leading to hyperpolarization and relaxation of vascular smooth muscle cells [347].

In vivo studies in rats with accelerated atherosclerosis and metabolic diseases showed improved endothelial dysfunction [57]. In another study using obese Zucker rats, a model for metabolic syndrome, chronic resveratrol administration increased eNOS expression and reduced dyslipidemia and hypertension [348]. Similarly, in hypercholesterolemic rabbits treated with resveratrol, plasma NO concentration increased, and endothelial dysfunction was reduced [349]. However, in estrogen receptor-deficient mice, resveratrol showed no vasorelaxation effects compared to normal mice, demonstrating that estrogen receptors mediate the intracellular effects of resveratrol. Activation of these receptors leads to phosphorylation of Src, Erk ½, and eNOS, resulting in NO synthesis activation [342].

Stilbenes, such as resveratrol found in grapes, need further testing in humans. In a 12-week study where 75 mg/day of resveratrol was administrated to 15 healthy non-obese postmenopausal women, no changes were detected in metabolic functions, such as AMPK signaling, inflammatory markers, insulin, and mitochondrial functions [350]. The absence of effects in humans aligns with previous results in healthy rodents [351], suggesting that resveratrol's effects are limited to conditions like type II diabetes, dyslipidemia, and obesity [117].

A study on resveratrol's role in primary and secondary cardiovascular disease prevention involved a small sample size (n = 75) over one year, limiting clinical relevance. Participants were divided into three groups: one received grape extract with 8.1 mg/day of resveratrol for six months followed by an increase in the stilbene's concentration to 16.2 mg/day for another six months, a second group received grape extract without resveratrol, and the third received a placebo. All participants were on statins and treated per cardiovascular disease prevention guidelines [352]. Results were promising, with the resveratrol-treated group showing improved inflammatory and fibrinolytic profiles in comparison to the remaining groups [352]. Resveratrol provided additional benefits for patients at high risk of cardiovascular disease beyond other phenolic compounds in grape extract, complementing the current primary prevention guidelines [353].

In animal models, resveratrol and pterostilbene demonstrated anti-hyperlipidemic effects, but clinical trials in humans showed no significant changes in low-density lipoprotein (LDL)/high-density lipoprotein (HDL) ratios. A meta-analysis of randomized clinical trials revealed that the use of resveratrol as a supplement did not induce significant alterations in lipid parameters, including LDL, HDL, total cholesterol, and triglycerides [354]. Pterostilbene at high and low doses promotes an increase in LDL without affecting HDL or triglycerides in a placebo-controlled trial [355]. Similarly, grape extract administration did not influence LDL levels [117]. However, in a secondary prevention study of cardiovascular diseases, grape extract with resveratrol (8.1 mg/day for six months, then 16.2 mg/day) in patients with stable coronary artery disease and dietary restrictions showed increased anti-inflammatory adiponectin and reduced plasminogen activator inhibitor-1 (PAI-1) [356].

These findings suggest that resveratrol has cardioprotective effects, improving anti-inflammatory responses and preventing atherothrombotic signaling (Figure 3) [356].

Preclinical data have demonstrated that resveratrol can be utilized in diabetes management through correction of insulin signaling defects, improvement in insulin resistance, and prevention of pancreatic beta cells' dysfunction [357]. Moreover, resveratrol prevents hyperglycemia in animal models for diabetic pathology by promoting glucose uptake and GLUT4 translocation to the caveolar membrane within the diabetic myocardium [358]. It also improves glucose tolerance and reduces the expression of advanced glycation end-product (AGE) receptors in the liver and kidneys of diabetic rats [359].

Resveratrol exhibits actions that inhibit reactive oxygen and nitrogen species, such as superoxide anion ($O_2^{\bullet-}$), hydroxyl radical ($OH^{\bullet}$), H_2O_2, and malondialdehyde (MDA), while increasing levels of antioxidant enzymes like SOD, catalase, and glutathione peroxidase in diabetic animals [360]. It also inhibits pro-inflammatory signaling through NF-κB and reduces the production of inflammatory cytokines, including IL-1β, IL-4, and IL-6 and TNF-α, [360]. Additionally, resveratrol increases insulin sensitivity, glucose tolerance, and mitochondrial biogenesis via AMPK-dependent pathways [361]. A study showed that resveratrol did not achieve the same effects in AMPK-deficient mice, highlighting the essential role of this protein in the metabolic actions of resveratrol [361].

Resveratrol's role in glycemic control remains slightly uncertain, as some human studies lack significant evidence of the stilbene's effect on metabolic dysfunctions. Daily co-administration of metformin or glibenclamide and 250 mg of resveratrol for three months improves glycemic parameters in patients with type 2 diabetes, when compared with those receiving therapies lacking the stilbene [362]. As reported by Movahed and colleagues, 1 g/day of resveratrol for 45 days reduced systolic blood pressure, fasting glucose serum levels, and HbA1c [363]. Furthermore, a much lower daily 5 mg dose of resveratrol was found to reduce systolic blood pressure, HbA1c levels, and improve insulin sensitivity, while not affecting the homeostatic model assessment of insulin resistance, in a 28-day long treatment [363].

In contrast, a randomized clinical trial by Thazhath et al., with 500 mg of resveratrol twice daily for five weeks in diet-controlled type 2 diabetes patients, showed no significant effects on glycemic control [364]. No differences were observed in fasting glucose, postprandial glucose, HbA1c, gastric emptying, or glucagon-like peptide-1 (GLP-1) secretion between the resveratrol and placebo groups. Similarly, a six-month treatment did not show metabolic improvements in type 2 diabetes patients [365]. Therefore, the effects of resveratrol on human diabetes remain not fully understood [117].

Pterostilbene has shown promising results in glycemic control in obese rats with insulin resistance, enhancing hepatic glucokinase activity and glucose uptake in skeletal muscle [366]. In vitro, studies indicate that pterostilbene protects pancreatic beta cells from oxidative stress and apoptosis [367]. Pterostilbene, along with other constituents of *Pterocarpus marsupium*, has demonstrated anti-hyperglycemic properties [368,369]. However, human data on pterostilbene are still limited. One study administered sea buckthorn and blueberry extract to children with type 1 diabetes for two months, resulting in elevated SOD and glutathione peroxidase levels and reduced HbA1c levels [370].

Piceatannol also has glucose-regulating capabilities, increasing glucose disposal. Minakawa et al. observed that piceatannol promotes glucose uptake in cultured myotubes in a dose-dependent manner by activating AMPK and inducing glucose transporter type 4 (GLUT4) translocation [371]. AMPK activators are promising for type 2 diabetes treatment since AMPK stimulates GLUT4 translocation, with skeletal muscle being the primary site of glucose clearance [372,373]. Piceatannol was shown to lower fasting glucose in vivo by activating AMPK in myotubes and minimally impacting insulin secretion in vivo, establishing it as a glucose modulator [371,374,375]. Insulin signaling is modified by piceatannol in different tissues, for instance, it blocked insulin action in adipocytes but improved endothelial function under inflammatory stress [376–378].

3.6. Neuroprotective Effects and Cognitive Function

In neurodegenerative diseases (NDs), there is progressive damage to the structure and function of neurons. The three main NDs are Alzheimer's disease, Parkinson's disease, and amyotrophic lateral sclerosis. Each presents different clinical signs; however, the pathological processes appear similar, suggesting common biological pathways in the onset and progression of NDs. Factors involved in NDs include aging, lifestyle, and genetics [379]. Generally, neuroinflammation and oxidative stress are the primary causes of neuronal dysfunction and death, along with excitotoxicity, mitochondrial dysfunction, and apoptosis [380]. Oxidative stress in neurodegeneration is correlated with the progression of Parkinson's and Alzheimer's diseases [381]. One hypothesis involves the use of natural products due to their centuries-long application in human diseases [382–384]. In recent decades, many studies have described the protective effects of natural polyphenols and their active derivatives against various diseases, including cardiovascular diseases, diabetes, cancers, and NDs [385]. Additionally, natural compounds demonstrate great potential as neuroprotective agents for treating NDs due to their inherently multi-target profiles [386,387].

The neuroprotective effects of stilbenes include antioxidants and anti-inflammatory properties (Figure 3) [388–392]. Resveratrol protects neurons from ROS and improves motor coordination in 1-methyl-4-phenyl-1,2,3,6-tetrahydropyridine-induced parkinsonism in rats by neutralizing hydroxyl radicals [388]. Furthermore, resveratrol improves mitochondrial function, motor coordination, and neuronal survival, inhibits β-secretase, enhances non-amyloidogenic amyloid precursor protein (APP) cleavage, and increases clearance of beta-amyloid (Aβ) peptides [393]. Moreover, resveratrol protects against lipopolysaccharide (LPS)-induced dopaminergic neurodegeneration by inhibiting microglial activation and NF-κB signaling [394]. In Alzheimer's disease, resveratrol shows therapeutic potential by reducing amyloid plaques in the brain. Marambaud et al. demonstrated that resveratrol may not inhibit Aβ production but promotes proteasome-dependent Aβ degradation [395]. Winemaking residues are a source of piceatannol and vitisinol C, which inhibits Aβ aggregation [396,397].

Pre-treatment with resveratrol in rats subjected to cerebral ischemia reperfusion injury increased Nrf2 and HO-1 levels, reducing oxidative damage during ischemic events within the brain [398]. This pre-treatment also results in improved neurological scores with associated reductions in brain water content and infarct volume. In models of global cerebral ischemia, resveratrol blocked neuronal death by activating PI3K/Akt signaling, downregulating GSK-3β, and regulating CREB (cAMP response element-binding protein) [399].

Resveratrol improves cognition in animal models of vascular dementia, showing effects on cerebral ischemia [400]. In this study, vascular dementia was induced by bilateral common carotid artery occlusion for 8–12 weeks. Resveratrol treatment improved learning and memory scores by reducing malondialdehyde, a lipid peroxidation product, and increasing antioxidant enzyme levels, such as glutathione and SOD, in the cerebral cortex and hippocampus [117,400].

Pterostilbene acts as a neuroprotectant by counteracting glucose-induced damage in neuroblastoma cells, preventing the decline in viable cells and ROS generation in a concentration-dependent manner [389]. It also increased mitochondrial cytochrome C, mitochondrial complex I and III activities, and membrane potential, along with Nrf2, HO-1, and GST levels, offering protection against neuronal oxidative stress [389]. Pterostilbene improves memory by increasing RE1-silencing transcription factor (REST), postsynaptic density protein 95 (PSD-95), and mitochondrial porin 1 in the dentate gyrus of aged rats. Resveratrol improves memory in healthy and diabetic individuals with subclinical impairment but not in full Alzheimer's cases [393].

Oxyresveratrol has shown neuroprotective effects. Studies in rat cortical neurons demonstrated that it prevented Aβ (25–35)-induced damage by reducing cytosolic Ca^{2+} levels, inhibiting glutamate release, and lowering ROS generation [401]. Another study by Andrabi et al. demonstrated that, in a rat model of transient middle cerebral artery occlusion, oxyresveratrol significantly reduced cerebral infarct volume and improved subsequent neurological deficits through inhibition of cytochrome C release and caspase-3 activation [402].

In a 52-week long randomized placebo-controlled clinical trial, doses of resveratrol raging between 500 and 1000 mg administered twice daily were well tolerated by Alzheimer's patients, reducing Aβ40 levels in the cerebrospinal fluid and plasma, although other Alzheimer's biomarkers were unaffected [403]. Another trial with a daily supplementation of 200 mg of resveratrol and 320 mg of quercetin resulted in an improvement in memory performance when co-administrated in overweight elderly individuals [404].

A recent study showed that a double daily administration of 75 mg of resveratrol, for a period of 14 weeks, improved cerebrovascular function, cognition, and mood in postmenopausal women. Additionally, it enhanced cognitive performance and cerebral blood flow in individuals with type 2 diabetes [405,406]. In conclusion, human studies have demonstrated the beneficial effects of resveratrol in improving memory and cognition in healthy individuals and those with diabetes-related cognitive impairment, but not in Alzheimer's patients [117].

4. Conclusions and Future Perspectives

The wine industry represents a significant opportunity for obtaining high-value byproducts with potential industrial applications. Vineyard waste is a rich source of compounds with antioxidant effects (specifically stilbenes), which can act as potent reactive free radical scavengers, enzyme activators and/or inhibitors, antibacterial agents, anti-inflammatory agents, and anticancer agents, among other health-beneficial agents. For example, stilbenes inhibit platelet aggregation, possibly through the inhibition of COX-1. They also reduce markers of oxidative stress and inflammation, such as TNF-α and IL-1β, in various models of ischemia reperfusion injury, and inhibit cardiomyocyte hypertrophy through the activation of AMPK. In addition, stilbenes exert beneficial effects in cardiovascular conditions, such as improving glucose tolerance and insulin resistance in animal models of diabetes. Stilbenes exhibit anticancer effects through the inhibition of cancer cell proliferation, oxidative stress, and inflammation, engaging in regulatory roles of cell death mechanisms. From a neurological perspective, stilbenes can upregulate Nrf2 and HO-1 expression to mitigate oxidative damage during cerebral ischemia. Thus, their neuroprotective effects are largely attributed to their antioxidant activity. Stilbenes also exhibit a broad range of functions, including microbicidal and antifungal activities, various effects on the reversal of drug resistance, and biofilms and virulence factors, highlighting the potential of stilbenes as antimicrobial agents. Although the results are promising, most of the evidence comes from in vitro and in vivo studies, with limited clinical trials available to confirm the beneficial effects of stilbenes in humans.

From an industrial perspective, stilbenes have attracted growing interest not only in the pharmaceutical sector but also in the food and cosmetics industries. Their incorporation into biodegradable packaging, natural cosmetics, and functional supplements highlights their potential as multifunctional agents, aligning with the increasing demand for sustainable and bioactive products.

However, to allow large-scale applications, significant advancements are still required in extraction, purification, and formulation technologies. Low water solubility, limited bioavailability, and rapid metabolism remain technical challenges that must be addressed, potentially through innovative strategies such as nanoencapsulation, cyclodextrin complexation, or incorporation into polymeric matrices.

Future prospects for the utilization of stilbenes include the following: the optimization of extraction methods for improved efficiency and selectivity and reduced environmental impact including decreased energy and time consumption, the preservation of stilbene integrity, and the application of green chemistry principles; the development of controlled-release systems for pharmaceutical and cosmetic applications, including the formulation of nanoparticles with stilbenes to enhance bioabsorption and bioavailability for drug delivery; comprehensive toxicological and regulatory evaluations to ensure the safety of stilbene use in consumer products; the exploration of lesser-known stilbenes beyond resveratrol to broaden the range of applications and therapeutic potential, including resveratrol derivatives such as pterostilbene, pinosylvin, ε-viniferin, piceatannol, rhapontigenin, and isorhapontigenin, among others; and integration into circular production chains, contributing to the sustainability of the wine sector and fostering the bioeconomy, thereby fostering eco-design, one of the seven pillars of the circular economy.

In conclusion, stilbenes demonstrate great potential for multiple human health benefits. However, further studies are needed due to the number and complexity of the cellular processes involved. Additionally, because of their low concentrations in foods and associated rapid metabolism and excretion in mammalian organisms, improvements in stability, solubility, and delivery systems are required to enable clinical application. Moreover, there is limited information with regard to the potential beneficial properties of less-studied stilbenes. Therefore, more research is still necessary to uncover the impact of lesser-known stilbenes on health.

Author Contributions: Conceptualization, L.P.B., G.C., S.S. and A.P.D.; investigation, L.P.B.; writing—original draft preparation, L.P.B. and Â.L.; writing—review and editing, L.P.B., Â.L., G.C., S.S. and A.P.D.; supervision, G.C., S.S. and A.P.D.; project administration, S.S. and A.P.D. All authors have read and agreed to the published version of the manuscript.

Funding: This work was developed within the scope of the CICS-UBI (Portugal) core grant (https://doi.org/10.54499/UIDB/00709/2020) and the CICS-UBI programmatic grant (https://doi.org/10.54499/UIDP/00709/2020) with national funds registered in the budget of the FCT. Ângelo Luís acknowledges the Scientific Employment contract (Microbiology), funded by Portuguese Foundation for Science and Technology (FCT), within the scope of DL57/2016 (Reference: DL57/2016/CP1354/CT0010). Luis Brás acknowledges funding from individual PhD fellowships from FCT (UNITA) by project PRT/BD/154423/2022 (DOI https://doi.org/10.54499/PRT/BD/154423/2022).

Conflicts of Interest: The authors declare no conflicts of interest.

Abbreviations

The following abbreviations are used in this manuscript:

UV	Ultraviolet
STS	Stilbene synthase
DW	Dry weight
MAE	Microwave-assisted extraction
UAE	Ultrasound-assisted extraction
PSE	Pressurized solvent extraction
COX	Cyclooxygenase
Aβ	Beta-amyloid
ROS	Reactive oxygen species
RNS	Reactive nitrogen species
SOD	Superoxide dismutase
NADPH	Nicotinamide adenine dinucleotide phosphate
GPx	Glutathione peroxidase
GST	Glutathione-S-transferase
NF-κB	Nuclear factor κB
eNOS	Endothelial nitric oxide synthase
LDL	Low-density lipoprotein
SIRT1	Sirtuin 1
H_2O_2	Hydrogen peroxide
TNF-α	Tumor necrosis factor-α
AMPK	Adenosine monophosphate-activated protein kinase
LPS	Lipopolysaccharide
HO-1	Heme oxygenase-1
Nrf2	Nuclear factor erythroid 2-related factor 2
STAT3	Signal transducer and activator of transcription 3
NRU	Neutral red uptake
BH4	Tetrahydrobiopterin
cAMP	Cyclic adenosine monophosphate
cGMP	Cyclic guanosine monophosphate
NDs	Neurodegenerative diseases
CCR1	Chemokine receptor
PGE2	Prostaglandin E2
GSH	Glutathione
PTEN	Phosphatase and TENsin homolog
PKB	Protein kinase B
GSK-3β	Glycogen synthase kinase-3β
VEGF	Vascular endothelial growth factor
CSC	Cancer stem cell
VECs	Vascular endothelial cells
PI3K	Phosphoinositide 3-kinase
Akt	Protein kinase B
HDL	High-density lipoprotein

References

1. International Organisation of Vine and Wine. 2023 World Wine Production—Oiv First Estimates. Available online: https://www.fao.org/3/i7042e/i7042e.pdf (accessed on 11 August 2025).
2. OIV. *State of the World Vine and Wine Sector in 2023*; International Organisation of Vine and Wine: Paris, France, 2023.
3. Bustamante, M.A.; Moral, R.; Paredes, C.; Pérez-Espinosa, A.; Moreno-Caselles, J.; Pérez-Murcia, M.D. Agrochemical characterisation of the solid by-products and residues from the winery and distillery industry. *Waste Manag.* **2008**, *28*, 372–380. [CrossRef]

4. Bordiga, M.; Travaglia, F.; Locatelli, M.; Arlorio, M.; Coïsson, J.D. Spent grape pomace as a still potential by-product. *Int. J. Food Sci. Technol.* **2015**, *50*, 2022–2031. [CrossRef]
5. Yu, J.; Ahmedna, M. Functional components of grape pomace: Their composition, biological properties and potential applications. *Int. J. Food Sci. Technol.* **2013**, *48*, 221–237. [CrossRef]
6. Broome, J.C.; Warner, K.D. Agro-environmental partnerships facilitate sustainable wine-grape production and assessment. *Calif. Agric.* **2008**, *62*, 133–141. [CrossRef]
7. Zacharof, M.-P. Grape Winery Waste as Feedstock for Bioconversions: Applying the Biorefinery Concept. *Waste Biomass Valorization* **2017**, *8*, 1011–1025. [CrossRef]
8. Prozil, S.O.; Evtuguin, D.V.; Lopes, L.P.C. Chemical composition of grape stalks of *Vitis vinifera* L. from red grape pomaces. *Ind. Crops Prod.* **2012**, *35*, 178–184. [CrossRef]
9. Kalli, E.; Lappa, I.; Bouchagier, P.; Tarantilis, P.A.; Skotti, E. Novel application and industrial exploitation of winery by-products. *Bioresour. Bioprocess.* **2018**, *5*, 46. [CrossRef]
10. Chowdhary, P.; Gupta, A.; Gnansounou, E.; Pandey, A.; Chaturvedi, P. Current trends and possibilities for exploitation of Grape pomace as a potential source for value addition. *Environ. Pollut.* **2021**, *278*, 116796. [CrossRef] [PubMed]
11. Toma, D.-I.; Baroi, A.M.; Din, A.; Vizitiu, D.E.; Fierascu, I.; Fierascu, R.C. Grapevine Plant Waste Utilization in Nanotechnology. *AgroLife Sci. J.* **2024**, *13*, 203–216. [CrossRef]
12. Wei, M.; Ma, T.; Ge, Q.; Li, C.; Zhang, K.; Fang, Y.; Sun, X. Challenges and opportunities of winter vine pruning for global grape and wine industries. *J. Clean. Prod.* **2022**, *380*, 135086. [CrossRef]
13. Ilyas, T.; Chowdhary, P.; Chaurasia, D.; Gnansounou, E.; Pandey, A.; Chaturvedi, P. Sustainable green processing of grape pomace for the production of value-added products: An overview. *Environ. Technol. Innov.* **2021**, *23*, 101592. [CrossRef]
14. Khan, Z.A.; Iqbal, A.; Shahzad, S.A. Synthetic approaches toward stilbenes and their related structures. *Mol. Divers.* **2017**, *21*, 483–509. [CrossRef] [PubMed]
15. Baroi, A.M.; Popitiu, M.; Fierascu, I.; Sărdărescu, I.-D.; Fierascu, R.C. Grapevine Wastes: A Rich Source of Antioxidants and Other Biologically Active Compounds. *Antioxidants* **2022**, *11*, 393. [CrossRef]
16. Beres, C.; Costa, G.N.S.; Cabezudo, I.; da Silva-James, N.K.; Teles, A.S.C.; Cruz, A.P.G.; Mellinger-Silva, C.; Tonon, R.V.; Cabral, L.M.C.; Freitas, S.P. Towards integral utilization of grape pomace from winemaking process: A review. *Waste Manag.* **2017**, *68*, 581–594. [CrossRef]
17. Cuccia, P. Ethics+ economy+ environment= sustainability: Gambero Rosso on the front lines with a new concept of sustainability. *Wine Econ. Policy* **2015**, *4*, 69–70. [CrossRef]
18. Iuga, M.; Mironeasa, S. Potential of grape byproducts as functional ingredients in baked goods and pasta. *Compr. Rev. Food Sci. Food Saf.* **2020**, *19*, 2473–2505. [CrossRef]
19. Contreras, M.d.M.; Romero-García, J.M.; López-Linares, J.C.; Romero, I.; Castro, E. Residues from grapevine and wine production as feedstock for a biorefinery. *Food Bioprod. Process.* **2022**, *134*, 56–79. [CrossRef]
20. Bensid, A.; El Abed, N.; Houicher, A.; Regenstein, J.M.; Özogul, F. Antioxidant and antimicrobial preservatives: Properties, mechanism of action and applications in food—A review. *Crit. Rev. Food Sci. Nutr.* **2022**, *62*, 2985–3001. [CrossRef]
21. Vilela, A.; Pinto, T. Grape Infusions: Between Nutraceutical and Green Chemistry. *Sustain. Chem.* **2021**, *2*, 441–466. [CrossRef]
22. Teixeira, A.; Baenas, N.; Dominguez-Perles, R.; Barros, A.; Rosa, E.; Moreno, D.A.; Garcia-Viguera, C. Natural Bioactive Compounds from Winery By-Products as Health Promoters: A Review. *Int. J. Mol. Sci.* **2014**, *15*, 15638–15678. [CrossRef] [PubMed]
23. Ferrer-Gallego, R.; Silva, P. The Wine Industry By-Products: Applications for Food Industry and Health Benefits. *Antioxidants* **2022**, *11*, 2025. [CrossRef]
24. Nunes, M.A.; Rodrigues, F.; Oliveira, M.B.P.P. 11—Grape Processing By-Products as Active Ingredients for Cosmetic Proposes. In *Handbook of Grape Processing By-Products*; Galanakis, C.M., Ed.; Academic Press: Cambridge, MA, USA, 2017; pp. 267–292. [CrossRef]
25. Silva, A.; Silva, V.; Igrejas, G.; Aires, A.; Falco, V.; Valentão, P.; Poeta, P. Phenolic compounds classification and their distribution in winemaking by-products. *Eur. Food Res. Technol.* **2023**, *249*, 207–239. [CrossRef]
26. Constantin, O.E.; Stoica, F.; Rațu, R.N.; Stănciuc, N.; Bahrim, G.E.; Râpeanu, G. Bioactive Components, Applications, Extractions, and Health Benefits of Winery By-Products from a Circular Bioeconomy Perspective: A Review. *Antioxidants* **2024**, *13*, 100. [CrossRef]
27. Ianni, A.; Di Maio, G.; Pittia, P.; Grotta, L.; Perpetuini, G.; Tofalo, R.; Cichelli, A.; Martino, G. Chemical–nutritional quality and oxidative stability of milk and dairy products obtained from Friesian cows fed with a dietary supplementation of dried grape pomace. *J. Sci. Food Agric.* **2019**, *99*, 3635–3643. [CrossRef] [PubMed]
28. Troilo, M.; Difonzo, G.; Paradiso, V.M.; Summo, C.; Caponio, F. Bioactive Compounds from Vine Shoots, Grape Stalks, and Wine Lees: Their Potential Use in Agro-Food Chains. *Foods* **2021**, *10*, 342. [CrossRef]

29. Ahmad, B.; Yadav, V.; Yadav, A.; Rahman, M.U.; Yuan, W.Z.; Li, Z.; Wang, X. Integrated biorefinery approach to valorize winery waste: A review from waste to energy perspectives. *Sci. Total Environ.* **2020**, *719*, 137315. [CrossRef]
30. Bordiga, M.; Montella, R.; Travaglia, F.; Arlorio, M.; Coïsson, J.D. Characterization of polyphenolic and oligosaccharidic fractions extracted from grape seeds followed by the evaluation of prebiotic activity related to oligosaccharides. *Int. J. Food Sci. Technol.* **2019**, *54*, 1283–1291. [CrossRef]
31. Gabur, G.-D.; Teodosiu, C.; Fighir, D.; Cotea, V.V.; Gabur, I. From Waste to Value in Circular Economy: Valorizing Grape Pomace Waste through Vermicomposting. *Agriculture* **2024**, *14*, 1529. [CrossRef]
32. Vitalini, S.; Gardana, C.; Zanzotto, A.; Simonetti, P.; Faoro, F.; Fico, G.; Iriti, M. The presence of melatonin in grapevine (*Vitis vinifera* L.) berry tissues. *J. Pineal Res.* **2011**, *51*, 331–337. [CrossRef] [PubMed]
33. Moreira, M.M.; Barroso, M.F.; Porto, J.V.; Ramalhosa, M.J.; Švarc-Gajić, J.; Estevinho, L.; Morais, S.; Delerue-Matos, C. Potential of Portuguese vine shoot wastes as natural resources of bioactive compounds. *Sci. Total Environ.* **2018**, *634*, 831–842. [CrossRef]
34. Peña-Portillo, G.-C.; Acuña-Nelson, S.-M.; Bastías-Montes, J.-M. From Waste to Wealth: Exploring the Bioactive Potential of Wine By-Products—A Review. *Antioxidants* **2024**, *13*, 992. [CrossRef]
35. Goufo, P.; Singh, R.K.; Cortez, I. A Reference List of Phenolic Compounds (Including Stilbenes) in Grapevine (*Vitis vinifera* L.) Roots, Woods, Canes, Stems, and Leaves. *Antioxidants* **2020**, *9*, 398. [CrossRef] [PubMed]
36. Zwingelstein, M.; Draye, M.; Besombes, J.-L.; Piot, C.; Chatel, G. trans-Resveratrol and trans-ε-Viniferin in Grape Canes and Stocks Originating from Savoie Mont Blanc Vineyard Region: Pre-extraction Parameters for Improved Recovery. *ACS Sustain. Chem. Eng.* **2019**, *7*, 8310–8316. [CrossRef]
37. Zagoskina, N.V.; Zubova, M.Y.; Nechaeva, T.L.; Kazantseva, V.V.; Goncharuk, E.A.; Katanskaya, V.M.; Baranova, E.N.; Aksenova, M.A. Polyphenols in Plants: Structure, Biosynthesis, Abiotic Stress Regulation, and Practical Applications (Review). *Int. J. Mol. Sci.* **2023**, *24*, 13874. [CrossRef]
38. Quideau, S.; Deffieux, D.; Douat-Casassus, C.; Pouységu, L. Plant Polyphenols: Chemical Properties, Biological Activities, and Synthesis. *Angew. Chem. Int. Ed.* **2011**, *50*, 586–621. [CrossRef]
39. Manach, C.; Scalbert, A.; Morand, C.; Rémésy, C.; Jiménez, L. Polyphenols: Food sources and bioavailability. *Am. J. Clin. Nutr.* **2004**, *79*, 727–747. [CrossRef]
40. Bravo, L. Polyphenols: Chemistry, Dietary Sources, Metabolism, and Nutritional Significance. *Nutr. Rev.* **1998**, *56*, 317–333. [CrossRef]
41. Zhang, Y.; Li, P.; Cheng, L. Developmental changes of carbohydrates, organic acids, amino acids, and phenolic compounds in 'Honeycrisp'apple flesh. *Food Chem.* **2010**, *123*, 1013–1018. [CrossRef]
42. Dai, J.; Mumper, R.J. Plant Phenolics: Extraction, Analysis and Their Antioxidant and Anticancer Properties. *Molecules* **2010**, *15*, 7313–7352. [CrossRef]
43. Barry, T.N.; Forss, D.A. The condensed tannin content of vegetative Lotus pedunculatus, its regulation by fertiliser application, and effect upon protein solubility. *J. Sci. Food Agric.* **1983**, *34*, 1047–1056. [CrossRef]
44. Sies, H. Polyphenols and health: Update and perspectives. *Arch. Biochem. Biophys.* **2010**, *501*, 2–5. [CrossRef]
45. Kampa, M.; Nifli, A.-P.; Notas, G.; Castanas, E. Polyphenols and cancer cell growth. In *Reviews of Physiology, Biochemistry and Pharmacology*; Amara, S.G., Bamberg, E., Fleischmann, B., Gudermann, T., Hebert, S.C., Jahn, R., Lederer, W.J., Lill, R., Miyajima, A., Offermanns, S., et al., Eds.; Springer: Berlin/Heidelberg, Germany, 2007; pp. 79–113.
46. Guo, W.; Kong, E.; Meydani, M. Dietary Polyphenols, Inflammation, and Cancer. *Nutr. Cancer* **2009**, *61*, 807–810. [CrossRef]
47. Korkina, L.G.; De Luca, C.; Kostyuk, V.A.; Pastore, S. Plant Polyphenols and Tumors: From Mechanisms to Therapies, Prevention, and Protection Against Toxicity of Anti-Cancer Treatments. *Curr. Med. Chem.* **2009**, *16*, 3943–3965. [CrossRef]
48. Zhao, B. Natural Antioxidants Protect Neurons in Alzheimer's Disease and Parkinson's Disease. *Neurochem. Res.* **2009**, *34*, 630–638. [CrossRef]
49. Joseph, J.; Cole, G.; Head, E.; Ingram, D. Nutrition, Brain Aging, and Neurodegeneration. *J. Neurosci.* **2009**, *29*, 12795–12801. [CrossRef] [PubMed]
50. Gutierrez-Merino, C.; Lopez-Sanchez, C.; Lagoa, R.; Samhan-Arias, A.K.; Bueno, C.; Garcia-Martinez, V. Neuroprotective Actions of Flavonoids. *Curr. Med. Chem.* **2011**, *18*, 1195–1212. [CrossRef]
51. Michalska, M.; Gluba, A.; Mikhailidis, D.P.; Nowak, P.; Bielecka-Dabrowa, A.; Rysz, J.; Banach, M. The role of polyphenols in cardiovascular disease. *Med. Sci. Monit.* **2010**, *16*, RA110–RA119.
52. Grassi, D.; Desideri, G.; Croce, G.; Tiberti, S.; Aggio, A.; Ferri, C. Flavonoids, Vascular Function and Cardiovascular Protection. *Curr. Pharm. Des.* **2009**, *15*, 1072–1084. [CrossRef]
53. Milne, J.C.; Lambert, P.D.; Schenk, S.; Carney, D.P.; Smith, J.J.; Gagne, D.J.; Jin, L.; Boss, O.; Perni, R.B.; Vu, C.B.; et al. Small molecule activators of SIRT1 as therapeutics for the treatment of type 2 diabetes. *Nature* **2007**, *450*, 712–716. [CrossRef] [PubMed]
54. Zunino, S.J. Type 2 Diabetes and Glycemic Response to Grapes or Grape Products1,2. *J. Nutr.* **2009**, *139*, 1794S–1800S. [CrossRef] [PubMed]

55. Borriello, A.; Cucciolla, V.; Della Ragione, F.; Galletti, P. Dietary polyphenols: Focus on resveratrol, a promising agent in the prevention of cardiovascular diseases and control of glucose homeostasis. *Nutr. Metab. Cardiovasc. Dis.* **2010**, *20*, 618–625. [CrossRef]
56. Ferrari, C.K.B. Functional foods, herbs and nutraceuticals: Towards biochemical mechanisms of healthy aging. *Biogerontology* **2004**, *5*, 275–289. [CrossRef]
57. Pearson, K.J.; Baur, J.A.; Lewis, K.N.; Peshkin, L.; Price, N.L.; Labinskyy, N.; Swindell, W.R.; Kamara, D.; Minor, R.K.; Perez, E.; et al. Resveratrol Delays Age-Related Deterioration and Mimics Transcriptional Aspects of Dietary Restriction without Extending Life Span. *Cell Metab.* **2008**, *8*, 157–168. [CrossRef]
58. Queen, B.L.; Tollefsbol, T.O. Polyphenols and Aging. *Curr. Aging Sci.* **2010**, *3*, 34–42. [CrossRef]
59. Agouni, A.; Lagrue-Lak-Hal, A.-H.; Mostefai, H.A.; Tesse, A.; Mulder, P.; Rouet, P.; Desmoulin, F.; Heymes, C.; Martínez, M.C.; Andriantsitohaina, R. Red Wine Polyphenols Prevent Metabolic and Cardiovascular Alterations Associated with Obesity in Zucker Fatty Rats (Fa/Fa). *PLoS ONE* **2009**, *4*, e5557. [CrossRef] [PubMed]
60. Cherniack, E.P. Polyphenols: Planting the seeds of treatment for the metabolic syndrome. *Nutrition* **2011**, *27*, 617–623. [CrossRef] [PubMed]
61. Rahman, I.; Chung, S. Dietary polyphenols, deacetylases and chromatin remodeling in inflammation. *Pers. Nutr.* **2010**, *101*, 84–94.
62. Kostyuk, V.; Potapovich, A.; De Luca, C. The Promise of Plant Polyphenols as the Golden Standard Skin Anti-Inflammatory Agents. *Curr. Drug Metab.* **2010**, *11*, 415–424. [CrossRef] [PubMed]
63. Acquaviva, R.; Iauk, L. Natural polyphenols as anti-inflammatory agents. *J. Front. Biosci.* **2010**, *2*, 318–331.
64. Sáez, V.; Gayoso, C.; Riquelme, S.; Pérez, J.; Vergara, C.; Mardones, C.; von Baer, D. C18 core-shell column with in-series absorbance and fluorescence detection for simultaneous monitoring of changes in stilbenoid and proanthocyanidin concentrations during grape cane storage. *J. Chromatogr. B* **2018**, *1074–1075*, 70–78. [CrossRef]
65. Waffo-Teguo, P.; Krisa, S.; Richard, T.; Mérillon, J.-M. Grapevine Stilbenes and Their Biological Effects. In *Bioactive Molecules and Medicinal Plants*; Ramawat, K.G., Merillon, J.M., Eds.; Springer: Berlin/Heidelberg, Germany, 2008; pp. 25–54.
66. Mattivi, F.; Vrhovsek, U.; Malacarne, G.; Masuero, D.; Zulini, L.; Stefanini, M.; Moser, C.; Velasco, R.; Guella, G. Profiling of Resveratrol Oligomers, Important Stress Metabolites, Accumulating in the Leaves of Hybrid *Vitis vinifera* (Merzling × Teroldego) Genotypes Infected with Plasmopara viticola. *J. Agric. Food Chem.* **2011**, *59*, 5364–5375. [CrossRef]
67. Jiang, Y.L.; Liu, Z.P. Natural Products as Anti-Invasive and Anti-Metastatic Agents. *Curr. Med. Chem.* **2011**, *18*, 808–829. [CrossRef]
68. Park, E.-J.; Pezzuto, J.M. The pharmacology of resveratrol in animals and humans. *Biochim. Biophys. Acta (BBA)—Mol. Basis Dis.* **2015**, *1852*, 1071–1113. [CrossRef]
69. Seyed, M.A.; Jantan, I.; Bukhari, S.N.A.; Vijayaraghavan, K. A Comprehensive Review on the Chemotherapeutic Potential of Piceatannol for Cancer Treatment, with Mechanistic Insights. *J. Agric. Food Chem.* **2016**, *64*, 725–737. [CrossRef] [PubMed]
70. Billet, K.; Houillé, B.; Besseau, S.; Mélin, C.; Oudin, A.; Papon, N.; Courdavault, V.; Clastre, M.; Giglioli-Guivarc'h, N.; Lanoue, A. Mechanical stress rapidly induces E-resveratrol and E-piceatannol biosynthesis in grape canes stored as a freshly-pruned byproduct. *Food Chem.* **2018**, *240*, 1022–1027. [CrossRef] [PubMed]
71. Guerrero, R.F.; Biais, B.; Richard, T.; Puertas, B.; Waffo-Teguo, P.; Merillon, J.-M.; Cantos-Villar, E. Grapevine cane's waste is a source of bioactive stilbenes. *Ind. Crops Prod.* **2016**, *94*, 884–892. [CrossRef]
72. Navarro-Orcajada, S.; Conesa, I.; Vidal-Sánchez, F.J.; Matencio, A.; Albaladejo-Maricó, L.; García-Carmona, F.; López-Nicolás, J.M. Stilbenes: Characterization, bioactivity, encapsulation and structural modifications. A review of their current limitations and promising approaches. *Crit. Rev. Food Sci. Nutr.* **2023**, *63*, 7269–7287. [CrossRef]
73. Chong, J.; Poutaraud, A.; Hugueney, P. Metabolism and roles of stilbenes in plants. *Plant Sci.* **2009**, *177*, 143–155. [CrossRef]
74. Bábíková, P.; Vrchotová, N.; Tříska, J.; Kyseláková, M. Content of trans-resveratrol in leaves and berries of interspecific grapevine (*Vitis* sp.) varieties. *Czech J. Food Sci.* **2008**, *26*, S13–S17. [CrossRef]
75. Rusjan, D.; Halbwirth, H.; Stich, K.; Mikulič-Petkovšek, M.; Veberič, R. Biochemical response of grapevine variety 'Chardonnay' (*Vitis vinifera* L.) to infection with grapevine yellows (*Bois noir*). *Eur. J. Plant Pathol.* **2012**, *134*, 231–237. [CrossRef]
76. Parage, C.; Tavares, R.; Réty, S.; Baltenweck-Guyot, R.; Poutaraud, A.; Renault, L.; Heintz, D.; Lugan, R.; Marais, G.A.B.; Aubourg, S.; et al. Structural, Functional, and Evolutionary Analysis of the Unusually Large Stilbene Synthase Gene Family in Grapevine. *Plant Physiol.* **2012**, *160*, 1407–1419. [CrossRef]
77. Çetin, E.S.; Altinöz, D.; Tarçan, E.; Göktürk Baydar, N. Chemical composition of grape canes. *Ind. Crops Prod.* **2011**, *34*, 994–998. [CrossRef]
78. Ewald, P.; Delker, U.; Winterhalter, P. Quantification of stilbenoids in grapevine canes and grape cluster stems with a focus on long-term storage effects on stilbenoid concentration in grapevine canes. *Food Res. Int.* **2017**, *100*, 326–331. [CrossRef]
79. Cebrián, C.; Sánchez-Gómez, R.; Salinas, M.R.; Alonso, G.L.; Zalacain, A. Effect of post-pruning vine-shoots storage on the evolution of high-value compounds. *Ind. Crops Prod.* **2017**, *109*, 730–736. [CrossRef]
80. Ben Mohamed, H.; Vadel, A.M.; Geuns, J.M.C.; Khemira, H. Biochemical changes in dormant grapevine shoot tissues in response to chilling: Possible role in dormancy release. *Sci. Hortic.* **2010**, *124*, 440–447. [CrossRef]

81. Aaviksaar, A.; Haga, M.; Pussa, T.; Roasto, M.; Tsoupras, G. Purification of resveratrol from vine stems. *Proc.-Est. Acad. Sci. Chem.* **2003**, *52*, 155–164. [CrossRef]
82. Püssa, T.; Floren, J.; Kuldkepp, P.; Raal, A. Survey of Grapevine Vitis vinifera Stem Polyphenols by Liquid Chromatography−Diode Array Detection−Tandem Mass Spectrometry. *J. Agric. Food Chem.* **2006**, *54*, 7488–7494. [CrossRef]
83. Cruz, S.; Raposo, R.; Ruiz-Moreno, M.J.; Garde-Cerdán, T.; Puertas, B.; Gonzalo-Diago, A.; Moreno-Rojas, J.M.; Cantos-Villar, E. Grapevine-shoot stilbene extract as a preservative in white wine. *Food Packag. Shelf Life* **2018**, *18*, 164–172. [CrossRef]
84. Cebrián-Tarancón, C.; Sánchez-Gómez, R.; Salinas, M.R.; Alonso, G.L.; Oliva, J.; Zalacain, A. Toasted vine-shoot chips as enological additive. *Food Chem.* **2018**, *263*, 96–103. [CrossRef]
85. Rätsep, R.; Karp, K.; Maante-Kuljus, M.; Aluvee, A.; Kaldmäe, H.; Bhat, R. Recovery of Polyphenols from Vineyard Pruning Wastes—Shoots and Cane of Hybrid Grapevine (*Vitis* sp.) Cultivars. *Antioxidants* **2021**, *10*, 1059. [CrossRef]
86. Pietarinen, S.P.; Willför, S.M.; Ahotupa, M.O.; Hemming, J.E.; Holmbom, B.R. Knotwood and bark extracts: Strong antioxidants from waste materials. *J. Wood Sci.* **2006**, *52*, 436–444. [CrossRef]
87. García-Pérez, M.-E.; Royer, M.; Herbette, G.; Desjardins, Y.; Pouliot, R.; Stevanovic, T. Picea mariana bark: A new source of trans-resveratrol and other bioactive polyphenols. *Food Chem.* **2012**, *135*, 1173–1182. [CrossRef]
88. Rajbhar, K.; Dawda, H.; Mukundan, U. Polyphenols: Methods of extraction. *Sci. Revs. Chem. Commun.* **2015**, *5*, 1–6.
89. Zwingelstein, M.; Draye, M.; Besombes, J.-L.; Piot, C.; Chatel, G. Viticultural wood waste as a source of polyphenols of interest: Opportunities and perspectives through conventional and emerging extraction methods. *Waste Manag.* **2020**, *102*, 782–794. [CrossRef]
90. Angelov, G.; Boyadzhieva, S.; Georgieva, S. Rosehip extraction: Process optimization and antioxidant capacity of extracts. *Open Chem.* **2014**, *12*, 502–508. [CrossRef]
91. Angelov, G.; Georgieva, S.; Boyadzhieva, S.; Boyadzhiev, L. Optimizing the extraction of globe artichoke wastes. *Comptes Rendus De. L'Academie Bulg. Des. Sci.* **2015**, *68*, 1235–1240.
92. Kumar, B.; Bhardwaj, N.; Agrawal, K.; Chaturvedi, V.; Verma, P. Current perspective on pretreatment technologies using lignocellulosic biomass: An emerging biorefinery concept. *Fuel Process. Technol.* **2020**, *199*, 106244. [CrossRef]
93. Carvalho, T.M.J. Extraction of Raw Plant Material Using Supercritical Carbon Dioxide. Master's Thesis, Warsaw University of Technology, Warsaw, Poland, 2016.
94. Mahindrakar, K.V.; Rathod, V.K. Chapter 5—Ultrasound-assisted extraction of lipids, carotenoids, and other compounds from marine resources. In *Innovative and Emerging Technologies in the Bio-Marine Food Sector*; Garcia-Vaquero, M., Rajauria, G., Eds.; Academic Press: Cambridge, MA, USA, 2022; pp. 81–128.
95. Šimat, V.; Frleta, R.; Čagalj, M.; Skroza, D. Technological and Analytical Aspects of Bioactive Compounds and Nutraceuticals from Marine Algae. In *Bioactive Compounds and Nutraceuticals from Dairy, Marine, and Nonconventional Sources*; Apple Academic Press: Palm Bay, FL, USA, 2024; pp. 159–196.
96. Maddaloni, M.; Vassalini, I.; Alessandri, I. Green Routes for the Development of Chitin/Chitosan Sustainable Hydrogels. *Sustain. Chem.* **2020**, *1*, 325–344. [CrossRef]
97. Schnee, S.; Queiroz, E.F.; Voinesco, F.; Marcourt, L.; Dubuis, P.-H.; Wolfender, J.-L.; Gindro, K. Vitis vinifera Canes, a New Source of Antifungal Compounds against Plasmopara viticola, Erysiphe necator, and Botrytis cinerea. *J. Agric. Food Chem.* **2013**, *61*, 5459–5467. [CrossRef]
98. Gabaston, J.; Leborgne, C.; Waffo-Teguo, P.; Valls, J.; Palos Pinto, A.; Richard, T.; Cluzet, S.; Mérillon, J.-M. Wood and roots of major grapevine cultivars and rootstocks: A comparative analysis of stilbenes by UHPLC-DAD-MS/MS and NMR. *Phytochem. Anal.* **2019**, *30*, 320–331. [CrossRef]
99. Bavaresco, L.; Fregoni, C. Physiological Role and Molecular Aspects of Grapevine Stilbenic Compounds. In *Molecular Biology & Biotechnology of the Grapevine*; Roubelakis-Angelakis, K.A., Ed.; Springer: Dordrecht, The Netherlands, 2001; pp. 153–182.
100. El Khawand, T.; Courtois, A.; Valls, J.; Richard, T.; Krisa, S. A review of dietary stilbenes: Sources and bioavailability. *Phytochem. Rev.* **2018**, *17*, 1007–1029. [CrossRef]
101. Kapetanovic, I.M.; Muzzio, M.; Huang, Z.; Thompson, T.N.; McCormick, D.L. Pharmacokinetics, oral bioavailability, and metabolic profile of resveratrol and its dimethylether analog, pterostilbene, in rats. *Cancer Chemother. Pharmacol.* **2011**, *68*, 593–601. [CrossRef]
102. Marier, J.-F.; Vachon, P.; Gritsas, A.; Zhang, J.; Moreau, J.-P.; Ducharme, M.P. Metabolism and Disposition of Resveratrol in Rats: Extent of Absorption, Glucuronidation, and Enterohepatic Recirculation Evidenced by a Linked-Rat Model. *J. Pharmacol. Exp. Ther.* **2002**, *302*, 369–373. [CrossRef]
103. Walle, T. Bioavailability of resveratrol. *Ann. N. Y. Acad. Sci.* **2011**, *1215*, 9–15. [CrossRef] [PubMed]
104. Das, S.; Lin, H.-S.; Ho, P.C.; Ng, K.-Y. The Impact of Aqueous Solubility and Dose on the Pharmacokinetic Profiles of Resveratrol. *Pharm. Res.* **2008**, *25*, 2593–2600. [CrossRef] [PubMed]
105. Teplova, V.V.; Isakova, E.P.; Klein, O.I.; Dergachova, D.I.; Gessler, N.N.; Deryabina, Y.I. Natural Polyphenols: Biological Activity, Pharmacological Potential, Means of Metabolic Engineering (Review). *Appl. Biochem. Microbiol.* **2018**, *54*, 221–237. [CrossRef]

106. Schouten, A.; Wagemakers, L.; Stefanato, F.L.; Kaaij, R.M.v.d.; Kan, J.A.L.v. Resveratrol acts as a natural profungicide and induces self-intoxication by a specific laccase. *Mol. Microbiol.* **2002**, *43*, 883–894. [CrossRef]
107. Seppänen, S.K.; Syrjälä, L.; von Weissenberg, K.; Teeri, T.H.; Paajanen, L.; Pappinen, A. Antifungal activity of stilbenes in in vitro bioassays and in transgenic Populus expressing a gene encoding pinosylvin synthase. *Plant Cell Rep.* **2004**, *22*, 584–593. [CrossRef]
108. Liu, Q.; Yeo, W.S.; Bae, T. The SaeRS Two-Component System of Staphylococcus aureus. *Genes.* **2016**, *7*, 81. [CrossRef]
109. Duan, J.; Li, M.; Hao, Z.; Shen, X.; Liu, L.; Jin, Y.; Wang, S.; Guo, Y.; Yang, L.; Wang, L.; et al. Subinhibitory concentrations of resveratrol reduce alpha-hemolysin production in Staphylococcus aureus isolates by downregulating saeRS. *Emerg. Microbes Infect.* **2018**, *7*, 1–10. [CrossRef]
110. Flemming, H.-C.; Wuertz, S. Bacteria and archaea on Earth and their abundance in biofilms. *Nat. Rev. Microbiol.* **2019**, *17*, 247–260. [CrossRef]
111. Kattke, M.D.; Gosschalk, J.E.; Martinez, O.E.; Kumar, G.; Gale, R.T.; Cascio, D.; Sawaya, M.R.; Philips, M.; Brown, E.D.; Clubb, R.T. Structure and mechanism of TagA, a novel membrane-associated glycosyltransferase that produces wall teichoic acids in pathogenic bacteria. *PLoS Pathog.* **2019**, *15*, e1007723. [CrossRef]
112. Rabin, N.; Zheng, Y.; Opoku-Temeng, C.; Du, Y.; Bonsu, E.; Sintim, H.O. Biofilm Formation Mechanisms and Targets for Developing Antibiofilm Agents. *Future Med. Chem.* **2015**, *7*, 493–512. [CrossRef]
113. Ramadan, H.H. Chronic rhinosinusitis and bacterial biofilms. *Curr. Opin. Otolaryngol. Head. Neck Surg.* **2006**, *14*, 183–186. [CrossRef] [PubMed]
114. Hall-Stoodley, L.; Costerton, J.W.; Stoodley, P. Bacterial biofilms: From the Natural environment to infectious diseases. *Nat. Rev. Microbiol.* **2004**, *2*, 95–108. [CrossRef] [PubMed]
115. Wang, W.-B.; Lai, H.-C.; Hsueh, P.-R.; Chiou, R.Y.-Y.; Lin, S.-B.; Liaw, S.-J. Inhibition of swarming and virulence factor expression in Proteus mirabilis by resveratrol. *J. Med. Microbiol.* **2006**, *55*, 1313–1321. [CrossRef]
116. Fernández-Alvarez, A.; Llorente-Izquierdo, C.; Mayoral, R.; Agra, N.; Boscá, L.; Casado, M.; Martín-Sanz, P. Evaluation of epigenetic modulation of cyclooxygenase-2 as a prognostic marker for hepatocellular carcinoma. *Oncogenesis* **2012**, *1*, e23. [CrossRef]
117. Akinwumi, B.C.; Bordun, K.-A.M.; Anderson, H.D. Biological Activities of Stilbenoids. *Int. J. Mol. Sci.* **2018**, *19*, 792. [CrossRef] [PubMed]
118. Remsberg, C.M.; Yáñez, J.A.; Ohgami, Y.; Vega-Villa, K.R.; Rimando, A.M.; Davies, N.M. Pharmacometrics of pterostilbene: Preclinical pharmacokinetics and metabolism, anticancer, antiinflammatory, antioxidant and analgesic activity. *Phytother. Res.* **2008**, *22*, 169–179. [CrossRef]
119. Kim, J.; Min, J.S.; Kim, D.; Zheng, Y.F.; Mailar, K.; Choi, W.J.; Lee, C.; Bae, S.K. A simple and sensitive liquid chromatography–tandem mass spectrometry method for trans-ε-viniferin quantification in mouse plasma and its application to a pharmacokinetic study in mice. *J. Pharm. Biomed. Anal.* **2017**, *134*, 116–121. [CrossRef]
120. Santucci, C.; Mignozzi, S.; Levi, F.; Malvezzi, M.; Boffetta, P.; Negri, E.; La Vecchia, C. European cancer mortality predictions for the year 2025 with focus on breast cancer. *Ann. Oncol.* **2025**, *36*, 460–468. [CrossRef]
121. Pereira, M.; Peleteiro, B.; Capewell, S.; Bennett, K.; Azevedo, A.; Lunet, N. Changing patterns of cardiovascular diseases and cancer mortality in Portugal, 1980–2010. *BMC Public Health* **2012**, *12*, 1126. [CrossRef] [PubMed]
122. Siegel, R.; Naishadham, D.; Jemal, A. Cancer statistics for Hispanics/Latinos, 2012. *CA Cancer J. Clin.* **2012**, *62*, 283–298. [CrossRef] [PubMed]
123. Kang, N.J.; Shin, S.H.; Lee, H.J.; Lee, K.W. Polyphenols as small molecular inhibitors of signaling cascades in carcinogenesis. *Pharmacol. Ther.* **2011**, *130*, 310–324. [CrossRef]
124. Mena, S.; Ortega, A.; Estrela, J.M. Oxidative stress in environmental-induced carcinogenesis. *Mutat. Res./Genet. Toxicol. Environ. Mutagen.* **2009**, *674*, 36–44. [CrossRef] [PubMed]
125. Jang, M.; Cai, L.; Udeani, G.O.; Slowing, K.V.; Thomas, C.F.; Beecher, C.W.W.; Fong, H.H.S.; Farnsworth, N.R.; Kinghorn, A.D.; Mehta, R.G.; et al. Cancer Chemopreventive Activity of Resveratrol, a Natural Product Derived from Grapes. *Science* **1997**, *275*, 218–220. [CrossRef]
126. Chatterjee, A.; Ronghe, A.; Singh, B.; Bhat, N.K.; Chen, J.; Bhat, H.K. Natural Antioxidants Exhibit Chemopreventive Characteristics through the Regulation of CNC b-Zip Transcription Factors in Estrogen-Induced Breast Carcinogenesis. *J. Biochem. Mol. Toxicol.* **2014**, *28*, 529–538. [CrossRef] [PubMed]
127. Sirerol, J.A.; Feddi, F.; Mena, S.; Rodriguez, M.L.; Sirera, P.; Aupí, M.; Pérez, S.; Asensi, M.; Ortega, A.; Estrela, J.M. Topical treatment with pterostilbene, a natural phytoalexin, effectively protects hairless mice against UVB radiation-induced skin damage and carcinogenesis. *Free Radic. Biol. Med.* **2015**, *85*, 1–11. [CrossRef]
128. Alfaras, I.; Juan, M.E.; Planas, J.M. trans-Resveratrol Reduces Precancerous Colonic Lesions in Dimethylhydrazine-Treated Rats. *J. Agric. Food Chem.* **2010**, *58*, 8104–8110. [CrossRef]

129. Azorín-Ortuño, M.; Yáñez-Gascón, M.J.; Vallejo, F.; Pallarés, F.J.; Larrosa, M.; Lucas, R.; Morales, J.C.; Tomás-Barberán, F.A.; García-Conesa, M.T.; Espín, J.C. Metabolites and tissue distribution of resveratrol in the pig. *Mol. Nutr. Food Res.* **2011**, *55*, 1154–1168. [CrossRef]
130. Roubille, C.; Martel-Pelletier, J.; Davy, J.-M.; Haraoui, B.; Pelletier, J.-P. Cardiovascular adverse effects of anti-inflammatory drugs. *Anti-Inflamm. Anti-Allergy Agents Med. Chem.* **2013**, *12*, 55–67. [CrossRef]
131. Levy, G.N. Prostaglandin H synthases, nonsteroidal antiinflammatory drugs, and colon cancer. *FASEB J.* **1997**, *11*, 234–247. [CrossRef]
132. Kitasato, A.; Kuroki, T.; Adachi, T.; Ono, S.; Tanaka, T.; Tsuneoka, N.; Hirabaru, M.; Takatsuki, M.; Eguchi, S. A Selective Cyclooxygenase-2 Inhibitor (Etodolac) Prevents Spontaneous Biliary Tumorigenesis in a Hamster Bilioenterostomy Model. *Eur. Surg. Res.* **2014**, *52*, 73–82. [CrossRef]
133. Cianciulli, A.; Calvello, R.; Cavallo, P.; Dragone, T.; Carofiglio, V.; Panaro, M.A. Modulation of NF-κB activation by resveratrol in LPS treated human intestinal cells results in downregulation of PGE2 production and COX-2 expression. *Toxicol. Vitr.* **2012**, *26*, 1122–1128. [CrossRef]
134. Schwartz, S.A.; Hernandez, A.; Mark Evers, B. The role of NF-κB/IκB proteins in cancer: Implications for novel treatment strategies. *Surg. Oncol.* **1999**, *8*, 143–153. [CrossRef]
135. Potter, G.A.; Patterson, L.H.; Wanogho, E.; Perry, P.J.; Butler, P.C.; Ijaz, T.; Ruparelia, K.C.; Lamb, J.H.; Farmer, P.B.; Stanley, L.A.; et al. The cancer preventative agent resveratrol is converted to the anticancer agent piceatannol by the cytochrome P450 enzyme CYP1B1. *Br. J. Cancer* **2002**, *86*, 774–778. [CrossRef] [PubMed]
136. Hart, J.H. Role of Phytostilbenes in Decay and Disease Resistance. *Annu. Rev. Phytopathol.* **1981**, *19*, 437–458. [CrossRef]
137. Adrian, M.; Jeandet, P. Effects of resveratrol on the ultrastructure of Botrytis cinerea conidia and biological significance in plant/pathogen interactions. *Fitoterapia* **2012**, *83*, 1345–1350. [CrossRef]
138. Collado-González, M.; Guirao-Abad, J.P.; Sánchez-Fresneda, R.; Belchí-Navarro, S.; Argüelles, J.-C. Resveratrol lacks antifungal activity against Candida albicans. *World J. Microbiol. Biotechnol.* **2012**, *28*, 2441–2446. [CrossRef] [PubMed]
139. Jeandet, P.; Douillet-Breuil, A.-C.; Bessis, R.; Debord, S.; Sbaghi, M.; Adrian, M. Phytoalexins from the Vitaceae: Biosynthesis, Phytoalexin Gene Expression in Transgenic Plants, Antifungal Activity, and Metabolism. *J. Agric. Food Chem.* **2002**, *50*, 2731–2741. [CrossRef] [PubMed]
140. Lee, K.; Lee, J.-H.; Ryu, S.Y.; Cho, M.H.; Lee, J. Stilbenes Reduce Staphylococcus aureus Hemolysis, Biofilm Formation, and Virulence. *Foodborne Pathog. Dis.* **2014**, *11*, 710–717. [CrossRef]
141. Beaumont, P.; Courtois, A.; Atgié, C.; Richard, T.; Krisa, S. In the shadow of resveratrol: Biological activities of epsilon-viniferin. *J. Physiol. Biochem.* **2022**, *78*, 465–484. [CrossRef]
142. Cho, H.S.; Lee, J.-H.; Ryu, S.Y.; Joo, S.W.; Cho, M.H.; Lee, J. Inhibition of Pseudomonas aeruginosa and Escherichia coli O157:H7 Biofilm Formation by Plant Metabolite ε-Viniferin. *J. Agric. Food Chem.* **2013**, *61*, 7120–7126. [CrossRef]
143. Frazzi, R.; Valli, R.; Tamagnini, I.; Casali, B.; Latruffe, N.; Merli, F. Resveratrol-mediated apoptosis of hodgkin lymphoma cells involves SIRT1 inhibition and FOXO3a hyperacetylation. *Int. J. Cancer* **2013**, *132*, 1013–1021. [CrossRef]
144. Shankar, S.; Chen, Q.; Siddiqui, I.; Sarva, K.; Srivastava, R.K. Sensitization of TRAIL-resistant LNCaP cells by resveratrol (3, 4′, 5 tri-hydroxystilbene): Molecular mechanisms and therapeutic potential. *J. Mol. Signal.* **2007**, *2*, 7. [CrossRef]
145. van Ginkel, P.R.; Yan, M.B.; Bhattacharya, S.; Polans, A.S.; Kenealey, J.D. Natural products induce a G protein-mediated calcium pathway activating p53 in cancer cells. *Toxicol. Appl. Pharmacol.* **2015**, *288*, 453–462. [CrossRef] [PubMed]
146. Lin, H.-S.; Yue, B.-D.; Ho, P.C. Determination of pterostilbene in rat plasma by a simple HPLC-UV method and its application in pre-clinical pharmacokinetic study. *Biomed. Chromatogr.* **2009**, *23*, 1308–1315. [CrossRef] [PubMed]
147. Singh, N.; Nigam, M.; Ranjan, V.; Sharma, R.; Balapure, A.K.; Rath, S.K. Caspase Mediated Enhanced Apoptotic Action of Cyclophosphamide- and Resveratrol-Treated MCF-7 Cells. *J. Pharmacol. Sci.* **2009**, *109*, 473–485. [CrossRef] [PubMed]
148. Kai, L.; Samuel, S.K.; Levenson, A.S. Resveratrol enhances p53 acetylation and apoptosis in prostate cancer by inhibiting MTA1/NuRD complex. *Int. J. Cancer* **2010**, *126*, 1538–1548. [CrossRef]
149. Rimando, A.M.; Cuendet, M.; Desmarchelier, C.; Mehta, R.G.; Pezzuto, J.M.; Duke, S.O. Cancer Chemopreventive and Antioxidant Activities of Pterostilbene, a Naturally Occurring Analogue of Resveratrol. *J. Agric. Food Chem.* **2002**, *50*, 3453–3457. [CrossRef]
150. Stivala, L.A.; Savio, M.; Carafoli, F.; Perucca, P.; Bianchi, L.; Maga, G.; Forti, L.; Pagnoni, U.M.; Albini, A.; Prosperi, E.; et al. Specific Structural Determinants Are Responsible for the Antioxidant Activity and the Cell Cycle Effects of Resveratrol. *J. Biol. Chem.* **2001**, *276*, 22586–22594. [CrossRef]
151. McCormack, D.; McFadden, D. Pterostilbene and Cancer: Current Review. *J. Surg. Res.* **2012**, *173*, e53–e61. [CrossRef]
152. McCormack, D.E.; Mannal, P.; McDonald, D.; Tighe, S.; Hanson, J.; McFadden, D. Genomic Analysis of Pterostilbene Predicts Its Antiproliferative Effects Against Pancreatic Cancer In Vitro and In Vivo. *J. Gastrointest. Surg.* **2012**, *16*, 1136–1143. [CrossRef] [PubMed]
153. Ferrer, P.; Asensi, M.; Segarra, R.; Ortega, A.; Benlloch, M.; Obrador, E.; Varea, M.T.; Asensio, G.; Jordá, L.; Estrela, J.M. Association between Pterostilbene and Quercetin Inhibits Metastatic Activity of B16 Melanoma. *Neoplasia* **2005**, *7*, 37–47. [CrossRef]

154. Tolomeo, M.; Grimaudo, S.; Cristina, A.D.; Roberti, M.; Pizzirani, D.; Meli, M.; Dusonchet, L.; Gebbia, N.; Abbadessa, V.; Crosta, L.; et al. Pterostilbene and 3′-hydroxypterostilbene are effective apoptosis-inducing agents in MDR and BCR-ABL-expressing leukemia cells. *Int. J. Biochem. Cell Biol.* **2005**, *37*, 1709–1726. [CrossRef]
155. Alosi, J.A.; McDonald, D.E.; Schneider, J.S.; Privette, A.R.; McFadden, D.W. Pterostilbene Inhibits Breast Cancer In Vitro Through Mitochondrial Depolarization and Induction of Caspase-Dependent Apoptosis1. *J. Surg. Res.* **2010**, *161*, 195–201. [CrossRef]
156. Wang, Y.; Ding, L.; Wang, X.; Zhang, J.; Han, W.; Feng, L.; Sun, J.; Jin, H.; Wang, X.J. Pterostilbene simultaneously induces apoptosis, cell cycle arrest and cyto-protective autophagy in breast cancer cells. *Am. J. Transl. Res.* **2012**, *4*, 44.
157. Moon, D.; McCormack, D.; McDonald, D.; McFadden, D. Pterostilbene induces mitochondrially derived apoptosis in breast cancer cells in vitro. *J. Surg. Res.* **2013**, *180*, 208–215. [CrossRef] [PubMed]
158. Chen, R.-J.; Tsai, S.-J.; Ho, C.-T.; Pan, M.-H.; Ho, Y.-S.; Wu, C.-H.; Wang, Y.-J. Chemopreventive Effects of Pterostilbene on Urethane-Induced Lung Carcinogenesis in Mice via the Inhibition of EGFR-Mediated Pathways and the Induction of Apoptosis and Autophagy. *J. Agric. Food Chem.* **2012**, *60*, 11533–11541. [CrossRef] [PubMed]
159. Chakraborty, A.; Gupta, N.; Ghosh, K.; Roy, P. In vitro evaluation of the cytotoxic, anti-proliferative and anti-oxidant properties of pterostilbene isolated from Pterocarpus marsupium. *Toxicol. Vitr.* **2010**, *24*, 1215–1228. [CrossRef]
160. Chiou, Y.-S.; Tsai, M.-L.; Wang, Y.-J.; Cheng, A.-C.; Lai, W.-M.; Badmaev, V.; Ho, C.-T.; Pan, M.-H. Pterostilbene Inhibits Colorectal Aberrant Crypt Foci (ACF) and Colon Carcinogenesis via Suppression of Multiple Signal Transduction Pathways in Azoxymethane-Treated Mice. *J. Agric. Food Chem.* **2010**, *58*, 8833–8841. [CrossRef] [PubMed]
161. Paul, S.; Rimando, A.M.; Lee, H.J.; Ji, Y.; Reddy, B.S.; Suh, N. Anti-inflammatory Action of Pterostilbene Is Mediated through the p38 Mitogen-Activated Protein Kinase Pathway in Colon Cancer Cells. *Cancer Prev. Res.* **2009**, *2*, 650–657. [CrossRef]
162. Hardie, D.G.; Ross, F.A.; Hawley, S.A. AMPK: A nutrient and energy sensor that maintains energy homeostasis. *Nat. Rev. Mol. Cell Biol.* **2012**, *13*, 251–262. [CrossRef]
163. Reinisalo, M.; Kårlund, A.; Koskela, A.; Kaarniranta, K.; Karjalainen, R.O. Polyphenol Stilbenes: Molecular Mechanisms of Defence against Oxidative Stress and Aging-Related Diseases. *Oxidative Med. Cell. Longev.* **2015**, *2015*, 340520. [CrossRef]
164. Cantó, C.; Gerhart-Hines, Z.; Feige, J.N.; Lagouge, M.; Noriega, L.; Milne, J.C.; Elliott, P.J.; Puigserver, P.; Auwerx, J. AMPK regulates energy expenditure by modulating NAD+ metabolism and SIRT1 activity. *Nature* **2009**, *458*, 1056–1060. [CrossRef]
165. Wilson, B.J.; Tremblay, A.M.; Deblois, G.v.; Sylvain-Drolet, G.; Giguère, V. An Acetylation Switch Modulates the Transcriptional Activity of Estrogen-Related Receptor α. *Mol. Endocrinol.* **2010**, *24*, 1349–1358. [CrossRef]
166. Annunziata, G.; Maisto, M.; Schisano, C.; Ciampaglia, R.; Narciso, V.; Tenore, G.C.; Novellino, E. Resveratrol as a Novel Anti-Herpes Simplex Virus Nutraceutical Agent: An Overview. *Viruses* **2018**, *10*, 473. [CrossRef]
167. Kolouchová, I.; Maťátková, O.; Paldrychová, M.; Kodeš, Z.; Kvasničková, E.; Sigler, K.; Čejková, A.; Šmidrkal, J.; Demnerová, K.; Masák, J. Resveratrol, pterostilbene, and baicalein: Plant-derived anti-biofilm agents. *Folia Microbiol.* **2018**, *63*, 261–272. [CrossRef] [PubMed]
168. Martelli, G.; Giacomini, D. Antibacterial and antioxidant activities for natural and synthetic dual-active compounds. *Eur. J. Med. Chem.* **2018**, *158*, 91–105. [CrossRef]
169. Kumar, S.N.; Nambisan, B. Antifungal Activity of Diketopiperazines and Stilbenes Against Plant Pathogenic Fungi In Vitro. *Appl. Biochem. Biotechnol.* **2014**, *172*, 741–754. [CrossRef] [PubMed]
170. Li, X.; Li, Y.; Xiong, B.; Qiu, S. Progress of Antimicrobial Mechanisms of Stilbenoids. *Pharmaceutics* **2024**, *16*, 663. [CrossRef]
171. Mattio, L.M.; Dallavalle, S.; Musso, L.; Filardi, R.; Franzetti, L.; Pellegrino, L.; D'Incecco, P.; Mora, D.; Pinto, A.; Arioli, S. Antimicrobial activity of resveratrol-derived monomers and dimers against foodborne pathogens. *Sci. Rep.* **2019**, *9*, 19525. [CrossRef] [PubMed]
172. Gerits, E.; Defraine, V.; Vandamme, K.; Cremer, K.D.; Brucker, K.D.; Thevissen, K.; Cammue, B.P.A.; Beullens, S.; Fauvart, M.; Verstraeten, N.; et al. Repurposing Toremifene for Treatment of Oral Bacterial Infections. *Antimicrob. Agents Chemother.* **2017**, *61*, 10.1128/aac.01846-01816. [CrossRef]
173. Lim, Y.R.I.; Preshaw, P.M.; Lim, L.P.; Ong, M.M.A.; Lin, H.-S.; Tan, K.S. Pterostilbene complexed with cyclodextrin exerts antimicrobial and anti-inflammatory effects. *Sci. Rep.* **2020**, *10*, 9072. [CrossRef]
174. Yang, B.; Yao, H.; Li, D.; Liu, Z. The phosphatidylglycerol phosphate synthase PgsA utilizes a trifurcated amphipathic cavity for catalysis at the membrane-cytosol interface. *Curr. Res. Struct. Biol.* **2021**, *3*, 312–323. [CrossRef]
175. Hrast, M.; Rožman, K.; Ogris, I.; Škedelj, V.; Patin, D.; Sova, M.; Barreteau, H.; Gobec, S.; Grdadolnik, S.G.; Zega, A. Evaluation of the published kinase inhibitor set to identify multiple inhibitors of bacterial ATP-dependent mur ligases. *J. Enzym. Inhib. Med. Chem.* **2019**, *34*, 1010–1017. [CrossRef] [PubMed]
176. Wu, X.-Z.; Cheng, A.-X.; Sun, L.-M.; Lou, H.-X. Effect of plagiochin E, an antifungal macrocyclic bis(bibenzyl), on cell wall chitin synthesis in Candida albicans. *Acta Pharmacol. Sin.* **2008**, *29*, 1478–1485. [CrossRef] [PubMed]
177. Liu, Q.; Guo, X.; Jiang, G.; Wu, G.; Miao, H.; Liu, K.; Chen, S.; Sakamoto, N.; Kuno, T.; Yao, F.; et al. NADPH-Cytochrome P450 Reductase Ccr1 Is a Target of Tamoxifen and Participates in Its Antifungal Activity via Regulating Cell Wall Integrity in Fission Yeast. *Antimicrob. Agents Chemother.* **2020**, *64*, 10.1128/aac.00079-00020. [CrossRef]

178. Park, H.B.; Crawford, J.M. Lumiquinone A, an α-Aminomalonate-Derived Aminobenzoquinone from Photorhabdus luminescens. *J. Nat. Prod.* **2015**, *78*, 1437–1441. [CrossRef]
179. Houillé, B.; Papon, N.; Boudesocque, L.; Bourdeaud, E.; Besseau, S.; Courdavault, V.; Enguehard-Gueiffier, C.; Delanoue, G.; Guérin, L.; Bouchara, J.-P.; et al. Antifungal Activity of Resveratrol Derivatives against Candida Species. *J. Nat. Prod.* **2014**, *77*, 1658–1662. [CrossRef]
180. Li, D.-D.; Zhao, L.-X.; Mylonakis, E.; Hu, G.-H.; Zou, Y.; Huang, T.-K.; Yan, L.; Wang, Y.; Jiang, Y.-Y. In Vitro and In Vivo Activities of Pterostilbene against Candida albicans Biofilms. *Antimicrob. Agents Chemother.* **2014**, *58*, 2344–2355. [CrossRef]
181. De Filippis, B.; Ammazzalorso, A.; Amoroso, R.; Giampietro, L. Stilbene derivatives as new perspective in antifungal medicinal chemistry. *Drug Dev. Res.* **2019**, *80*, 285–293. [CrossRef]
182. Mora-Pale, M.; Bhan, N.; Masuko, S.; James, P.; Wood, J.; McCallum, S.; Linhardt, R.J.; Dordick, J.S.; Koffas, M.A.G. Antimicrobial mechanism of resveratrol-trans-dihydrodimer produced from peroxidase-catalyzed oxidation of resveratrol. *Biotechnol. Bioeng.* **2015**, *112*, 2417–2428. [CrossRef]
183. Kim, S.; Lee, D.G. Oxyresveratrol-induced DNA cleavage triggers apoptotic response in Candida albicans. *Microbiology* **2018**, *164*, 1112–1121. [CrossRef]
184. Lee, W.; Lee, D.G. Resveratrol induces membrane and DNA disruption via pro-oxidant activity against Salmonella typhimurium. *Biochem. Biophys. Res. Commun.* **2017**, *489*, 228–234. [CrossRef] [PubMed]
185. Bock, F.J.; Tait, S.W.G. Mitochondria as multifaceted regulators of cell death. *Nat. Rev. Mol. Cell Biol.* **2020**, *21*, 85–100. [CrossRef]
186. Lee, J.; Lee, D.G. Novel Antifungal Mechanism of Resveratrol: Apoptosis Inducer in Candida albicans. *Curr. Microbiol.* **2015**, *70*, 383–389. [CrossRef] [PubMed]
187. Dadi, P.K.; Ahmad, M.; Ahmad, Z. Inhibition of ATPase activity of Escherichia coli ATP synthase by polyphenols. *Int. J. Biol. Macromol.* **2009**, *45*, 72–79. [CrossRef]
188. Hwang, D.; Lim, Y.-H. Resveratrol antibacterial activity against Escherichia coli is mediated by Z-ring formation inhibition via suppression of FtsZ expression. *Sci. Rep.* **2015**, *5*, 10029. [CrossRef] [PubMed]
189. Barrows, J.M.; Goley, E.D. FtsZ dynamics in bacterial division: What, how, and why? *Curr. Opin. Cell Biol.* **2021**, *68*, 163–172. [CrossRef]
190. Lou, F.; Wang, K.; Hou, Y.; Shang, X.; Tang, F. Inhibitory effect of resveratrol on swimming motility and adhesion ability against Salmonella enterica serovar Typhimurium infection. *Microb. Pathog.* **2023**, *184*, 106323. [CrossRef] [PubMed]
191. Lee, S.A.; Wallis, C.M.; Rogers, E.E.; Burbank, L.P. Grapevine phenolic compounds influence cell surface adhesion of Xylella fastidiosa and bind to lipopolysaccharide. *PLoS ONE* **2020**, *15*, e0240101. [CrossRef]
192. Kugaji, M.S.; Kumbar, V.M.; Peram, M.R.; Patil, S.; Bhat, K.G.; Diwan, P.V. Effect of Resveratrol on biofilm formation and virulence factor gene expression of Porphyromonas gingivalis in periodontal disease. *APMIS* **2019**, *127*, 187–195. [CrossRef]
193. Uppuluri, P.; Chaturvedi, A.K.; Srinivasan, A.; Banerjee, M.; Ramasubramaniam, A.K.; Köhler, J.R.; Kadosh, D.; Lopez-Ribot, J.L. Dispersion as an Important Step in the Candida albicans Biofilm Developmental Cycle. *PLoS Pathog.* **2010**, *6*, e1000828. [CrossRef]
194. Rosman, C.W.K.; van der Mei, H.C.; Sjollema, J. Influence of sub-inhibitory concentrations of antimicrobials on micrococcal nuclease and biofilm formation in Staphylococcus aureus. *Sci. Rep.* **2021**, *11*, 13241. [CrossRef]
195. Siegel, R.L.; Kratzer, T.B.; Giaquinto, A.N.; Sung, H.; Jemal, A. Cancer statistics, 2025. *CA Cancer J. Clin.* **2025**, *75*, 10–45. [CrossRef]
196. Jones, S.B. Cancer in the developing world: A call to action. *BMJ* **1999**, *319*, 505–508. [CrossRef]
197. Siegel, R.; Ma, J.; Zou, Z.; Jemal, A. Cancer statistics, 2014. *CA Cancer J. Clin.* **2014**, *64*, 9–29. [CrossRef]
198. Malhotra, A.; Nair, P.; Dhawan, D.K. Study to Evaluate Molecular Mechanics behind Synergistic Chemo-Preventive Effects of Curcumin and Resveratrol during Lung Carcinogenesis. *PLoS ONE* **2014**, *9*, e93820. [CrossRef] [PubMed]
199. Mazué, F.; Delmas, D.; Murillo, G.; Saleiro, D.; Limagne, E.; Latruffe, N. Differential protective effects of red wine polyphenol extracts (RWEs) on colon carcinogenesis. *Food Funct.* **2014**, *5*, 663–670. [CrossRef] [PubMed]
200. Back, J.H.; Zhu, Y.; Calabro, A.; Queenan, C.; Kim, A.S.; Arbesman, J.; Kim, A.L. Resveratrol-Mediated Downregulation of Rictor Attenuates Autophagic Process and Suppresses UV-Induced Skin Carcinogenesis. *Photochem. Photobiol.* **2012**, *88*, 1165–1172. [CrossRef]
201. Hsieh, T.-c.; Yang, C.-J.; Lin, C.-Y.; Lee, Y.-S.; Wu, J.M. Control of stability of cyclin D1 by quinone reductase 2 in CWR22Rv1 prostate cancer cells. *Carcinogenesis* **2012**, *33*, 670–677. [CrossRef]
202. Lin, C.; Crawford, D.R.; Lin, S.; Hwang, J.; Sebuyira, A.; Meng, R.; Westfall, J.E.; Tang, H.-Y.; Lin, S.; Yu, P.-Y.; et al. Inducible COX-2-dependent apoptosis in human ovarian cancer cells. *Carcinogenesis* **2011**, *32*, 19–26. [CrossRef]
203. Rajasekaran, D.; Elavarasan, J.; Sivalingam, M.; Ganapathy, E.; Kumar, A.; Kalpana, K.; Sakthisekaran, D. Resveratrol interferes with N-nitrosodiethylamine-induced hepatocellular carcinoma at early and advanced stages in male Wistar rats. *Mol. Med. Rep.* **2011**, *4*, 1211–1217.
204. Shrotriya, S.; Tyagi, A.; Deep, G.; Orlicky, D.J.; Wisell, J.; Wang, X.-J.; Sclafani, R.A.; Agarwal, R.; Agarwal, C. Grape seed extract and resveratrol prevent 4-nitroquinoline 1-oxide induced oral tumorigenesis in mice by modulating AMPK activation and associated biological responses. *Mol. Carcinog.* **2015**, *54*, 291–300. [CrossRef] [PubMed]

205. Kang, H.J.; Youn, Y.-K.; Hong, M.-K.; Kim, L.S. Antiproliferation and Redifferentiation in Thyroid Cancer Cell Lines by Polyphenol Phytochemicals. *JKMS* **2011**, *26*, 893–899. [CrossRef]
206. Tsan, M.-f.; White, J.E.; Maheshwari, J.G.; Chikkappa, G. Anti-leukemia Effect of Resveratrol. *Leuk. Lymphoma* **2002**, *43*, 983–987. [CrossRef]
207. Temraz, S.; Mukherji, D.; Shamseddine, A. Potential Targets for Colorectal Cancer Prevention. *Int. J. Mol. Sci.* **2013**, *14*, 17279–17303. [CrossRef] [PubMed]
208. Wang, D.; DuBois, R.N. The role of COX-2 in intestinal inflammation and colorectal cancer. *Oncogene* **2010**, *29*, 781–788. [CrossRef]
209. Mosalpuria, K.; Hall, C.; Krishnamurthy, S.; Lodhi, A.; Hallman, D.M.; Baraniuk, M.S.; Bhattacharyya, A.; Lucci, A. Cyclooxygenase-2 expression in non-metastatic triple-negative breast cancer patients. *Mol. Clin. Oncol.* **2014**, *2*, 845–850. [CrossRef] [PubMed]
210. Wang, D.; Guo, X.-Z.; Li, H.-Y.; Zhao, J.-J.; Shao, X.-D.; Wu, C.-Y. Prognostic significance of cyclooxygenase-2 protein in pancreatic cancer: A meta-analysis. *Tumor Biol.* **2014**, *35*, 10301–10307. [CrossRef]
211. Jiao, J.; Ishikawa, T.-O.; Dumlao, D.S.; Norris, P.C.; Magyar, C.E.; Mikulec, C.; Catapang, A.; Dennis, E.A.; Fischer, S.M.; Herschman, H.R. Targeted Deletion and Lipidomic Analysis Identify Epithelial Cell COX-2 as a Major Driver of Chemically Induced Skin Cancer. *Mol. Cancer Res.* **2014**, *12*, 1677–1688. [CrossRef]
212. Marnett, L.J. Aspirin and the potential role of prostaglandins in colon cancer. *Cancer Res.* **1992**, *52*, 5575.
213. Yi, C.-O.; Jeon, B.T.; Shin, H.J.; Jeong, E.A.; Chang, K.C.; Lee, J.E.; Lee, D.H.; Kim, H.J.; Kang, S.S.; Cho, G.J.; et al. Resveratrol activates AMPK and suppresses LPS-induced NF-κB-dependent COX-2 activation in RAW 264.7 macrophage cells. *ACB* **2011**, *44*, 194–203. [CrossRef]
214. Han, Y.; Jiang, C.; Tang, J.; Wang, C.; Wu, P.; Zhang, G.; Liu, W.; Jamangulova, N.; Wu, X.; Song, X. Resveratrol reduces morphine tolerance by inhibiting microglial activation via AMPK signalling. *Eur. J. Pain* **2014**, *18*, 1458–1470. [CrossRef] [PubMed]
215. GÜROCAK, Ş.; Karabulut, E.; KARADAĞ SOYLU, N.; ÖZGÖR, D.; Ozkeles, N.; Karabulut, A. Preventive effects of resveratrol against azoxymethane induced damage in rat liver. *Asian Pac. J. Cancer Prev.* **2013**, *14*. [CrossRef] [PubMed]
216. Akca, H.; Demiray, A.; Aslan, M.; Acikbas, I.; Tokgun, O. Tumour suppressor PTEN enhanced enzyme activity of GPx, SOD and catalase by suppression of PI3K/AKT pathway in non-small cell lung cancer cell lines. *J. Enzym. Inhib. Med. Chem.* **2013**, *28*, 539–544. [CrossRef] [PubMed]
217. Mena, S.; Rodríguez, M.L.; Ponsoda, X.; Estrela, J.M.; Jäättela, M.; Ortega, A.L. Pterostilbene-Induced Tumor Cytotoxicity: A Lysosomal Membrane Permeabilization-Dependent Mechanism. *PLoS ONE* **2012**, *7*, e44524. [CrossRef]
218. Casanova, F.; Quarti, J.; da Costa, D.C.F.; Ramos, C.A.; da Silva, J.L.; Fialho, E. Resveratrol chemosensitizes breast cancer cells to melphalan by cell cycle arrest. *J. Cell. Biochem.* **2012**, *113*, 2586–2596. [CrossRef]
219. Yu, X.-D.; Yang, J.-l.; Zhang, W.-L.; Liu, D.-X. Resveratrol inhibits oral squamous cell carcinoma through induction of apoptosis and G2/M phase cell cycle arrest. *Tumor Biol.* **2016**, *37*, 2871–2877. [CrossRef]
220. Gokbulut, A.A.; Apohan, E.; Baran, Y. Resveratrol and quercetin-induced apoptosis of human 232B4 chronic lymphocytic leukemia cells by activation of caspase-3 and cell cycle arrest. *Hematology* **2013**, *18*, 144–150. [CrossRef] [PubMed]
221. McCormack, D.; McFadden, D. A Review of Pterostilbene Antioxidant Activity and Disease Modification. *Oxidative Med. Cell. Longev.* **2013**, *2013*, 575482. [CrossRef]
222. Schneider, J.G.; Alosi, J.A.; McDonald, D.E.; McFadden, D.W. Pterostilbene Inhibits Lung Cancer Through Induction of Apoptosis1. *J. Surg. Res.* **2010**, *161*, 18–22. [CrossRef]
223. Suh, N.; Paul, S.; Hao, X.; Simi, B.; Xiao, H.; Rimando, A.M.; Reddy, B.S. Pterostilbene, an Active Constituent of Blueberries, Suppresses Aberrant Crypt Foci Formation in the Azoxymethane-Induced Colon Carcinogenesis Model in Rats. *Clin. Cancer Res.* **2007**, *13*, 350–355. [CrossRef]
224. Cichocki, M.; Paluszczak, J.; Szaefer, H.; Piechowiak, A.; Rimando, A.M.; Baer-Dubowska, W. Pterostilbene is equally potent as resveratrol in inhibiting 12-O-tetradecanoylphorbol-13-acetate activated NFκB, AP-1, COX-2, and iNOS in mouse epidermis. *Mol. Nutr. Food Res.* **2008**, *52*, S62–S70. [CrossRef] [PubMed]
225. Dhar, S.; Kumar, A.; Rimando, A.M.; Zhang, X.; Levenson, A.S. Resveratrol and pterostilbene epigenetically restore PTEN expression by targeting oncomiRs of the miR-17 family in prostate cancer. *Oncotarget* **2015**, *6*, 27214. [CrossRef]
226. Paul, S.; DeCastro, A.J.; Lee, H.J.; Smolarek, A.K.; So, J.Y.; Simi, B.; Wang, C.X.; Zhou, R.; Rimando, A.M.; Suh, N. Dietary intake of pterostilbene, a constituent of blueberries, inhibits the β-catenin/p65 downstream signaling pathway and colon carcinogenesis in rats. *Carcinogenesis* **2010**, *31*, 1272–1278. [CrossRef] [PubMed]
227. McCormack, D.; Schneider, J.; McDonald, D.; McFadden, D. The antiproliferative effects of pterostilbene on breast cancer in vitro are via inhibition of constitutive and leptin-induced Janus kinase/signal transducer and activator of transcription activation. *Am. J. Surg.* **2011**, *202*, 541–544. [CrossRef]
228. Liu, Y.; Wang, L.; Wu, Y.; Lv, C.; Li, X.; Cao, X.; Yang, M.; Feng, D.; Luo, Z. Pterostilbene exerts antitumor activity against human osteosarcoma cells by inhibiting the JAK2/STAT3 signaling pathway. *Toxicology* **2013**, *304*, 120–131. [CrossRef]

229. Chen, R.-J.; Ho, C.-T.; Wang, Y.-J. Pterostilbene induces autophagy and apoptosis in sensitive and chemoresistant human bladder cancer cells. *Mol. Nutr. Food Res.* **2010**, *54*, 1819–1832. [CrossRef]
230. Chakraborty, A.; Bodipati, N.; Demonacos, M.K.; Peddinti, R.; Ghosh, K.; Roy, P. Long term induction by pterostilbene results in autophagy and cellular differentiation in MCF-7 cells via ROS dependent pathway. *Mol. Cell. Endocrinol.* **2012**, *355*, 25–40. [CrossRef] [PubMed]
231. Magee, J.A.; Piskounova, E.; Morrison, S.J. Cancer Stem Cells: Impact, Heterogeneity, and Uncertainty. *Cancer Cell* **2012**, *21*, 283–296. [CrossRef]
232. Nguyen, L.V.; Vanner, R.; Dirks, P.; Eaves, C.J. Cancer stem cells: An evolving concept. *Nat. Rev. Cancer* **2012**, *12*, 133–143. [CrossRef]
233. Visvader, J.E.; Lindeman, G.J. Cancer Stem Cells: Current Status and Evolving Complexities. *Cell Stem Cell* **2012**, *10*, 717–728. [CrossRef]
234. Hagiwara, K.; Kosaka, N.; Yoshioka, Y.; Takahashi, R.-u.; Takeshita, F.; Ochiya, T. Stilbene derivatives promote Ago2-dependent tumour-suppressive microRNA activity. *Sci. Rep.* **2012**, *2*, 314. [CrossRef]
235. Mak, K.-K.; Wu, A.T.H.; Lee, W.-H.; Chang, T.-C.; Chiou, J.-F.; Wang, L.-S.; Wu, C.-H.; Huang, C.-Y.F.; Shieh, Y.-S.; Chao, T.-Y.; et al. Pterostilbene, a bioactive component of blueberries, suppresses the generation of breast cancer stem cells within tumor microenvironment and metastasis via modulating NF-κB/microRNA 448 circuit. *Mol. Nutr. Food Res.* **2013**, *57*, 1123–1134. [CrossRef]
236. Lee, C.-M.; Su, Y.-H.; Huynh, T.-T.; Lee, W.-H.; Chiou, J.-F.; Lin, Y.-K.; Hsiao, M.; Wu, C.-H.; Lin, Y.-F.; Wu, A.T.H.; et al. BlueBerry Isolate, Pterostilbene, Functions as a Potential Anticancer Stem Cell Agent in Suppressing Irradiation-Mediated Enrichment of Hepatoma Stem Cells. *Evid.-Based Complement. Altern. Med.* **2013**, *2013*, 258425. [CrossRef]
237. Hsieh, T.-C.; Bennett, D.J.; Lee, Y.-S.; Wu, E.; Wu, J.M. In Silico and Biochemical Analyses Identify Quinone Reductase 2 as a Target of Piceatannol. *Curr. Med. Chem.* **2013**, *20*, 4195–4202. [CrossRef] [PubMed]
238. Murias, M.; Jäger, W.; Handler, N.; Erker, T.; Horvath, Z.; Szekeres, T.; Nohl, H.; Gille, L. Antioxidant, prooxidant and cytotoxic activity of hydroxylated resveratrol analogues: Structure–activity relationship. *Biochem. Pharmacol.* **2005**, *69*, 903–912. [CrossRef]
239. Piotrowska, H.; Kucinska, M.; Murias, M. Biological activity of piceatannol: Leaving the shadow of resveratrol. *Mutat. Res./Rev. Mutat. Res.* **2012**, *750*, 60–82. [CrossRef] [PubMed]
240. Azmi, A.S.; Bhat, S.H.; Hadi, S.M. Resveratrol–Cu(II) induced DNA breakage in human peripheral lymphocytes: Implications for anticancer properties. *FEBS Lett.* **2005**, *579*, 3131–3135. [CrossRef] [PubMed]
241. Sirerol, J.A.; Rodríguez, M.L.; Mena, S.; Asensi, M.A.; Estrela, J.M.; Ortega, A.L. Role of Natural Stilbenes in the Prevention of Cancer. *Oxidative Med. Cell. Longev.* **2016**, *2016*, 3128951. [CrossRef] [PubMed]
242. Ashikawa, K.; Majumdar, S.; Banerjee, S.; Bharti, A.C.; Shishodia, S.; Aggarwal, B.B. Piceatannol Inhibits TNF-Induced NF-κB Activation and NF-κB-Mediated Gene Expression Through Suppression of IκBα Kinase and p65 Phosphorylation1. *J. Immunol.* **2002**, *169*, 6490–6497. [CrossRef]
243. Islam, S.; Hassan, F.; Mu, M.M.; Ito, H.; Koide, N.; Mori, I.; Yoshida, T.; Yokochi, T. Piceatannol Prevents Lipopolysaccharide (LPS)-Induced Nitric Oxide (NO) Production and Nuclear Factor (NF)-κB Activation by Inhibiting IκB Kinase (IKK). *Microbiol. Immunol.* **2004**, *48*, 729–736. [CrossRef]
244. Jin, C.-Y.; Moon, D.-O.; Lee, K.-J.; Kim, M.-O.; Lee, J.-D.; Choi, Y.H.; Park, Y.-M.; Kim, G.-Y. Piceatannol attenuates lipopolysaccharide-induced NF-κB activation and NF-κB-related proinflammatory mediators in BV2 microglia. *Pharmacol. Res.* **2006**, *54*, 461–467. [CrossRef]
245. Xueyan, R.; Jia, Y.; Xuefeng, Y.; Lidan, T.; Qingjun, K. Isolation and purification of five phenolic compounds from the Xinjiang wine grape (Vitis Vinifera) and determination of their antioxidant mechanism at cellular level. *Eur. Food Res. Technol.* **2018**, *244*, 1569–1579. [CrossRef]
246. Sharma, A.; Boise, L.H.; Shanmugam, M. Cancer Metabolism and the Evasion of Apoptotic Cell Death. *Cancers* **2019**, *11*, 1144. [CrossRef]
247. Barjot, C.; Tournaire, M.; Castagnino, C.; Vigor, C.; Vercauteren, J.; Rossi, J.-F. Evaluation of antitumor effects of two vine stalk oligomers of resveratrol on a panel of lymphoid and myeloid cell lines: Comparison with resveratrol. *Life Sci.* **2007**, *81*, 1565–1574. [CrossRef]
248. Billard, C.; Izard, J.-C.; Roman, V.; Kern, C.; Mathiot, C.; Mentz, F.; Kolb, J.-P. Comparative Antiproliferative and Apoptotic Effects of Resveratrol, ϵ-viniferin and Vine-shots Derived Polyphenols (Vineatrols) on Chronic B Lymphocytic Leukemia Cells and Normal Human Lymphocytes. *Leuk. Lymphoma* **2002**, *43*, 1991–2002. [CrossRef]
249. Nivelle, L.; Aires, V.; Rioult, D.; Martiny, L.; Tarpin, M.; Delmas, D. Molecular analysis of differential antiproliferative activity of resveratrol, epsilon viniferin and labruscol on melanoma cells and normal dermal cells. *Food Chem. Toxicol.* **2018**, *116*, 323–334. [CrossRef] [PubMed]

250. Özdemir, F.; Akalın, G.; Şen, M.; Önder, N.I.; Işcan, A.; Kutlu, H.M.; Incesu, Z. Towards Novel Anti-tumor Strategies for Hepatic Cancer: ε-Viniferin in Combination with Vincristine Displays Pharmacodynamic Synergy at Lower Doses in HepG2 Cells. *OMICS J. Integr. Biol.* **2014**, *18*, 324–334. [CrossRef]
251. Özdemir, F.; İncesu, Z.; Şena, M.; Öndera, N.İ.; Dikme, M. Implications of Enhanced Effectiveness of Vincristine Sulfate/ε-Viniferin Combination Compared to Vincristine Sulfate Only on HepG2 Cells. *Dicle Tıp Derg.* **2016**, *43*, 534–541.
252. Tarhan, S.; Özdemir, F.; İncesu, Z.; Demirkan, E.S. Direct and protective effects of single or combined addition of vincristine and ε-viniferin on human HepG2 cellular oxidative stress markers in vitro. *Cytotechnology* **2016**, *68*, 1081–1094. [CrossRef]
253. Özdemir, F.; Apaydın, E.; Önder, N.İ.; Şen, M.; Ayrım, A.; Öğünç, Y.; İncesu, Z. Apoptotic effects of ε-viniferin in combination with cis-platin in C6 cells. *Cytotechnology* **2018**, *70*, 1061–1073. [CrossRef]
254. Ito, T.; Akao, Y.; Yi, H.; Ohguchi, K.; Matsumoto, K.; Tanaka, T.; Iinuma, M.; Nozawa, Y. Antitumor effect of resveratrol oligomers against human cancer cell lines and the molecular mechanism of apoptosis induced by vaticanol C. *Carcinogenesis* **2003**, *24*, 1489–1497. [CrossRef]
255. Muhtadi; Hakim, E.H.; Juliawaty, L.D.; Syah, Y.M.; Achmad, S.A.; Latip, J.; Ghisalberti, E.L. Cytotoxic resveratrol oligomers from the tree bark of Dipterocarpus hasseltii. *Fitoterapia* **2006**, *77*, 550–555. [CrossRef]
256. Nur Ainaa, A.; Norizan Ahmat, N.A.; Mashita Abdullah, M.A.; Norrizah Jaafar Sidik, N.J.S.; Syed Ahmad, T. Antioxidant, antimicrobial and cytotoxic activities of resveratrol oligomers of Shorea macroptera dyer. *Aust. J. Basic. Appl. Sci.* **2012**, *6*, 431–436.
257. Wibowo, A.; Ahmat, N.; Hamzah, A.S.; Latif, F.A.; Norrizah, J.S.; Khong, H.Y.; Takayama, H. Identification and biological activity of secondary metabolites from Dryobalanops beccarii. *Phytochem. Lett.* **2014**, *9*, 117–122. [CrossRef]
258. Colin, D.; Lancon, A.; Delmas, D.; Lizard, G.; Abrossinow, J.; Kahn, E.; Jannin, B.; Latruffe, N. Antiproliferative activities of resveratrol and related compounds in human hepatocyte derived HepG2 cells are associated with biochemical cell disturbance revealed by fluorescence analyses. *Biochimie* **2008**, *90*, 1674–1684. [CrossRef] [PubMed]
259. Aja, I.; Ruiz-Larrea, M.B.; Courtois, A.; Krisa, S.; Richard, T.; Ruiz-Sanz, J.-I. Screening of Natural Stilbene Oligomers from Vitis vinifera for Anticancer Activity on Human Hepatocellular Carcinoma Cells. *Antioxidants* **2020**, *9*, 469. [CrossRef] [PubMed]
260. Ames, B.N.; Shigenaga, M.K.; Hagen, T.M. Oxidants, antioxidants, and the degenerative diseases of aging. *Proc. Natl. Acad. Sci. USA* **1993**, *90*, 7915–7922. [CrossRef]
261. Pintea, A.; Rugină, D.; Pop, R.; Bunea, A.; Socaciu, C.; Diehl, H.A. Antioxidant Effect of Trans-Resveratrol in Cultured Human Retinal Pigment Epithelial Cells. *J. Ocul. Pharmacol. Ther.* **2011**, *27*, 315–321. [CrossRef]
262. Ksila, M.; Vejux, A.; Prost-Camus, E.; Durand, P.; Ghzaiel, I.; Nury, T.; Duprey, D.; Meziane, S.; Masmoudi-Kouki, O.; Latruffe, N.; et al. Cytotoxic and Antioxidant Activities of Imine Analogs of Trans-Resveratrol towards Murine Neuronal N2a Cells. *Molecules* **2022**, *27*, 4713. [CrossRef]
263. Almeida, T.C.; Seibert, J.B.; Almeida, S.H.d.S.; Amparo, T.R.; Teixeira, L.F.d.M.; Barichello, J.M.; Postacchini, B.B.; Santos, O.D.H.d.; Silva, G.N.d. Polymeric micelles containing resveratrol: Development, characterization, cytotoxicity on tumor cells and antimicrobial activity. *Braz. J. Pharm. Sci.* **2020**, *56*, e18411. [CrossRef]
264. Selvaraj, S.; Mohan, A.; Narayanan, S.; Sethuraman, S.; Krishnan, U.M. Dose-Dependent Interaction of trans-Resveratrol with Biomembranes: Effects on Antioxidant Property. *J. Med. Chem.* **2013**, *56*, 970–981. [CrossRef]
265. Khayoon, H.A.; Al-Rekabi, F.M. Cytotoxic effect of resveratrol on colorectal cancer cell line. *Iraqi J. Vet. Med.* **2020**, *44*, 68–74. [CrossRef]
266. Osman, A.-M.M.; Bayoumi, H.M.; Al-Harthi, S.E.; Damanhouri, Z.A.; ElShal, M.F. Modulation of doxorubicin cytotoxicity by resveratrol in a human breast cancer cell line. *Cancer Cell Int.* **2012**, *12*, 47. [CrossRef]
267. Jeon, D.; Jo, M.; Lee, Y.; Park, S.-H.; Phan, H.T.L.; Nam, J.H.; Namkung, W. Inhibition of ANO1 by Cis- and Trans-Resveratrol and Their Anticancer Activity in Human Prostate Cancer PC-3 Cells. *Int. J. Mol. Sci.* **2023**, *24*, 1186. [CrossRef]
268. Fernández-Pérez, F.; Belchí-Navarro, S.; Almagro, L.; Bru, R.; Pedreño, M.A.; Gómez-Ros, L.V. Cytotoxic Effect of Natural trans-Resveratrol Obtained from Elicited Vitis vinifera Cell Cultures on Three Cancer Cell Lines. *Plant Foods Hum. Nutr.* **2012**, *67*, 422–429. [CrossRef]
269. Anisimova, N.Y.U.; Kiselevsky, M.V.; Sosnov, A.V.; Sadovnikov, S.V.; Stankov, I.N.; Gakh, A.A. Trans-, cis-, and dihydro-resveratrol: A comparative study. *Chem. Cent. J.* **2011**, *5*, 88. [CrossRef]
270. Siddiqui, M.; Saquib, Q.; Ahamed, M.; Ahmad, J.; Al-Khedhairy, A.; Abou-Tarboush, F.; Musarrat, J. Effect of trans-resveratrol on rotenone-induced cytotoxicity in human breast adenocarcinoma cells. *Toxicol. Int.* **2011**, *18*, 105.
271. Kocsis, Z.; Marcsek, Z.L.; Jakab, M.G.; Szende, B.; Tompa, A. Chemopreventive properties of trans—Resveratrol against the cytotoxicity of chloroacetanilide herbicides in vitro. *Int. J. Hyg. Environ. Health* **2005**, *208*, 211–218. [CrossRef] [PubMed]
272. Siddiqui, M.A.; Kashyap, M.P.; Kumar, V.; Al-Khedhairy, A.A.; Musarrat, J.; Pant, A.B. Protective potential of trans-resveratrol against 4-hydroxynonenal induced damage in PC12 cells. *Toxicol. Vitr.* **2010**, *24*, 1592–1598. [CrossRef]
273. Leischner, C.; Burkard, M.; Michel, A.; Berchtold, S.; Niessner, H.; Marongiu, L.; Busch, C.; Frank, J.; Lauer, U.M.; Venturelli, S. Comparative Analysis of the Antitumor Activity of Cis- and Trans-Resveratrol in Human Cancer Cells with Different p53 Status. *Molecules* **2021**, *26*, 5586. [CrossRef] [PubMed]

274. Miksits, M.; Wlcek, K.; Svoboda, M.; Kunert, O.; Haslinger, E.; Thalhammer, T.; Szekeres, T.; Jäger, W. Antitumor activity of resveratrol and its sulfated metabolites against human breast cancer cells. *Planta Medica* **2009**, *75*, 1227–1230. [CrossRef] [PubMed]
275. Marel, A.-K.; Lizard, G.; Izard, J.-C.; Latruffe, N.; Delmas, D. Inhibitory effects of trans-resveratrol analogs molecules on the proliferation and the cell cycle progression of human colon tumoral cells. *Mol. Nutr. Food Res.* **2008**, *52*, 538–548. [CrossRef]
276. Filomeni, G.; Graziani, I.; Rotilio, G.; Ciriolo, M.R. trans-Resveratrol induces apoptosis in human breast cancer cells MCF-7 by the activation of MAP kinases pathways. *Genes. Nutr.* **2007**, *2*, 295–305. [CrossRef]
277. Nashine, S.; Nesburn, A.B.; Kuppermann, B.D.; Kenney, M.C. Role of Resveratrol in Transmitochondrial AMD RPE Cells. *Nutrients* **2020**, *12*, 159. [CrossRef]
278. El-Melegy, M.G.; Eltaher, H.M.; Gaballah, A.; El-Kamel, A.H. Enhanced oral permeability of Trans-Resveratrol using nanocochleates for boosting anticancer efficacy; in-vitro and ex-vivo appraisal. *Eur. J. Pharm. Biopharm.* **2021**, *168*, 166–183. [CrossRef]
279. Tsang, S.W.; Zhang, H.; Lin, Z.; Mu, H.; Bian, Z.-X. Anti-fibrotic effect of trans-resveratrol on pancreatic stellate cells. *Biomed. Pharmacother.* **2015**, *71*, 91–97. [CrossRef]
280. dos Santos Moreno, C.; Rogero, S.O.; Ikeda, T.I.; Cruz, Á.S.; Rogero, J.R. Resveratrol and radiation biological effects. *Int. J. Nutrology* **2012**, *5*, 28–33. [CrossRef]
281. Lucas, I.K.; Kolodziej, H. Trans-resveratrol induces apoptosis through ROS-triggered mitochondria-dependent pathways in A549 human lung adenocarcinoma epithelial cells. *Planta Medica* **2015**, *81*, 1038–1044. [CrossRef]
282. Mannal, P.; McDonald, D.; McFadden, D. Pterostilbene and tamoxifen show an additive effect against breast cancer in vitro. *Am. J. Surg.* **2010**, *200*, 577–580. [CrossRef]
283. Elsherbini, A.M.; Sheweita, S.A.; Sultan, A.S. Pterostilbene as a Phytochemical Compound Induces Signaling Pathways Involved in the Apoptosis and Death of Mutant P53-Breast Cancer Cell Lines. *Nutr. Cancer* **2021**, *73*, 1976–1984. [CrossRef]
284. Wakimoto, R.; Ono, M.; Takeshima, M.; Higuchi, T.; Nakano, S. Differential Anticancer Activity of Pterostilbene Against Three Subtypes of Human Breast Cancer Cells. *Anticancer Res.* **2017**, *37*, 6153–6159.
285. Shin, H.J.; Han, J.M.; Choi, Y.S.; Jung, H.J. Pterostilbene Suppresses both Cancer Cells and Cancer Stem-Like Cells in Cervical Cancer with Superior Bioavailability to Resveratrol. *Molecules* **2020**, *25*, 228. [CrossRef]
286. Chatterjee, K.; Mukherjee, S.; Vanmanen, J.; Banerjee, P.; Fata, J.E. Dietary Polyphenols, Resveratrol and Pterostilbene Exhibit Antitumor Activity on an HPV E6-Positive Cervical Cancer Model: An in vitro and in vivo Analysis. *Front. Oncol.* **2019**, *9*, 352. [CrossRef]
287. Zhang, Y.; Li, Y.; Sun, C.; Chen, X.; Han, L.; Wang, T.; Liu, J.; Chen, X.; Zhao, D. Effect of Pterostilbene, a Natural Derivative of Resveratrol, in the Treatment of Colorectal Cancer through Top1/Tdp1-Mediated DNA Repair Pathway. *Cancers* **2021**, *13*, 4002. [CrossRef] [PubMed]
288. Wawszczyk, J.; Jesse, K.; Smolik, S.; Kapral, M. Mechanism of Pterostilbene-Induced Cell Death in HT-29 Colon Cancer Cells. *Molecules* **2022**, *27*, 369. [CrossRef]
289. Hsiao, Y.-H.; Chen, N.-C.; Koh, Y.-C.; Nagabhushanam, K.; Ho, C.-T.; Pan, M.-H. Pterostilbene Inhibits Adipocyte Conditioned-Medium-Induced Colorectal Cancer Cell Migration through Targeting FABP5-Related Signaling Pathway. *J. Agric. Food Chem.* **2019**, *67*, 10321–10329. [CrossRef] [PubMed]
290. Wen, W.; Lowe, G.; Roberts, C.M.; Finlay, J.; Han, E.S.; Glackin, C.A.; Dellinger, T.H. Pterostilbene, a natural phenolic compound, synergizes the antineoplastic effects of megestrol acetate in endometrial cancer. *Sci. Rep.* **2017**, *7*, 12754. [CrossRef]
291. Wen, W.; Lowe, G.; Roberts, C.M.; Finlay, J.; Han, E.S.; Glackin, C.A.; Dellinger, T.H. Pterostilbene Suppresses Ovarian Cancer Growth via Induction of Apoptosis and Blockade of Cell Cycle Progression Involving Inhibition of the STAT3 Pathway. *Int. J. Mol. Sci.* **2018**, *19*, 1983. [CrossRef]
292. Lin, V.C.-H.; Tsai, Y.-C.; Lin, J.-N.; Fan, L.-L.; Pan, M.-H.; Ho, C.-T.; Wu, J.-Y.; Way, T.-D. Activation of AMPK by Pterostilbene Suppresses Lipogenesis and Cell-Cycle Progression in p53 Positive and Negative Human Prostate Cancer Cells. *J. Agric. Food Chem.* **2012**, *60*, 6399–6407. [CrossRef]
293. Mannal, P.W.; Alosi, J.A.; Schneider, J.G.; McDonald, D.E.; McFadden, D.W. Pterostilbene Inhibits Pancreatic Cancer In Vitro. *J. Gastrointest. Surg.* **2010**, *14*, 873–879. [CrossRef] [PubMed]
294. Bracht, J.W.P.; Karachaliou, N.; Berenguer, J.; Pedraz-Valdunciel, C.; Filipska, M.; Codony-Servat, C.; Codony-Servat, J.; Rosell, R. Osimertinib and pterostilbene in EGFR-mutation-positive non-small cell lung cancer (NSCLC). *Int. J. Biol. Sci.* **2019**, *15*, 2607–2614. [CrossRef]
295. Guo, L.; Tan, K.; Wang, H.; Zhang, X. Pterostilbene inhibits hepatocellular carcinoma through p53/SOD2/ROS-mediated mitochondrial apoptosis. *Oncol. Rep.* **2016**, *36*, 3233–3240. [CrossRef]
296. Mori, S.; Kishi, S.; Honoki, K.; Fujiwara-Tani, R.; Moriguchi, T.; Sasaki, T.; Fujii, K.; Tsukamoto, S.; Fujii, H.; Kido, A.; et al. Anti-Stem Cell Property of Pterostilbene in Gastrointestinal Cancer Cells. *Int. J. Mol. Sci.* **2020**, *21*, 9347. [CrossRef]

297. Chang, H.-P.; Lu, C.-C.; Chiang, J.-H.; Tsai, F.-J.; Juan, Y.-N.; Tsao, J.-W.; Chiu, H.-Y.; Yang, J.-S. Pterostilbene modulates the suppression of multidrug resistance protein 1 and triggers autophagic and apoptotic mechanisms in cisplatin-resistant human oral cancer CAR cells via AKT signaling. *Int. J. Oncol.* **2018**, *52*, 1504–1514. [CrossRef] [PubMed]
298. Aly, S.; El-Kamel, A.H.; Sheta, E.; El-Habashy, S.E. Chondroitin/Lactoferrin-dual functionalized pterostilbene-solid lipid nanoparticles as targeted breast cancer therapy. *Int. J. Pharm.* **2023**, *642*, 123163. [CrossRef] [PubMed]
299. Chen, R.-J.; Lyu, Y.-J.; Chen, Y.-Y.; Lee, Y.-C.; Pan, M.-H.; Ho, Y.-S.; Wang, Y.-J. Chloroquine Potentiates the Anticancer Effect of Pterostilbene on Pancreatic Cancer by Inhibiting Autophagy and Downregulating the RAGE/STAT3 Pathway. *Molecules* **2021**, *26*, 6741. [CrossRef]
300. Kumar, V.; Haldar, S.; Das, N.S.; Ghosh, S.; Dhankhar, P.; Sircar, D.; Roy, P. Pterostilbene-isothiocyanate inhibits breast cancer metastasis by selectively blocking IKK-β/NEMO interaction in cancer cells. *Biochem. Pharmacol.* **2021**, *192*, 114717. [CrossRef]
301. Tong, C.; Wang, Y.; Li, J.; Cen, W.; Zhang, W.; Zhu, Z.; Yu, J.; Lu, B. Pterostilbene inhibits gallbladder cancer progression by suppressing the PI3K/Akt pathway. *Sci. Rep.* **2021**, *11*, 4391. [CrossRef] [PubMed]
302. Gao, H.; Liu, Z.; Xu, W.; Wang, Q.; Zhang, C.; Ding, Y.; Nie, W.; Lai, J.; Chen, Y.; Huang, H. Pterostilbene promotes mitochondrial apoptosis and inhibits proliferation in glioma cells. *Sci. Rep.* **2021**, *11*, 6381. [CrossRef]
303. Sáez, V.; Pastene, E.; Vergara, C.; Mardones, C.; Hermosín-Gutiérrez, I.; Gómez-Alonso, S.; Gómez, M.V.; Theoduloz, C.; Riquelme, S.; von Baer, D. Oligostilbenoids in Vitis vinifera L. Pinot Noir grape cane extract: Isolation, characterization, in vitro antioxidant capacity and anti-proliferative effect on cancer cells. *Food Chem.* **2018**, *265*, 101–110. [CrossRef] [PubMed]
304. Kuo, P.-L.; Hsu, Y.-L. The grape and wine constituent piceatannol inhibits proliferation of human bladder cancer cells via blocking cell cycle progression and inducing Fas/membrane bound Fas ligand-mediated apoptotic pathway. *Mol. Nutr. Food Res.* **2008**, *52*, 408–418. [CrossRef] [PubMed]
305. Morales, P.; Haza, A.I. Selective apoptotic effects of piceatannol and myricetin in human cancer cells. *J. Appl. Toxicol.* **2012**, *32*, 986–993. [CrossRef]
306. Du, M.; Zhang, Z.; Gao, T. Piceatannol induced apoptosis through up-regulation of microRNA-181a in melanoma cells. *Biol. Res.* **2017**, *50*, 36. [CrossRef]
307. Dias, S.J.; Li, K.; Rimando, A.M.; Dhar, S.; Mizuno, C.S.; Penman, A.D.; Levenson, A.S. Trimethoxy-Resveratrol and Piceatannol Administered Orally Suppress and Inhibit Tumor Formation and Growth in Prostate Cancer Xenografts. *Prostate* **2013**, *73*, 1135–1146. [CrossRef]
308. Yokozawa, T.; Kim, Y.J. Piceatannol Inhibits Melanogenesis by Its Antioxidative Actions. *Biol. Pharm. Bull.* **2007**, *30*, 2007–2011. [CrossRef]
309. Liu, W.-H.; Chang, L.-S. Piceatannol induces Fas and FasL up-regulation in human leukemia U937 cells via Ca2+/p38α MAPK-mediated activation of c-Jun and ATF-2 pathways. *Int. J. Biochem. Cell Biol.* **2010**, *42*, 1498–1506. [CrossRef]
310. Takasawa, R.; Akahane, H.; Tanaka, H.; Shimada, N.; Yamamoto, T.; Uchida-Maruki, H.; Sai, M.; Yoshimori, A.; Tanuma, S.-i. Piceatannol, a natural trans-stilbene compound, inhibits human glyoxalase I. *Bioorganic Med. Chem. Lett.* **2017**, *27*, 1169–1174. [CrossRef] [PubMed]
311. Wolter, F.; Clausnitzer, A.; Akoglu, B.; Stein, J. Piceatannol, a Natural Analog of Resveratrol, Inhibits Progression through the S Phase of the Cell Cycle in Colorectal Cancer Cell Lines. *J. Nutr.* **2002**, *132*, 298–302. [CrossRef]
312. Ovesná, Z.; Kozics, K.; Bader, Y.; Saiko, P.; Handler, N.; Erker, T.; Szekeres, T. Antioxidant activity of resveratrol, piceatannol and 3,3′,4,4′,5,5′-hexahydroxy-trans-stilbene in three leukemia cell lines. *Oncol. Rep.* **2006**, *16*, 617–624. [CrossRef]
313. Duarte, N.; Kayser, O.; Abreu, P.; Ferreira, M.-J.U. Antileishmanial activity of piceatannol isolated from Euphorbia lagascae seeds. *Phytother. Res.* **2008**, *22*, 455–457. [CrossRef]
314. Chowdhury, S.A.; Kishino, K.; Satoh, R.; Hashimoto, K.; Kikuchi, H.; Nishikawa, H.; Shirataki, Y.; Sakagami, H. Tumor-specificity and Apoptosis-inducing Activity of Stilbenes and Flavonoids. *Anticancer Res.* **2005**, *25*, 2055–2063.
315. Zhang, Y.; Gu, Y.; Xie, J.; Hu, Y. Anti-tumor effect of piceatannol through induction of cell apoptosis via up-regulation of microRNA-125b expression on pancreatic cancer. *Int. J. Clin. Exp. Med.* **2017**, *10*, 14495–14502.
316. Siedlecka-Kroplewska, K.; Ślebioda, T.; Kmieć, Z. Induction of autophagy, apoptosis and aquisition of resistance in response to piceatannol toxicity in MOLT-4 human leukemia cells. *Toxicol. Vitr.* **2019**, *59*, 12–25. [CrossRef] [PubMed]
317. Liu, T.; Liu, M.; Guo, Q.; Liu, Y.; Zhao, Y.; Wu, Y.; Sun, B.; Wang, Q.; Liu, J.; Han, J. Investigation of binary and ternary systems of human serum albumin with oxyresveratrol/piceatannol and/or mitoxantrone by multipectroscopy, molecular docking and cytotoxicity evaluation. *J. Mol. Liq.* **2020**, *311*, 113364. [CrossRef]
318. Arai, D.; Kataoka, R.; Otsuka, S.; Kawamura, M.; Maruki-Uchida, H.; Sai, M.; Ito, T.; Nakao, Y. Piceatannol is superior to resveratrol in promoting neural stem cell differentiation into astrocytes. *Food Funct.* **2016**, *7*, 4432–4441. [CrossRef]
319. Rüweler, M.; Gülden, M.; Maser, E.; Murias, M.; Seibert, H. Cytotoxic, cytoprotective and antioxidant activities of resveratrol and analogues in C6 astroglioma cells in vitro. *Chem.-Biol. Interact.* **2009**, *182*, 128–135. [CrossRef]
320. Billack, B.; Radkar, V.; Adiabouah, C. In Vitro evaluation of the cytotoxic and anti-proliferative properties of resveratrol and several of its analogs. *Cell. Mol. Biol. Lett.* **2008**, *13*, 553–569. [CrossRef]

321. Kang, C.-H.; Moon, D.-O.; Choi, Y.H.; Choi, I.-W.; Moon, S.-K.; Kim, W.-J.; Kim, G.-Y. Piceatannol enhances TRAIL-induced apoptosis in human leukemia THP-1 cells through Sp1- and ERK-dependent DR5 up-regulation. *Toxicol. Vitr.* **2011**, *25*, 605–612. [CrossRef] [PubMed]
322. Rohaiza, S.; Yaacob, W.; Din, L.; Nazlina, I. Cytotoxic oligostilbenes from Shorea hopeifolia. *Afr. J. Pharm. Pharmacol.* **2011**, *5*, 1272–1277. [CrossRef]
323. Chang, C.-I.; Chien, W.-C.; Huang, K.-X.; Hsu, J.-L. Anti-Inflammatory Effects of Vitisinol A and Four Other Oligostilbenes from Ampelopsis brevipedunculata var. Hancei. *Molecules* **2017**, *22*, 1195. [CrossRef]
324. Ha, D.T.; Chen, Q.C.; Hung, T.M.; Youn, U.J.; Ngoc, T.M.; Thuong, P.T.; Kim, H.J.; Seong, Y.H.; Min, B.S.; Bae, K. Stilbenes and oligostilbenes from leaf and stem of Vitis amurensis and their cytotoxic activity. *Arch. Pharmacal Res.* **2009**, *32*, 177–183. [CrossRef]
325. Vion, E.; Page, G.; Bourdeaud, E.; Paccalin, M.; Guillard, J.; Rioux Bilan, A. Trans ε-viniferin is an amyloid-β disaggregating and anti-inflammatory drug in a mouse primary cellular model of Alzheimer's disease. *Mol. Cell. Neurosci.* **2018**, *88*, 1–6. [CrossRef]
326. Tian, X.; Guo, S.; Zhang, S.; Li, P.; Wang, T.; Ho, C.-T.; Pan, M.-H.; Bai, N. Chemical characterization of main bioactive constituents in Paeonia ostii seed meal and GC-MS analysis of seed oil. *J. Food Biochem.* **2020**, *44*, e13088. [CrossRef] [PubMed]
327. Medrano-Padial, C.; Puerto, M.; Merchán-Gragero, M.d.M.; Moreno, F.J.; Richard, T.; Cantos-Villar, E.; Pichardo, S. Cytotoxicity studies of a stilbene extract and its main components intended to be used as preservative in the wine industry. *Food Res. Int.* **2020**, *137*, 109738. [CrossRef] [PubMed]
328. Kim, H.J.; Chang, E.J.; Bae, S.J.; Shim, S.M.; Park, H.D.; Rhee, C.H.; Park, J.H.; Choi, S.W. Cytotoxic and antimutagenic stilbenes from seeds ofPaeonia lactiflora. *Arch. Pharmacal Res.* **2002**, *25*, 293–299. [CrossRef]
329. Kang, J.H.; Park, Y.H.; Choi, S.W.; Yang, E.K.; Lee, W.J. Resveratrol derivatives potently induce apoptosis in human promyelocytic leukemia cells. *Exp. Mol. Med.* **2003**, *35*, 467–474. [CrossRef]
330. Zghonda, N.; Yoshida, S.; Araki, M.; Kusunoki, M.; Mliki, A.; Ghorbel, A.; Miyazaki, H. Greater Effectiveness of ε-Viniferin in Red Wine Than Its Monomer Resveratrol for Inhibiting Vascular Smooth Muscle Cell Proliferation and Migration. *Biosci. Biotechnol. Biochem.* **2011**, *75*, 1259–1267. [CrossRef]
331. Courtois, A.; Garcia, M.; Krisa, S.; Atgié, C.; Sauvant, P.; Richard, T.; Faure, C. Encapsulation of ε-viniferin in onion-type multi-lamellar liposomes increases its solubility and its photo-stability and decreases its cytotoxicity on Caco-2 intestinal cells. *Food Funct.* **2019**, *10*, 2573–2582. [CrossRef]
332. Zghonda, N.; Yoshida, S.; Ezaki, S.; Otake, Y.; Murakami, C.; Mliki, A.; Ghorbel, A.; Miyazaki, H. ε-Viniferin Is More Effective Than Its Monomer Resveratrol in Improving the Functions of Vascular Endothelial Cells and the Heart. *Biosci. Biotechnol. Biochem.* **2012**, *76*, 954–960. [CrossRef]
333. Wu, C.W.; Nakamoto, Y.; Hisatome, T.; Yoshida, S.; Miyazaki, H. Resveratrol and its dimers ε-viniferin and δ-viniferin in red wine protect vascular endothelial cells by a similar mechanism with different potency and efficacy. *Kaohsiung J. Med. Sci.* **2020**, *36*, 535–542. [CrossRef]
334. Huang, C.; Lin, Z.-J.; Lee, C.-J.; Lai, W.-H.; Chen, J.-C.; Huang, H.-C. ε-Viniferin and α-viniferin alone or in combination induced apoptosis and necrosis in osteosarcoma and non-small cell lung cancer cells. *Food Chem. Toxicol.* **2021**, *158*, 112617. [CrossRef]
335. Richard, T.; Poupard, P.; Nassra, M.; Papastamoulis, Y.; Iglésias, M.-L.; Krisa, S.; Waffo-Teguo, P.; Mérillon, J.-M.; Monti, J.-P. Protective effect of ε-viniferin on β-amyloid peptide aggregation investigated by electrospray ionization mass spectrometry. *Bioorganic Med. Chem.* **2011**, *19*, 3152–3155. [CrossRef]
336. Frombaum, M.; Therond, P.; Djelidi, R.; Beaudeux, J.-L.; Bonnefont-Rousselot, D.; Borderie, D. Piceatannol is more effective than resveratrol in restoring endothelial cell dimethylarginine dimethylaminohydrolase expression and activity after high-glucose oxidative stress. *Free Radic. Res.* **2011**, *45*, 293–302. [CrossRef] [PubMed]
337. Woo, A.; Min, B.; Ryoo, S. Piceatannol-3′-O-β-D-glucopyranoside as an active component of rhubarb activates endothelial nitric oxide synthase through inhibition of arginase activity. *Exp. Mol. Med.* **2010**, *42*, 524–532. [CrossRef] [PubMed]
338. Yuan, Q.; Peng, J.; Liu, S.-Y.; Wang, C.-J.; Xiang, D.-X.; Xiong, X.-M.; Hu, C.-P.; Li, Y.-J. Inhibitory effect of resveratrol derivative BTM-0512 on high glucose-induced cell senescence involves dimethylaminohydrolase/asymmetric dimethylarginine pathway. *Clin. Exp. Pharmacol. Physiol.* **2010**, *37*, 630–635. [CrossRef] [PubMed]
339. Gómez-Zorita, S.; Fernández-Quintela, A.; Lasa, A.; Aguirre, L.; Rimando, A.M.; Portillo, M.P. Pterostilbene, a Dimethyl Ether Derivative of Resveratrol, Reduces Fat Accumulation in Rats Fed an Obesogenic Diet. *J. Agric. Food Chem.* **2014**, *62*, 8371–8378. [CrossRef]
340. Spanier, G.; Xu, H.; Xia, N.; Tobias, S.; Deng, S.; Wojnowski, L.; Forstermann, U.; Li, H. Resveratrol reduces endothelial oxidative stress by modulating the gene expression of superoxide dismutase 1 (SOD1), glutathione peroxidase 1 (GPx1) and NADPH oxidase subunit (Nox4). *J. Physiol. Pharmacol.* **2009**, *60*, 111–116.
341. Wallerath, T.; Deckert, G.; Ternes, T.; Anderson, H.; Li, H.; Witte, K.; Förstermann, U. Resveratrol, a Polyphenolic Phytoalexin Present in Red Wine, Enhances Expression and Activity of Endothelial Nitric Oxide Synthase. *Circulation* **2002**, *106*, 1652–1658. [CrossRef]

342. Frombaum, M.; Le Clanche, S.; Bonnefont-Rousselot, D.; Borderie, D. Antioxidant effects of resveratrol and other stilbene derivatives on oxidative stress and NO bioavailability: Potential benefits to cardiovascular diseases. *Biochimie* **2012**, *94*, 269–276. [CrossRef]
343. Yang, J.; Wang, N.; Li, J.; Zhang, J.; Feng, P. Effects of resveratrol on NO secretion stimulated by insulin and its dependence on SIRT1 in high glucose cultured endothelial cells. *Endocrine* **2010**, *37*, 365–372. [CrossRef]
344. Arunachalam, G.; Yao, H.; Sundar, I.K.; Caito, S.; Rahman, I. SIRT1 regulates oxidant- and cigarette smoke-induced eNOS acetylation in endothelial cells: Role of resveratrol. *Biochem. Biophys. Res. Commun.* **2010**, *393*, 66–72. [CrossRef]
345. Gresele, P.; Pignatelli, P.; Guglielmini, G.; Carnevale, R.; Mezzasoma, A.M.; Ghiselli, A.; Momi, S.; Violi, F. Resveratrol, at Concentrations Attainable with Moderate Wine Consumption, Stimulates Human Platelet Nitric Oxide Production123. *J. Nutr.* **2008**, *138*, 1602–1608. [CrossRef]
346. Nicholson, S.K.; Tucker, G.A.; Brameld, J.M. Physiological concentrations of dietary polyphenols regulate vascular endothelial cell expression of genes important in cardiovascular health. *Br. J. Nutr.* **2009**, *103*, 1398–1403. [CrossRef] [PubMed]
347. Klinger, J.R. The Nitric Oxide/cGMP Signaling Pathway in Pulmonary Hypertension. *Clin. Chest Med.* **2007**, *28*, 143–167. [CrossRef] [PubMed]
348. Rivera, L.; Morón, R.; Zarzuelo, A.; Galisteo, M. Long-term resveratrol administration reduces metabolic disturbances and lowers blood pressure in obese Zucker rats. *Biochem. Pharmacol.* **2009**, *77*, 1053–1063. [CrossRef] [PubMed]
349. Zou, J.-G.; Wang, Z.-R.; Huang, Y.-Z.; Cao, K.-J.; Wu, J.M. Effect of red wine and wine polyphenol resveratrol on endothelial function in hypercholesterolemic rabbits. *Int. J. Mol. Med.* **2003**, *11*, 317–320. [CrossRef]
350. Yoshino, J.; Conte, C.; Fontana, L.; Mittendorfer, B.; Imai, S.I.; Schechtman, K.B.; Gu, C.; Kunz, I.; Fanelli, F.R.; Patterson, B.W.; et al. Resveratrol Supplementation Does Not Improve Metabolic Function in Nonobese Women with Normal Glucose Tolerance. *Cell Metab.* **2012**, *16*, 658–664. [CrossRef]
351. Juan, M.E.; Vinardell, M.P.; Planas, J.M. The Daily Oral Administration of High Doses of trans-Resveratrol to Rats for 28 Days Is Not Harmful. *J. Nutr.* **2002**, *132*, 257–260. [CrossRef]
352. Tomé-Carneiro, J.; Gonzálvez, M.; Larrosa, M.; Yáñez-Gascón, M.J.; García-Almagro, F.J.; Ruiz-Ros, J.A.; García-Conesa, M.T.; Tomás-Barberán, F.A.; Espín, J.C. One-Year Consumption of a Grape Nutraceutical Containing Resveratrol Improves the Inflammatory and Fibrinolytic Status of Patients in Primary Prevention of Cardiovascular Disease. *Am. J. Cardiol.* **2012**, *110*, 356–363. [CrossRef]
353. Tomé-Carneiro, J.; Gonzálvez, M.; Larrosa, M.; García-Almagro, F.J.; Avilés-Plaza, F.; Parra, S.; Yáñez-Gascón, M.J.; Ruiz-Ros, J.A.; García-Conesa, M.T.; Tomás-Barberán, F.A.; et al. Consumption of a grape extract supplement containing resveratrol decreases oxidized LDL and ApoB in patients undergoing primary prevention of cardiovascular disease: A triple-blind, 6-month follow-up, placebo-controlled, randomized trial. *Mol. Nutr. Food Res.* **2012**, *56*, 810–821. [CrossRef]
354. Sahebkar, A. Effects of resveratrol supplementation on plasma lipids: A systematic review and meta-analysis of randomized controlled trials. *Nutr. Rev.* **2013**, *71*, 822–835. [CrossRef] [PubMed]
355. Riche, D.M.; Riche, K.D.; Blackshear, C.T.; McEwen, C.L.; Sherman, J.J.; Wofford, M.R.; Griswold, M.E. Pterostilbene on Metabolic Parameters: A Randomized, Double-Blind, and Placebo-Controlled Trial. *Evid.-Based Complement. Altern. Med.* **2014**, *2014*, 459165. [CrossRef] [PubMed]
356. Tomé-Carneiro, J.; Gonzálvez, M.; Larrosa, M.; Yáñez-Gascón, M.J.; García-Almagro, F.J.; Ruiz-Ros, J.A.; Tomás-Barberán, F.A.; García-Conesa, M.T.; Espín, J.C. Grape Resveratrol Increases Serum Adiponectin and Downregulates Inflammatory Genes in Peripheral Blood Mononuclear Cells: A Triple-Blind, Placebo-Controlled, One-Year Clinical Trial in Patients with Stable Coronary Artery Disease. *Cardiovasc. Drugs Ther.* **2013**, *27*, 37–48. [CrossRef]
357. Szkudelski, T.; Szkudelska, K. Resveratrol and diabetes: From animal to human studies. *Biochim. Biophys. Acta (BBA)—Mol. Basis Dis.* **2015**, *1852*, 1145–1154. [CrossRef] [PubMed]
358. Penumathsa, S.V.; Thirunavukkarasu, M.; Zhan, L.; Maulik, G.; Menon, V.P.; Bagchi, D.; Maulik, N. Resveratrol enhances GLUT-4 translocation to the caveolar lipid raft fractions through AMPK/Akt/eNOS signalling pathway in diabetic myocardium. *J. Cell. Mol. Med.* **2008**, *12*, 2350–2361. [CrossRef]
359. Moridi, H.; Karimi, J.; Sheikh, N.; Goodarzi, M.T.; Saidijam, M.; Yadegarazari, R.; Khazaei, M.; Khodadadi, I.; Tavilani, H.; Piri, H.; et al. Resveratrol-Dependent Down-regulation of Receptor for Advanced Glycation End-products and Oxidative Stress in Kidney of Rats With Diabetes. *Int. J. Endocrinol. Metab.* **2015**, *13*, e23542. [CrossRef]
360. Soufi, F.G.; Mohammad-nejad, D.; Ahmadieh, H. Resveratrol improves diabetic retinopathy possibly through oxidative stress—Nuclear factor κB—Apoptosis pathway. *Pharmacol. Rep.* **2012**, *64*, 1505–1514. [CrossRef]
361. Um, J.-H.; Park, S.-J.; Kang, H.; Yang, S.; Foretz, M.; McBurney, M.W.; Kim, M.K.; Viollet, B.; Chung, J.H. AMP-Activated Protein Kinase–Deficient Mice Are Resistant to the Metabolic Effects of Resveratrol. *Diabetes* **2009**, *59*, 554–563. [CrossRef] [PubMed]
362. Bhatt, J.K.; Thomas, S.; Nanjan, M.J. Resveratrol supplementation improves glycemic control in type 2 diabetes mellitus. *Nutr. Res.* **2012**, *32*, 537–541. [CrossRef] [PubMed]

363. Movahed, A.; Nabipour, I.; Lieben Louis, X.; Thandapilly, S.J.; Yu, L.; Kalantarhormozi, M.; Rekabpour, S.J.; Netticadan, T. Antihyperglycemic Effects of Short Term Resveratrol Supplementation in Type 2 Diabetic Patients. *Evid.-Based Complement. Altern. Med.* **2013**, *2013*, 851267. [CrossRef] [PubMed]
364. Thazhath, S.S.; Wu, T.; Bound, M.J.; Checklin, H.L.; Standfield, S.; Jones, K.L.; Horowitz, M.; Rayner, C.K. Administration of resveratrol for 5 wk has no effect on glucagon-like peptide 1 secretion, gastric emptying, or glycemic control in type 2 diabetes: A randomized controlled trial12. *Am. J. Clin. Nutr.* **2016**, *103*, 66–70. [CrossRef]
365. Bo, S.; Ponzo, V.; Ciccone, G.; Evangelista, A.; Saba, F.; Goitre, I.; Procopio, M.; Pagano, G.F.; Cassader, M.; Gambino, R. Six months of resveratrol supplementation has no measurable effect in type 2 diabetic patients. A randomized, double blind, placebo-controlled trial. *Pharmacol. Res.* **2016**, *111*, 896–905. [CrossRef]
366. Gómez-Zorita, S.; Fernández-Quintela, A.; Aguirre, L.; Macarulla, M.T.; Rimando, A.M.; Portillo, M.P. Pterostilbene improves glycaemic control in rats fed an obesogenic diet: Involvement of skeletal muscle and liver. *Food Funct.* **2015**, *6*, 1968–1976. [CrossRef]
367. Bhakkiyalakshmi, E.; Shalini, D.; Sekar, T.V.; Rajaguru, P.; Paulmurugan, R.; Ramkumar, K.M. Therapeutic potential of pterostilbene against pancreatic beta-cell apoptosis mediated through Nrf2. *Br. J. Pharmacol.* **2014**, *171*, 1747–1757. [CrossRef]
368. Manickam, M.; Ramanathan, M.; Farboodniay Jahromi, M.A.; Chansouria, J.P.N.; Ray, A.B. Antihyperglycemic Activity of Phenolics from Pterocarpus marsupium. *J. Nat. Prod.* **1997**, *60*, 609–610. [CrossRef]
369. Grover, J.K.; Vats, V.; Yadav, S.S. Pterocarpus marsupium extract (Vijayasar) prevented the alteration in metabolic patterns induced in the normal rat by feeding an adequate diet containing fructose as sole carbohydrate. *Diabetes Obes. Metab.* **2005**, *7*, 414–420. [CrossRef]
370. Nemes-Nagy, E.; Szőcs-Molnár, T.; Dunca, I.; Balogh-Sămărghiţan, V.; Hobai, Ş.; Morar, R.; Pusta, D.L.; Crăciun, E.C. Effect of a dietary supplement containing blueberry and sea buckthorn concentrate on antioxidant capacity in type 1 diabetic children. *Acta Physiol. Hung.* **2008**, *95*, 383–393. [CrossRef]
371. Minakawa, M.; Miura, Y.; Yagasaki, K. Piceatannol, a resveratrol derivative, promotes glucose uptake through glucose transporter 4 translocation to plasma membrane in L6 myocytes and suppresses blood glucose levels in type 2 diabetic model db/db mice. *Biochem. Biophys. Res. Commun.* **2012**, *422*, 469–475. [CrossRef] [PubMed]
372. Towler, M.C.; Hardie, D.G. AMP-Activated Protein Kinase in Metabolic Control and Insulin Signaling. *Circ. Res.* **2007**, *100*, 328–341. [CrossRef]
373. Saltiel, A.R.; Kahn, C.R. Insulin signalling and the regulation of glucose and lipid metabolism. *Nature* **2001**, *414*, 799–806. [CrossRef]
374. Hijona, E.; Aguirre, L.; Pérez-Matute, P.; Villanueva-Millán, M.J.; Mosqueda-Solis, A.; Hasnaoui, M.; Nepveu, F.; Senard, J.M.; Bujanda, L.; Aldámiz-Echevarría, L.; et al. Limited beneficial effects of piceatannol supplementation on obesity complications in the obese Zucker rat: Gut microbiota, metabolic, endocrine, and cardiac aspects. *J. Physiol. Biochem.* **2016**, *72*, 567–582. [CrossRef]
375. Oritani, Y.; Okitsu, T.; Nishimura, E.; Sai, M.; Ito, T.; Takeuchi, S. Enhanced glucose tolerance by intravascularly administered piceatannol in freely moving healthy rats. *Biochem. Biophys. Res. Commun.* **2016**, *470*, 753–758. [CrossRef]
376. Kershaw, J.; Kim, K.-H. The Therapeutic Potential of Piceatannol, a Natural Stilbene, in Metabolic Diseases: A Review. *J. Med. Food* **2017**, *20*, 427–438. [CrossRef]
377. Spalding, K.L.; Arner, E.; Westermark, P.O.; Bernard, S.; Buchholz, B.A.; Bergmann, O.; Blomqvist, L.; Hoffstedt, J.; Näslund, E.; Britton, T.; et al. Dynamics of fat cell turnover in humans. *Nature* **2008**, *453*, 783–787. [CrossRef]
378. Jeong, S.O.; Son, Y.; Lee, J.H.; Cheong, Y.K.; Park, S.H.; Chung, H.T.; Pae, H.O. Resveratrol analog piceatannol restores the palmitic acid-induced impairment of insulin signaling and production of endothelial nitric oxide via activation of anti-inflammatory and antioxidative heme oxygenase-1 in human endothelial cells. *Mol. Med. Rep.* **2015**, *12*, 937–944. [CrossRef] [PubMed]
379. Jellinger, K.A. *General Aspects of Neurodegeneration*; Spring: Vienna, Austria, 2003; pp. 101–144.
380. de A Boleti, A.P.; Almeida, J.A.; Migliolo, L. Impact of the metabolic syndrome on the evolution of neurodegenerative diseases. *Neural Regen. Res.* **2021**, *16*, 688–689. [CrossRef] [PubMed]
381. Pisoschi, A.M.; Pop, A.; Iordache, F.; Stanca, L.; Predoi, G.; Serban, A.I. Oxidative stress mitigation by antioxidants-an overview on their chemistry and influences on health status. *Eur. J. Med. Chem.* **2021**, *209*, 112891. [CrossRef]
382. de Araújo, F.F.; de Paulo Farias, D.; Neri-Numa, I.A.; Pastore, G.M. Polyphenols and their applications: An approach in food chemistry and innovation potential. *Food Chem.* **2021**, *338*, 127535. [CrossRef] [PubMed]
383. Olajide, O.A.; Sarker, S.D. Alzheimer's disease: Natural products as inhibitors of neuroinflammation. *Inflammopharmacology* **2020**, *28*, 1439–1455. [CrossRef] [PubMed]
384. Carradori, S.; D'Ascenzio, M.; Chimenti, P.; Secci, D.; Bolasco, A. Selective MAO-B inhibitors: A lesson from natural products. *Mol. Divers.* **2014**, *18*, 219–243. [CrossRef]
385. Rahman, M.M.; Rahaman, M.S.; Islam, M.R.; Rahman, F.; Mithi, F.M.; Alqahtani, T.; Almikhlafi, M.A.; Alghamdi, S.Q.; Alruwaili, A.S.; Hossain, M.S.; et al. Role of Phenolic Compounds in Human Disease: Current Knowledge and Future Prospects. *Molecules* **2022**, *27*, 233. [CrossRef]

386. Fantacuzzi, M.; Amoroso, R.; Carradori, S.; De Filippis, B. Resveratrol-based compounds and neurodegeneration: Recent insight in multitarget therapy. *Eur. J. Med. Chem.* **2022**, *233*, 114242. [CrossRef] [PubMed]
387. Mohd Sairazi, N.S.; Sirajudeen, K.N.S. Natural Products and Their Bioactive Compounds: Neuroprotective Potentials against Neurodegenerative Diseases. *Evid.-Based Complement. Altern. Med.* **2020**, *2020*, 6565396. [CrossRef]
388. Lu, K.-T.; Ko, M.-C.; Chen, B.-Y.; Huang, J.-C.; Hsieh, C.-W.; Lee, M.-C.; Chiou, R.Y.Y.; Wung, B.-S.; Peng, C.-H.; Yang, Y.-L. Neuroprotective Effects of Resveratrol on MPTP-Induced Neuron Loss Mediated by Free Radical Scavenging. *J. Agric. Food Chem.* **2008**, *56*, 6910–6913. [CrossRef]
389. Yang, Y.; Fan, C.; Wang, B.; Ma, Z.; Wang, D.; Gong, B.; Di, S.; Jiang, S.; Li, Y.; Li, T.; et al. Pterostilbene attenuates high glucose-induced oxidative injury in hippocampal neuronal cells by activating nuclear factor erythroid 2-related factor 2. *Biochim. Biophys. Acta (BBA)—Mol. Basis Dis.* **2017**, *1863*, 827–837. [CrossRef]
390. Wang, B.; Liu, H.; Yue, L.; Li, X.; Zhao, L.; Yang, X.; Wang, X.; Yang, Y.; Qu, Y. Neuroprotective effects of pterostilbene against oxidative stress injury: Involvement of nuclear factor erythroid 2-related factor 2 pathway. *Brain Res.* **2016**, *1643*, 70–79. [CrossRef]
391. Su, Q.; Pu, H.; Hu, C. Neuroprotection by combination of resveratrol and enriched environment against ischemic brain injury in rats. *Neurol. Res.* **2016**, *38*, 60–68. [CrossRef]
392. Fang, L.; Gao, H.; Zhang, W.; Zhang, W.; Wang, Y. Resveratrol alleviates nerve injury after cerebral ischemia and reperfusion in mice by inhibiting inflammation and apoptosis. *Int. J. Clin. Exp. Med.* **2015**, *8*, 3219.
393. Teka, T.; Zhang, L.; Ge, X.; Li, Y.; Han, L.; Yan, X. Stilbenes: Source plants, chemistry, biosynthesis, pharmacology, application and problems related to their clinical Application-A comprehensive review. *Phytochemistry* **2022**, *197*, 113128. [CrossRef]
394. Zhang, F.; Shi, J.-S.; Zhou, H.; Wilson, B.; Hong, J.-S.; Gao, H.-M. Resveratrol Protects Dopamine Neurons Against Lipopolysaccharide-Induced Neurotoxicity through Its Anti-Inflammatory Actions. *Mol. Pharmacol.* **2010**, *78*, 466–477. [CrossRef]
395. Marambaud, P.; Zhao, H.; Davies, P. Resveratrol promotes clearance of Alzheimer's disease amyloid-beta peptides. *J. Biol. Chem.* **2005**, *280*, 37377–37382. [CrossRef] [PubMed]
396. Biais, B.; Krisa, S.; Cluzet, S.; Da Costa, G.; Waffo-Teguo, P.; Mérillon, J.-M.; Richard, T. Antioxidant and Cytoprotective Activities of Grapevine Stilbenes. *J. Agric. Food Chem.* **2017**, *65*, 4952–4960. [CrossRef] [PubMed]
397. Chaher, N.; Arraki, K.; Dillinseger, E.; Temsamani, H.; Bernillon, S.; Pedrot, E.; Delaunay, J.-C.; Mérillon, J.-M.; Monti, J.-P.; Izard, J.-C.; et al. Bioactive stilbenes from Vitis vinifera grapevine shoots extracts. *J. Sci. Food Agric.* **2014**, *94*, 951–954. [CrossRef]
398. Ren, J.; Fan, C.; Chen, N.; Huang, J.; Yang, Q. Resveratrol Pretreatment Attenuates Cerebral Ischemic Injury by Upregulating Expression of Transcription Factor Nrf2 and HO-1 in Rats. *Neurochem. Res.* **2011**, *36*, 2352–2362. [CrossRef]
399. Simão, F.; Matté, A.; Pagnussat, A.S.; Netto, C.A.; Salbego, C.G. Resveratrol prevents CA1 neurons against ischemic injury by parallel modulation of both GSK-3β and CREB through PI3-K/Akt pathways. *Eur. J. Neurosci.* **2012**, *36*, 2899–2905. [CrossRef]
400. Ma, X.; Sun, Z.; Liu, Y.; Jia, Y.; Zhang, B.; Zhang, J. Resveratrol improves cognition and reduces oxidative stress in rats with vascular dementia. *Neural Regen. Res.* **2013**, *8*, 2050–2059.
401. Ban, J.Y.; Jeon, S.-Y.; Nguyen, T.T.H.; Bae, K.; Song, K.-S.; Seonga, Y.H. Neuroprotective Effect of Oxyresveratrol from *Smilacis Chinae* Rhizome on Amyloid Beta Protein (25–35)-Induced Neurotoxicity in Cultured Rat Cortical Neurons. *Biol. Pharm. Bull.* **2006**, *29*, 2419–2424. [CrossRef]
402. Andrabi, S.A.; Spina, M.G.; Lorenz, P.; Ebmeyer, U.; Wolf, G.; Horn, T.F.W. Oxyresveratrol (trans-2,3′,4,5′-tetrahydroxystilbene) is neuroprotective and inhibits the apoptotic cell death in transient cerebral ischemia. *Brain Res.* **2004**, *1017*, 98–107. [CrossRef]
403. Turner, R.S.; Thomas, R.G.; Craft, S.; van Dyck, C.H.; Mintzer, J.; Reynolds, B.A.; Brewer, J.B.; Rissman, R.A.; Raman, R.; Aisen, P.S.; et al. A randomized, double-blind, placebo-controlled trial of resveratrol for Alzheimer disease. *Neurology* **2015**, *85*, 1383–1391. [CrossRef] [PubMed]
404. Witte, A.V.; Kerti, L.; Margulies, D.S.; Flöel, A. Effects of Resveratrol on Memory Performance, Hippocampal Functional Connectivity, and Glucose Metabolism in Healthy Older Adults. *J. Neurosci.* **2014**, *34*, 7862–7870. [CrossRef] [PubMed]
405. Wong, R.H.X.; Raederstorff, D.; Howe, P.R.C. Acute Resveratrol Consumption Improves Neurovascular Coupling Capacity in Adults with Type 2 Diabetes Mellitus. *Nutrients* **2016**, *8*, 425. [CrossRef]
406. Wong, R.H.X.; Nealon, R.S.; Scholey, A.; Howe, P.R.C. Low dose resveratrol improves cerebrovascular function in type 2 diabetes mellitus. *Nutr. Metab. Cardiovasc. Dis.* **2016**, *26*, 393–399. [CrossRef] [PubMed]

International Journal of
Molecular Sciences

MDPI

Review

Salvinorin A and *Salvia divinorum*: Toxicology, Pharmacological Profile, and Therapeutic Potential

Sara Calado [1,2], Bruno Pires [1,2], Luana M. Rosendo [1,2], Tiago Rosado [1,2,3], Eugenia Gallardo [1,2,3] and Ana Paula Duarte [1,2,*]

1 RISE-Health, Departamento de Ciências Médicas, Faculdade de Ciências da Saúde, Universidade da Beira Interior, Av. Infante D. Henrique, 6200-506 Covilhã, Portugal; sara.calado@ubi.pt (S.C.); bruno.pinheiro.pires@ubi.pt (B.P.); may.rosendo@ubi.pt (L.M.R.); tiago.rosado@fcsaude.ubi.pt (T.R.); egallardo@fcsaude.ubi.pt (E.G.)

2 Laboratório de Fármaco-Toxicologia, UBIMedical, Universidade da Beira Interior, Estrada Municipal 506, 6200-284 Covilhã, Portugal

3 Grupo de Investigação Sobre Problemas Relacionados Com Toxicofilias, Centro Académico Clínico das Beiras (CACB), Universidade da Beira Interior, Av. Infante D. Henrique, 6200-506 Covilhã, Portugal

* Correspondence: apduarte@fcsaude.ubi.pt

Abstract: *Salvia divinorum* is a psychoactive plant presenting a complex pharmacological profile, attracting significant scientific interest due to its potential therapeutic applications and associated health risks. This review provides a comprehensive analysis of the toxic and therapeutic effects of *S. divinorum*, evaluating its potential medical applications while highlighting the risks associated with its consumption. Additionally, the review examines the plant's recreational use, global consumption trends, and legal status. By synthesising current research, this article aims to clarify the implications of *S. divinorum* use and inform future studies on its pharmacological potential and regulatory considerations.

Keywords: *Salvia divinorum*; salvinorin A; toxicity; therapeutic effects

1. Introduction

Salvia divinorum is a perennial member of the *Lamiaceae* (mint) family and it is native to the Sierra of Oaxaca, Mexico. It is also referred to as "diviner's sage," "magic mint," "mystic sage," "ska Maria," and "ska Pastora". The Mazatecs used this psychotropic herb to treat various ailments, including rheumatism, headaches, diarrhoea, and abdominal swelling, as well as for shamanic and divinatory purposes [1–4]. The plant was initially recognised as a distinct species of *salvia* in 1962 by Carl Epling and Carlos D. Játiva of the University of California [5].

In recent years, the potent hallucinogenic effects of *S. divinorum* have led to its use as a substance of abuse, particularly by young people in the United States and Europe. The consumption of this herb has increased rapidly, particularly among adolescents and young adults, due to its widespread availability both online and in smartshops, where its distribution remains widely unregulated in most countries [6]. In general, the hallucinogenic efficacy of *S. divinorum* is comparable to that of other classical hallucinogens such as lysergic acid diethylamide (LSD). However, although both substances may be equally potent, *S. divinorum* has shorter-lasting effects than LSD [7].

At least 22 diterpene-type compounds have been identified in the phytochemistry of *S. divinorum*. Ten of them are classified as salvinorins (salvinorins A, B, C, D, E, F, G, H, I, and J—with salvinorin J occurring in two configurations: α and β), six as divinatorins

(divinatorins A, B, C, D, E, and F), four as salvidivins (salvidivins A, B, C, and D), and two as salvinicins (salvinicins A and B) (Figure 1) [8,9].

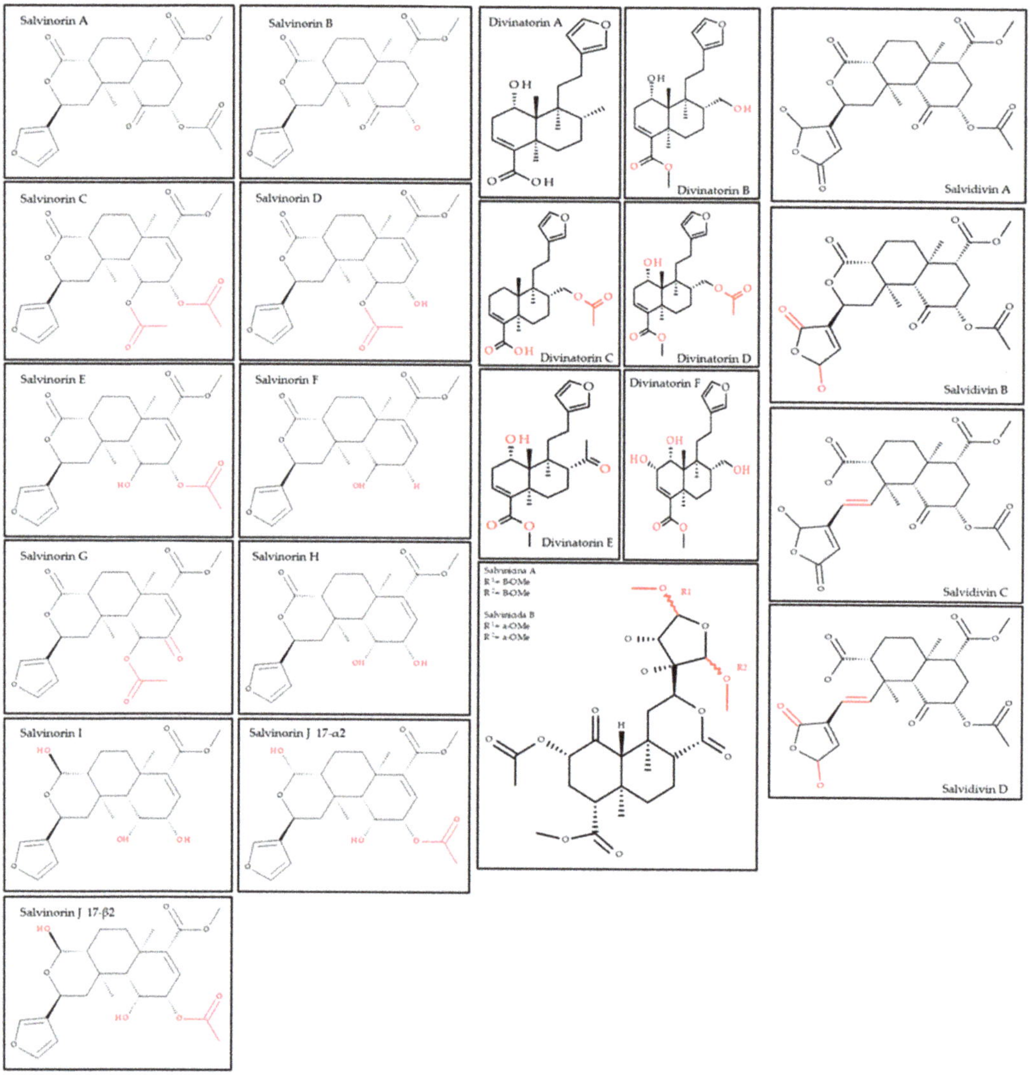

Figure 1. Chemical structures of the 22 diterpene-type compounds identified in *S. divinorum*.

The main bioactive compound of this plant is salvinorin A (SA), which belongs to the salvinorins group and is formed in glandular trichomes on the surface of the leaves of the plant. SA is responsible for the hallucinogenic effects [10–12]. SA has also been linked to the release and modification of monoaminergic neurotransmitters such as noradrenaline, dopamine, and serotonin. Clinical trials have demonstrated the beneficial effects of terpenes, with little to no adverse effects when used to treat pain [13].

SA is a neoclerodane diterpene that was first isolated and identified by Ortega et al. [14] and later by Valdes III et al. [15]. It is recognised as a selective and potent κ-opioid receptor agonist [14–16]. Classic hallucinogenic molecules, such as LSD, psilocin, DMT, and mescaline, activate serotonin-2A (5-HT_2A) receptors, while SA is unique among psychoactive substances due to its selective agonism at the kappa-opioid receptor (KOR) [10,17,18].

The most common methods of administration for recreational users are smoking the leaves, crushing them to extract juice for oral consumption, or masticating and swallowing the whole leaves [19]. Two early reports indicate that the hallucinogenic effects of SA can last for up to an hour following conventional oral dosing [20]. Depending on the dosage, vaporisation or inhalation of SA results in an almost immediate experience of the full effects (30 s after consumption), which can quickly subside after intake [6,17,21–23]. Inhaling doses of 200–500 μg can result in effects lasting up to two hours [6]. In contrast, the effects of oral consumption take longer to manifest; they usually begin 5–10 min after ingestion and last longer, with the intensity of the effects increasing rapidly. This suggests that the oral mucosa acts as a release buffer, gradually releasing SA into the bloodstream [6,21].

SA exhibits numerous potential metabolic sites, including a lactone ring, a C_2 acetyl group, and a C_4 ester. The C_2 acetate undergoes ester hydrolysis to produce salvinorin B, which has lower KOR activity, while the lactone group is known to be labile. The C4 ester has been proven to be enzymatically stable. Overall, it is believed that no active metabolites contribute to the effects of SA (Figure 2) [24].

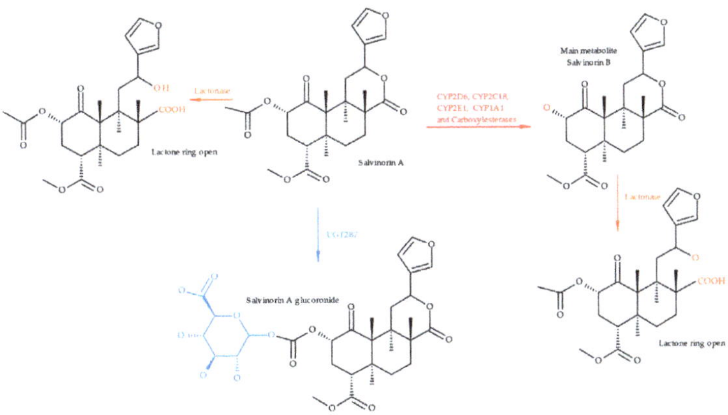

Figure 2. Metabolism of salvinorin A (SA).

Currently, few studies provide a complete in vivo toxicokinetic analysis of SA in humans, though some significant findings must be considered. Lipophilicity and molecular weight have long been thought to be the primary factors influencing drug permeation across the blood–brain barrier (BBB) [25]. SA is a highly lipophilic molecule, which may account for its rapid absorption in the central nervous system and its apparent ease of BBB penetration. In vitro studies suggest that SA is metabolised in humans by CYP450 enzymes (notably CYP2D6, CYP1A1, CYP2C18, and CYP2E1 isoforms), with glucuronidation likely serving as the drug's primary metabolic pathway (Figure 2) [25]. Studies conducted in non-human primates have revealed interesting findings regarding SA toxicokinetics. According to Schmidt et al. [26], following intravenous administration of SA, effects appeared immediately and dissipated within 15 min, which is consistent with SA's short duration of action. The total elimination half-life was 56.6 ± 24.8 min, with potential gender-specific variations in SA elimination, suggesting that its effects may differ between males and females.

This review aims to explore various aspects of *S. divinorum* with an emphasis on the toxic and therapeutic effects, addressing possible medical applications and health risks arising from its consumption. Furthermore, the plant's recreational use, consumption statistics, and legal status will be discussed.

2. Toxicology and Pharmacokinetics

As previously stated, psychedelic drugs can cause a remarkable variety of changes in perception, emotion, and thought processes, with effects that are highly influenced by the user and the setting in which they are used [27]. SA (Figure 3) is a non-nitrogenous diterpenoid molecule that is the primary psychoactive ingredient in *S. divinorum*, although fewer active compounds have been isolated. SA, like ketamine, is regarded as a dissociative drug [6,20–23,27–30].

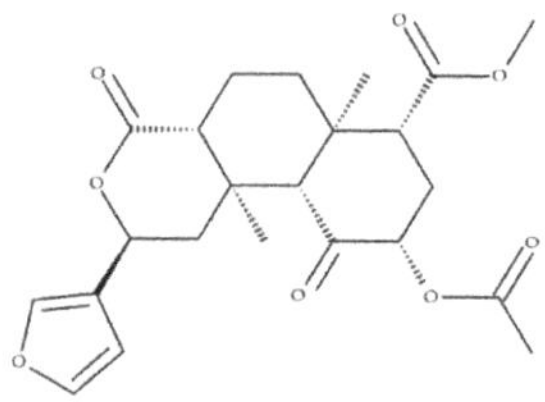

Figure 3. Chemical structure of salvinorin A (SA).

SA does not act on serotonin or dopamine receptors but targets KOR with high potency and selectivity. Notably, it is the first naturally occurring compound of its kind that lacks a basic nitrogen atom, diverging structurally from typical opioid ligands. Upon binding to KOR, SA initiates a cascade of intracellular events via Gi/o protein coupling, leading to the inhibition of adenylyl cyclase and a subsequent decrease in cyclic AMP (cAMP) levels. Additionally, the released $\beta\gamma$ subunits modulate ion channels, inhibiting voltage-gated calcium channels and activating G protein-coupled inwardly rectifying potassium (GIRK) channels, ultimately promoting neuronal hyperpolarisation and reduced neurotransmitter release. Beyond these classical mechanisms, KOR activation by SA can also lead to extracellular signal-regulated kinase (ERK1/2) phosphorylation, which influences dopamine transporter function and may underlie some of the compound's effects on mood, perception, and reward processing [10,17,18].

These intracellular pathways are critical not only for SA's acute psychoactive effects, but also for its safety profile. By selectively engaging the KOR and avoiding interaction with mu-opioid (MOR) and serotonergic receptors, SA elicits atypical dissociative states without producing euphoria or cardiorespiratory depression. The downstream modulation of dopaminergic and glutamatergic systems may account for its hallucinogenic and anxiolytic properties, while also contributing to its low addictive potential [18,20]. Altogether, the molecular signalling pathways activated by SA, particularly its selective KOR engagement and associated intracellular cascades, are integral to understanding both its short-acting psychoactive profile and its pharmacological distinctiveness. These mechanisms not only distinguish SA from serotonergic hallucinogens but also support its emerging role as a scaffold for the development of functionally selective KOR ligands with potential clinical applications [18,31].

This lack of interaction with MORs and specific KOR modulation contributes to a lower risk of respiratory depression and other side effects often related to MOR modulation while also reducing the risk of addiction [21,28].

Additionally, SA does not affect 5-hydroxytryptamine receptor subtype 2A (5-HT_{2A}R), the main molecular target of classical hallucinogenic drugs such as mescaline, psilocybin, and LSD [6,20–23,30]. It also exhibits a different behavioural profile from that of cannabinoids and has no affinity for the cannabinoid CB1 receptors in vitro [7,32]. Widely distributed throughout the central nervous system (CNS) and even in some peripheral tissues such as the intestines, lungs, kidneys, adrenals, spleen, stomach, testis, uterus, and ovaries, KORs are essential for controlling several physiological functions, including stress, pain, mood, reward, inflammation, and motor function [6,22].

This specific KOR affinity leads to a plethora of effects, with individuals usually reporting a disconnection from reality, visual and auditory hallucinations accompanied by object distortions, loss of contact with the body or depersonalisation, dizziness, unconsciousness, and motor discoordination [6,21–23,29]. Nevertheless, SA does not provide a sense of exhilaration or excitement, and it does not affect vital signs or cognitive impairments [21,23,33].

SA is primarily absorbed by the respiratory system and the oral mucosa, being quickly degraded in the gastrointestinal tract and metabolised in the liver and gallbladder, producing salvinorin B, its main inactive metabolite [6,20–23,32].

It is rapidly dispersed throughout the body, readily crossing the blood–brain barrier, where it accumulates and exerts its psychoactive effects before being quickly eliminated [6,17,21,22]. Its short psychoactive half-life (around 8 min) is closely associated with its pharmacokinetic characteristics and well correlated with its psychological effects [6,20–23].

Acute toxicity (1600 μg/kg) of SA in rats showed no impact on temperature, galvanic response, or cardiac conduction, while simultaneously being significantly less hazardous than morphine in humans (LD_{50} = 0.78 μg/mL for males and 0.98 μg/mL for females, respectively) [17]. Pure SA inhaled at concentrations as high as 12 mg has been shown to cause no harm or adverse effects in humans [17]. While most of the in vivo research indicates that SA and *S. divinorum* are relatively safe, in vitro toxicity in various cell lines (N 27, A 549, Caco 2, HepG2, Cos7, and Hek 293 cells) has shown dose- and time-dependent effects [6,17]. Additionally, Hep G2 and, to a lesser degree, Caco-2 cells have shown reduced susceptibility to *S. divinorum* and SA in contrast to N27, A549, COS-7, and HEK-293 cell lines [6,17]. Nevertheless, further knowledge regarding the complete scope of *S. divinorum* and SA toxicity in vivo, including both acute and chronic ingestion, is necessary to ensure its safe use.

3. Recreational Use and Legislation

The most popular methods of administration for recreational use include chewing and smoking the leaves [19,23,34]. Usually, a dose between 0.25 and 0.75 g of leaf material is smoked, producing effects within a minute and lasting up to 20 min [34]. A longer-lasting, mild to moderate effect is obtained by chewing the bitter leaves, typically 10–30 g of fresh leaves or 2–5 g of dried leaves. Drinking tea is uncommon since SA is easily broken down in the gastrointestinal tract [34].

Despite the potential dangers and laws enforced by each nation, the recreational use of this plant persists due to its extensive psychoactive properties. Recreational products, such as *S. divinorum* leaves and seeds, are easily accessible and can be purchased online and in physical stores that sell drug paraphernalia, such as head shops or smart shops, in countries or territories where their consumption is not prohibited [2,23]. These products are typically sold in packages containing leaves or another absorbent material impregnated with an extract of SA [2,23]. Although it is known by many street names, it is commonly marketed under the name "Purple Sticky™" in smoke shops [2,23,29,32].

The use of *S. divinorum* is strongly associated with alcohol, tobacco, cannabis, prescription and abused drugs, and other illegal substances, including LSD, ecstasy, heroin, phencyclidine, and cocaine [23]. Epidemiological evidence suggests that young adult white men with higher-income parents are more likely to use this substance. Additionally, a significant proportion of consumers report using *Salvia* primarily in their homes or apartments, highlighting the need for a secure setting to facilitate a smoother hallucinatory experience [23,29]. Individuals use *S. divinorum* for a variety of reasons, including curiosity and the desire to experiment with new substances. Others seek altered states of consciousness for spiritual experiences, self-discovery, or personal growth. Some consume it for meditation, relaxation, or enjoyment, while others use it recreationally to experience a high or an enhanced altered state. Additionally, some individuals turn to *S. divinorum* as a means of coping with psychological distress or to explore drug-induced states of altered consciousness [23].

Given that *S. divinorum* is often acquired through unregulated herbal products, special attention must be paid to the authentication of the plant material. Several morphologically similar *Salvia* species—particularly from the mint family—have been found to contain hepatotoxic compounds such as salviarin and rhyacophiline. Therefore, misidentification or contamination of commercial preparations may pose serious toxicological risks. Ensuring botanical accuracy is essential to avoid inadvertent exposure to harmful substances and to support the safe use of *S. divinorum*, whether for recreational or potential therapeutic purposes [4,35,36].

The scientific community remains divided concerning *S. divinorum*'s toxicity, potential for addiction, and medical applications, making its legal status a contentious issue [32]. In 2002, Australia became the first country to prohibit the sale and possession of *S. divinorum*, with several other nations following [32].

Currently, the legal status of *S. divinorum* varies by location and intended use. While it is not classified as a controlled substance at the federal level in the United States, the Drug Enforcement Administration (DEA) has identified it as a "drug of concern" due to its potential for misuse and abuse. It remains legal at the federal level but is restricted in eighteen states [37]. Some states impose no specific regulations, allowing legal use, possession, and sale, whereas others restrict their sales to individuals over 18 [37]. In Georgia, SA is classified as a dangerous drug, making its sale, distribution, or possession illegal, except when cultivated purely for decorative, landscaping, or aesthetic purposes [37]. Several European and Asian countries, including Belgium, Denmark, Italy, Latvia, Lithuania, Romania, Sweden, Australia, and Japan, have laws regulating *S. divinorum*. In Estonia, Finland, and Norway, it is allowed only for medicinal use, whereas in Croatia, Germany, Poland, and Spain, only the plant itself is regulated. In Canada, selling *Salvia* without authorisation is prohibited under the Natural Health Products Regulation [34].

4. Potential Therapeutic Uses

Historically, it has been employed in small doses to treat ailments such as diarrhoea, bloating, headaches, and rheumatism. The plant's infusion or juice has also been used as a palliative remedy, helping to alleviate discomfort and promoting well-being [37].

SA is a potent and selective KOR agonist, which distinguishes it from classical psychedelics such as LSD and psilocybin, which primarily target the serotonin 2^a ($5\text{-}HT_{2A}$) receptor. The unique pharmacological profile of SA has spurred interest in its potential therapeutic applications, ranging from pain relief to the treatment of mental health disorders. The following sections outline the therapeutic potentials described in the literature.

4.1. Pain Management

Chronic pain poses a significant medical challenge due to the side effects and dependency risks associated with conventional analgesics. Research has highlighted *S. divinorum* as a promising alternative for pain relief. SA's interaction with KOR plays a crucial role in modulating pain perception, mood regulation, and motor control [37].

Studies have shown that SA produces antinociceptive effects without activating mu-opioid receptors, which are commonly associated with opioid addiction [37]. Pharmacological research demonstrates its dual agonist action on KOR and CB1 cannabinoid receptors, enhancing its analgesic potential [30,37,38].

In formalin-induced pain models, SA effectively reduced inflammatory pain and hyperalgesia. Aviello et al. [39] reported decreased paw oedema and nociceptive behaviour in mice, confirming its anti-inflammatory and pain-relieving properties [39]. Capasso et al. [40] further demonstrated that SA mitigated ileitis-induced hypermotility by modulating KOR and CB1 receptor interactions [40].

Recent studies suggest that SA may serve as a basis for developing short-term anaesthetics or non-addictive painkillers, given its fast onset and brief duration of action [37].

4.2. Mental Health Applications

Depression and Anxiety

The KOR agonism of SA has opened new avenues for treating mood disorders. Unlike serotonergic psychedelics, SA's distinct mechanism offers an alternative approach to managing depression and anxiety.

A well-documented case study by Hanes [41] described a 26-year-old woman with chronic depression who achieved full remission after sublingual administration of *S. divinorum* leaves (0.5–0.75 g) three times per week for six months. Her progress, assessed using the Hamilton Depression Scale, revealed significant improvements.

Animal models have provided additional support. Rodents administered SA (0.001–1000 μg/kg, s.c.) exhibited reduced anxiety-like behaviour and increased exploration, indicating anxiolytic effects [19]. However, high doses (0.125–2.0 mg/kg) produced depressive-like symptoms and decreased locomotor activity, highlighting the importance of dose regulation [42].

Recent research also suggests that SA influences neuroplasticity by regulating brain-derived neurotrophic factor (BDNF) expression, which may underlie its antidepressant properties [43].

Despite promising preclinical data, clinical trials exploring *S. divinorum*'s therapeutic effects remain scarce. A Phase 1/2 trial registered on ClinicalTrials.gov investigated its impact on major depressive disorder but faced challenges with participant recruitment [44]. The open-label design and small sample size (only one participant at the time of review) limited the study's conclusions.

4.3. Substance Use Disorders

SA has shown promising results in addressing substance use disorders by targeting the KOR-dopaminergic system. KOR activation reduces dopamine release in the nucleus accumbens, a brain region central to reward processing.

Preclinical studies by Glick et al. [45] and Appel and Kim-Appel [42] demonstrated that SA dose-dependently decreased cocaine and remifentanil self-administration in rhesus monkeys. This suggests that SA may diminish drug reward responses, offering a pharmacological route for treating addiction.

Rothman et al. [46] also found that SA disrupted drug-associated memory reconsolidation, reducing the risk of relapse. Furthermore, recent findings highlight its role in microglial polarisation—shifting from a pro-inflammatory M1 to an anti-inflammatory M2 phenotype—which could mitigate neuroinflammation linked to substance abuse [47].

4.4. Neuroprotection

Emerging research highlights SA's neuroprotective potential, particularly in the context of cerebral ischaemia and brain injury. Studies have shown that SA can preserve vascular integrity and reduce inflammation.

In a middle cerebral artery occlusion (MCAO) mouse model, intranasal administration of SA (12.5–50 μg/kg) significantly reduced infarct volume, improved neurological outcomes, and increased IL-10 levels, supporting its anti-inflammatory action [48].

Furthermore, in models of hypoxic-ischemic injury in neonatal pigs, SA enhanced cerebral vasodilation and prevented apoptosis through nitric oxide synthase (NOS) and ATP-sensitive potassium channel (KATP channel) activation. This suggests it could play a therapeutic role in conditions involving cerebral vascular dysfunction [49].

4.5. Anti-Inflammatory and Antinociceptive Effects

SA's anti-inflammatory properties extend to various models of inflammatory disease. Aviello et al. [39] found that doses of 0.5–2 mg/kg reduced carrageenan-induced paw oedema and formalin-induced inflammatory pain via KOR and CB1 receptor pathways. Additionally, SA inhibited leukotriene biosynthesis in activated rat peritoneal macrophages, reducing cell infiltration and vascular permeability. This suggests its potential for treating leukotriene-related conditions, such as asthma and cardiovascular diseases [50].

While *S. divinorum*'s hallucinogenic properties present certain challenges, its bioactive compound, SA, holds considerable therapeutic potential. From pain management and mood disorder treatments to neuroprotection and anti-inflammatory applications, its KOR agonism and interaction with cannabinoid receptors highlight a diverse pharmacological profile.

However, advancing its medical use hinges on rigorous clinical trials to balance therapeutic benefits with psychoactive risks. The ongoing exploration of SA and its analogues may pave the way for innovative, non-addictive treatments for pain, addiction, and mental health disorders.

4.6. Microdosing

Microdosing is the practice of consuming psychoactive substances in sub-perceptual dosages to acquire potential therapeutic advantages while avoiding full hallucinogenic or dissociative effects [51].

This approach has been extensively studied with classic psychedelics such as LSD and psilocybin, which have demonstrated promising results in improving mood, cognitive flexibility, and emotional resilience [51,52]. However, there has been little research into microdosing *S. divinorum*, particularly its primary active ingredient, SA.

Microdosing SA differs significantly from LSD and psilocybin, which act on 5-HT2A receptors and are commonly used for cognitive enhancement and mood regulation [51,52]. A study on microdosing found that while many users report benefits such as increased focus and creativity, a significant portion discontinued due to a lack of efficacy. Unlike LSD and psilocybin, SA targets the KOR receptor, producing short-lived effects that may require frequent dosing [51].

Ketamine, another KOR-interacting compound, provides prolonged antidepressant effects, whereas SA's transient action presents challenges for sustained therapeutic use [51]. Similarly, ibogaine's long-lasting influence on addiction pathways contrasts with SA's rapid metabolism. Ayahuasca, a serotonergic psychedelic, and kratom, a partial opioid agonist, offer more sustained benefits and are more frequently used in microdosing practices [51].

While SA's potential in mood regulation, pain management, and addiction treatment warrants further exploration, its distinct pharmacology and short duration necessitate targeted research to determine optimal dosing strategies and clinical viability.

Limited reports suggest that microdosing SA may influence mood, stress resilience, and pain perception [53,54]. Some users describe mild anxiolytic and antidepressant effects, while others report increased introspection and altered cognitive processing [54]. However, systematic studies are lacking, and concerns exist regarding KOR activation potentially inducing dysphoria. Unlike classical psychedelics, which promote sustained neuroplasticity, SA's transient nature may require frequent administration to maintain any potential benefits.

Further research is needed to evaluate the efficacy and safety of SA microdosing, particularly its potential in mood disorders, pain management, and addiction treatment. Establishing optimal dosing strategies and understanding its long-term effects will be crucial for determining the viability in therapeutic applications.

5. Analysis of Salvinorin A in Biological Specimens

Given the significant role of *S. divinorum* and its primary active compound, SA, in both clinical and forensic toxicology, it is crucial to explore the methodologies available for their detection, the biological matrices that can be utilised, and the analytical techniques applicable to their identification. This section is based on a literature review conducted using the PubMed database with the search terms "*Salvia divinorum* and biological matrix" and "salvinorin A and biological matrix," considering only human biological samples. The conventional matrices examined include whole blood, plasma, serum, and urine, while alternative specimens considered are oral fluid/saliva, hair, vitreous humour, sweat, and pericardial fluid. The selection process involved three independent reviewers, and only studies identified as relevant by at least two of them were incorporated into this analysis, covering publications from 2004 to 2025.

Despite the relevance of identifying these psychoactive substances, there are relatively few studies available, and the conclusions regarding the effects of their consumption remain limited. A comprehensive review by Margalho et al. [55] summarised the benefits and limitations of different biological samples and the analytical methods used for detection.

Table 1 presents an overview of the methodologies developed for detecting SA in biological samples. Among conventional matrices, plasma and urine are the most frequently used.

Table 1. Summary of methodologies for the detection of salvinorin A (SA) in biological samples.

Compounds	Sample/ Volume	Extraction Procedure	Analytical	LOD (ng/mL)	LOQ (ng/mL)	Recoveries (%)	Reference
Salvinorin A	Urine (20 mL)	LLE (chloroform). SPME (85 μm polyacrylate fibre)	GC × GC-ToFMS	200 for LLE; 4 for SPME	300 for LLE; 8 for SPME	n.s.	[56]
Salvinorin A	Urine (0.2 mL)	MEPS (C_{18})	GC-MS/MS (EI)	5	20	71.91–80.15	[57]
Salvinorin A Salvinorin B	Plasma and Urine (0.5 mL)	SPE (Oasis HLB (30 mg, 1 mL))	LC-MS (APCI)	2 (Salvinorin A in Plasma)	2 (Salvinorin A in Plasma) 50 (Salvinorin B in Plasma)	104.00–106.00 (Urine)	[58]
Salvinorin A	Plasma (0.25 mL)	SPE (Waters Oasis SPE cartridges)	LC-MS (APCI)	n.s.	1	n.s.	[59]
Salvinorin A	Plasma (0.4 mL)	SPE (Strata C_{18}-E 55 micron, 70 Å)	LC-MS/MS (ESI)	n.s.	0.05	93.00–114.00	[60]
Salvinorin A	Plasma, Urine, Saliva, and Sweat (1 mL)	LLE (chloroform/ isopropanol (9:1))	GC-MS (EI)	5 (Plasma, Urine and Saliva); 0.003 μg/patch (Sweat)	15 (Plasma, Urine and Saliva); 0.01 μg/patch (Sweat)	84.60 (Plasma); 93.70 (Urine); 84.20 (Saliva); 77.10 (Sweat)	[61]
Salvinorin A	Hair (25 mg)	LLE (Diethyl ether)	UHPLC-MS/MS (ESI)	0.02 ng/mg	0.05 ng/mg	79.60–97.40	[62]

Table 1. *Cont.*

Compounds	Sample/ Volume	Extraction Procedure	Analytical	LOD (ng/mL)	LOQ (ng/mL)	Recoveries (%)	Reference
Salvinorin A	Vitreous humour, Pericardial fluid, Blood, and Plasma (0.1–0.25 mL)	SPE (Oasis® HLB (3 mL, 60 mg))	GC-MS (EI)	5	5	79.60–100.60 (Vitreous humour); 93.40–100.20 (Pericardial fluid); 88.80–99.10 (Blood); 88.30–98.00 (Plasma)	[63]
Salvinorin A Salvinorin B	Urine and Plasma (1 mL)	LLE (DMSO/ PEG-400, (25:75, *v*/*v*)	LC-MS/MS (APCI)	n.s.	0.5	n.s.	[64]
Salvinorin A	Urine and Blood (1 mL)	SPE (Oasis® HLB (3 mL, 60 mg))	LC-MS (ESI)	2.5	5	71.70	[65]

APCI: Atmospheric-pressure chemical ionisation; DMSO: dimethyl sulfoxide; EI: Electron impact; ESI: Electrospray ionisation; GC: Gas chromatography; GC × GC-ToFMS: Two-dimensional gas chromatography–time of flight mass spectrometry; LC: Liquid chromatography; LLE: Liquid–liquid extraction; LOD: Limit of detection; LOQ: Limit of quantification; MEPS: Microextraction by packed sorbent; MS: Mass spectrometry; MS-MS: Tandem mass spectrometry; n.s.: Not specified; PEG-400: polyethylene glycol 400; SPE: Solid-phase extraction; SPME: Solid-phase microextraction; UHPLC: Ultra-high-pressure liquid chromatography.

A noteworthy study by Barnes and Snow [56] investigated urine samples using both liquid–liquid extraction (LLE) and solid-phase microextraction (SPME) coupled with two-dimensional gas chromatography–time of flight mass spectrometry. The microextraction approach provided a detection limit (LOD) of 4 ng/mL and a quantification limit (LOQ) of 8 ng/mL, though it required a substantial sample volume (20 mL). Conversely, Moreno et al. [57] developed a method employing microextraction by packed sorbent with gas chromatography–tandem mass spectrometry using only 200 μL of urine, obtaining an LOD of 5 ng/mL and an LOQ of 20 ng/mL, with recoveries ranging from 72% to 80%.

Several studies have examined plasma and urine matrices using LC coupled to mass spectrometry. Schmidt and collaborators [58] utilised 500 μL of these samples, while Tidgewell et al. [59] developed a method for plasma analysis using SPE extraction, only using 250 μL. More recently, Caspers et al. [60] employed SPE extraction and LC-tandem mass spectrometry for 400 μL of plasma, achieving an LOQ of 0.05 ng/mL and recoveries between 93% and 114% [60].

Alternative biological samples have also been explored. Pichini et al. [61] assessed plasma, urine, saliva (1 mL), and sweat patches for SA detection. Using gas chromatography–mass spectrometry (GC-MS), they reported LODs of 5 ng/mL and LOQs of 15 ng/mL for saliva, along with an LOD of 0.003 μg/patch and an LOQ of 0.01 μg/patch for sweat, achieving recoveries above 77%. Later, the same research group [61] developed a method for detecting the compound in 25 mg of hair using ultra-high-pressure liquid chromatography–tandem mass spectrometry, with an LOD of 0.02 ng/mg and an LOQ of 0.05 ng/mg [62].

From a *postmortem* perspective, Margalho et al. [63] established a method for detecting SA in vitreous humour (100 μL) and pericardial fluid (250 μL), alongside blood and plasma samples. Using SPE extraction and GC-MS analysis, they reported excellent LOD and LOQ values of 5 ng/mL, with recoveries exceeding 79% [63].

Among the different approaches employed for the analysis of SA, SPE emerges as the most widely utilised technique, particularly due to its efficiency in selectively isolating analytes from complex biological matrices. Within SPE protocols, Oasis® HLB (hydrophilic lipophilic balance) sorbents have demonstrated superior performance, likely attributed to their balanced retention of both polar and non-polar compounds, facilitating enhanced recovery rates and reproducibility. LLE remains a conventional alternative, particularly in studies involving urine and plasma. A critical assessment of LLE methodologies reveals that the most employed solvents include chloroform and binary solvent mixtures such as ethyl acetate–heptane and chloroform–isopropanol. The choice of solvent system is crucial, as it influences both extraction efficiency and the elimination of interfering compounds.

Despite the growing interest in miniaturised extraction techniques, their practical application in forensic and clinical toxicology remains relatively limited. However, some methodologies, particularly solid-phase microextraction (SPME) and microextraction by packed sorbent (MEPS), have gained recognition due to their potential for reducing solvent consumption and sample volume requirements. While these techniques align with green chemistry principles, their adoption has been constrained by factors such as lower sensitivity compared to conventional SPE and LLE methods.

Among the available instrumental techniques for detecting SA, liquid chromatography coupled to mass spectrometry (LC-MS and LC-MS/MS) is the most widely used approach. This preference is primarily due to its high sensitivity, specificity, and the fact that SA determination is often part of a multi-analyte method that enables the detection of other drugs of abuse, particularly those that are more thermally unstable, such as LSD. Although gas chromatography coupled to mass spectrometry (GC-MS and GC-MS/MS) is also used, especially in urine and hair analysis (Table 1), its application is comparatively less frequent than LC-MS-based methods.

In cases where the LOQ value was not specified, the lowest calibration curve point was considered.

6. General Discussion and Future Directions

Although mechanistic, toxicological, and therapeutic aspects of *S. divinorum* and SA have been discussed throughout this review, this final section aims to synthesise key insights and highlight future directions based on current gaps in the literature. SA remains a pharmacologically unique compound due to its potent and selective agonism of KOR, without significant interaction with serotonergic, dopaminergic, or cannabinoid systems [20,23,31]. This positions it outside the classical hallucinogen framework and underscores its relevance for studying alternative pathways of perception and consciousness.

The ability of SA to activate both G protein–dependent and β-arrestin–dependent pathways opens new opportunities for biased agonism research. Such functional selectivity could, in principle, allow for therapeutic benefits such as analgesia, anti-inflammatory, or antidepressant effects, while reducing the risk of adverse events like hallucinations or dysphoria. These pharmacological dimensions are still poorly explored in vivo, particularly in models that assess circuit-level changes or behavioural outcomes linked to specific signalling biases [16,17,31].

There is also a need for translational research bridging molecular pharmacology and clinical application. While preclinical studies suggest potential roles for SA in pain management, mood disorders, and substance use disorders, no placebo-controlled clinical trials have yet been conducted. Critical knowledge gaps remain regarding dose–response relationships, safety in humans, sex-related pharmacokinetic variability, and possible long-term neuroadaptations. Likewise, the impact of genetic variability in KOR expression or function warrants further investigation [19,21,26,33].

Future research should also explore the development of analogues or derivatives with improved pharmacokinetic profiles, such as extended half-life, enhanced bioavailability, and optimised receptor selectivity [11,18,22]. Finally, given the increasing non-medical use of S. *divinorum*, regulatory, ethical, and public health implications must be addressed alongside scientific developments [1,34].

7. Conclusions

S. divinorum and its active compound, SA, represent a pharmacologically distinct class of psychoactive agents due to their selective agonism of kappa opioid receptors (KOR). Unlike classical serotonergic hallucinogens, SA induces short-lasting dissociative effects through a unique mechanism of action. This review outlined current knowledge on its pharmacodynamics, toxicology, and potential therapeutic applications. While preclinical data indicate promise in areas such as pain, addiction, and mood disorders, translational and clinical research remains limited. Continued investigation is essential to clarify its medical value, address safety concerns, and explore its regulatory and public health implications.

Author Contributions: Conceptualization, E.G., B.P., T.R. and A.P.D.; methodology, S.C., B.P. and L.M.R.; formal analysis, E.G., B.P., T.R. and A.P.D.; investigation, S.C., B.P. and L.M.R.; writing—original draft preparation, S.C., B.P., L.M.R. and E.G.; writing—review and editing, S.C., B.P., L.M.R., E.G., T.R. and A.P.D.; supervision, E.G., B.P., T.R. and A.P.D. All authors have read and agreed to the published version of the manuscript.

Funding: Luana Rosendo and Bruno Pires acknowledge the PhD fellowship from FCT (References: (2024.00478.BD and 2024.02748.BD, respectively).

Conflicts of Interest: The authors declare no conflicts of interest.

References

1. Maqueda, A.E. The Use of *Salvia divinorum* from a Mazatec Perspective. In *Plant Medicines, Healing and Psychedelic Science: Cultural Perspectives*; Springer International Publishing: Cham, Switzerland, 2018; pp. 55–70. ISBN 9783319767208.
2. Vohra, R.; Seefeld, A.; Cantrell, F.L.; Clark, R.F. *Salvia divinorum*: Exposures Reported to a Statewide Poison Control System over 10 Years. *J. Emerg. Med.* **2011**, *40*, 643–650. [CrossRef] [PubMed]
3. Mahendran, R.; Lim, H.A.; Tan, J.Y.S.; Chua, S.M.; Winslow, M. *Salvia divinorum*: An Overview of the Usage, Misuse, and Addiction Processes. *Asia-Pac. Psychiatry* **2016**, *8*, 23–31. [CrossRef] [PubMed]
4. Faudree, P. "Making Medicine" with *Salvia divinorum*: Competing Approaches and Their Implications. *Med. Anthropol. Cross Cult. Stud. Health Illn.* **2020**, *39*, 582–596. [CrossRef]
5. Kowalczuk, A.P.; Raman, V.; Galal, A.M.; Khan, I.A.; Siebert, D.J.; Zjawiony, J.K. Vegetative Anatomy and Micromorphology of *Salvia divinorum* (Lamiaceae) from Mexico, Combined with Chromatographic Analysis of Salvinorin A. *J. Nat. Med.* **2014**, *68*, 63–73. [CrossRef]
6. Martinho, A.; Silva, S.M.; Gallardo, E. Cytotoxic Effects of Salvinorin A, A Major Constituent of *Salvia divinorum*. *Med. Chem.* **2016**, *12*, 432–440. [CrossRef]
7. Listos, J.; Merska, A.; Fidecka, S. Pharmacological Activity of Salvinorin A, the Major Component of *Salvia divinorum*. *Pharmacol. Rep.* **2011**, *63*, 1305–1309. [CrossRef]
8. Hernández-Bello, R.; García-Rodríguez, R.V.; García-Sosa, K.; Peña-Rodríguez, L.M.; Vázquez-Hernández, M.; Ramos-Morales, F.R.; Corcoran, O.; Sánchez-Medina, A. Salvinorin A Content in Legal High Products of *Salvia divinorum* Sold in Mexico. *Forensic Sci. Int.* **2015**, *249*, 197–201. [CrossRef]
9. Kutrzeba, L.M.; Ferreira, D.; Zjawiony, J.K. Salvinorins J from *Salvia divinorum*: Mutarotation in the Neoclerodane System. *J. Nat. Prod.* **2009**, *72*, 1361–1363. [CrossRef]
10. Wang, Y.; Metz, P. Total Synthesis of the Neoclerodane Diterpene Salvinorin A via an Intramolecular Diels-Alder Strategy. *Org. Lett.* **2018**, *20*, 3418–3421. [CrossRef]
11. Hill, S.J.; Brion, A.U.C.M.; Shenvi, R.A. Chemical Syntheses of the Salvinorin Chemotype of KOR Agonist. *Nat. Prod. Rep.* **2020**, *37*, 1478–1496. [CrossRef]
12. Ford, S.A.; Ness, R.W.; Kwon, M.; Ro, D.K.; Phillips, M.A. A Chromosome Level Reference Genome of Diviner's Sage (*Salvia divinorum)* Provides Insight into Salvinorin A Biosynthesis. *BMC Plant Biol.* **2024**, *24*, 914. [CrossRef] [PubMed]

13. Tlacomulco-Flores, L.L.; Déciga-Campos, M.; González-Trujano, M.E.; Carballo-Villalobos, A.I.; Pellicer, F. Antinociceptive Effects of *Salvia divinorum* and Bioactive Salvinorins in Experimental Pain Models in Mice. *J. Ethnopharmacol.* **2020**, *248*, 112276. [CrossRef] [PubMed]
14. Ortega, A.; Blount, J.F.; Manchand, P.S. Salvinorin, a New Trans-Neoclerodane Diterpene from *Salvia divinorum* (Labiatae). *J. Chem. Soc. Perkin 1* **1982**, *17*, 2505–2508. [CrossRef]
15. Valdes III, L.; Butler, W.; Hatfield, G.; Paul, A.; Koreeda, M. Divinorin A, a Psychotropic Terpenoid, and Divinorin B from the Hallucinogenic Mexican Mint *Salvia divinorum*. *J. Org. Chem.* **1984**, *49*, 4716–4720. [CrossRef]
16. Prisinzano, T.E. Neoclerodanes as Atypical Opioid Receptor Ligands. *J. Med. Chem.* **2013**, *56*, 3435–3443. [CrossRef]
17. Hernández-Alvarado, R.B.; Madariaga-Mazón, A.; Ortega, A.; Martinez-Mayorga, K. DARK Classics in Chemical Neuroscience: Salvinorin A. *ACS Chem. Neurosci.* **2020**, *11*, 3979–3992. [CrossRef]
18. Roach, J.J.; Shenvi, R.A. A Review of Salvinorin Analogs and Their Kappa-Opioid Receptor Activity. *Bioorg. Med. Chem. Lett.* **2018**, *28*, 1436–1445. [CrossRef]
19. Grundmann, O.; Phipps, S.M.; Zadezensky, I.; Butterweck, V. *Salvia divinorum* and Salvinorin A: An Update on Pharmacology and Analytical Methodology. *Planta Med.* **2007**, *73*, 1039–1046. [CrossRef]
20. Cunningham, C.W.; Rothman, R.B.; Prisinzano, T.E. Neuropharmacology of the Naturally Occurring κ-Opioid Hallucinogen Salvinorin A. *Pharmacol. Rev.* **2011**, *63*, 316–347. [CrossRef]
21. Quiñonez-Bastidas, G.N.; Grijalva-Contreras, L.E.; Patiño-Camacho, S.I.; Navarrete, A. Emerging Psychotropic Drug for the Treatment of Trigeminal Pain: Salvinorin A. *Pharmaceuticals* **2024**, *17*, 1619. [CrossRef]
22. Tibouti, S.; Barriault, L. An Overview of Syntheses of Salvinorin A and Its Analogues. *ChemCatChem* **2024**, *16*, e202400182. [CrossRef]
23. Zawilska, J.B.; Wojcieszak, J. *Salvia divinorum*: From Mazatec Medicinal and Hallucinogenic Plant to Emerging Recreational Drug. *Hum. Psychopharmacol.* **2013**, *28*, 403–412. [CrossRef] [PubMed]
24. Chakraborty, S.; Majumdar, S. Natural Products for the Treatment of Pain: Chemistry and Pharmacology of Salvinorin A, Mitragynine, and Collybolide. *Biochemistry* **2021**, *60*, 1381–1400. [CrossRef]
25. Teksin, Z.S.; Lee, I.J.; Nemieboka, N.N.; Othman, A.A.; Upreti, V.V.; Hassan, H.E.; Syed, S.S.; Prisinzano, T.E.; Eddington, N.D. Evaluation of the Transport, *In Vitro* Metabolism and Pharmacokinetics of Salvinorin A, a Potent Hallucinogen NIH Public Access. *Eur. J. Pharm. Biopharm.* **2009**, *72*, 471–477. [CrossRef]
26. Schmidt, M.D.; Schmidt, M.S.; Butelman, E.R.; Harding, W.W.; Tidgewell, K.; Murry, D.J.; Kreek, M.J.; Prisinzano, T.E. Pharmacokinetics of the Plant-Derived κ-Opioid Hallucinogen Salvinorin A in Nonhuman Primates. *Synapse* **2005**, *58*, 208–210. [CrossRef]
27. Calder, A.E.; Qualls, C.; Hasler, G.; Elmiger, D.; Strassman, R. The Hallucinogen Rating Scale: Updated Factor Structure in a Large, Multistudy Sample. *Biol. Psychiatry Glob. Open Sci.* **2025**, *5*, 100436. [CrossRef]
28. Ngo, I.; Kumar, R.; Li, L.; Kim, S.W.; Kwon, M.; Ro, D.K. Identification of Clerodane Diterpene Modifying Cytochrome P450 (CYP728D26) in *Salvia divinorum*—En Route to Psychotropic Salvinorin A Biosynthesis. *Physiol. Plant* **2024**, *176*, e14569. [CrossRef]
29. Wu, L.-T.; Woody, G.; Yang, C.; Li, J.-H. Blazer Recent National Trends in *Salvia divinorum* Use and Substance-Use Disorders among Recent and Former *Salvia divinorum* Users Compared with Nonusers. *Subst. Abuse Rehabil.* **2011**, *2*, 53–68. [CrossRef]
30. Prisinzano, T.E. Psychopharmacology of the Hallucinogenic Sage *Salvia divinorum*. *Life Sci.* **2005**, *78*, 527–531. [CrossRef]
31. Butelman, E.R.; Kreek, M.J. Salvinorin A, a Kappa-Opioid Receptor Agonist Hallucinogen: Pharmacology and Potential Template for Novel Pharmacotherapeutic Agents in Neuropsychiatric Disorders. *Front. Pharmacol.* **2015**, *6*, 190. [CrossRef]
32. Silva, A.; Simão, A.Y.; Gonçalves, J.; Caramelo, D.; Soares, S.; Luís, Â.; Rosado, T.; Barroso, M.; Duarte, A.P.; Gallardo, E. Salvia divinorum: Beyond Its Therapeutic Uses. In *Salvia: Chemistry and Effects*; Rendón Villalobos, J.R., Ed.; Nova Science Publishers, Inc.: Hauppauge, NY, USA, 2020; Chapter 5; pp. 1–33.
33. Bagdasarian, F.A.; Hansen, H.D.; Chen, J.; Yoo, C.H.; Placzek, M.S.; Hooker, J.M.; Wey, H.Y. Acute Effects of Hallucinogens on Functional Connectivity: Psilocybin and Salvinorin-A. *ACS Chem. Neurosci.* **2024**, *15*, 2654–2661. [CrossRef] [PubMed]
34. European Monitoring Centre for Drugs and Drug Addiction *Salvia divinorum* Drug Profile. Available online: https://www.euda.europa.eu/publications/drug-profiles/salvia_en#control (accessed on 16 February 2025).
35. Fotovvat, M.; Radjabian, T.; Saboora, A. HPLC Fingerprint of Important Phenolic Compounds in Some *Salvia* L. Species from Iran. *Rec. Natl. Prod.* **2019**, *13*, 37–49. [CrossRef]
36. Bodnar Willard, M.A.; McGuffin, V.L.; Waddell Smith, R. Forensic Analysis of *Salvia divinorum* Using Multivariate Statistical Procedures. Part I: Discrimination from Related *Salvia* Species. *Anal. Bioanal. Chem.* **2012**, *402*, 833–842. [CrossRef] [PubMed]
37. José Petit-Rodríguez, M. Salvia Legality by State: In What States Is Salvia Legal? Available online: https://addictionresource.com/drugs/salvia/legality/ (accessed on 16 March 2025).
38. Coffeen, U.; Pellicer, F. *Salvia divinorum*: From Recreational Hallucinogenic Use to Analgesic and Anti-Inflammatory Action. *J. Pain. Res.* **2019**, *12*, 1069–1076. [CrossRef]

39. Aviello, G.; Borrelli, F.; Guida, F.; Romano, B.; Lewellyn, K.; De Chiaro, M.; Luongo, L.; Zjawiony, J.K.; Maione, S.; Izzo, A.A.; et al. Ultrapotent Effects of Salvinorin A, a Hallucinogenic Compound from *Salvia divinorum*, on LPS-Stimulated Murine Macrophages and Its Anti-Inflammatory Action in vivo. *J. Mol. Med.* **2011**, *89*, 891–902. [CrossRef]
40. Capasso, R.; Borrelli, F.; Cascio, M.G.; Aviello, G.; Huben, K.; Zjawiony, J.K.; Marini, P.; Romano, B.; Di Marzo, V.; Capasso, F.; et al. Inhibitory Effect of Salvinorin A, from *Salvia divinorum*, on Ileitis-Induced Hypermotility: Cross-Talk between κ-Opioid and Cannabinoid CB 1 Receptors. *Br. J. Pharmacol.* **2008**, *155*, 681–689. [CrossRef]
41. Hanes, K.R.P. Antidepressant Effects of the Herb *Salvia divinorum*: A Case Report. *J. Clin. Psychopharmacol.* **2001**, *21*, 634–635. [CrossRef]
42. Appel, J.; Kim-Appel, D. The Rise of a New Psychoactive Agent: *Salvia divinorum*. *Int. J. Ment. Health Addict.* **2007**, *5*, 248–253. [CrossRef]
43. Reiff, C.M.; Richman, E.E.; Nemeroff, C.B.; Carpenter, L.L.; Widge, A.S.; Rodriguez, C.I.; Kalin, N.H.; McDonald, W.M. Psychedelics and Psychedelic-Assisted Psychotherapy. *Am. J. Psychiatry* **2020**, *177*, 391–410. [CrossRef]
44. Siegel, A.N.; Meshkat, S.; Benitah, K.; Lipsitz, O.; Gill, H.; Lui, L.M.W.; Teopiz, K.M.; McIntyre, R.S.; Rosenblat, J.D. Registered Clinical Studies Investigating Psychedelic Drugs for Psychiatric Disorders. *J. Psychiatr. Res.* **2021**, *139*, 71–81. [CrossRef]
45. Glick, S.D.; Maisonneuve, I.M.; Raucci, J.; Sydney, A. Kappa Opioid Inhibition of Morphine and Cocaine Self-Administration in Rats. *Brain Res.* **1995**, *681*, 147–152. [CrossRef]
46. Rothman, R.B.; Gorelick, D.A.; Heishman, S.J.; Eichmiller, P.R.; Hill, B.H.; Norbeck, J.; Liberto, J.G. An Open-Label Study of a Functional Opioid κ Antagonist in the Treatment of Opioid Dependence. *J. Subst. Abuse Treat.* **2000**, *18*, 277–281. [CrossRef] [PubMed]
47. Yang, J.; Cai, J.H.; Wu, T.X.; Gao, Z.Q.; Zhou, C.; Wu, Q.; Ji, M.J. Salvinorin A Ameliorates Pilocarpine-Induced Seizures by Regulating Hippocampal Microglia Polarization. *J. Ethnopharmacol.* **2024**, *335*, 118697. [CrossRef] [PubMed]
48. Su, D.; Riley, J.; Kiessling, W.J.; Armstead, W.M.; Liu, R. Salvinorin A Produces Cerebrovasodilation through Activation of Nitric Oxide Synthase, κ Receptor, and Adenosine Triphosphate–Sensitive Potassium Channel. *Anesthesiology* **2011**, *114*, 374. [CrossRef]
49. Xin, J.; Zhang, Y.; He, Z.; Wang, Z. Highly Selective Non-Opioid Kappa Opioid Receptor (KOR) Agonist Salvinorin A Protects against Forebrain Ischemia-Induced Brain Injury in Rats. *Brain Res.* **2016**, *1637*, 168–176. [CrossRef]
50. Rossi, A.; Caiazzo, E.; Bilancia, R.; Riemma, M.A.; Pagano, E.; Cicala, C.; Ialenti, A.; Zjawiony, J.K.; Izzo, A.A.; Capasso, R.; et al. Salvinorin A Inhibits Airway Hyperreactivity Induced by Ovalbumin Sensitization. *Front. Pharmacol.* **2017**, *7*, 233757. [CrossRef]
51. Polito, V.; Liknaitzky, P. The Emerging Science of Microdosing: A Systematic Review of Research on Low Dose Psychedelics (1955–2021) and Recommendations for the Field. *Neurosci. Biobehav. Rev.* **2022**, *139*, 104706. [CrossRef]
52. Hutten, N.R.P.W.; Mason, N.L.; Dolder, P.C.; Kuypers, K.P.C. Motives and Side-Effects of Microdosing With Psychedelics Among Users. *Int. J. Neuropsychopharmacol.* **2019**, *22*, 426–434. [CrossRef]
53. Fichna, J.; Dicay, M.; Lewellyn, K.; Janecka, A.; Zjawiony, J.K.; MacNaughton, W.K.; Storr, M.A. Salvinorin A Has Antiinflammatory and Antinociceptive Effects in Experimental Models of Colitis in Mice Mediated by KOR and CB1 Receptors. *Inflamm Bowel Dis.* **2012**, *18*, 1137–1145. [CrossRef]
54. Kaertner, L.S.; Steinborn, M.B.; Kettner, H.; Spriggs, M.J.; Roseman, L.; Buchborn, T.; Balaet, M.; Timmermann, C.; Erritzoe, D.; Carhart-Harris, R.L. Positive Expectations Predict Improved Mental-Health Outcomes Linked to Psychedelic Microdosing. *Sci Rep.* **2021**, *11*, 1941. [CrossRef]
55. Margalho, C.; Corte-Real, F.; López-Rivadulla, M.; Gallardo, E. *Salvia divinorum*: Toxicological Aspects and Analysis in Human Biological Specimens. *Bioanalysis* **2016**, *8*, 1415–1425. [CrossRef] [PubMed]
56. Barnes, B.B.; Snow, N.H. Analysis of Salvinorin A in Plants, Water, and Urine Using Solid-Phase Microextraction-Comprehensive Two-Dimensional Gas Chromatography–Time of Flight Mass Spectrometry. *J. Chromatogr. A* **2012**, *1226*, 110–115. [CrossRef] [PubMed]
57. Moreno, I.; Da Fonseca, B.; Oppolzer, D.; Martinho, A.; Barroso, M.; Cruz, A.; Queiroz, J.A.; Gallardo, E. Analysis of Salvinorin A in Urine Using Microextraction in Packed Syringe and GC–MS/MS. *Bioanalysis* **2013**, *5*, 661–668. [CrossRef] [PubMed]
58. Schmidt, M.S.; Prisinzano, T.E.; Tidgewell, K.; Harding, W.; Butelman, E.R.; Kreek, M.J.; Murry, D.J. Determination of Salvinorin A in Body Fluids by High Performance Liquid Chromatography–Atmospheric Pressure Chemical Ionization. *J. Chromatogr. B* **2005**, *818*, 221–225. [CrossRef]
59. Tidgewell, K.; Harding, W.W.; Schmidt, M.; Holden, K.G.; Murry, D.J.; Prisinzano, T.E. A Facile Method for the Preparation of Deuterium Labeled Salvinorin A: Synthesis of [2,2,2-2H3]-Salvinorin A. *Bioorg. Med. Chem. Lett.* **2004**, *14*, 5099–5102. [CrossRef]
60. Caspers, M.J.; Williams, T.D.; Lovell, K.M.; Lozama, A.; Butelman, E.R.; Kreek, M.J.; Johnson, M.; Griffiths, R.; MacLean, K.; Prisinzano, T.E. LC-MS/MS Quantification of Salvinorin A from Biological Fluids. *Anal. Methods* **2013**, *5*, 7042–7048. [CrossRef]
61. Pichini, S.; Abanades, S.; Farré, M.; Pellegrini, M.; Marchei, E.; Pacifici, R.; De La Torre, R.; Zuccaro, P. Quantification of the Plant-Derived Hallucinogen Salvinorin A in Conventional and Non-Conventional Biological Fluids by Gas Chromatography/Mass Spectrometry after *Salvia divinorum* Smoking. *Rapid Commun. Mass Spectrom.* **2005**, *19*, 1649–1656. [CrossRef]

62. Pichini, S.; Marchei, E.; García-Algar, O.; Gomez, A.; Di Giovannandrea, R.; Pacifici, R. Ultra-High-Pressure Liquid Chromatography Tandem Mass Spectrometry Determination of Hallucinogenic Drugs in Hair of Psychedelic Plants and Mushrooms Consumers. *J. Pharm. Biomed. Anal.* **2014**, *100*, 284–289. [CrossRef]
63. Margalho, C.; Gallardo, E.; Castanheira, A.; Vieira, D.N.; López-Rivadulla, M.; Real, F.C. A Validated Procedure for Detection and Quantitation of Salvinorin a in Pericardial Fluid, Vitreous Humor, Whole Blood and Plasma Using Solid Phase Extraction and Gas Chromatography–Mass Spectrometry. *J. Chromatogr. A* **2013**, *1304*, 203–210. [CrossRef]
64. Mendelson, J.E.; Coyle, J.R.; Lopez, J.C.; Baggott, M.J.; Flower, K.; Everhart, E.T.; Munro, T.A.; Galloway, G.P.; Cohen, B.M. Lack of Effect of Sublingual Salvinorin A, a Naturally Occurring Kappa Opioid, in Humans: A Placebo-Controlled Trial. *Psychopharmacology* **2011**, *214*, 933–939. [CrossRef]
65. McDonough, P.C.; Holler, J.M.; Vorec, S.P.; Bosy, T.Z.; Magluilo, J.; Past, M.R. The Detection and Quantitative Analysis of the Psychoactive Component of *Salvia divinorum*, Salvinorin A, in Human Biological Fluids Using Liquid Chromatography-Mass Spectrometry. *J. Anal. Toxicol.* **2008**, *32*, 417–421. [CrossRef]

MDPI AG
Grosspeteranlage 5
4052 Basel
Switzerland
Tel.: +41 61 683 77 34

International Journal of Molecular Sciences Editorial Office
E-mail: ijms@mdpi.com
www.mdpi.com/journal/ijms